THE ANCIENT EGYPTIANS & THE NATURAL WORLD

sidestonepress

THE ANCIENT EGYPTIANS & THE NATURAL WORLD

FLORA, FAUNA, & SCIENCE

edited by
SALIMA IKRAM, JESSICA KAISER & STÉPHANIE PORCIER

Published by Sidestone Press, Leiden
www.sidestone.com

Lay-out & cover design: Sidestone Press

Photograph cover:

- Front cover: Cat mummy with a rare occurrence of a cartonnage mask, from the Egyptian Museum (Cairo), c. 300 BC-AD 100 – Anna-Marie Kellen, Salima Ikram and Francis Dzikowski.
- Back cover: Radiograph of a wooden statue of Anubis that contains a disarticulated dog mummy, from the Egyptian Museum (Cairo), c. 320 BC-30 BC – Salima Ikram and Francis Dzikowski

ISBN 978-94-6426-035-9 (softcover)
ISBN 978-94-6426-036-6 (hardcover)
ISBN 978-94-6426-037-3 (PDF e-book)

archimede
ARCHÉOLOGIE ET HISTOIRE DE LA MÉDITERRANÉE ET DE L'ÉGYPTE ANCIENNES
LabEx ANR-11-LABX-0032-01

ARCE
AMERICAN RESEARCH CENTER IN EGYPT

Institute for Bioarchaeology

The American University in Cairo

Contents

Preface

Salima Ikram, Jessica Kaiser, Stéphanie Porcier

Although different aspects of bioarchaeology have played a varying role in Egyptian archaeology over time, with physical anthropology being most consistently and frequently being integrated into excavations, it was not until the latter half of the 20th century that the other various branches of bioarchaeology (archaeozoology, archaeobotany, and related fields) became a more regular component of the greater archaeological picture. Over time, the diverse aspects of bioarchaeology have become increasingly integrated into every excavation, as is attested by the steadily growing number of publications in all these fields, together with their presence in site reports. The Bioarchaeology in Egypt conference has contributed to the growth in, and integration of, bioarchaeology into mainstream Egyptian archaeology.

The first conference relating to bioarchaeology in Egypt that was sponsored by the American Research Center in Egypt, the American University in Egypt and the Institute of Bioarchaeology. It took place in 2010, and focused on human remains, with the understanding that in future years it would become a more inclusive bioarchaeological conference. This was indeed the case with the next event that took place in 2013, sponsored by the same group, and supported by the Wenner-Gren foundation, with a broad spectrum of papers being offered. As a result of the conference, new collaborations were born, and fresh initiatives for intellectual exchange in different branches of bioarchaeology were created, including an archaeobotanical working group focussing on Egyptian and Nubian remains.

Additionally, training sessions in all aspects of Bioarchaeology were engendered. In conjunction with the Ministry of Antiquities, training sessions were hosted, not just by institutions in Cairo, such as the American University in Cairo, Ancient Egypt Research Associates, the French Institute, and the German Institute, but these were joined by other institutions, such as the Institute for Bioarchaeology, UNESCO, and the University of Zurich, to name but a few. Soon, the students became teachers, and established their own training programs all over Egypt, with a dedicated facility for training and research being established by the Ministry at Saqqara. Increasing numbers of graduates of training programmes started working in different aspects of bioarchaeology within the Ministry and on different excavations, both national and international.

In 2016, the First International Symposium on Animals in Ancient Egypt was held in Lyon, mainly under the auspices of the Musée des Confluences and the Agence Nationale de la Recherche. The symposium was unlike the Bioarchaeology conference in that its focus was on non-human animals, not just through the lens of bioarchaeology, but through traditional Egyptology, thus including animals in texts and images. There, it was agreed that the second symposium should take place in Cairo.

For 2019 we hit upon the happy idea of combining the Bioarchaeology in Egypt conference with the Second International Symposium on Animals in Ancient Egypt.

This has been generously supported by the American University in Cairo, the American Research Center in Egypt, the Institute of Bioarchaeology, the Wenner-Gren Foundation, and the Musée des Confluences. The overall theme, interpreted in its broadest sense, was the interaction between humans and the natural world. This has been remarkably successful, bringing together more than 320 individuals from at least 17 different countries.

This volume brings together the results of studies using not only traditional methodologies such as texts, images, metrics, and macroscopic examinations of the remains, but also new and advanced technologies, including different sorts of radiography, radiocarbon and isotope analyses, as well as in-depth studies of tool-use, and ways of applying scientific methodology. These studies have deepned our knowledge of ancient Egypt, its inhabitants, and their interaction with their environment in both religious and daily life. The diverse topics that are covered include Egyptian burial customs, religion, diet, health, and diseases of the ancient Egyptians, animal mummification, butchery, food production, the ancient environment and its flora and fauna.

While the majority of presentations focus on humans, there is a marked increase in lectures on other animals; archaeobotany is sadly under-represented, but this is because there is a separate Egyptian archaeobotany conference, which is more commonly attended by archaeobotanists working in Egypt. Hopefully, in future years they will present more frequently at the general Bioarchaeology conferences.

Now, with bioarchaeology firmly established as part of 'usual excavation activities', the next challenge is the fuller integration of archaeometry with both archaeology and bioarchaeology, which will provide evidence that will lead to a more profound understanding of the ancient Egyptians and their interactions with the natural world. But that is for the next conference!

We are very grateful to the different institutions that supported this conference, in particular the American University in Cairo, the American Research Center in Egypt, the Institute of Bioarchaeology, the Wenner-Gren Foundation, and the Musée des Confluences. We are grateful to Mary Sadek and Djodi Deutsch for logistical help in organising the conference, and Lubna Ali, Salah Sarouk, Hazem Hassib, Khadiga Sayed, and their teams for facilitating the conference at the American University in Cairo. Additionally, we would like to thank André J. Veldmeijer, Mariam Ayad, Lisa Sabbahy, Ariel Singer, Katherine Bateman, Emmy Malak, Yasmin el Shazly, Sara Abd el Aziz, Hassan El Zawy, Jayme Reichart, Hayley Goddard, Bianca von Sittert, Omar Kamel, Matthew Treasure, Nadine Issa, and our volunteers for the time and effort that they put into running the conference, and without whose help the conference would not have been possible.

Abstracts

Curatorial Training in Human Remains for the Egyptian Museum, Cairo

Hasnaa Askalany[1] & Gehad Shawky Ibrahem[2]

[1] Archaeological inspector at the Center of Documentation of Egyptian Antiquities-MSA, hasnaa.askalany@gmail.com

[2] Museum Curator at Egyptian Museum in Cairo, Gehad_shawky@yahoo.com

During the practical part of the Egyptian Museum Curators Human Remains Training Course that was carried out by the Institute for Bioarchaeology, the American University in Cairo, the Egyptian Museum, and the (then) Ministry of Antiquities in 2018, we started the documentation of one box TR 28-4-26-27, which contained skeletal and mummified remains. Our initial information stated that all these skeletal material within the box originated from the site of Bab al-Gusus, which is located in the Theban necropolis near Deir al-Bahari and dates to the 21st Dynasty (ca. 1069-945 BC). However, upon closer examination we discovered that the contents were unprovenanced. Our research showed that the material came from Kawmil, Al-Aamra, Naqada north and south, Beit Allam, Gebel Silsilah, and Deir al-Bahari. The report will investigate the links between materials of this box and their archaeological contexts and follow the history of these skeletal materials post-excavation.

Keywords: Egyptian Museum, human remains, mummified remains

Hyperostosis Frontalis Interna in the Early Dynastic Period at Abydos, Egypt

Brenda J. Baker[1] & Ahmed Gabr[2]

[1] Center for Bioarchaeological Research, School of Human Evolution and Social Change, Arizona State University, Tempe, AZ, 85287-2402, BrendaJ.Baker@asu.edu

[2] Ministry of Antiquities, 3 Adel Abou Bakr, Zamalek, Cairo, 11211, ahmedgabr85@gmail.com

Hyperostosis frontalis interna (HFI) is a condition that involves thickening of the internal aspect of the frontal bone. It is of unknown cause and is much more common today in older adult females than in males and is typically associated with hormonal changes related to menopause and diet. At Abydos, four individuals with HFI have been identified in Early Dynastic remains from two cemetery contexts that date to approximately 3100-3000 BC. These examples of HFI are among the earliest yet recognized. A mostly complete middle adult female, age 45-49, is from one of five excavated subsidiary graves of King Aha's main funerary enclosure in the North Cemetery. Eight large, dense nodules are located on either side of the frontal crest and the entire area is thickened. Three affected individuals are from the Early Dynastic South Abydos cemetery, where

16 tombs have been excavated and 17 individuals are represented. Those with HFI include a probable female middle adult represented only by skull fragments, with thickening and nodules located on the internal aspect of the frontal bone. A more complete middle adult female exhibits similar involvement. A male from one of the largest tombs, between 20 and 25 years old, shows characteristic changes of HFI despite his young age at death. The rate and distribution of HFI in these two cemeteries is discussed in relation to the high status of these individuals.

Keywords: bioarchaeology, elite status, Early Dynastic Period, endocranial lesions, paleopathology, Upper Egypt

Humans and Animals Together in the Journey to the Afterlife. The Burial in Area R11 under the Temple of Millions of Years of Amenhotep II, Luxor, West Thebes – Italian Archaeological Project

Fabio Bona*, Giovanna Bellandi, Letizia Cavallini, Anna Consonni, Tommaso Quirino & Angelo Sesana
Italian Archaeological Project at the Temple of Amenhotep II in Western Thebes – CEFB (Centro di Egittologia Francesco Ballerini, Como – Italy)
* Corresponding author: fabgeo@libero.it

We present an analysis of the interaction between humans and animals in a funerary context, based on bones recovered from tomb R11 at the Temple of Millions of Years of Amenhotep II. Tomb R11 dates to the Third Intermediate Period-early Late Period (9th to 7th centuries BC) and consists of a shaft with two irregularly shaped chambers (A and B) at the bottom. The tomb had been disturbed in antiquity. In chamber B, we excavated one articulated skeleton (an older woman), while in chamber A we found the scattered fragmentary remains of four individuals: a man, two women and an adult of indeterminable sex. The faunal remains from the fill levels of the shaft and chamber A are abundant and date to the periods from the original tomb-use through the Ptolemaic era and into modern times. Taphonomic and osteological observations and the ossification levels of the bones suggest the presence of two groups of remains. Firstly, ancient bones that relate to activities connected to the original burials or to other more recent ritual or funerary activities. Secondly, bones deposited in the shaft in modern times. The first group consists of complete skeletons of cows and sheep from the shaft, and the rear half of a cow and a complete sheep from chamber A. The second group includes a recent camel and fragments of a dog and a bird. The presence of the remains of a crocodile, frequently found in fill levels of local tombs, is currently not clearly understood. This paper works toward an interpretation of the finds in Chamber A.

Keywords: Amenhotep II, ancient Egypt, anthropology, archaeozoology, cemetery, Luxor, temple

To Be or Not to Be a Dog Mummy: How a Metric Study of the Skull Can Inform on Selection Practices Pertaining to Canid Mummification in Ancient Egypt

Colline Brassard[1,2,*], Stéphanie Porcier[2,3], Cécile Callou[1]
[1] Archéozoologie, Archéobotanique : sociétés, pratiques et environnements (UMR 7209), Sorbonne Universités, Muséum national d'Histoire naturelle, CNRS, Département Homme et Environnement, 55 rue Buffon, 75005 Paris, France
[2] Laboratoire CNRS "Archéologie des Sociétés Méditerranéennes" (ASM_UMR 5140), Université Paul-Valéry Montpellier 3, Labex ARCHIMEDE, France.
[3] Laboratoire CNRS "Histoire et Sources des Mondes Antiques" (HiSoMA_UMR 5189), Maison de l'Orient et de la Méditerranée, Lyon, France.
* Corresponding author: colline.brassard@mnhn.fr

Egyptian animal cemeteries contain millions of canine mummies dating from the Late Period to the Roman era. Mostly dogs, but also jackals and foxes were mummified and offered to deities like Anubis or Wepwawet, attesting to unique cultural and religious practices in the history of Egyptian civilization. This phenomenon raises questions pertaining to the selection of animals for votive mummies through time and space. To what extent were domestic and/or wild canids involved? Was a specific morphotype or sex privileged for the mummification process? Did these choices change depending on geographical and chrono-cultural contexts? Iconography and texts provide some answers to understand the place of these animals in the pantheon and the mental space of the ancient Egyptians. However, bioarchaeological studies of the remains can provide a complementary and important outlook on the practical application of these considerations. Most of the time, the cranium (the skull without the mandible) is well preserved, enabling descriptive and qualitative studies, yet metric investigations remain relatively uncommon. The present study focuses on a multivariate analysis of measurements taken on the crania of ancient and modern Egyptian canids from the Musée des Confluences in Lyon (France), the Museums of Natural History in Paris (France) and Geneva (Switzerland), and dogs of the archaeological site of Kerma. Using log-shape ratios, principal components analyses, ascendant hierarchical classification, and discriminant analyses we demonstrate that multivariate statistics is an excellent tool to investigate which animals were selected for the preparation of votive mummies. We further provide comparative data that can be used to integrate individuals of different sites or data of scanned complete mummies into future studies.

Keywords: ancient Egypt, animal mummies, dog, morphometrics, multivariate analyses

Newcomers in the Bestiary. A Review of the Presence of *Lycaon pictus* in Late Predynastic and Early Dynastic Environment and Iconography

Axelle Brémont
Sorbonne University & CNRS UMR 8167 "Orient et Méditerranée", 1 rue Victor Cousin, 75005 Paris, France, axelle.bremont@paris-sorbonne.fr

Because it is unmistakably represented on several palettes from the late Predynastic to the early Protodynastic period, it has been assumed that the African wild dog (*Lycaon pictus*) was present in Egypt at least during the Middle Holocene. However, no osteological evidence, either from anthropic or geological assemblages, supports its presence anywhere in the Egyptian territory. Instead, examination of the rest of the depicted bestiary from this period shows that many animals that were represented were indigenous to Egypt (or at least not anymore), such as fallow deer, baboon and elephant. A re-examination of iconographic details and the associated complete lack of faunal remains suggests the lycaon was represented because it was valued as an exotic animal, rather than as a part of the Egyptian biotope.

Keywords: African wild dog, exoticism, iconography, Predynastic, zooarchaeology

Dévots et animaux sacrés

Alain Charron
Musée départemental Arles antique, Presqu'île du cirque romain, BP 205, 13635 Arles cedex, France
CNRS, UMR 5140, Archéologie des sociétés méditerranéennes, Montpellier
alain.charron@departement13.fr

Le monde des animaux liés aux cultes recèle encore bien des ombres et l'attitude des anciens Égyptiens à leur encontre n'est quasiment pas étudiée. Nous savons qu'ils n'accordaient pas la même valeur à toutes les bêtes ; certaines, comme le taureau Apis, bénéficiaient d'un régime très favorable alors que bien d'autres, élevées en batteries, pourrait-on dire, ne servaient que de support à l'occasion d'une offrande faite en l'honneur d'une divinité.

Dans les faits, nous ignorons quasiment tout de ce qu'il advenait quand un dévot se rendait dans un sanctuaire où un animal était associé à un culte, à l'exception notable des témoignages laissés par ceux qui prenaient part à l'enterrement du taureau Apis. L'animal est souvent mis en avant dans nos études alors que c'est la divinité qui était invoquée par les croyants. Il s'agit de s'interroger sur ce qui se produisait lors des fêtes religieuses, connaître les participants. Doit-on d'ailleurs parler de pèlerins, alors que les témoignages montreraient plutôt des honneurs rendus essentiellement par les populations locales ? Le troc établi entre une personne et la divinité est mieux connu. En revanche, il n'est pas facile de savoir quand et quel personnel pénétrait dans les nécropoles. Il y avait certainement peu de monde en dehors des prêtres et des serviteurs attachés aux animaux, notamment les membres des associations chargées de placer les momies dans les souterrains. Le but de cette communication est de faire le point sur les connaissances et d'inciter à réaliser de nouvelles recherches.

Mots-clés: animaux sacrés, cultes animaliers, dévots, momies animales, piété, religion, taureau Apis, taureau Buchis, taureau Mnevis

Lions and Science and Whorls, Oh My!

Karen Polinger Foster
Yale University, karen.foster@yale.edu

The interaction theme of this conference afforded a welcome opportunity to apply the results of recent scientific studies on lions to the matter of lion representations in Egyptian art, especially the depictions of hair whorls on their shoulders. The scientific work includes research on lion genetics and phylogeographic traits; phenotype plasticity and the secondary sexual character of manes in the context of lion group-living; thermographic imaging of lions; and ecological niche modeling of lion ranges. With regard to the hair whorl regularly shown on lion shoulders in Egyptian and other ancient art, the author herself examined 55 skins and stuffed specimens in museum collections. Taking all this into consideration, the present paper holds that traditional interpretations of the lions in Predynastic and pharaonic art, such as a long-maned/short-maned subspecies dichotomy, or their presence as proof that wild prides were still prevalent in the Nile Valley, can no longer be sustained. Instead, the conclusions offer a fresh look at how and why Egyptian artists rendered particular aspects of lion behavior and appearance.

Keywords: Egyptian art, hair whorl, lions, lion phylogeography, lion genetics

Tuberculosis at Tell-el Amarna: A Theoretical Exercise in the Economic and Social Effects of Chronic, Terminal Disease in Ancient Egypt

Gretchen R. Dabbs
Department of Anthropology, Southern Illinois University, 1000 Faner Dr. Mail Code 4502, Carbondale, IL 62901, gdabbs@siu.edu

This article presents the skeletal evidence of tuberculosis in the South Tombs Cemetery at Tell al-Amarna (ca. 1353-1332 BC). A single, adult female exhibits lytic lesions of the thoracic and lumbar vertebral column with associated kyphotic collapse of the column. A differential diagnosis is presented and the lesions are suggested to be tuberculosis after eliminating fungal, bacterial, and parasitic infections, traumatic injury, internal dysfunction, and external conditions known to produce morphologically similar lesions. The presence of tuberculosis at the ancient capital city is then used to discuss broader economic and social aspects of potentially wide-spread disease within the ancient city and its effect on the total economic output. It is suggested as many as one in seven households could have had an individual with active tuberculosis infection. The impact of the loss of economic output of that individual plus that individual's caretaker would have substantially reduced the overall economic standing of the household by as much as 20%, culminating in a collective loss of as much as 2.8% of the local economic activity. This value is in the range of the economic losses observed in modern African nations, where the World Health Organization has identified tuberculosis as a "sizeable economic burden."

Keywords: Amarna Period, bioarchaeology, disease related income loss, differential diagnosis, disability, endemic disease, familial caregiver stress, New Kingdom, osteological paradox

Burial Practies in the West Delta: Cases from Kom Aziza

Shereen El-Morsi & Aya M. Salem
Ministry of Tourism and Antiquities, Egypt, shereenelmorsi9@gmail.com, ayah.salem86@yahoo.com

This paper presents the preliminary results from the excavation of a cemetery at Kom Aziza in the Delta that dates primarily to the Roman period.

Keywords: amphorae, ceramic coffins, Roman burials

A Structure-from-Motion Pipeline for Bone Morphology 3D Analysis

Margaret Farmer & Angelique Corthals
John Jay College of Criminal Justice, City University of New York, 524 West 59th Street, New York, NY 10019, margaret.farmer@jjay.cuny.edu; acorthals@jjay.cuny.edu

Structure from Motion (SfM), the process of creating 3D models from photographs of objects at multiple viewpoints, is a promising method for comparative morphology in osteology. The point cloud data generated from an SfM 3D model can be compared to other models and analyzed statistically. SfM can be combined with machine-learning algorithms to classify specimens with complex features, such as skeletonized human remains that have been subjected to trauma or disease. Several samples of human remains from the Theban Tomb Complex 29 in the Valley of the Nobles, Luxor, Egypt, were photographed with both a consumer-grade smartphone camera and high-resolution digital camera for reconstruction in 3D. The images were uploaded to proprietary and open-source SfM software in order to compare the renderings for accuracy and quantifiable error. We found that the surface detail of a bone photographed with different devices will exhibit the same unique physical features that can be extracted and measured accurately by machine learning algorithms. When two 3D point clouds, rendered from the same bone but photographed with different devices, were digitally superimposed, the mean distance between the points in the two models was ≤ 0.11 mm ($s \leq 1.0$ mm). These features can be used for automatic classification of 3D data and comparison with existing 3D models of similar specimens. We also developed a protocol for simplifying and measuring point clouds despite differences in file size and resolution. As part of a pipeline in progress, the 3D data collected from TTC 29 is being added to a cloud-based repository that will be publicly available. Machine learning software will be implemented in order to develop a web-based application, available for mobile devices, that can be used to classify unique physical traits in bones of pathological interest. Ultimately, this technology will be deployed in both archaeological and forensic settings to scan cheaply and efficiently entire contexts and recreate them in 3D for later interpretation. This method will be particularly helpful in situations where the material cannot be transported for further studies.

Keywords: application programming interface, artificial intelligence, machine learning, osteomorphology, photogrammetry, pipeline

Human and Faunal Remains in Egypt: A New Department and a New Approach

Zeinab Hashesh[1] & Ahmed Gabr[2]
[1] Beni Suef University, zeinab.hashesh@gmail.com
[2] Ministry of Tourism and Antiquities, ahmedgabr85@gmail.com

Although the term bioarchaeology has been used in Egypt since the 1970s, the methodology used has often been ineffective. To mitigate the lack of expertise, and to increase the awareness about human and animal remains, the Ministry of Tourism and Antiquities (MoTA) has therefore collaborated with various foreign missions and organizations to provide training for MoTA archaeologists, who have an important role in ministry sponsored excavations. The authors of this paper have participated in several such programs since 2006, and have since then attempted to pass on what we learnt not only to other MoTA inspectors, but also to the public. This has been achieved by organizing human remains workshops and training programs in different inspectorates around Egypt and with different cultural authorities. While these measures have been successful in building an understanding among MoTA staff about the importance of bioarcheological research, the lack of care taken with such materials during earlier excavations remains problematic. Further, the training so far provided has mainly concentrated on the excavation of human and faunal remains, rather than documentation and curating of the materials. For that reason, a department within the MoTA for studying, documentation and storing of human and faunal remains was established in 2017, led by the first author of this paper. To date, the department has surveyed the bioarcheological remains in 34 MoTA magazines, in addition to 147 magazines of foreign missions. After completing the country-wide survey, the authors plan to document the results in a custom-made database. This paper will provide an overview of the methodology of the current survey, as well as an outline of the plans for the project going forward with the long and crucial process of documenting and curating human and faunal remains in Egypt.

Keywords: Atfih, cultural heritage management, human remains, Maadi, storage

Venerunt, Viderunt, Vicerunt: The Roman Conquest and the Non-Elite

Jessica Kaiser
Department of Near Eastern Studies, University of California Berkeley, jessicakaiser@berkeley.edu

In 31 BC, Octavian famously defeated the combined forces of Cleopatra and Mark Anthony at Actium, and Egypt became the breadbasket of Rome. Contemporary written sources suggest that this change in Egypt's status, from an independent power to a province of the Roman Empire, had far-reaching consequences for the Egyptian population. Taxes were raised and more efficiently collected, and the Romans instituted a more or less segregationist policy towards native Egyptians. Women in particular became increasingly marginalized when the Egyptian laws were abandoned in favor of the more patriarchal Roman legal system. However, a narrative based on textual sources generally paints a picture seen from the vantage point of the elite. Skeletal remains, in contrast, can provide insight into the quality of life among population groups rarely heard from in the literary record. The Wall of the Crow Cemetery in Giza presents an opportunity to examine how the Roman conquest affected a non-elite population from the Memphite region. The majority of the human remains from the cemetery fall into two distinct phases: an earlier period of use in the Saite period (664-525 BC) that was relatively prosperous and stable; and a later period of occupation in the early to mid-Roman period (1st-2nd century AD). Higher frequencies of markers of skeletal stress markers in the Roman sample suggest that the effects of the Roman conquest were felt not only by the elite, but also by the less fortunate.

Keywords: Bioarchaeology, Egyptian archaeology, Egyptology, markers of skeletal stress, Roman period, Saite period

Women's Health Issues Reflected in Case Studies from Theban Tomb 16

Suzanne Onstine[1], Jesus Herrerín,[2] Miguel Sanchez[3] & Rosa Dinarès[4]
[1] Deptartment of History, 101 Mitchell Hall, University of Memphis. Memphis, TN 38152 USA sonstine@memphis.edu
[2] Universidad Autónoma de Madrid, Fac. Ciencias, Darwin s/n 28049 Madrid, Spain jesus.herrerin@uam.es
[3] MD Chief, Deptartment of Pathology, Englewood Hospital 350 Engle St, Englewood, NJ 07631 USA miguel.sanchez@ehmchealth.org
[4] MD (retired General University Hospital of Catalonia) rdinares@gmail.com

University of Memphis fieldwork in TT16 at Thebes has yielded numerous human remains. Within this corpus of material there are several interesting cases related to the health of women in ancient Egypt. This article describes cases including death due to complications during childbirth, arthritis, parietal thinning, and *Hyperostosis frontalis interna.*

Keywords: arthritis, childbirth, Hyperostosis frontalis *interna, metabolic conditions, parietal thinning, secondary burials, women's health*

Creatures of the Sun, Creatures of the Moon: Animal Mummies from Lisbon's National Archaeological Museum

Salima Ikram[1], Carlos Prates[2], Sandra Sousa[2], and Carlos Oliveira[2]
[1] American University in Cairo
[2] IMI-art/Affidea

Although now scholars are increasing the attention paid to animal mummies, these artefacts are still less studied than their human counterparts. This study presents a group of mummies held by the Museu Nacional de Arqueologia (National Archaeological Museum) in Lisbon, and highlights the importance of examining them through various technologies. The study underscores the importance of using computer tomography (CT) in addition to traditional X-rays as the former reveals details that are not visible in the latter.

Keywords: ancient Egypt, animal mummies, crocodile, CT imaging, falcon, ibis

Brief Notes about a Mummified Crocodile from the National Archaeological Museum (MANN) of Naples, Italy

Ilaria Incordino[1] & Cinzia Oliva[2]
[1] University of Naples "L'Orientale", iincordino@unior.it
[2] oliva.c@libero.it

This paper deals with the ongoing analysis of an unusual mummified crocodile, currently on display in the Archaeological Museum of Naples (MANN, Room XXII, Hogg Collection 1883, Inv. No. 2338). The 2.54 m long mummy still preserves its almost complete original wrapping of linen bandages, palm leaves and ropes, and two small crocodiles deposed next to the adult one, showing remains of linen bandage covering the rear part of the bodies. On the occasion of the reopening on the Egyptian section of the museum, the crocodile was the subject of a conservation project, during which the original outer wrapping that had been cut along the length on the back at the time of the object's entrance to the collection, was studied and properly conserved. The outer, almost intact shroud, bandages, fragments of tunics and decorated textiles were also studied and analyzed during the course of the restoration work, with special attention to any technical data (fibers, dyes, weaving, fringe, stitches) that could help in dating the complete artefact. The mummy has been acquired by the Museum through a donation by Dr. Hogg in 1833, who traveled in Egypt and Near East, but is actually without certain provenance. Its date is problematic too, even if clearly related to the Late Period, because no persuasive parallels have been found so far.

Keywords: crocodile mummy, conservation, Dr. Hoggs, Naples Archeological Museum (MANN), wrapping

Faunal Remains at the Causeway of Sahura

Mohamed Ismail Khaled[1] & Mohamed Hussein Ahmed[2]
[1]Ministry of Tourism and Antiquities, Egypt/University of Würzburg, Germany, mohamed.khaled@uni-wuerzburg.de
[2] Ministry of Tourism and Antiquities, Egypt, husseinabydos@gmail.com

In 2017, the excavation at the uppermost part of the northern side of the causeway of Sahura revealed a Late Period (664-332 BC) settlement. This is the first known Late Period settlement recorded from the Pyramid Complex of Sahura Abusir, Egypt. The faunal remains from the excavation were studied to widen our knowledge of the function of the pyramid complex of Sahura during the Late Period. Also, we sought the reason behind the existence of such a settlement, especially, as the local cult of Sekhmet-Sahura begins from the middle of the New Kingdom (1481-1425 BC) and lasts to the 1st millennium BC. The Faunal analysis gives us clear confirmation for the use of this place as a settlement.

Keywords: causeway, cult, entrance hall, faunal remains, Late Period Settlement, Sahura, Sekhmet Sahura

Interactions Between Teeth and Their Environment: A Study of the Effect on Adult Dental Age Estimation

Casey L. Kirkpatrick
Department of Anthropology, Social Science Centre Rm. 3326, University of Western Ontario, London, Ontario, Canada, N6A 3K7, ckirkpat@uwo.ca

Teeth are the only elements of the human skeleton that interact directly with the environment, often leaving evidence that can provide information about ancient people and the world in which they lived. Dental wear is one of the most commonly documented conditions resulting from the interaction of teeth and the environment because of its relationships with diet, behavior, and age-at-death. Nevertheless, there have been few large-scale studies of dental wear in ancient Egyptians. This paper presents some of the initial results from an ongoing large-scale study of dental wear aimed at investigating its aetiology in ancient Egypt and creating region-specific adult dental age estimation standards. This project is currently focused on a population sample from the Kellis 2 cemetery (al-Dakhla Oasis,

Egypt), with future plans to expand this sample within and beyond the Kellis 2 cemetery.

This paper briefly discusses the aetiology and mechanisms of dental wear, and considers how dental erosion may have contributed to the relatively severe dental wear observed in ancient Egyptian populations. Using photographs of occlusal dentition from the Kellis 2 cemetery, a relatively new method for the photogrammetric quantification of dental wear is subsequently explored and shown to have high rates of inter- and intra-observer reliability: ideal for scoring dental wear. Comparisons of data collected through this method are then used to identify significant dental wear asymmetry between first molar isomeres and between first molar antimeres. These results are discussed in relation to the design of new adult dental age estimation standards.

Keywords: age estimation, dental erosion, dental wear, al-Dakhla Oasis, intra-oral wear patterns, methodology, photogrammetry

Discovery of an Unexpected Textile Fiber in a Fish Mummy from the Musée des Conflunces (Lyon) Collection

Fleur Letellier-Willemin

Project MAHES, Mission of al-Deir Kharga Oasis, Western Egyptian Desert, CRIHAM EA 4270 University of Limoges France, f.letellier.willemin@free.fr

The study of the wrapping of a fish mummy from Esna, stored in the Lyon Musée des Confluences, showed something very surprising: the textile was woven with cotton threads. The exceptional use of cotton raises questions about how religion and daily business interact, especially because it all takes place near a great temple in the Nile Valley. This discovery underlines the importance of studying animal mummies' textiles because of their highly symbolic meaning as well as the practicalities associated with the textile industry.

Keywords: animal, animal mummy, cotton, mummy, textile

Analyse des gazelles momifiées de Kom Mereh/Komir (Haute Egypte) conservées au Musée des Confluences (Lyon, France).

Stéphanie Porcier[1] & Louis Chaix[2]

[1] Laboratoire CNRS "Archéologie des Sociétés Méditerranéennes" (ASM_UMR 5140), Université Paul-Valéry Montpellier 3; Labex ARCHIMEDE, Site Saint-Charles 2, Route de Mende, 34199 Montpellier cedex 05, France, stephanie.porcier@gmail.com

[2] Département d'archéozoologie, Muséum d'histoire naturelle, 1 route de malagnou, CH-1211 Genève, louis.chaix@bluewin.ch

La campagne de fouille menée en 1882 par Gaston Maspero près du village de Kom Mereh/Komir (13 km au SE d'Esna) a mis au jour de nombreuses gazelles momifiées datant de l'époque romaine. Certaines gisaient dans l'une des chambres du temple de Komir. D'autres étaient enterrées dans la plaine à l'ouest du village, à même le sable ou dans des fosses quadrangulaires de 3 à 4 mètres de profondeur. Le musée des Confluences de Lyon conserve aujourd'hui une vingtaine de spécimens provenant des inhumations de la plaine. Ce lot envoyé par Maspero à Louis Lortet a fait l'objet d'une étude pluri/interdisciplinaire dans le cadre du programme de recherche MAHES (Momies Animales et Humaines EgyptienneS). Après l'étude réalisée sur les textiles de ces momies par Fleur Letellier-Willemin (2019), nous nous intéressons ici non plus au contenant mais au contenu des momies. Sont présentés les résultats de l'analyse archéozoologique réalisée sur les restes osseux et les momies encore enveloppées.

Mots-clés : gazelles, Kom Mereh/Komir, mummies, Roman period

Did Egyptians Eat Donkeys? Reflections from Historical and Archaeological Data

Mathilde Prévost[1] & Joséphine Lesur[2]

[1] Sorbonne Université, Faculté des Lettres (École Doctorale 22 'Mondes anciens et médiévaux')/Resarch Unit 8167 'Orient et Méditerranée', team 'Mondes pharaoniques', prevostmathilde48@gmail.com

[2] Museum National d'Histoire Naturelle, in Research Unit 7209 'Archéozoologie et archéobotanique', jolesur@mnhn.fr.

The paucity of faunal remains and historical documents suggesting that the ancient Egyptians ate donkeys led to the assumption that there was a taboo prohibiting their consumption. However, as Y. Volokhine questioned the existence of such an interdiction that could apply to the entire Egyptian population, it seems relevant to review all the evidence, from archeozoological as well as historical data, and to re-evaluate to what extent donkeys were eaten and in which daily or ritual contexts.

Keywords: alimentation, daily life, donkey, history, ritual, taboo, zooarchaeology

What I Have Learned: Assumptions Bad, Intersections Good (keynote lecture)

Richard W. Redding
Kelsey Museum, University of Michigan, Ann Arbor, Michigan, 48104/Ancient Egypt Research Associates, 26 Lincoln St., Suite 5, Boston, Massachusetts, 02135, rredding@umich.edu

Based on 52 years of research in bioarchaeology, specifically, archaeozoology, I argue that assumptions create problems, while intersection between fields and colleagues are critical in advancing bioarchaeology. Here, I present three examples of invalid assumptions that have affected research. These include the origins of wet-cooking, the use of bone density as a proxy for bone survivability, and the use of limb bone fragments as indicators of butchering versus consumption. I go on to discuss how challenges to these assumptions are the result of intersections, sometimes between fields, sometimes between colleagues, and sometimes both.

Keywords: assumptions, bone density, Heit al-Ghurab, intersections, Shorbet Kawara, wet-cooking

Biomolecular Stable Isotope and Carbon-14 Dates of Ancient Egyptian Food Offerings: A Case Study from a Provincial Cemetery of Deir al-Ballas

Amr Khalaf Shahat[1] & Victoria Jensen[2]
[1] Cotsen Institute of Archaeology, A210 Fowler Building, Box 951510, 308 Charles E. Young Dr. North, University of California Los Angeles, Los Angeles, CA 90095, United States of America, akshahat@ucla.edu
[2] Independent researcher, Qurna, Egypt, vjensen@berkeley.edu

This paper presents baseline ^{13}C stable isotope data and AMS ^{14}C dates for five botanical samples excavated from Cemetery 1-200 at Deir al-Ballas, Upper Egypt. The site is located on the western bank of the Qena bend of the Nile, approximately 40 km north of Luxor, and was excavated by the Hearst Expedition of the University of California under the direction of George A. Reisner in 1900-1901. It yielded important archaeological features such as a palace used between the reigns of Seqenenre Taa (ca. 1553 BC) and Ahmose (ca. 1550-1525 BC), a monumental mudbrick two-level platform at the south end of the site, and numerous houses and administrative buildings. A workers' village consisting of tightly spaced mudbrick houses was subsequently reused as a cemetery when the floors of many rooms and alleyways were cut through to insert the graves of Cemetery 1-200. Based on the material culture, these graves date between the early 18th Dynasty to the early 19th Dynasty, with only a few tombs containing material that could possibly date as early as the late 17th Dynasty. While there was variability in the quantity and variety of grave goods as well as the subterranean tomb architecture indicating a range of socio-economic hierarchy in the local population, none of the tombs was decorated or provided evidence of elite titles. Archaeobotanical analysis of remains from Cemetery 1-200 yielded a total of 243 specimens of 13 species (Shahat & Jensen, in press). In this article we focus on five graves from this cemetery, reporting AMS^{14}C dates on botanical findings from the tombs to help determine the time span of the burials, ascertain that these species are not intrusive, and relate the archaeobotanical offerings to the material culture assemblages. Additionally, the carbon stable isotope measurements help identify local species grown with Nile water versus imported species based on water-use efficiency reflected by the ^{13}C. The botanical species of focus here include dom fruit from Tomb 105, juniper berries from Tomb 128, grapes from Tomb 163, the unusual finding of domesticated watermelon from Tomb 244 or 255, and pomegranate from Tomb 257. Although the samples presented here came from funerary contexts, they offer a baseline with significant implications for further applications of stable isotope in archaeobotany and bioarchaeology of diet in ancient Egypt. We recognize the caveat that the species found in funerary food offerings might not be representative of the diet of individuals. However, we argue that the mortuary offerings represent food items that were available to at least some members of the community during life. Furthermore, all of these fruit species except for pomegranate have been identified in non-elite settlement contexts, either in a recent archaeobotanical analysis by A. Shahat in Deir al-Ballas or in the Amarna Workmen's Village. This article aims to communicate the results of stable isotope analysis in plants to bioarcheologists to serve as a baseline on the diversity of plant sources contributing to isotopic input observed in bioarchaeological remains.

Keywords: archaeobotany, Deir al-Ballas, 18th Dynasty, food offerings, non-elite, radiocarbon dates, stable isotope

Animal Butchering Technology in Old and Middle Kingdom Egypt: Textual and Iconographic Evidence for the Shift from Stone to Metal Tools

Eleuterio Sousa* & Haskel J. Greenfield
University of Manitoba, Department of Anthropology and St. Paul's College, Winnipeg, MB, R3T5V5, Canada,
*Corresponding author: Eleuterio.sousa@icloud.com

While research has been carried out on meat consumption and its importance in secular and religious life in ancient Egyptian society, there is a gap in knowledge that coincides with the transition from a stone- to metal-based food processing technology. In this paper, the textual and iconographic evidence for the different technologies and their raw materials used in meat processing during the 3rd millennium BC in Egypt are discussed. The shift from stone to metal tool technology in mundane matters, such as meat processing, occurs long after the introduction of earliest metal tools. It is most likely that metal based butchering technology appears during the Middle Kingdom, based on the current state of knowledge.

Keywords: butchery, copper, Early Bronze Age, metal, metallurgy, technology

Anthropological Study of the Egyptian Mummy from the Pushkin State Museum of Fine Arts Using Computed Tomography

Sergey V. Vasilyev[1*], Ekaterina B. Yatsishina[2], Ravil M. Galeev[3], Svatlana B. Borutskaya[4], Mikhail V. Kovalchuk[2], Olga A. Vasilieva[5], Olga P. Dyuzheva[5] & Vadim L. Ushakov[2]
[1] Institute of Ethnology and Anthropology, Russian Academy of Sciences, Leninsky pr., 32A, 119991 Moscow, Russia, vasbor1@yandex.ru (* corresponding author)
[2] National Research Centre "Kurchatov Institute", Akademik Kurchatov Sq., 1, 123182 Moscow, Russia
[3] Institute of Ethnology and Anthropology, Russian Academy of Sciences, Leninsky pr., 32A, 119991 Moscow, Russia
[4] Lomonosov Moscow State University, Leninskie Gori 1, 119991 Moscow, Russia
[5] Pushkin State Museum of Fine Arts, Volkhonka 12, 119019 Moscow

This work presents the results of paleoanthropological research using the method of computed tomography of the ancient Egyptian mummy of Khor-ha. The mummy dates to the 7th to 4th centuries BC and is stored in the Pushkin State Museum of Fine Arts under inventory number №5301. For magnetic resonance scanning, Siemens Magnetom Verio magnetic field strength 3 Tesla magnetic resonance imaging (MRI) was used. X-ray computed tomography of the mummy was performed on a positron emission computed tomography (PET-CT) Biograph mCT40 by Siemens. This paper presents the results of this study, including a facial reconstruction, using the method of M.M. Gerasimov.

Keywords: ancient Egypt, anthropology, computed tomography, facial reconstruction, osteology

Intentionally Burnt Human Remains from Kom Ombo Temple Salvage Excavation

Afaf Wahba
Ministry of Tourism and Antiquities, Cairo
afafwahba4@yahoo.com

This paper presents preliminary anthropological research from the salvage excavations carried out in conjunction with the Ground Water Lowering Project (GWLP) at Kom Ombo temple (2018). The material came from salvage excavations, The archaeological monitoring revealed skeletal remains from two diffirent periods and two different areas within the temple, The material presented in this paper comes from the Ptolemaic period and is represented by numerous inhumations of what appears to be intentionally burnt human bones that were found inside two chambers. These chambers were designed originally for residential purposes and not as a burial place. Each room contained the remains of several individuals whose bones showed signs of in situ burning.

Keywords: burnt bone, cremation, Kom Ombo temple, Ptolemaic Period

Curatorial Training in Human Remains for the Egyptian Museum, Cairo

Hasnaa Askalany & Gehad Shawky Ibrahem

Introduction

A group of inspectors and museums curators of the Ministry of Antiquities and Tourism participated in a course on studying and curating human remains that was jointly organized by the Egyptian Museum, the Institute for Bioarchaeology, the American University in Cairo, and the Ministry of Antiquities, in 2018. The aim of the course was to train the group, which had no previous experience with human remains, to learn basic analytical methods, and to learn how to curate the remains in museums, archeological sites, and storage magazines. The program included a theoretical part, carried out at the American University in Cairo, and was followed by practical sessions at the Egyptian Museum (Figure 1). The theoretical component, which included some haptic learning, taught basic anatomy, how to age and sex, handle bones, pack skeletons, organize collections of human remains, and to identify major threats (environmental and verminous) to such collections and to safeguard against them. The goal of the practical portion of the course was to implement what had been learned during the theoretical portions.[1]

The Practicum: Box TR 28-4-26-27

The material provided for the practical part of the course was kept in a large wooden packing case in the Egyptian Museum (about 1.8 x 1 x 0.8 m), and it consisted of crania and other human remains. A label on the box suggested that the remains were from Bab al-Gusus, a group burial of the clergy of Amun of the 21st Dynasty (ca. 1069-945 BC), located outside the northeastern corner of the Hatshepsut temple (Figure 2). However, the contents of the box turned out to be from a variety of excavations, rather than from a single one. A total of 82 specimens were found, including some non-human animals; the majority of these were excavated from different sites by Jacques de Morgan. The specimens could be identified based on information put on them, in pencil or ink, by the excavators, such as site name, excavation number, and other markings. After investigating the material, we found that many of the specimens appeared in his 1896 book, *Recherches sur les Origines de l'Égypte*, in which Dr Fouquet provided the analyses of the skeletal material, while de

---

1 We would like to thank Madam Sabah Abelrazzek, Director of the Egyptian Museum, for providing the opportunity to participate in the course and for facilitating the practical section, our teachers, Roxie Walker, Afaf Wahba, Zeinab Hashesh, Ahmed Gabr, Amira Shahin, and Salima Ikram for their help and encouragement, and the Ministry for supporting this course. We are also indebted to the other curators and colleagues who formed part of this 'rescue' team and participated in the course. In addition to the authors, they were Roqaya Ali, Sally Hosney, Asmaa Ahmed, Ahmed Abdel-Latif, and Enas Ahmed.

In: Ikram, S., J. Kaiser & S. Porcier. Eds. 2021. *The Ancient Egyptians and the Natural World. Flora, Fauna, and Science.* – Leiden, Sidestone Press: 19-22.

Figure 1. Roxie Walker, the manager of the project with the trainees in the practical training that place in the Egyptian Museum. Photograph by H. Askalany.

Figure 2. The opened box. Photography by H. Askalany.

Morgan focused on the material culture. It was rewarding to find the raw material that yielded the data for Fouquet and de Morgan's contributions to and interpretations of the physical anthropology in Egypt during the late 19th century, albeit controversial, and, in light of modern science, largely incorrect. These materials would certainly benefit from further study.

The team was able in some cases to match cranial and post cranial material that was found in different parts of the large packing case as the excavators had written on some of the bones, particularly once we had learned to decode the excavator's abbreviations (NS is Naqada South, K for Kawamil, etc.) (Figures 4, 5). The sites from which the contents of the box originated, insofar as could be identified, were: Bab el Gasus, al-Amra, Gebel el Silsila, Naqada North, Naqada South, Kawamil, Deir el Bahari, and Beit Alam (Figure 3).

We first removed all the material from the packing case, which was thoroughly vacuumed. Then each body-part was cleaned using brushes and blowers such as are used to clean camera lenses. Any papers (wrapping or identification slips) associated with each anatomical element was also cleaned and conserved by the paper conservators of the museum, and notes made on these materials. These included the French-Arabic (Sada Al-Sharq – صدى الشرق) newspaper dated to 1895. After cleaning, each element was documented. We recorded: any information concerning the provenance, excavator,

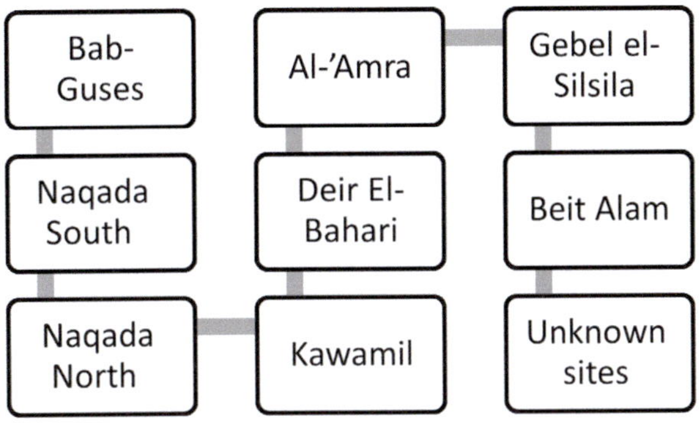

Figure 3. Distribution of skeletal elements from identifiable sites.

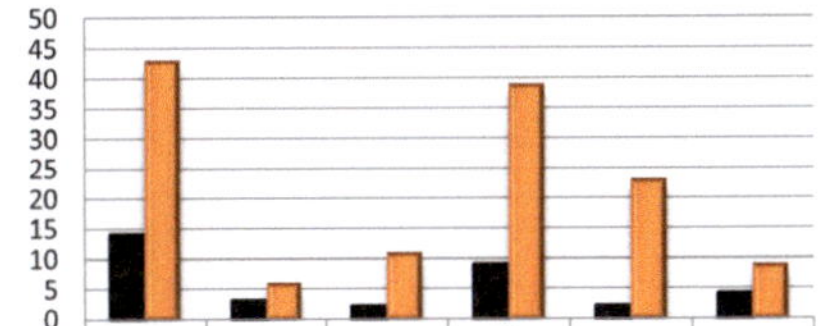

	NAQADA SOUth	NAQADA NORTH	ALAMRA	E-KAWAMIL	BEIT ALLAM	GABEL EL SILSILA
■ Black column Skelteal Material of De Morgan in 28-4-26-27	14	3	2	9	2	4
■ the orange colomn total Skelteal Material of De Morgan	43	6	11	39	23	9

Figure 4. Materials recovered from TR 28-4-26-27 that show the number of anatomical elements mentioned in de Morgan's publication in contrast with what we found.

date excavated, described the element and when possible determined sex, age, and tried to identify the presence of any pathologies. These will be the focus of a future study. After this, each element was wrapped in acid-free tissue, provided with supports, and re-housed, following best practice.

The materials found in the packing case were certainly not all of those used in de Morgan's book, and some were probably not from his excavations. This is particularly true of the mummified remains (Figure 4).

Many of the unprovenanced materials consisted of mummified fragments, which might have come from Bab el Gassus. These included diverse body parts, none of which seem to belong together. These were also catalogued and examined and were used for training on how to study mummification and wrappings.

Conclusion

This course has illustrated the importance of creating a group of curators who have a basic knowledge of human and animal anatomy, as well as of best practice in terms of handling, cleaning, and storing bones and mummified remains, enabling them to deal with these materials in museums and on-site storage facilities. If some (all) of these individuals have more knowledge of such remains, it would be beneficial, but these curators can work in tandem with specialists. The combination of teachers, each with different expertise, who gave diverse lectures and led different laboratory sections, was beneficial and stimulating. Most useful was the chance to implement the theoretical part of the course. In addition to providing actual experience handling such remains, it gave us a tremendous sense of pride and accomplishment to 'rescue' skeletal and mummified remains, to clean them, re-house them, and to demonstrate that we had learnt the basics of anatomical identification, ageing and sexing of humans, and examining mummies. The Egyptian Museum and the many on-site storage areas spread throughout the country are filled with skeletal and mummified material that is in need of cataloguing, conserving, and studying. We hope that the training project that we participated in is the first of many so that a larger group of specifically trained curators can be created to deal specifically with such remains, so that they can be preserved and studied, and trust that the Ministry of Tourism and Antiquities and the Museum Sector will support similar programs in the future.

Hyperostosis Frontalis Interna in the Early Dynastic Period at Abydos, Egypt

Brenda J. Baker & Ahmed Gabr

Introduction

Hyperostosis frontalis interna (HFI) is a pathological condition of unknown etiology that affects the internal surface of the frontal bone. The prevalence of HFI in past populations is poorly known. The discovery of HFI in four individuals from two Early Dynastic cemetery contexts at Abydos, Egypt, suggests that the frequency of this condition may be much greater in past populations than has been generally acknowledged.

Hyperostosis Frontalis Interna in Modern and Past Populations

HFI manifests as nodular lesions and thickening of the endocranial (internal) surface of the frontal bone. If other cranial vault bones are involved, the condition is called *hyperostosis calvaria interna* (HCI) or *hyperostosis cranialis diffusa* (HCD) and should be considered an independent process with a different etiology even if it co-occurs with HFI (Hershkovitz *et al.*, 1999). In modern postindustrial populations, HFI is far more common in females than in males, particularly in postmenopausal females (Hershkovitz *et al.*, 1999; Western & Bekvalac, 2016). With a reported sex ratio of up to 10:1 affected females to males in clinical studies (Hershkovitz *et al.*, 1999: 318), it has been suggested that the presence of HFI can "*serve as a means of sexing and aging in archaeological material*" (Ortner, 2003: 416). Such an application in archaeological samples is inadvisable, however, as the condition seems to have affected males at a higher rate in antiquity than today (Rühli *et al.*, 2004: 95; Watrous *et al.*, 1993). Even in cadaver studies, it has been demonstrated that males with less severe manifestations are often overlooked and a more realistic sex ratio of 5:1 has been proposed for those under age 60 and 3:1 in those age 60 and older (Hershkovitz *et al.*, 1999: 318; *cf.* Raikos *et al.*, 2011: 457).

Although the specific cause of HFI is undetermined, the condition has been shown to be independent of any syndrome, such as Morgagni-Stewart-Morel-Moore syndrome, to which it was initially linked (Hershkovitz *et al.*, 1999; Raikos *et al.*, 2011: 456). HFI generally is associated with hormonal changes related to menopause in females and hypogonadism in males (Hershkovitz *et al.*, 2009: 322-324; see Western & Bekvalac [2016] for a review of the literature). Obesity and diabetes also are frequently linked with HFI today (Raikos *et al.*, 2011: 457; Verdy *et al.*, 1978; Western & Bekvalac, 2016). Estrogen, androgen, leptin, resisten, and dietary phytoestrogens have all been suggested as factors affecting the development of HFI (May *et al.*, 2010; Raikos *et al.*, 2011: 457-458; Rühli & Henneberg, 2002; Rühli *et al.*, 2004; Western & Bekvalac, 2016).

HFI is thought to be infrequent in the past, becoming much more common only after the Industrial Revolution (*e.g.*, see Hajdu *et al.*, 2009; Hershkovitz *et al.*, 1999; Rühli *et al.*, 2004). The modern prevalence of HFI is generally reported to range from 5-12% (*e.g.*,

In: Ikram, S., J. Kaiser & S. Porcier. Eds. 2021. *The Ancient Egyptians and the Natural World. Flora, Fauna, and Science.* – Leiden, Sidestone Press: 23-30.

Figure 1. Satellite view of Abydos showing the areas of Early Dynastic mortuary activity. Aha's funerary enclosures (1) are in the North Cemetery (2), where enclosures of other Dynasty 1 and 2 rulers are also located. The mudbrick enclosure of Khasekhemwy (3), is the only royal funerary enclosure still standing. The royal tombs (4) are toward the western cliffs, while the South Cemetery (6) is south-southeast of the New Kingdom temple of Seti I (5). Map by A. Gabr., based on GoogleEarth.

May *et al.*, 2011: 392; She & Szakacs, 2004: 207). In modern samples, frequency of HFI typically ranges from 18-25% in females, with only 1-5% of males affected (Cvetković *et al.*, 2018: 174; May *et al.*, 2011: 392; Nikolić *et al.*, 2010: 206). Barber *et al.* (1997:158) suggested an overall prevalence of 1-4% in archaeological samples, a rate supported by some systematic studies of skeletal collections (*e.g.*, Hajdu *et al.*, 2009; Sperduti & Manzi, 1990) but not by others (*e.g.*, Flohr & Witzel, 2010; Lazer, 1996: 622-623; Mulhern *et al.*, 2006; Szeniczey *et al.*, 2018) that show rates within, or even higher than, the reported range in modern populations. Studies of historical collections and cadaver series have demonstrated a correlation with age and longevity (*e.g.*, Hershkovitz *et al.*, 1999; Nikolić *et al.*, 2010; Western & Bekvalac, 2016), with the condition becoming increasingly common in younger age cohorts in recent decades (May *et al.*, 2011). What is clear from such studies is that HFI distribution has altered from the 19th to 21st centuries. Its distribution in populations that predate industrialization, therefore, should not be based on modern patterns of HFI and merits increased attention in past populations.

In paleopathology, scoring for HFI is tied to stages of severity and has been evaluated through macroscopic observation, radiography, and computed tomography (*e.g.*, Barber *et al.*, 1997; Hershokvitz *et al.*, 1999; May *et al.*, 2011; Western & Bekvalac, 2017). Hershkovitz *et al.* (1999: 306-309) provided a classification scheme for macroscopic observation with four types that has become the standard method of scoring HFI. Type A consists of raised single or multiple "bony islands" with discrete margins, generally less than 10 mm in size, and either unilateral or bilateral and most commonly on the anteromedial aspect of the frontal bone. Type B shows nodular "bony overgrowths" that lack discrete margins and have slight or no elevation and cover less than 25% of the frontal bone. Type C shows more extensive "nodular bony overgrowth" and irregular thickening on up to half of the endocranial surface, with type D consisting of an irregularly elevated and clearly demarcated area covering more than half of the surface. In this study, individuals with observable complete or partial frontal bones who were at least 15 years of age at death were included.

Early Dynastic Abydos

Early Dynastic mortuary activity at Abydos, in Upper Egypt, is located in the area of the royal tombs at Umm al-Qa'ab, near the base of the western cliffs, and their associated funerary enclosures in the North Cemetery, approximately 1.6 km to the north (Figure 1). The South Cemetery is an elite cemetery approximately 1.5 km northeast of the royal tombs and 1.7 km southeast of the funerary enclosures (Figure 1). Skeletal remains examined for this article include only those from recent excavations of funerary enclosures in the North Cemetery and tombs in the South Cemetery.

In the North Cemetery, excavation of three funerary enclosures associated with King Aha of Dynasty 1 were conducted in 2002-2003 (in which BB participated) and in 2004-2005 as part of the University of Pennsylvania Museum-Yale University-Institute of Fine Arts, New York University Expedition to Abydos (PYIFA; Adams & O'Connor, 2003; Bestock, 2008; 2009; O'Connor, 2009). Five of the six subsidiary graves around the larger, main enclosure were excavated, along with five of the six graves surrounding the smaller two enclosures associated with Aha. The subsidiary tombs around all three of the Aha enclosures were single chambers with walls of mudbrick that were plastered with mud. Tombs around the main enclosure had floors of limestone cobbles. The graves around the two smaller enclosures had only sand floors with reed matting on the floor of at least two of them (Bestock, 2009: 73-86). Bodies were placed in wooden coffins. One grave around the smaller enclosure was intact (Bestock, 2008: 48), while all others were badly disturbed. Despite looting, it is evident that furnishings for the subsidiary burials associated with the main enclosure were much richer than those of the smaller enclosures, including wine jars and other ceramic and stone vessels, jewelry of ivory, lapis lazuli, and carnelian, and animal bone indicative of food offerings. Interments around the two smaller enclosures were accompanied only by ceramic vessels and jar stoppers (Bestock, 2009: 73-86). The 10 excavated individuals (Table 1) include one young child, a young adult male, four young adult females, three middle adult females, and one old adult female (Baker, 2013).

The South Cemetery is located approximately 435 m south-southeast of the New Kingdom temple of Seti I at

	Female (n = 14)				Male (n = 7)					Unknown (n = 6)		Total
	YA	MA	OA	Adult	Adol.	YA	MA	OA	Adult	Preadult	Adult	
Aha enclo-sures	4	3	1	0	0	1	0	0	0	1	0	10
South Cemetery	2	1	0	3	1	2	2	0	1	4	1	17
Total	**6**	**4**	**1**	**3**	**1**	**3**	**2**	**0**	**1**	**5**	**1**	**27**

Table 1. Sex and age distribution of the Early Dynastic samples from the Aha enclosures and South Cemetery. Age ranges for Preadult = 0-14 years, Adolescent (Adol). = 15-19 years, Young Adult (YA) = 20-34 years, Middle Adult (MA) = 35-49 years, Old Adult (OA) = 50+.

the edge of the modern village of al-Araba al-Madfuna (Figure 1). Excavation of this cemetery (in which AG has participated) was initiated by the Supreme Council of Antiquities in 2007 (Gabr, 2011; Hossein, 2011) and has continued by the Ministry of Antiquities to 2019. Although these graves have all been badly disturbed, enough remains to conclude that the tombs were constructed of mudbrick with mud-plastered walls and floors (Hossein, 2011). Most tombs were multichambered (Figure 2), and the bodies within were all interred in wooden coffins. Accompaniments found in these tombs include ceramic and stone vessels, a wooden box, and faunal remains (Hossein, 2011). The grave construction and inclusions indicate that the people buried in the South Cemetery during the Early Dynastic period were of very high status. The human remains include 17 individuals (Table 1). Of the six adult females, two are young adults, one is a middle adult, and three are of unknown age. Of the six males, one is an adolescent, two are young adults, two are middle adults, and one is of unknown age. One adult of unknown sex is present, as well as one young child and three older children.

HFI from the Aha Enclosures

From the Aha enclosure burials excavated in the North Cemetery, only the skeletal remains of the individual buried in the southeast subsidiary grave of the main enclosure show evidence of HFI (Table 2). This grave, like the others around the main enclosure, was a separate mudbrick tomb for a single individual, with mud-plastered walls and a floor of limestone cobbles. The grave was looted in antiquity but retained portions of the wooden coffin, ceramic and calcite vessels, and a ceramic palette. The skeleton was disturbed and largely disarticulated, with only the cranium, legs and feet left *in situ*. Although the upper body was incomplete, bones present are well preserved. The individual is estimated to be a female based upon a sharp orbital margin, lack of supraorbital ridges and glabellar development, extremely vertical forehead, and extremely broad greater sciatic notch with a large preauricular sulcus (Buikstra & Ubelaker, 1994). Although neither pubis was present, age was estimated primarily from the auricular surface, which showed no transverse organization and a mostly dense surface with slight coarse granularity near the apical area and other characteristics corresponding with Lovejoy *et al.* (1985) phase 6, age 45-49.

The internal table of the frontal bone of this female exhibits multiple bony nodules to the right of the frontal crest and three to the left (Figure 3). The largest of the nodules is 4.0 mm wide by 7.1 mm long and is raised 3.2 mm above the endocranial surface. The lesions and extent of involvement correspond to Hershkovitz *et al.* (1999) type A (Table 2).

Figure 2. Abydos South Cemetery Tomb 3 during excavation. Photograph by Mohamed Samah. Courtesy of Yasser Hossein.

Cemetery	Tomb	Skeleton No.	Sex	Age	HFI Type
North	Aha encl. SE	ANC2002.57	Female	45-49 y	A
South	3	121	Female	40-49 y	C
South	4	126	Male	20-25 y	C
South	Surface cleaning	136	Female	Middle Adult	D

Table 2. Sex, age, and HFI classification for affected Early Dynastic individuals from the Aha enclosures and the South Cemetery.

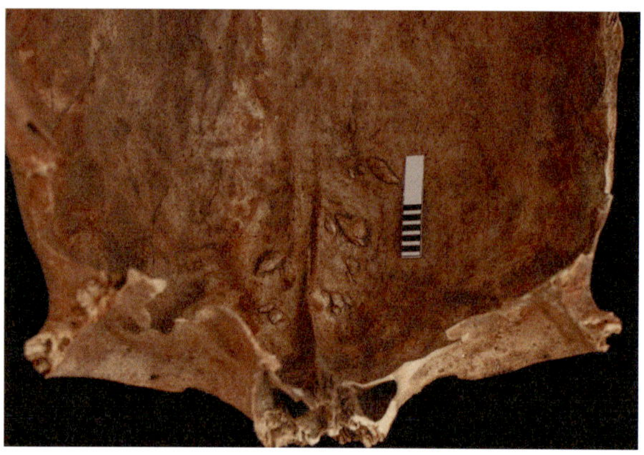

Figure 3. The endocranial surface of the frontal bone of a 45-49-year-old female (ANC2002.57) from the southeast subsidiary grave of the main Aha enclosure in the North Cemetery. The raised bony nodules on either side of the frontal crest are indicative of HFI type A. Photograph by B. Baker.

South Cemetery HFI

Three of the 13 adolescent-adult individuals from the South Cemetery display lesions characteristic of HFI, including a middle adult female, a female in her 40s, and a male age 20-25 (Table 1). Tomb 3 (Figure 2) was excavated in April 2019. The skeleton was disturbed and the bones in fragile condition. Sex was estimated principally from features of the skull, including a very sharp orbital margin, and slight development of glabella and the mental eminence (Buikstra & Ubelaker, 1994). The greater sciatic notch was also judged as female. Age was based on the auricular surface. The surface is granular with islands of densification and some activity at the apex and in the retroauricular area, corresponding with Lovejoy *et al.* (1985) phases 5-6, age 40-49. Cranial vault sutures are all completely closed. The frontal bone is fragmentary, with only about half of it remaining (Figure 4). The bone displays thickening with bony overgrowth lacking a discrete margin. Despite the incomplete condition, it is estimated that more than 25% of the bone surface was involved. These characteristics are most consistent with type C (Table 2).

Tomb 4 was also disturbed in antiquity. This tomb is the one of the largest to have been excavated to date in the South Cemetery. The skeleton is incomplete, but all of the frontal bone is preserved. Sex was estimated on the basis of cranial and pubis morphology (Buikstra & Ubelaker, 1994; Phenice, 1969). The ischiopubic ramus is wide with a V-shaped subpubic concavity and there is no ventral arc. The mandibular third molar is erupted. A fusion line is evident on the distal fibula, while the iliac crests, medial clavicle, and vertebral ring epiphyses are unfused. The auricular surface and pubic symphysis are observable. Estimated age at death based on all these factors is

Figure 4. The endocranial surface of the fragmentary frontal bone from the South Cemetery Tomb 3 female (Skeleton 121) who died in her 40s. Thickening and bony overgrowth is consistent with HFI type C. Photograph by A. Gabr.

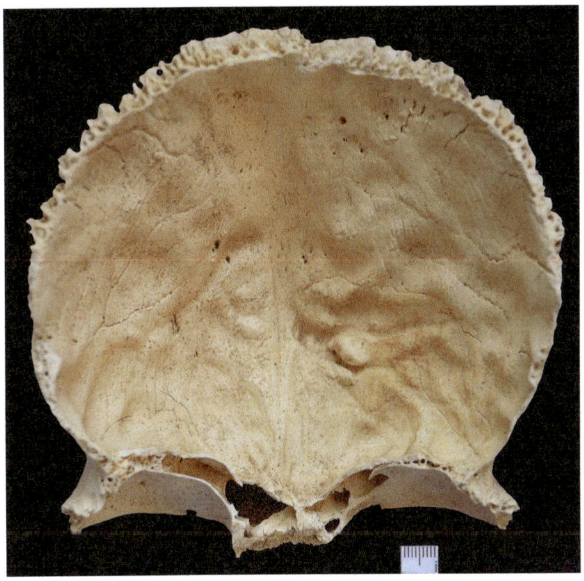

Figure 5. The internal aspect of the complete frontal bone of a 20-25-year-old male (Skeleton 126) from the South Cemetery Tomb 4. Photograph by A. Gabr.

20-25 years. The endocranial surface of the frontal bone shows nodular overgrowth and thickening on both sides of the frontal crest and sagittal sinus that covers 25-50% of the surface (Figure 5). The changes are consistent with type C (Table 2).

Fragments of a disturbed skull from a probable middle adult female were found on the surface outside of Tomb 4 (Figure 6). Also represented are part of the right parietal, the right temporal, and a fragment of right maxilla with empty sockets for the incisors, canine, and premolars. Sex was estimated from both the mastoid process and orbital margin. The midcoronal suture and area of bregma are completely closed. Less than half of the frontal bone is present, but the endocranial surface shows extensive thickening with undulating bony overgrowths and vascular tracks covering the surface as far as the groove for the middle meningeal artery, which usually is not crossed (Figure 5). The extent of involvement, therefore, is in keeping with HFI type D (Table 2).

Discussion

The frequency of HFI in these small samples from royal and elite Early Dynastic contexts at Abydos is high. One of nine adults from the subsidiary burials of the Aha enclosures shows HFI (11.1%), while three of 13 in the South Cemetery are affected (23.1%). The overall rate of HFI in the combined sample (four of 22 individuals) is 18.2%, which is considerably higher than rates reported for many modern samples. This finding may be due in part to the age and sex composition of the sample (Table 1) in which there are 14 females (six young adults, four middle adults, one old adult, and three adults of unknown age) and seven males (one adolescent, three young adults, two middle adults, no old adults, and one adult of unknown age). Although the ratio of females to males is 2:1 in this sample, only one old adult is represented (although the four adults with an unknown age could also be older adults). Three of the 14 females are affected (21.4%) and all are middle adults. This rate is comparable to that found in modern females. One of the seven males is affected (14.3%), which is higher than expected. This individual is a young adult with moderate involvement that is not frequently seen in males, particularly in this age group.

These results support the postulate that lifestyle and diet may be important factors in the development of HFI (*e.g.*, Flohr & Witzel, 2010; Szeniczey *et al.*, 2018), rather

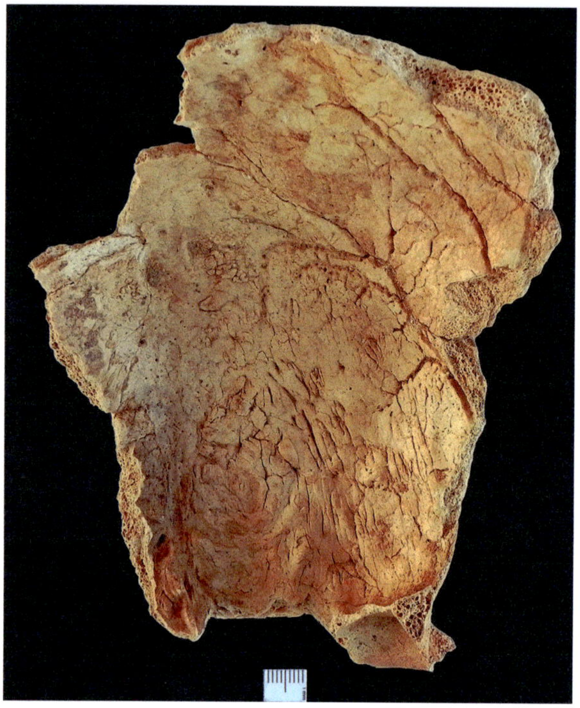

Figure 6. The endocranial surface of a fragmentary frontal bone from a female middle adult (Skeleton 136) found in surface cleaning around Tomb 4 shows severe manifestations of HFI including thickening and extensive bony overgrowth typical of HFI type D. Photograph by A. Gabr.

than being related only to age and the increased longevity found in modern postindustrial populations. Where archaeological skeletal collections have been examined systematically, HFI appears to occur at high frequency (comparable to modern rates) in samples comprised of high-status individuals (*e.g.*, Flohr & Witzel, 2010; Mulhern *et al.*, 2006). Flohr & Witzel (2010) suggest that the individuals in their sample from a Bronze Age palace in Syria would have had a high-calorie diet and low physical activity, increasing their risk of obesity and metabolic and endocrine disorders. They report nine affected in a sample of at least 70 individuals (yielding a frequency of 12.8%). Szeniczey *et al.* (2018) examined 4668 crania from 57 sites in the Carpathian basin across different time periods from 4900 BC to the 1600s AD and found a higher frequency of HFI in pastoralist versus sedentary groups. HFI prevalence in the Migration and Hungarian Conquest periods of 14.4% and 14.3%, respectively, was linked to the lifestyle of mobile pastoralists with diets rich in meat and dairy products and low in carbohydrates that reduced glucose levels and elevated insulin levels, the latter affecting sex hormones implicated in the development of obesity, type 2 diabetes, and likely HFI as well (Szeniczey *et al.*, 2018: 116-117).

The diet of those buried at Abydos during the Early Dynastic period may have differed from non-royal interments of later periods (especially Middle Kingdom and later) based on dental indicators (see Baker, 1997: 110-111). Attrition in the Early Dynastic sample from the Aha enclosures is mild compared with individuals of similar age groups who died during later periods, although it is more severe in the older adults from the South Cemetery compared with those from the Aha enclosures. Frequencies of dental enamel hypoplasia, caries, and abscesses among the courtiers are also quite low compared with later groups interred at Abydos, although abscesses were noted in some older individuals. These Early Dynastic courtiers and elite, therefore, likely were better nourished with diets containing less starch and more meat than lower status people of later periods. Future analysis of stable isotopes may help elucidate the relative proportion of protein in the diet of these individuals compared with lower status people interred at Abydos in subsequent periods.

Alternatively, the high rate of HFI seen in the Early Dynastic Abydos sample may reflect genetic factors, particularly given observations that some archaeological samples in which HFI has been found may consist of closely related individuals (*e.g.*, Glab *et al.*, 2005; Hajdu *et al.*, 2009: 201; Mulhern *et al.*, 2006: Watrous *et al.*, 1993). It is not unreasonable to assume that the individuals buried around royal funerary monuments and in a nearby elite cemetery may have come from closely related families. Further research using nonmetric skeletal traits and metric data may help elucidate relatedness in the Abydos Early Dynastic sample, as would analysis of DNA if it becomes possible in the future.

Few reports of HFI in ancient Egyptians have been published, with only eight others documented to date (Bebel & Golijewskaja, 2015; Mant, 2014; Rösing, 1990; Shahin *et al.*, 2013; Watrous *et al.*, 1993). The oldest individual with HFI that has been reported previously is from Dynasty 2 at Tarkhan (Shahin *et al.*, 2013). As with most archaeological examples from other regions, HFI reported in Egypt tends to be from incidental discoveries in fragmented crania rather than systematic studies that include examination of intact skulls. This study extends the presence of HFI back to the very beginning of the 1st Dynasty, ca. 3100-3000 BC. HFI clearly has great time depth in Egypt but we still know very little about its frequency over time, its distribution by sex and age, or across socioeconomic groups. Contextual information concerning the samples in which HFI is found is crucial to understanding the distribution of this condition. Expansion of this research, therefore, is highly recommended.

Conclusion

HFI in ancient Egyptians affected ancient Egyptians by the beginning of the 1st Dynasty based on its presence in a subsidiary burial from King Aha's enclosure. The study of this small sample of Early Dynastic skeletal remains from Abydos demonstrates the need for systematic investigations of both incomplete and intact crania from multiple samples through time, both within and among different socioeconomic groups and across different regions of the Nile, desert areas, and oases. Because it appears quite likely that a combination of factors relating to diet, physical activity, hormonal shifts over a lifetime, and genetics may underlie the development of HFI, greater attention to investigation of HFI in large samples is sorely needed to unravel the etiology and microevolution of this condition.

Acknowledgments

We thank Salima Ikram, Jessica Kaiser, and Stéphanie Porcier for organizing the Bioarchaeology of Ancient Egypt conference in January 2019 and inviting us to submit this work for publication. Access to the Abydos North Cemetery remains from the Aha enclosures was provided by Matthew Adams and the University of Pennsylvania Museum-Yale University-Institute of Fine Arts, New York University Expedition to Abydos. Yasser Hossein, director of the South Cemetery excavation for Egypt's Ministry of Antiquities, facilitated work on the human remains during excavation and study seasons.

Bibliography

Adams, M. & D. O'Connor. 2003. The Royal Mortuary Enclosures of Abydos and Hierakonpolis. In: Hawass, Z. Ed. The Treasures of the Pyramids. – Cairo, American University in Cairo Press: 78-85.

Baker, B.J. 1997. Contributions of Biological Anthropology to the Understanding of Ancient Egyptian and Nubian Societies. In: Lustig, J. Ed. Anthropology and Egyptology: A Developing Dialogue. – Sheffield, England, Sheffield Academic Press (Monographs in Mediterranean Archaeology 8): 106-116.

Baker, B.J. 2013. Sacrifices for the State? The Subsidiary Burials from Aha's Funerary Enclosures at Abydos. Paper Presented at the 2nd Bioarchaeology of Ancient Egypt conference, Cairo Egypt. Abstract. – PalArch's Journal of Archaeology of Egypt/Egyptology 10, 1: 11-12.

Bebel, A. & A. Golijewskaja. A Case of *Hyperostosis Frontalis Interna* from Deir el-Bahari, Egypt. – Bioarchaeology of the Near East 9: 45-54.

Bestock, L.D. 2008. The Early Dynastic Funerary Enclosures of Abydos. – Archéo-Nil 18: 42-59.

Bestock, L. 2009. The Development of Royal Funerary Cult at Abydos: Two Funerary Enclosures from the Reign of Aha (MENES, Studien zur Kultur und Sprache der ägyptischen Frühzeit und des Alten Reiches, Band 6). – Wiesbaden, Harrassowitz.

Buikstra, J.E. & D.H. Ubelaker. Eds. 1994. Standards for Data Collection from Human Skeletal Remains. – Fayetteville, Arkansas Archeological Survey (Research Series 44).

Cvetković, D., S. Nikolić, V. Brković, & V. Živković. 2018. Hyperostosis Frontalis Interna as an Age-Related Phenomenon – Differences between Males and Females and Possible Use in Identification. – Science & Justice 59 (2019): 172-176.

Flohr, S. & C. Witzel. 2010. *Hyperostosis Frontalis Interna* – A Marker of Social Status? Evidence from the Bronze-Age "high society" of Qatna, Syria. – HOMO – Journal of Comparative Human Biology 62: 30-43.

Gabr, A.M. 2011. The New Archaic Period Cemetery at Abydos: Osteological Report. In: Friedman, R.F & P.N. Fiske. Eds. Egypt at Its Origins 3. – Leuven, Peeters (Orientalia Lovaniensia Analecta 205): 281-291.

Glab, H., K. Szostek & K. Kaczanowski. 2005. *Hyperostosis Frontalis Interna*, a Genetic Disease? Two Medieveal Cases from Southern Poland. – HOMO – Journal of Comparative Human Biology 57 (2006): 19-27.

Hajdu, T., E. Fóthi, Zs. Bernert, E. Molnár, G. Lovász, I. Kővári, K. Köhler & A. Marcxik. 2009. Appearance of *Hyperostosis Frontalis Interna* in Some Osteoarchaeological Series from Hungary. – HOMO: Journal of Comparative Human Biology 60: 185-205.

Hershkovitz, I., C. Greenwald, B.M. Rothschild, B. Latimer, O. Dutour, L.M. Jellema & S. Wish-Baratz. 1999. Hyperostosis Frontalis Interna: An Anthropological Perspective. – American Journal of Physical Anthropology 109: 303-325.

Hossein, Y.M. 2011. A New Archaic Period Cemetery at Abydos. In: Friedman, R.F. & P.N. Fiske. Eds. Egypt at Its Origins 3. – Leuven, Peeters (Orientalia Lovaniensia Analecta 205): 269-280.

Lazer, E. 1996. Revealing Secrets of a Lost City. An Archaeologist Examines Skeletal Remains from the Ruins of Pompeii. – Medical Journal of Australia 165: 620-623.

Lovejoy, C.O., R.S. Meindl, T.R. Pryzbeck & R.P. Mensforth. 1985. Chronological Metamorphosis of the Auricular Surface of the Ilium: A New Method for the Determination of Age at Death. – American Journal of Physical Anthropology 68: 15-28.

Mant, M. 2014. Palaeopathology of Human Remains at Ancient Medes (Tell er-Rub'a), Egypt. – Bioarchaeology of the Near East 8: 1-27.

May, H., N. Peled, G. Dar, J. Abbas, B. Medlej, Y. Masharawi & I. Hershkovitz. 2010. Hyperostosis Frontalis Interna and Androgen Suppression. – The Anatomical Record 293: 1333-1336.

May, H., N. Peled, G. Dar, J. Abbas & I. Hershkovitz. 2011. Hyperostosis Frontalis Interna: What Does it Tell Us

About our Health? – American Journal of Human Biology 23: 392-397.

Mulhern, D.M., C.A. Wilczak & J.C. Dudar. 2006. Brief Communication: Unusual Finding at Pueblo Bonito: Multiple Cases of Hyperostosis Frontalis Interna. – American Journal of Physical Anthropology 130: 480-484.

Nikolić, S., D. Djonić, V. Živković, D. Babić, F. Juković & M. Djurić. 2010. Rate of Occurrence, Gross Appearance, and Age Relation of Hyperostosis Frontalis Interna in Females: A Prospective Autopsy Study. – American Journal of Forensic Medicine & Pathology 31: 205-207.

O'Connor, D. 2009. Abydos: Egypt's First Pharaohs and the Cult of Osiris. – London, Thames & Hudson.

Ortner, D.J. 2003. Identification of Pathological Conditions in Human Skeletal Remains (2nd ed.). – San Diego, Academic Press.

Phenice, T. 1969. A Newly Developed Visual Method of Sexing in the Os Pubis. – American Journal of Physical Anthropology 30: 297-301.

Raikos, A., G.K. Paraskevas, F. Yusuf, P. Kordali, S. Meditskou, A. Al-Haj & B. Brand-Saberi. 2011. Etiopathogenesis of Hyperostosis Frontalis Interna: A Mystery Still. – Annals of Anatomy 193: 453-458.

Rösing, F.W. 1990. Qubbet el-Hawa und Elephantine. Zur Bevölkerungsgeschichte von Ägypten. – Stuttgart & New York, Gustav Fischer.

Rühli, F.J., T. Böni & M. Henneberg. 2004. *Hyperostosis Frontalis Interna*: Archaeological Evidence of Possible Microevolution of Human Sex Steroids? – HOMO – Journal of Comparative Human Biology 55: 91-99.

Shahin, A.A. S. Alhoseiny & M. Aldali. 2013. Hyperostosis Frontalis Interna: An Egyptian Case Referred to the Second Dynasty (2890-2650 BC) from Tarkhan-Egypt. – The Egyptian Rhuematologist 36: 41-45.

She, R. & J. Szakacs. 2004. Hyperostosis Frontalis Interna: Case Report and Review of Literature. – Annals of Clinical & Laboratory Science 34: 206-208.

Sperduti, A. & G. Manzi. Hyperostosis Frontalis Interna in a Cranial Sample from the Roman Population of Portus. – Rivista di Antropologia (Roma) 68: 279-286.

Szeniczey, S., A. Marcsik, Z. Ács, T. Balassa, Z. Bernert, K. Bakó, T. Czuppon, A. Endrődi, S. Évinger, Z. Farkas, L. Hlavenková, K. Hoppál, C. Kálmán Kiss, K. Kiss, K. Kocsis, L O. Kovács, P.F. Kovács, K. Köhler, L. Költő, I. Kővári, O. László, G. Lovász, J. Lovranits, J. Lukács, Z. Masek, M. Merczi, E. Molnár, C.E. Németh, J.G. Ódor, L. Paja, I. Pap, R. Patay, I. Rácz, Z. Rácz, Á. Ritoók, G. Szenthe, G. Szilas, B.M. Szőke, Z. Tóth, T. Vidaz, K. Wolff, M. Finnegan & T. Hajdua. 2018. *Hyperostosis Frontalis Interna* in Ancient Populations from the Carpathian Basin – A Possible Relationship between Lifestyle and Risk of Development. – International Journal of Paleopathology 24 (2019): 108-118.

Verdy, M., J. Guimond, P. Fauteux & M. Aube. 1978. Prevalence of Hyperostosis Frontalis Interna in Relation to Body Weight. – The American Journal of Clinical Nutrition 31: 2002-2004.

Watrous, A.C., S.C. Anton & A.M. Plourde. 1993. Hyperostosis Frontalis Interna in Ancient Egyptians. – American Journal of Physical Anthropology 36 (S16): 205.

Western, A.G. & J.J. Bekvalac. 2016. Hyperostosis Frontalis Interna in Female Historic Skeletal Populations: Age, Sex, Hormones and the Impact of Industrialization. – American Journal of Physical Anthropology 162: 501-515.

Humans and Animals Together in the Journey to the Afterlife

The Burial in Area R11 under the Temple of Millions of Years of Amenhotep II, Luxor, West Thebes – Italian Archaeological Project

Fabio Bona, Giovanna Bellandi, Letizia Cavallini, Anna Consonni, Tommaso Quirino & Angelo Sesana

General Introduction (AS)

The CEFB (Francesco Ballerini Centre for Egyptology) Italian Archaeological Project, directed by Angelo Sesana, began its work at the Temple of Millions of Years of pharaoh Amenhotep II (18th Dynasty, 1428-1397 BC) in 1997/1998.[1] The 21st field season at the ruins of the temple located in Western Thebes, just north of the Ramesseum, was completed in January 2019. Most excavation activity occurred during the earlier field seasons, with more recent years seeing a focus on stabilisation, restoration and the analysis of materials excavated in prior years.

The excavations have revealed, since 2007, the intensive funerary use of the area, both before the construction of the Temple and again after its abandonment (Consonni *et al.*, 2017; Sesana, 2010; Sesana & Quirino, 2010; Sesana *et al.*, 2018). We documented many funerary structures and complexes that largely belong to three main phases between the Middle Kingdom (ca. 2055-1650 BC) and the Ptolemaic Period (305-30 BC) (Consonni, 2016a; 2016b; Consonni & Sesana, 2016; Sesana & Consonni, 2013).

The first phase dates from the Middle Kingdom to the beginning of the New Kingdom, prior to the construction of the temple. These burials are characterised by a range of forms, from complex structures to shallow shafts and simple pits. The burials of the second phase belong to the Third Intermediate Period-Late Period, long after the temple was abandoned, and consist of 25 shaft tombs, sometimes provided with a mud-brick chapel on the surface.

The third and latest evidence of organised activity dates to the Ptolemaic Period, and is characterised primarily by pottery, spread throughout the upper levels in different sectors and found in large quantities in the fill of the funerary shafts, mainly in the area of the Temple courtyard. There are scattered remains of human bone in the upper levels of shaft fill that may well be related to Ptolemaic burials. In addition to these finds, two *in situ* Ptolemaic deposits were discovered in 2004/2005 and 2007/2008.

1 Field seasons occur in December/January and hence each falls in two years.

In: Ikram, S., J. Kaiser & S. Porcier. Eds. 2021. *The Ancient Egyptians and the Natural World. Flora, Fauna, and Science.* – Leiden, Sidestone Press: 31-44.

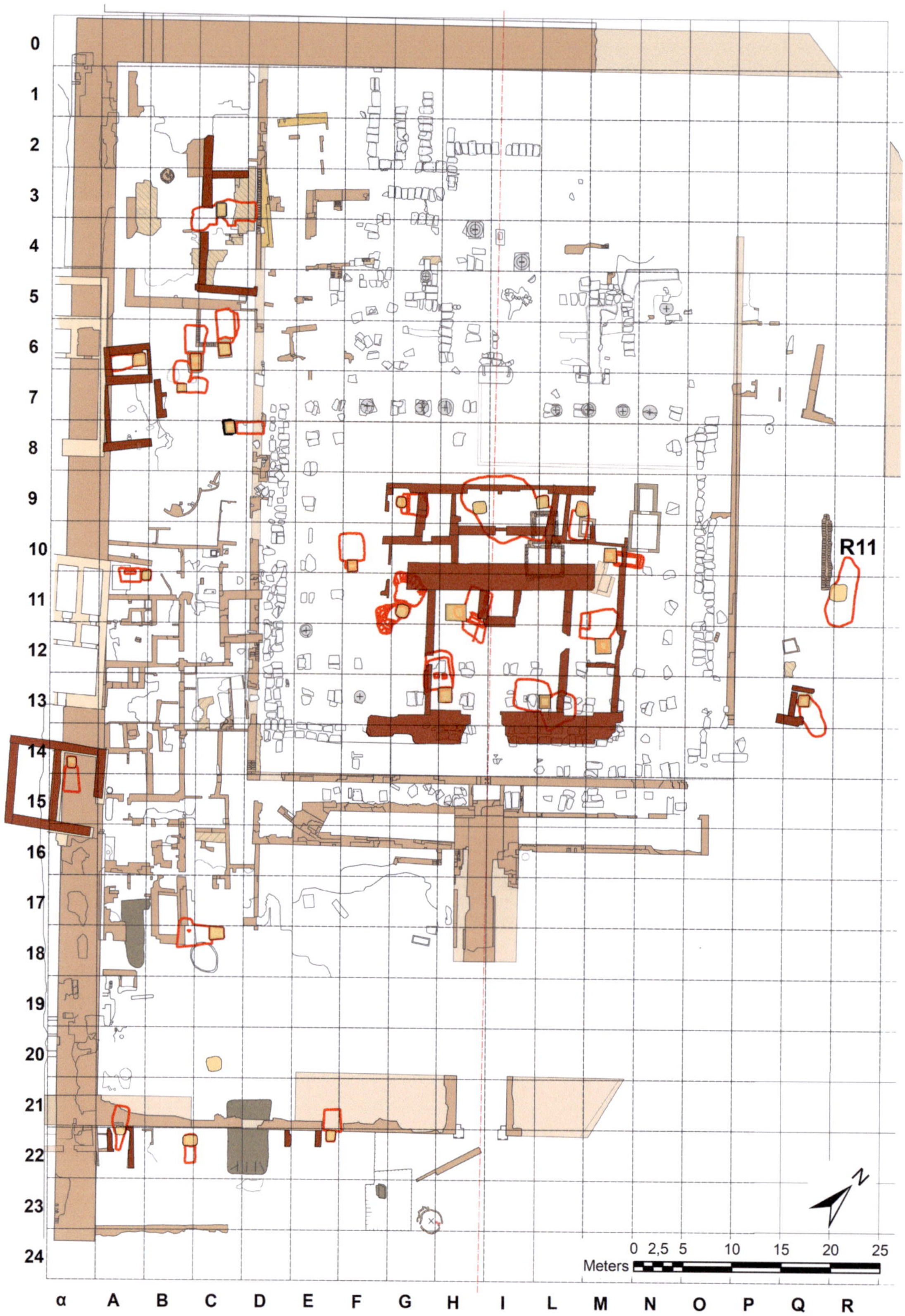

Figure 1. Plan of the Temple of Millions of Years of Amenhotep II, showing the position of tomb R11 and of the other Third Intermediate- early Late Period tombs. Plan: arch. E. Negri; GIS and data processing: T. Quirino.

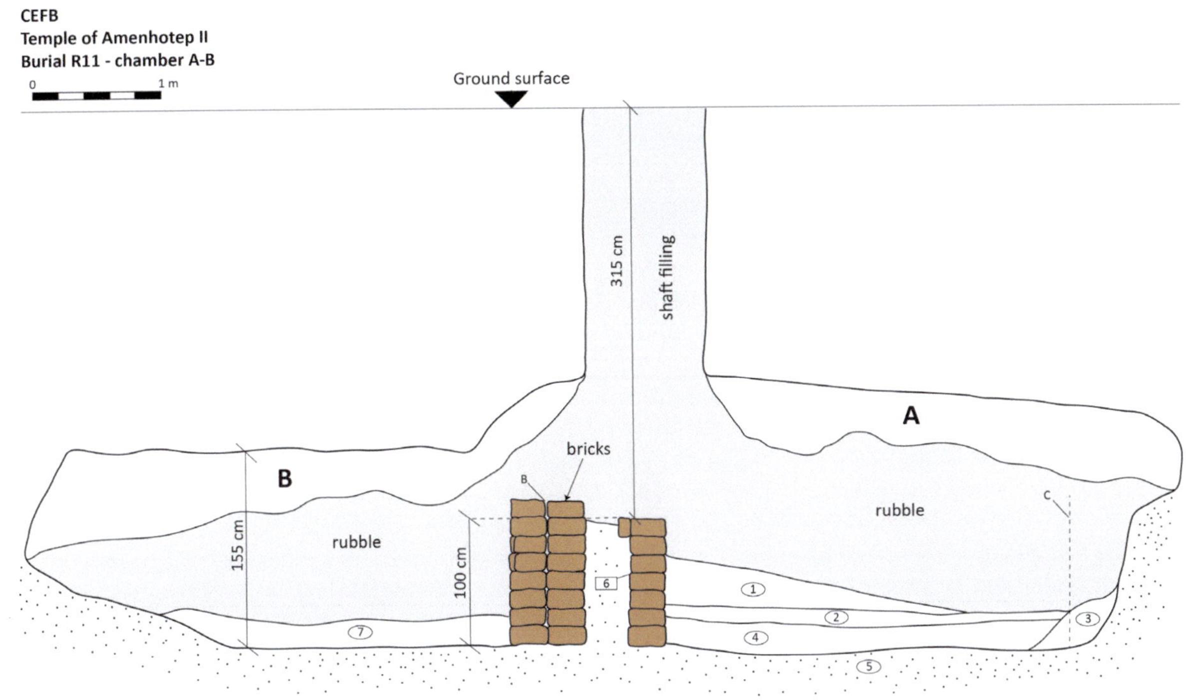

Figure 2. Section of tomb R11. Drawing by F. Longhi, F. Redolfi Riva.

In these tombs, we documented more than 100 fully or partially articulated human skeletons and numerous scattered human remains (Bellandi *et al.*, 2015: 22-23). For each, where possible, we recorded taphonomic observations, determined the age and sex, and collected anthropometric data and observations for the reconstruction of health status and any evidence of functional stress. We have yet to complete a full study of the necropolis population across the different phases of use. In some tombs, we found not only human remains but also those of animals. The shaft tomb in area R11[2] that is the focus of the present article is one such example (Figure 1).

R11 – Funerary Context (AC, TQ)

The tomb in area R11 has a square shaft 3.5 m in depth with two irregularly-shaped chambers opening off it at the bottom, one – named A – toward the east and the other – named B – toward the west (Sesana *et al.*, 2018: 15-16, fig. 11-12). Their original dimensions are difficult to determine due to the partial collapse, in particular in chamber B, of the walls, which were partly cut into the breccia (desert conglomerate) stratum and partly into the compacted sand (number 5 in the section drawing; Figure 2). At the time of discovery, the entrances of both chambers were partially sealed by mud-brick walls.

The shaft fill consisted of a fine grey sand, mixed with many fragments of pottery (sometimes reassemblable, and dating mainly from the Third Intermediate Period to Ptolemaic times[3]), animal bones and a few fragments of sandstone from the temple architecture, all of which had been thrown into the pit.[4] In the upper levels inside the chambers, the fill was also characterized by the presence of rubble that fell, perhaps at different times, from the ceiling. In Chamber A, level 1 consisted of greyish fine sand with fragments of pottery, mud bricks and animal bones and represented the interface between the shaft fill and the archaeological levels found within the chamber.[5]

Only the lowest stratigraphic levels (levels 2 and 4 in chamber A and level 7 in chamber B) could be directly related to the original burial activities or to subsequent looting. Levels 2 and 4, very similar in terms of composition, are reddish compacted sand layers with animal bones and fragments of pottery. The boundary between these two

2 The tomb, excavated between 2012/2013 and 2014/2015 under the supervision of Francesco Longhi, Flavio Redolfi Riva and Tommaso Quirino, is named with the number of the square where its entrance was discovered.

3 The pottery from this context is under study by A. Consonni.

4 In the upper part of the shaft fill there were also modern materials, mostly rubbish from the nearby cafeteria.

5 Remains found in this level have been studied together with the material collected in the shaft fill and in the rubble level. Only the fill of the shaft, the rubble level and chamber A yielded faunal remains.

levels is determined by the presence of remains of a coffin and other finds. Items placed along the perimeter of the chambers were well preserved thanks to the collapse of the walls, which covered them with a sandy layer (level 3) the composition of which is indistinguishable from that of the walls of the chambers.

Tomb R11 (Figure 3), as with other tombs excavated in this area, was robbed in antiquity as well as possibly more recently. This destructive activity denies us the possibility of fully understanding the original placement of the finds and their relationships within the burial, thus impeding the clear interpretation of the whole funerary (and possibly post-funerary) context. Despite this, we can suggest the following reconstruction.

Chamber B, explored in 2012/2013, contained the burial of an older woman lying with her head oriented toward the east, inside the remains of an outer anthropoid coffin that had totally vanished and of an inner wooden anthropoid coffin and a cartonnage, preserved only as traces on the lower part of the body. It is only thanks to our application of colour transform techniques to these very scanty remains[6] that it was possible to recognize some details of the cartonnage (blue wings with red details on a white background, crossing each other behind the central band that was no longer visible), and of the inner coffin (the traces of an Abydos fetish and, in correspondence with the feet, of a yellow central vertical band flanked by a reticulate network of rhomboidal pattern painted in blue in imitation of bead-netting on a red background). The mummy's chest sported two small clay udjat eyes with a hole for suspension.

The funerary assemblage consisted of four limestone uninscribed canopic jars lying in pairs, two on either side of the woman's head, at the entrance of the chamber. The cavity inside the jars is hollowed out only so far as to house the stopper (for parallels see Consonni *et al.*, 2018: 116), rather than the viscera. On the right side of the coffin we found a total of about 400 sun-dried clay ushabtis of the mummiform type, with light blue paint on the surface, possibly once contained in a box of which no trace remains. The general character of this assemblage fits well with the Aston (2011: 16-21) Theban burial assemblages dated from 890 to 750 BC.

Chamber A, excavated in 2014/2015, had been extensively looted. At the time of discovery, it contained the badly preserved remains of two wooden rectangular coffins of the *ḳrsw* type. Each of them contained at least one inner anthropomorphic wooden coffin, of which only a pair of bronze inlaid eyes and eyebrows were preserved. Human remains attributed to three individuals were lying in the chamber (a fourth individual, in a very fragmentary form, was found in the shaft fill), but not inside the coffins and clearly not in their original place of deposition. The funerary assemblage, also partly displaced, was composed of a set of four uninscribed limestone canopic jars of the dummy type, made of one of single pieces of stone (Figure 4). One of the jars was found in the upper archaeological level of chamber B, very close to the entrance. The remains of three small wooden boxes contained a total of about 660 ushabtis of the mummiform type, all made of sun-dried clay, blue-washed, only crudely modelled, and often provided with a beard.

Many faience beads, mainly light blue, but also red, black and white, were recovered from the lowest archaeological levels of the chamber. The presence of bead nets (Aston, 2009: 292, in use from the second half of the 8th century BC), the type of ushabtis, provided with a beard (but also the possible type of shabti boxes that can be deduced from the dimensions of their traces) and the presence of rectangular coffins of the *ḳrsw* type suggest a dating to the 25th or 26th Dynasty for these burials (Aston, 2011: 23-30). This date can be further supported by the study of a set of pottery vases found *in situ* inside the chamber.

At the feet of the southernmost coffin, two superimposed square chests, preserved only as traces, were found. The upper one was divided into four square compartments. Each of them contained a black mass with traces of textiles on the surface, probably the remains of mummified internal organs. Originally, a lid was present, but in the upper chest it had totally vanished and only a white trace of the lower remained. No traces of colours or decorations, but just the traces of the white stucco that probably covered all the surfaces, were preserved.

Inside the lower chest, a large storage jar decorated with white bands – completely reconstructed from its fragments – along with a deep bowl with simple rim and a flattened cup, all in Nile clay, were recovered: a set datable to the 7th century BC (Boulet, 2018: 344, fig. 6j-k, Theban ceramic production phase IV – from the end of the 25th Dynasty to the beginning of the 26th Dynasty).

From the study of the preserved funerary assemblages we can hypothesize that, in tomb R11, chamber B was the first to be used to house burials. Chamber A was used – and perhaps even excavated from the breccia and compacted sand – later. Probably after a relatively short period of time, sufficient for the partial collapse of the sand walls that covered part of its contents, chamber A was looted.

Human remains – R11 (GB; LC)

Methods

Taphonomic observations, that is information useful for the reconstruction of the original position of the body at the time of deposition, were recorded during the

6 For the application of colour transformation techniques to coffins found at the Temple see Consonni & Quirino (2019).

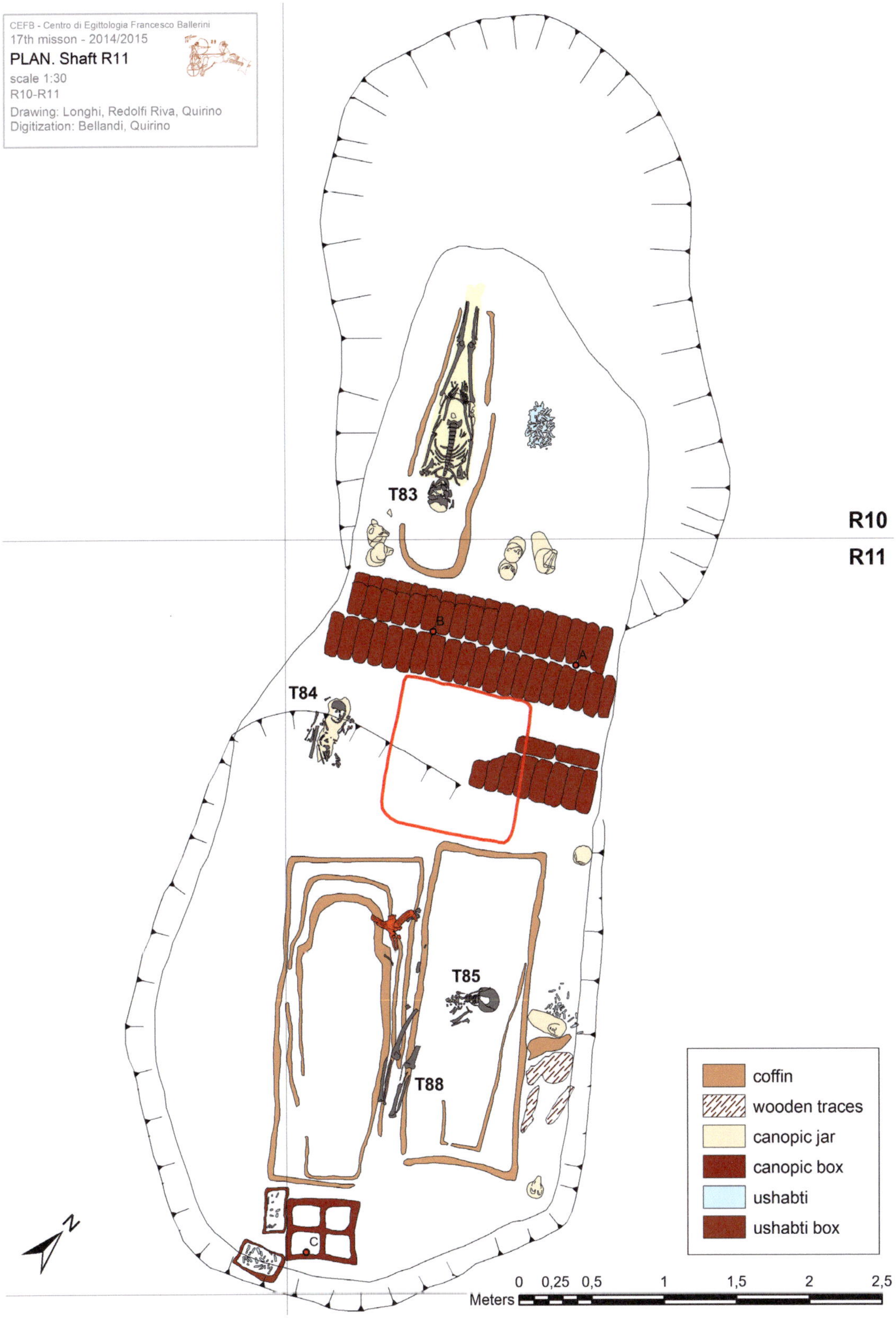

Figure 3. Plan of tomb R11, with the positions of the burials and of the funerary assemblage. Plan: F. Longhi, F. Redolfi Riva, T. Quirino; GIS and data processing: G. Bellandi, T. Quirino.

Figure 4. Two of the four canopic jars found in R11, chamber B. One of them was thrown by looters onto a *shabti* box, preserved only as traces. Photograph by T. Quirino © CEFB.

excavation of the burials using standard archaeological practice and the approach of Duday (2005). All finds underwent macroscopic observation and were measured with classical metric instruments: curved calipers, straight-bladed calipers, osteometric board etc., following standard physical anthropology protocols (Bass, 2005; Basili Gualandi, 1995; White & Folkens, 2005). Determination of sex was based on discriminating characteristics of the skull and pelvis (Acsadi & Nemeskeri, 1970; but see also Buikstra & Ubelaker, 1994 and Cattaneo & Grandi, 2004). Estimation of age was based on several methods (summarised in Buikstra & Ubelaker, 1994) depending on the extent of preservation of the finds: degree of tooth wear (Lovejoy, 1985); ectocranial suture closure (Meindl & Lovejoy, 1985); morphology of the auricular surface of the ilium (Lovejoy *et al.*, 1985); age-related modification of the pubic symphysis (Suchey & Brooks, 1988; Todd, 1921); variations of the surface and margins of the sternal ends of the ribs (Iscan *et al.* 1984; Iscan & Loth, 1985).

Sub-adults were aged according to the level of epiphyseal closure and/or fusion of ossification centres (Scheuer & Black, 2000), dental eruption and development (Ubelaker, 1989) and long-bone length (Scheuer & Black, 2000; Stloukal & Hanáková, 1978).

Anthropometric measures were taken using the guidelines established by Martin & Saller (1957). Stature was estimated using the method of Trotter & Gleser (1977) for the black population, applied to adults. Non-metric traits were evaluated according to Buikstra & Ubelaker (1994). The various pathologies encountered were evaluated on the basis of the indications reported in Ortner (2003; see also Fornaciari & Giuffra, 2009). For the specific case of dental diseases, Hilson's (1997) approach was used.

ID	Area	Sex	Age	Stature	Horizontal cephalic index
T83	R11B	F	40/45	161	81.3 Brachycranial
T84	R11A	M?	>50	173	86.4 Hyperbrachycranial
T85	R11A	F	35/40	155	82.4 Brachycranial
T88	R11A	F?	Older adult	153	75.2 Mesocranial
ID 5	R11 shaft fill (scattered remains)	M?	Adult	not determinable	not determinable

Table 5. The human individuals in R11.

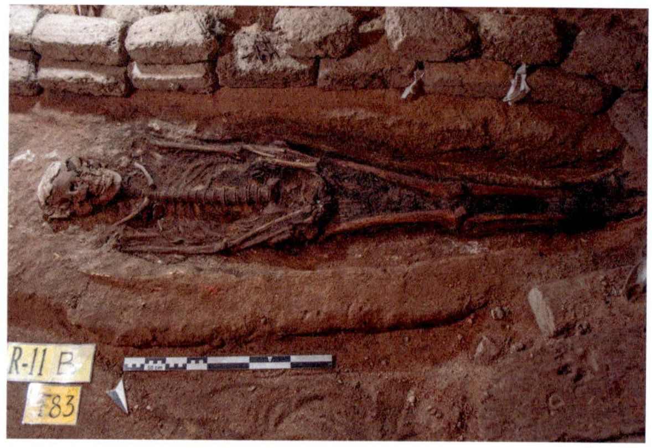

Figure 6. Chamber B. Individual T83. Photograph by T. Quirino © CEFB.

The Human Remains in R11

The analysis of the skeletal remains from tomb R11 returned a minimum number (MNI) of five individuals (Table 5, Figure 3). Chamber B contained only one skeleton, that of a woman 40-45 years old (T83; Figure 6), in her original burial position inside a cartonnage case and, probably, a set of two wooden anthropoid coffins, now visible only as traces. The skeleton showed evidence of having once been wrapped in bandages: the joints of the hands and feet, as well as general skeletal articulation, are preserved and there are suggestions of wrapping. The vertical position of the clavicles (which suggests a compression or bandaging of the chest) allows us to hypothesise that the body was originally buried supine, tightly bound, with the hands placed on the hips.

In Chamber A, looting precludes the reconstruction of the original positions of the bodies, though the remains of four individuals were noted here (probably two females and two males). However, there are traces of two coffins and it seems that each one contained a body. Indeed, individual T85 (Figure 7), an adult woman, found prone, retains only the upper portion of the skeleton, suggesting that this body was dragged out of its coffin by tomb robbers seeking to steal amulets and other grave goods. Human bones belonging to this individual were also present in the

Figure 7. Chamber A. Individual T85. Photograph by T. Quirino © CEFB.

Figure 8. Chamber B. Individual T88. Photograph by T. Quirino © CEFB.

Figure 9. Chamber B. Individual T84. Photograph by T. Quirino © CEFB.

upper layers of the chamber fill, which further suggests that the context was violated in the past.

Individual T88 (Figure 8), probably another woman, had been treated in the same way as the T85, and was preserved partially articulated outside her coffin. Another adult (T84; Figure 9), probably a man, of which, again, only the upper part of the skeleton remained, was found near the entrance of Chamber B. The remains of an adult male (ID 5) were found scattered in the upper fill of the shaft but the poor condition of his bones does not allow further conclusions or inferences.

Reconstruction of Health Condition and Functional Stress

Regarding health status, we observed that, as for most of the skeletons from the site (Bellandi *et al.*, 2015), the individuals from R11 have heavy dental wear (a phenomenon widely documented in skeletal remains from ancient Egypt: Forshaw, 2009; Harris *et al.*, 1998; Nerlich & Zink, 2003; Rose *et al.*, 1993) that makes the morphometric study of their teeth difficult. The teeth are also often very fragmentary and friable due to the environment in which they lay and the thermic variations that occurred during their burial.

We also recorded arthritis of the spine (*spondylosis*) in T83, which is associated with marked vertebral osteophytes on the cervical and lumbar vertebral bodies (Figure 10; this pathological condition is also often documented in ancient Egypt: Macke *et al.*, 2002; Nerlich & Zink, 2003; Rose, 2006).

The presence of porotic transformations of the orbital bone (*cribra orbitalia*), probably due to some form of chronic anaemia, and a case of benign tumours of the skull (*osteoma*) are visible in the same individual – T85.

The most interesting individual is T84. Unfortunately, the poor and very partial preservation of the skull, and especially the absence of the pelvis, makes the determination of both sex and age difficult. Furthermore, we note that, as in the case of other male skulls from the necropolis, this skull has "female" morphometric characteristics and so distinguishing male from female can be difficult. However, given the morphology of the mandible, we can hypothesise that we are talking about a mature male. The skull of this individual shows some interesting pathologies: on the inner surface of the frontal bone, we see a *Hyperostosis frontalis interna* (HFI) (Figure 11) that is visible as an overgrowth of bony tissue. This pathological condition manifested itself in the thickening of the inner surface of the frontal bone. HFI is much more common in females than in males (see Baker and Gabr, this volume). The etiology of HFI is not completely clear, but is most likely related to a hormonal imbalance: *"in males, HFI most probably emerges as a result of inadequate androgen stimulation, or rather as a result of the changes in the oestrogen/androgen ratio"* (Cvetkovićab

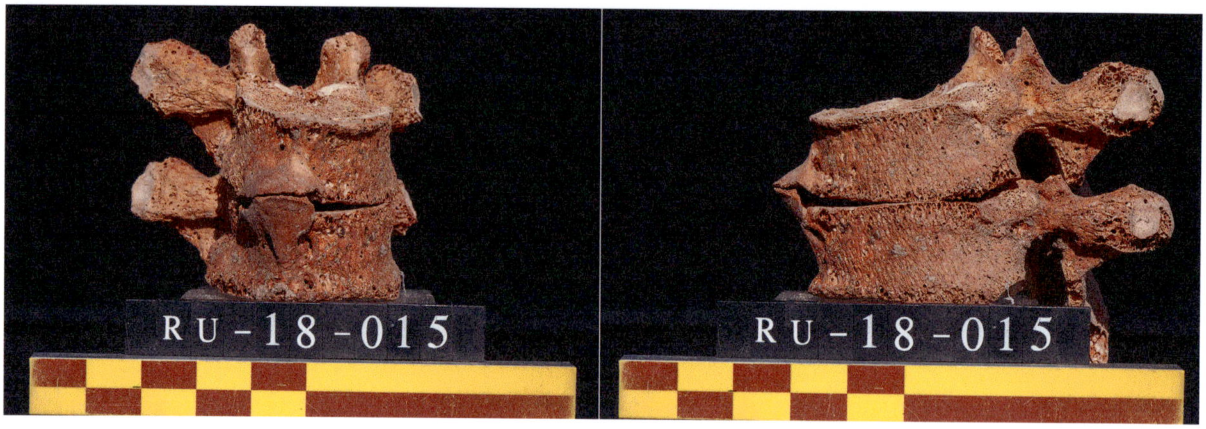

Figure 10. Arthritis of the spine (*spondylosis*) in T83, associated with marked vertebral osteophytes. Photograph by F.M. Giani © CEFB.

Figure 11. Frontal bone of T 84, *hyperostosis frontalis interna* (HFI). Photograph by F.M. Giani © CEFB.

et al., 2019: 172). However, many different hypotheses regarding the etiology of HFI have been proposed. Among the possible factors causing this lesion we can mention: genetic background, disorders of the skeletal system, and hormonal and metabolic disorders (Bebel & Golijewskaja, 2015). Any definitive conclusions in the current case would be premature without further analysis.

The same skull exhibits a regular circular depression on the external parietal bone (Figure 12). The bone is very thin in the middle of this depression and one might hypothesise that it represents the results of a cranial trauma that the subject survived for a considerable period. Unfortunately, without a radiological analysis it is not possible to confirm this hypothesis.

The sample of individuals from R11 is rather small for statistical comparison with anthropometric data for other skeletons from the necropolis (there are just four "complete" skeletons in R11). However, tombs L13 (five skeletons) and C6 (two skeletons) are of similar age and we have conducted (Dr. Craig Alexander undertook the statistical analyses, for which we thank him) simple analyses (parametric t-tests with appropriate corrections and non-parametric U-tests).

In this group of three tombs, the males are statistically significantly taller than the females – p-values are less than 0.05. Only L13 and R11 contain identifiable males – three in the first case and one in the second. All three tombs, however, contain at least two females. Comparing the statures of these, there are no conventionally statistically significant differences, although the females in L13 are, perhaps, a little taller while those in C6 are, perhaps, a little shorter. These latter inferences are, as noted, not statistically significant but might be regarded as "suggestive" and may indicate directions for future research.

Faunal Remains – R11 (FB)

Materials and Methods

The R11 animal bones consist of 1,252 specimens collected during field excavations in 2015/2016 and 2016/2017. The bones represent three classes of vertebrate: mammal, bird and reptile.

Figure 12. Regular circular depression on the external parietal bone of T84. Photograph by G. Bellandi.

The identification of the bones collected during the excavations was conducted in the field, mainly using published osteological images (Barone, 1995; Pales & Garcia, 1981; Pales & Lambert, 1971). In addition to identifying species, anatomical element, age, and sex, the bones were examined to identify traces of butchery, burning and various other taphonomic processes. The nomenclature used for domestic mammals follows Gentry *et al.* (2004). The following parameters were calculated: NISP (Number of Identifiable Specimens) and MNI (Minimum Number of Individuals) for each taxon. The MNI estimates were calculated following Bökönyi (1970), Klein and Cruz-Uribe (1984) and De Grossi Mazzorin (2008, with references).

Distinction between sheep and goats was made primarily using the criteria of Boessneck *et al.* (1964) and Prummel and Frisch (1986). Sexing was based on morphological characteristics of the horns in sheep (Boessneck *et al.*, 1964) and the presence of baculum in dogs. The tooth-wear stage of the domestic mammals was recorded following Payne (1973) and Grant (1982). The fusion stages of post-cranial bones were observed and related to age ranges following Barone (1995). Measurements of skeletal elements were taken following von den Driesch (1976). The sizes of the animals were estimated using the parameters proposed by Matolcsi (1970) for cattle (*Bos taurus*), Teichert (1975) for sheep (*Ovis aries*), and Koudelka (1884) and Harcourt (1974) for dogs (*Canis familiaris*).

Age of the Remains

As with other similar shaft tombs located in the area of the temple, on the basis of the archaeological evidence described above, the R11 animal remains can be dated between the Third Intermediate Period and the Ptolemaic age.

Some Taphonomic Remarks

The tomb sediments contained the full range of skeletal elements belonging to sheep and cows – the remains of a very young ovicaprid may be under-represented because the bones are so small and fragile and, hence, very difficult to collect during excavation.

No cut marks were detected on the bones (as is true for other tombs in the area, with the exception of a bovine radius from L13). Finally, few intrusive, recent, finds are present.

Taxon	R11 shaft fill				R11 levels 2 and 4				Total			
	n	MNI	%n	%MNI	n	MNI	%n	%MNI	n	MNI	%n	%MNI
Aves	1	1	0.1%	9.1%					1	1	0.1%	7.1%
Canis familiaris	1	1	0.1%	9.1%					1	1	0.1%	7.1%
Bos taurus	731	4	65.1%	36.4%	73	1	56.6%	33.3%	804	5	64.2%	35.8%
Caprovino	263	3	23.4%	27.3%	41	2	31.8%	66.7%	304	5	24.3%	35.8%
Ovis aries	19		1.7%		15		11.6%		34		2.7%	
Crocodylus niloticus	8	1	0.7%	9.1%					8	1	0.6%	7.1%
Camelus sp.	100	1	8.9%	9.1%					100	1	8.0%	7.1%
Total	1123	11	100.0%	100.0%	129	3	100.0%	100.0%	1252	14	100.0%	100.0%

Table 13. R11 faunal remains by stratigraphic context.

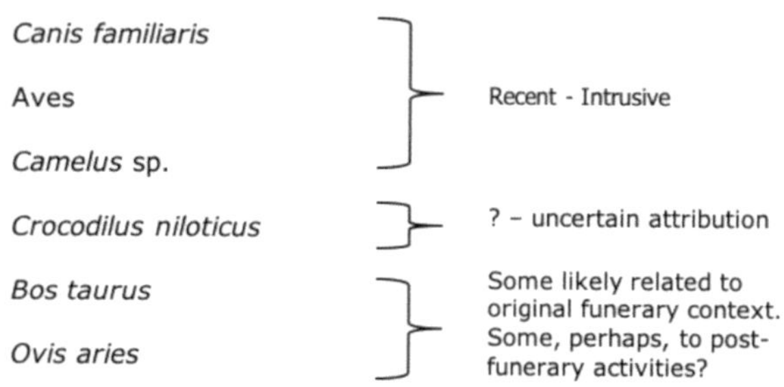

Figure 14. Schematic interpretation of the stratigraphic distribution of animal bones.

The Remains and their Stratigraphic Position

According to the stratigraphy (see the section "R11 – Funerary Context" above) we can recognize two main divisions: the remains collected from the shaft fill and level 1 and those collected from the lower chamber levels, mainly from levels 2 and 4. Bovines and sheep are the best-represented animals.

The stratigraphic positions, even if only indicative due to looting and taphonomic considerations, allow us to recognise three main assemblages (Table 13; Figure 14): i) Recent bones: an almost complete camel skeleton, one bird bone and one dog phalanx; ii) Bovine and sheep bones from the shaft fill and level 1 and iii) Bovine and sheep bones from the lowest levels (levels 2 and 4). It is also very important to emphasise the presence of scant remains of a very young ovicaprid in these levels. Are these remains direct evidence of the original funerary ritual conducted in the chamber?

In the shaft were found – with uncertain stratigraphic attribution – remains of a crocodile (parts of the skull and mandible, several vertebrae and parts of the forelimbs). It is difficult to understand the significance of the presence, in the archaeological context of R11, of this (possibly partial) crocodile. Although it is not rare and, indeed, is often seen in other shaft tombs in the area of the Temple of Millions of Years of Amenhotep II (Bona *et al.*, 2019), the crocodile's function remains unknown (Figure 15).

Results of the Zooarchaeological Analyses

Bos taurus

Upper levels (shaft fill and level 1):

The MNI is 4 individuals. The estimated height at the shoulder ranges between 1350 and 1500 mm. All the animals were adult.

Lower levels (levels 2 and 4):

Only the rear half of a bovine of 1410 mm height and age between 24 and 30 months was found.

Ovis aries

Upper levels (shaft fill and level 1):

The MNI is 3 individuals. The estimated height at the shoulder ranges between 700 and 780 mm. The ages of the sheep were, respectively: 8-10 years; 6-8 years; more than 36 months.

Lower levels (levels 2 and 4):

The remains suggest the presence of a single ram (identified by horns) of 750 mm height and aged 8-10 years. It is also important to note the presence of scattered remains of a very young ovicaprid, largely found in the lower levels.

Remarks on R11

The main question that the faunal remains raise is what relationship did these animals have to the burials in the tomb? We propose three possible scenarios: i) The remains in the lowest level of chamber A, if actually in their original positions, show the placing in the chamber of two haunches of beef (including the sacrum and tail) – meat of good quality (save for the tail, which is less meat yielding and of lower status; Ikram, 1995) – and an older ram, all directly connected to the burials as food offerings; ii) All the remains were connected with subsequent shaft deposit formation and are unconnected to the primary funerary

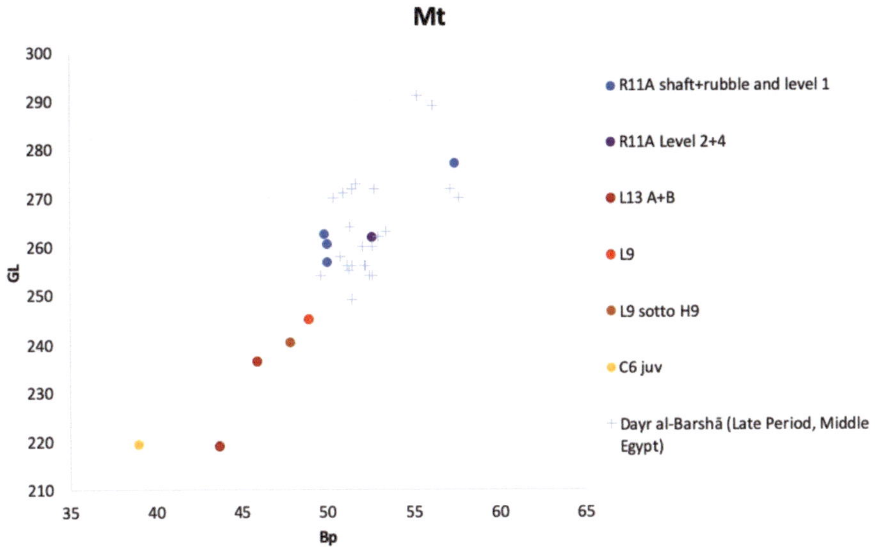

Figure 15. Dimensional comparison between metacarpal bones of cattle from various contexts in the area of Temple of Millions of Years of Amenhotep II and from Dayr al-Barshā (GL: greatest length; Bp: proximal breadth).

context; iii) Larger and older mammals were connected to the shaft deposit formation and the very young ovicaprid was deposited on the surface of the chamber as part of the funerary ritual, a situation observed, for example, in shaft tomb L13A (Bona *et al.*, 2019).

In our opinion, the data collected allow us to assume that the animal assemblage probably should be the result of a composite action of points i and iii, where some portions of cattle and ovicaprids were laid as funeral offerings but the rituals probably also included deposition, perhaps linked to some specific deceased, of a young ovicaprid as seen in other contexts in the area (Bona *et al.*, 2019). Point ii is connected to the presence of the recent remains (crocodile, dog, and bird).

Some Ideas regarding Bovine Remains

Some issues about which we might think further are: Can we recognise different bovine populations from metric data? Furthermore, can different bovine populations provide evidence of transitions between different rulers?

In figure 15 we can see some differences between cattle from different shaft tombs in the area of the temple and from Dayr al-Barshā (Linseele *et al.*, 2017). There are clearly visible distinctions but only better data will confirm or deny the validity or utility of this conclusion.

If size differences amongst the bovines linked to various groups that came under Egyptian control are confirmed, there would be the potential to use data on the average size of bovine remains from different contexts to assist in dating and to provide important information about the funerary contexts of both the Theban necropolis and others.

Conclusions

Interaction between human and animal in ancient Egypt is an interesting and important theme in understanding this ancient society. The passage from the world of the living to that of the dead sees a complex interaction between man and animal. Often the deeper meanings of these interactions tend to escape us. The study of the faunal remains related to funerary context R11 has provided us with important data to shed new light on this issue, or at least raise fresh questions concerning this relationship.

At this preliminary point in investigating these burials, there are at least two main depositional contexts. First, the remains of complete animals in the shaft and a few selected remains of portions of cows (rear legs) in the burial chamber were perhaps connected to some funeral ritual or given as offerings. Infant ovicaprids and raptors directly connected with a burial were certainly elements of a burial rite observed locally. See, for example, tomb L13A, and an as-yet unpublished new tomb in the area of the Ramesseum, at which excavation is still ongoing by the CEFB as part of the scientific collaboration with the Franco-Egyptian Archaeological Mission of West Thebes. In R11 no bird bones have been found, but we cannot exclude that their absence is a result of post-depositional events. Second, the animal bones in the shaft fill might be connected with some unknown post funeral rituals (see tomb C6 – Bona *et al.*, 2019 – or tomb M12).

Acknowledgements

We thank Craig Alexander for assistance with the English text.

Bibliography

Acsádi, G. & J. Nemeskéri. 1970. History of Human Life Span and Mortality – Budapest, Akadémiai Kiado.

Aston, D.A. 2011. Theban Funerary Customs ca. 950-650 BC. In: Hawass, Z., T.A. Bács & G. Schreiber. Eds. Proceedings of the colloquium in Theban Archaeology at the Supreme Council of Antiquities, November 5. – Cairo, Conseil Suprême des Antiquités: 9-32.

Aston, D.A. 2009. Burial Assemblages of Dynasty 21-25. Chronology – Typology – Developments. – Wien, Verlag der Österreichischen Akademie der Wissenschaften (Denkschriften der Gesamtakademie 54).

Barone, R. 1995. Anatomia comparata dei mammiferi domestici. I. Osteologia. – Calderini, Bologna.

Bass, W.M. 1995. Human Osteology. A Laboratory and Field Manual. – Columbia, Missouri Archaeological Society.

Basili Gualandi, P. 1995. Osteometria In: Facchini, F. Ed. Antropologia. Evoluzione uomo ambiente. – Torino, UTET Università: 568-590.

Bebel, A. & A. Golijewskaja. 2015. A Case of *Hyperostosis Frontalis Interna* from Deir el-Bahari, Egypt. – Bioarchaeology of the Near East, 9: 45-54.

Bellandi, G., R. De Marzo, S. Benazzi & A. Sesana. 2015. Burials under the Temple of Millions of Years of Amenhotep II – Luxor, West Thebes. In: Ikram, S., J. Kaiser & R. Walker. Eds. Egyptian Bioarchaeology: Humans, Animals, and the Environment. – Leiden, Sidestone Press: 19-32.

Boessneck, J., H-H. Müller & M. Teichert. 1964. Osteologische Unterscheidungsmerkmale zwischen Schaf (*Ovis aries* Linné) und Ziege (*Capra hircus* Linné). – Kühn-Archiv 78, 1/2: 5-126.

Bökönyi, S. 1970. A New Method for the Determination of the Number of Individuals in Animal Bone Material. – American Journal of Archaeology 74: 291-292.

Bona, F.A. Consonni, T. Quirino & A.Sesana. 2019. Interpreting the Faunal Remains from the Tombs at the Temple of Millions of Years of Amenhotep II in Western Thebes: Funerary Practices, Ritual Practices or, Perhaps, Something Else? In: Porcier S., S. Ikram & S. Pasquali. Eds. Creatures of Earth, Water and Sky. Essays on Animals in Ancient Egypt and Nubia. – Leiden, Sidestone Press: 99-107.

Boulet, S. 2018. Ceramic Industry Developments in the Theban Area during the Twenty-fifth Dynasty: Between Traditions and Innovations. In: Pischikova, E., J. Budka & K. Griffin. Eds. Thebes in the First Millennium BC: Art and Archaeology of the Kushite Period and Beyond. – London, Golden House Publications: 336-356.

Buikstra, J.E. & D.H. Ubelaker. 1994. Standards for Data Collection from Human Skeletal Remains. – Fayetteville, Arkansas Archaeological Survey.

Cattaneo, C. & M. Grandi. 2004. Antropologia e Odontoiatria Forense. Guida allo studio dei resti umani, testo atlante. – Bologna, Monduzzi editore.

Consonni, A. 2016a. Precious Finds from an Early Middle Kingdom Tomb in Thebes: Reconstructing Connections Between the Dead and Their Goods. In: Miniaci, G. & W. Grajetzki. Eds. Middle Kingdom Studies. The World of Middle Kingdom Egypt (2000-1550 BC). Vol. 2. Contributions on Archaeology, Art, Religion and Written Souces. – London, Golden House Publication: 13-26.

Consonni, A. 2016b. Ptolemaic Pottery from the Area of the Temple of Millions of Years of Amenhotep II: An Overview. In: David, R. Ed. Céramiques ptolémaïques de la région thébaine. – Le Caire, Institut français d'archéologie orientale (Cahiers de la Céramique Égyptienne 10, Numéro special): 193-213.

Consonni, A. & T. Quirino. 2019. Reconstructing the Perishable: Perspectives on Coffins at the Temple of Amenhotep II in Western Thebes. In: Strudwick, H. & J. Dawson. Eds. Ancient Egyptian Coffins: Past – Present – Future. – Oxford, Oxbow Books: 198.

Consonni, A., T. Quirino & A. Sesana. 2017. Before and After the Temple: The Long-Lived Necropolis in the Area of the Temple of Millions of Years of Amenhotep II – Western Thebes. In: Guidotti, M.C. & G. Rosati. Eds. Proceedings of the Eleventh International Congress of Egyptologists (Florence 23-30 August 2015). – Oxford, Archaeopress: 112-119.

Consonni, A. & A. Sesana. 2016. The Pottery from a Middle Kingdom Tomb at the Temple of Million of Years of Amenhotep II – Thebes. In: Bader, B., C.M. Knoblauch & E.C. Köhler. Eds. Vienna 2 – Ancient Egyptian Ceramics in the 21st Century, Proceedings of the International Conference held at the University of Vienna 14th-18th of May 2012. – Leuven, Peeters (Orientalia Lovaniensia Analecta 245): 103-118.

Cvetkovićab, D., S. Nikolićab, V. Brkovićbc & V. Živković. 2019. *Hyperostosis Frontalis Interna* as an Age-Related Phenomenon – Differences between Males and Females, and Possible Use in Identification. – Science & Justice 59 (2): 172-176.

De Grossi Mazzorin, J. 2008. Archeozoologia, lo studio dei resti animali in archeologia. – Bari, Laterza.

Duday, H. 2005. Lezioni di archeotanatologia. Archeologia funeraria e antropologia di campo, – Roma, Soprintendenza Archeologica editore.

Fornaciari, G. & V. Giuffra. 2009. Lezioni di Paleopatologia. – Genova, ECIG editore.

Forshaw, R.J. 2009. Dental Health and Disease in Ancient Egypt. – British Dental Journal 206: 421-424.

Gentry, A., J. Clutton-Brock & C.P. Groves. 2004. The Naming of Wild Animal Species and Their Domestic

Derivatives. – Journal of Archaeological Science 31: 645-651
Grant, A. 1982. The Use of Tooth Wear as a Guide to the Age of Domestic Ungulates. In: Wilson, B., C. Grigson & S. Payne. Eds. Aging and Sexing Animal Bones from Archaeological Sites. – Oxford, Archaeopress (British Archaeological Reports 109): 91-108.
Harris, J.H, P.V. Ponitz & B.K. Ingalls. 1998. Dental Health in Ancient Egypt. In: Cockburn A. & E. Cockburn. Eds. Mummies, Disease & Ancient Cultures. – Cambridge, Cambridge University Press: 59-68.
Harcourt, R.A. 1974. The Dog in Prehistoric and Early Historic Britain. – Journal of Archaeological Science 1 (2): 151-175.
Hillson, S. 1997. Dental Anthropology. – Cambridge, Cambridge University Press.
Ikram, S. 1995. Choice Cuts: Meat Production in Ancient Egypt. – Leuven, Peeters (Orientalia Lovaniensia Analecta 69).
Iscan, M.Y., S.R. Loth & R.K. Wright. 1984. Age Estimation from the Rib by Phase Analysis: White Males. – Journal of Forensic Science 29 (4): 1094-1104.
Iscan, M.Y. & S.R. Loth. 1985. Age Estimation from the Rib by Phase Analysis: White Females. – Journal of Forensic Science 30 (3): 853-863.
Klein, R.G. & K. Cruz-Uribe. 1984. The Analysis of Animal Bones from Archeological Sites. – Chicago, The University of Chicago Press.
Koudelka, F. 1884. Das Verhältnis der *ossa longa* zur skeletthohe bei den Saugetieren. – Verhandlung des Naturforschung Vereines in Brünn 24: 127-153.
Linseele, V., W. Van Neer, H. Willems & B. Vanthuyne. 2017. An Unusual Cattle Burial at Dayr al-Barshā. In: Mashkour M. & M. Beech. Ed. International Symposium on the Archaeozoology of Southwestern Asia and Adjacent Areas, Abu Dhabi. – Archaeozoology of the Near East 9: 353-377.
Lovejoy, C.O. 1985. Dental Wear in Libben Population: Its Functional Pattern and Role in the Determination of Adult Skeletal Age at Death. – American Journal of Physical Anthropology 68: 47-56.
Lovejoy, C.O., R.S. Meindl, T.R. Przybeck & R.P. Mensforth. 1985. Chronological Metamorphosis of the Auricular Surface of the Ilium. A New Method for the Determination of Adult Skeletal Age at Death. – American Journal of Physical Anthropology 68: 15-28.
Macke, A., C. Macke-Ribet & J. Connan. 2002. Ta set neferou. Une nécropole de Thèbes-ovest et son histoire 5. – Le Caire, Dar Namatallah Press.
Matolcsi, J. 1970. Historische Erforschung der Körpergröße des Rindes auf Grund von Ungarischem knocken material. – Zeitschrift für Tierzüchtung und Züchtungsbiologie 87: 89-137.
Martin, R. & K. Saller. 1957. Lehrbuch der Anthropologie in systematischer Darstellung 3. I- III. – Stuttgart.
Meindl, R.S. & C.O. Lovejoy. 1985. Ectocranial Suture Closure: A Revised Method for the Determination of Skeletal Age at Death Based on the Lateral – Anterior Sutures. – American Journal of Physical Anthropology 68: 57-66.
Nerlich, A.G. & A. Zink. 2003. Anthropological and Palaeopathological Analysis of the Human Remains from Three "Tombs of the Nobles" of the Necropolis of Thebes-West, Upper Egypt. In: Strudwick N. & J. Taylor. Eds. The Theban Necropolis: Past, Present and Future. – London, The British Museum Press: 218-228.
Ortner, D.J. 2003. Identification of Pathological Conditions in Human Skeletal Remains. – London, Academic Press.
Pales, L. & M.A. Garcia. 1981. Atlas ostéologique pour servir à l'identification des mammifères du Quaternaire. Vol. II. Tête, rachis, ceintures scapulaire et pelvienne (Carnivore et Homme, Herbivores). – Paris, CNRS.
Pales, L. & Ch. Lambert. 1971. Atlas ostéologique pour servir à l'identification des mammifères du Quaternaire I. Membres (Herbivores et Carnivores). – Paris, CNRS.
Payne, S. 1973. Kill-Off Patterns in Sheep and Goats: The Mandibles from Asvan Kale. – Anatolian Studies 23: 281-303.
Prummel, W. & H-J. Frisch. 1986. A Guide for the Distinction of Species, Sex and Body Size in Bones of Sheep and Goat. – Journal of Archaeological Science 13: 567-577.
Rose, J., G.J. Armelagos, & L.S. Perry. 1993. Dental Anthropology of the Nile Valley. In: Davies, W.V. & R. Walker. Eds. Biological Anthropology and the Study of Ancient Egypt. – London, The British Museum Press: 6-74.
Rose, J. 2006. Paleopathology of the Commoners at Tell Amarna, Egypt, Akhenaten's Capital City. – Memórias do Instituto Oswaldo Cruz 101(II): 73-76.
Scheuer, L. & S. Black. 2000. Developmental Juvenile Osteology. – San Diego, CA, Elsevier Academic Press.
Sesana, A. 2010. Le temple d'Amenhotep II à Thèbes-Ouest: du passé au present. In: Leblanc C. & G. Zahi. Eds. The Temples of Millions of Years and The Royal Power at Thebes, in the New Kingdom. Science and New Technologies Applied to Archaeology. International Symposium, Luxor, 3-5 January 2010. – Memnonia, Cahier Supplémentaire 2: 73-79.
Sesana, A. & T. Quirino. 2010. La Missione Italiana di scavo presso l'area del Tempio di Milioni di Anni di Amenhotep II – Luxor, West Bank, Campagne di scavo 1998-2009, Campagna di scavo 2009-2010. In: Pirelli, R. Ed. Ricerche Italiane e Scavi in Egitto IV. – Cairo, Centro Archeologico Italiano: 321-342.

Sesana, A. & A. Consonni 2013. Un sépulture de nouveau-nèé dans l'aire du Temple de millions d'années d'Amenhotep II à Thèbes-Ouest. – Memnonia XXIV: 185-198.
Sesana, A., A. Consonni & T. Quirino 2018. Il Tempio di Milioni di Anni di Amenhotep II a Luxor – West Bank. Risultati preliminari delle campagne di scavo 2012-2017. In: Capriotti, G. Ed. Ricerche Italiane e Scavi in Egitto VII. – Cairo, Centro Archeologico Italiano: 11-27.
Sesana, A. & T. Quirino. 2010. La Missione Italiana di scavo presso l'area del Tempio di Milioni di Anni di Amenhotep II – Luxor, West Bank, Campagne di scavo 1998-2009, Campagna di scavo 2009-2010. In Pirelli, R. Ed. Ricerche Italiane e Scavi in Egitto IV. – Cairo, Centro Archeologico Italiano: 321-342.
Stloukal, M. & H. Hanáková. 1978. Die lange der Langsknochen altslawischer Bovolkerungen Unter besonder Berucksichistigung von Wachstumsfragen. – HOMO: Journal of Comparative Human Biology 29: 53-69.
Suchey, J. M. & S.T. Brooks. 1988. Skeletal Age Determination Based on the Male *Os Pubis*. – Book of Abstracts 12th International Conference of Anthropological and Ethnological Sciences, Zagreb.
Teichert, M. 1975. Osteometrische Untersuchungen zur Berechnung der Widderisthöhe bei Schafen. In: Clason, A.T. Ed. Archaeozoological Studies. – New York, American Elsevier: 51-69.
Todd, T.W. 1921. Age Changes in the Pubic Bone. – American Journal of Physical Anthropology 4: 1-70.
Trotter, M. & G.C. Gleser. 1977. Corrigenda to "Estimation of Stature from Long Bones of American Whites and Negroes". – American Journal Physical Anthropology 47: 355-356.
Ubelaker, D.H. 1989. Human Skeletal Remains: Excavation, Analysis, Interpretation. – Washington, DC, Taraxacum.
von den Driesh, A. 1976. A Guide to the Measurement of Animal Bones from Archaeological Sites. – Cambridge, Harvard University.
White, T.D. & P.A. Folkens. 2005. The Human Bone Manual. – San Diego, CA, Academic Press.

To Be or Not to Be a Dog Mummy

How a Metric Study of the Skull Can Inform on Selection Practices Pertaining to Canid Mummification in Ancient Egypt

Colline Brassard, Stéphanie Porcier & Cécile Callou

Introduction

Animal mummification in Egypt is known from the Predynastic Period onwards (Ikram, 2013) and reached its peak during the Late Period. Studying this practice can provide important information allowing a better understanding of the attitude of ancient Egyptians towards the animals surrounding them. In addition to a few cases of burials of "pets", canids were also subjected to mass mummification. According to the literature, canid mummies were dedicated to gods such as Anubis or Wepwawet that provided protection in life or in the afterlife.

Thousands of canid remains of exceptional preservation are represented in dozens of cemeteries all over the Nile valley and the western desert (Ikram, 2013: figure 3). The most well documented dog necropolises are the ones from Saqqara gathering over 8 million mummies (Ikram *et al.*, 2013; Nicholson *et al.*, 2013; 2015), and Asyut (Kitagawa, 2013; 2016; 2019). According to bioarchaeological evidence and the interpretation of art and texts, the current consensus is that the mummified animals were dogs kept in captivity, and thus bred and killed at an early age for the purpose of mummification (Brassard, 2017; 2018; section by Callou in Dunand *et al.*, 2017; Dunand *et al.*, 2015). Several authors have suggested that the dogs used for mummification were most probably males, to reflect Anubis' own gender (Hartley, 2017; Ikram, 2013).

The extraordinarily good preservation of some mummies allows us to perceive the physical appearance of these ancient canids, even including the color of their fur. Unfortunately, these cases are rare. Most canid mummies are discovered as bony remains. Moreover, individuals are often scattered, making it often impossible to reconstruct the complete skeleton of each individual. Determining and interpreting isolated skeletal pieces is therefore crucial to answer questions surrounding mummification of wild canids. Although only a few foxes and jackals[1] were discovered among wrapped mummies, it is possible that wild canids were more frequently mummified than typically estimated.

---

1 Changes in the phylogeny have led to numerous issues about nomenclature of jackals (Gaubert *et al.*, 2012; Koepfli *et al.*, 2015; Rueness *et al.*, 2011; 2015). Here, the generic term "jackal" is used to refer to the species *Canis aureus* and *Canis anthus* (the two species that are currently recognized by phylogeny) as well as to animals previously referred to *Canis lupaster* or *Canis sacer*.

In: Ikram, S., J. Kaiser & S. Porcier. Eds. 2021. *The Ancient Egyptians and the Natural World. Flora, Fauna, and Science.* – Leiden, Sidestone Press: 45-60.

	Species/Breed	Provenance and sample size	Sample size
Wild canids	*Canis aureus*	Algeria (2), Syria, Ethiopia (6), Egypt: Cairo (*C. a. lupaster* + *C. sacer*)	11
	Vulpes rueppellii	Sudan, Tunisia (2), France (Ménagerie)	4
	Vulpes vulpes	Egypt, Marocco, unknown (2), Egypt (*V. v. aegyptica*): Assouan (4), Louqsor (1), Cairo (1)	10
	Vulpes zerda	unknown (1)	1
	Vulpes sp.	Egypt: Cairo (3)	3
	Canis dingo	Unknown	1
Domestic canid: *Canis familiaris*	German shepherd (1), boxer (2), braque allemand (1), bullmastiff (1), poodle (2), wolfhound (1), Alaskan (1), chow-chow (1), cocker (2), doberman (1), épagneul (2), wire-haired terrier (1), greyhound (1), afghan greyhound (2), sloughi (5), Pomeranian (3), malanese (1), setter (1), dachshund (2)		31
	Modern dogs from Sudan (6) and pariah from Constantinople (1)		7
TOTAL			68

Table 1. Modern specimens used for comparison with ancient remains in this study.

Indeed, researchers working on animal remains can easily miss information, especially if not very familiar with osteology and comparative anatomy of canids. In fact, very few studies are available to interpret the remains with regards to sex, species or morphotype (Brassard, 2017; Brassard & Callou, 2020; Dunand *et al.*, 2015; 2017; Hartley, 2017; Ikram *et al.*, 2020). Most often, they focus on qualitative morphological characteristics. However, the interpretation is not only subjective but also complex in big assemblages, and the relevance of these criteria has scarcely been tested on reference material. Even when metric data are used following the measurements recommended by von den Driesch (1976), most authors rely on univariate statistics. Yet, multivariate analyses are more powerful and allow a more precise description of shape (Reyment, 1982) and a better discrimination between groups as they take the multivariate nature of the biological object into account.

Expeditions in Egypt and the Egyptomania of the late 19th and the early 20th century resulted in the export of many archaeological remains to Europe. In addition to jewels, furniture and human mummies, animal mummies were considered fascinating objects. Moreover, they were even used as fertilizer or fuel (Baber, 2019; Herdman, 1889-1890). The Egyptian campaign of Napoleon Bonaparte brought much information about Egyptian treasures, thanks to the precise descriptions of the artists and scientists who accompanied him on his campaigns. Saint-Hilaire, and later Gaillard and Lortet were the first naturalists to focus on the fauna of modern and ancient Egypt. Lortet and Gaillard provided a full morphological and metric description of animals from modern and ancient Egypt in their famous publication called *La Faune momifiée de l'ancienne Egypte*, published between 1903 and 1909. They collected and brought back to France large quantities of mummies that were housed in the Musée Guimet in Lyon (now the Musée des Confluences). Maspero completed these collections with other remains creating the most important collection of animal mummies outside of Egypt (Porcier & Berthet, 2014).

The goal of the present study was to develop a multivariate approach for determining the species and morphotypes of canid remains based on the cranium (*i.e.* the skull without the mandible). Modern specimens of dogs from various breeds, jackals, and foxes from the Musée des Confluences (Lyon), the Museum of Natural History (Paris) and the museum of natural History in Geneva (Switzerland) were used as a reference sample to classify the ancient canids from the Musée des Confluences.

Material

The dataset is composed of 124 canid crania from the Musée des Confluences (Lyon, France), the Museum of Natural History (Paris, France) and the Museum of Geneva (Switzerland). The origin and inventory numbers are given in the supplementary data.

The modern comparative collection is composed of 38 domestic dogs from various breeds and 30 wild canids including a feral dog (*Canis dingo*). Most of the wild canids are from North Africa. Seven stray dogs from North Africa, called pariah dogs, were included in the domestic dogs. Details about the modern reference sample are provided in Table 1. The crania collected by Louis Lortet and Claude Gaillard in Cairo, Luxor and Aswan were identified as *Vulpes vulpes aegyptiaca* Sonnini, *Canis lupaster* Hemprich & Ehrenberg; *Canis sacer* Hemprich & Ehrenberg or *Canis dœderleini* Hilzheimer (Table 6).

We also studied the crania of 46 canid mummies at the state of isolated bones from the collection of the conservation and study center of the Musée des Confluences in Lyon. These crania were collected by Louis Lortet and Claude Gaillard or Maspéro in the early 20th century. No precise date is available for these specimens, but some other dog mummies from the collection were radiocarbon dated, and the oldest date to the 30th Dynasty, around 360 BC (Porcier *et al.*, 2019; Richardin *et al.*, 2017). Some identifications were proposed in *La Faune momifiée de l'ancienne Egypte* (1903; 1907; 1909). Drawings and precise descriptions enabled us to link this information to the crania we used in the present study. For others, the Museum's entry logbook provided little and often incomplete information on the arrival dates and provenance of these mummies. Finally, writings on some of the crania indicated the species, as well as provenance (Table 6).

Measurement	Definition	Measurability
cr 3	Basal length: Basion - Prosthion	+
cr 9	Facial length: Frontal midpoint - Prosthion	+
cr 14	Length of the horizontal part of the palatine: Staphylion - Palatinoorale	+
cr 15	Length of the cheek tooth row (measured along the alveoli on the buccal side)	-
cr 16	Length of the molar row (measured along the alveoli on the buccal side)	-
cr 24	Breadth dorsal to the external auditory meatus	+
cr 29	Greatest neurocranium breadth: Euryon - Euryon	-
cr 31	Least breadth of cranium: beadth at the postorbital constriction	+
cr 32	Frontal breadth: Ectorbitale - Ectorbitale	+
cr 33	Least breadth between the orbits: Entorbitale - Entorbitale	+
cr 36	Breadth at the canine alveoli	+
cr 38	Height of the cranium. The two pointers of the slide gauge are placed basally on the base of the cranium (on the basioccipital) and dorsally on the highest elevation of the sagittal crest	
cr 39	Height of the cranium without the sagittal crest. The lower pointer is placed on the basioccipital and the upper pointer is placed beside the sagittal crest on the highest point of the braincase	-

Table 2. Cranial measurements taken in this study following von den Driesch (1976).

To this sample we added 10 ancient pariah dogs from the archaeological site of Kerma (Sudan) collected by Louis Chaix, former curator at the Museum of Geneva.

Methods

We used used R 3.4.2. and Rstudio 0.99.903 for all the statistics analyses.

Measurements

To describe the form of the cranium, we chose 13 easy-to-take cranial measurements as proposed by von den Driesch (1976) (Table 2). We obtained a matrix of 13 variables and 124 individuals. Our objective was to consider all these measurements together to increase the accuracy of comparisons between individuals.

Data Conversion

To help interpret differences between individuals, the form of the cranium was characterized by its size and shape using log-shape ratios (Mosimann, 1970). From the raw data matrix, we obtained a new "form" matrix of 14 variables (13 variables of shape and 1 variable corresponding to the log size). The Euclidian distance between individuals was obtained using function dist().

Differences in Size and Shape between Extant Species

In the following statistical analyses, we considered 61 individuals of which the species identity is certain. This represents 9 *Canis aureus*, 38 *Canis familiaris*, 4 *Vulpes rueppellii* and 10 *Vulpes vulpes*.

Because of high disparity in variance between species, we performed one-way analyses of means, not assuming equal variances (function oneway.test() with the argument var.equal=FALSE) to look for differences in isometric size between extant species of known identity. The results indicate a highly significant difference in size between species ($p = 7.158e-12$; $F = 188.5$; $df=3$). To understand the origin of this difference, we performed a Games-Howell test, which represents an extension of the Tukey-Kramer post-hoc test for unequal variances. We used function posthocTGH() from package userfriendlyscience, with the parameter method="games-howell".

To compare form variables (shape and size taken together), we performed a Permutational Multivariate Analysis of Variance Using Distance Matrices. PERMANOVA is a nonparametric MANOVA, which uses a permutation test with pseudo-F ratios. We used the function adonis() from the vegan package because adonis is less sensitive to dispersion effects. We also performed post-hoc tests with pairwise.adonis(). We next performed a multivariate analogue of Levene's test for homogeneity of variances. We ran betadisper() and then permutest() from the vegan package. We specified PAIRWISE=TRUE to perform comparisons of group mean dispersions. A classical t-test was performed on the pairwise group dispersions, combined with a permutation test based on the t statistic calculated on pairwise group dispersions.

Unsupervised Analyses

In order to study the structure of the dataset without making any assumptions about species or morphotype membership, a hierarchical clustering was conducted based on the distance matrix. This allowed us to identify groups without specifying the number of clusters. We used the Ward's minimum variance method, which minimizes the total within-cluster variance (at each step of the process, the pair of clusters with minimum between-cluster distance are merged). It aims at finding compact, spherical clusters gathering morphologically similar individuals. The results are represented by a dendrogram.

Visualisation

A problem when dealing with sets of many measurements is the visualisation of the differences between individuals. We chose to perform a Principal Component Analysis (PCA) on the matrix combining shape and size. PCA produces linear combinations of the original variables to generate the principal components. As a result, the

PCA calculates uncorrelated variables and reduces the dimensionality of a data set with many variables. The first principal components (or axes) explain most of the variance of the dataset. Since all the log-shaped variables are on the same scale and have the same unit they were not scaled further and we performed a PCA on the covariance matrix allowing for the maximization of variation. We used the scores of non-zero components as new data for the following tests.

Supervised Analyses

To predict the species of ancient canids, we performed linear discriminant analyses (MASS package), partial least square discriminant analysis (mixOmics package) and neural networks (nnet package). These three methods are based on different principles, the most robust being neural networks, although its functioning is more obscure. The number of layers and the number of neurons per layer is determined using the function tune.model() that looks for parsimonious parameters maximizing cross-validation.

Rules are calculated from the extant individuals of known species (38 dogs, 9 jackals[2], 10 red foxes, 4 Rüppell's foxes) and then applied to the archaeological canids. The cross validation was performed on the same samples used for the construction of the model, because of the small number of reference specimens. As a consequence, the model depends on the reference sample and is subject to change if dog breeds are added or deleted.

Results

Differences in Size and Shape Between Extant Species

The analyses indicate a significant difference in size between species (p-value = 7.158e-12; F = 188.5; df=3). All species appear to have significantly different sizes (Table 3). Foxes and dogs are easily distinguishable using size (Figure 1). However, there is a strong overlap in the size of dogs and jackals (*Canis aureus*), partly due to great diversity between dog breeds. Moreover, the presence of dog-jackal hybrids cannot be ruled out (Galov *et al.*, 2015). As a consequence, size by itself is not a sufficient element to discriminate species and morphotypes. Moreover, the ancient Egyptian canids from the Musée des Confluences have sizes compatible with both dogs and the nine modern jackals, *Canis aureus*, of our sample.

2 Because of some uncertainty as to the identification of modern canids with writings mentioning ancient jackal species names, we did not take into account the modern canids that refer to *Canis sacer* or *Canis aureus lupaster* in the sample used for the establishment of the decision rules. We only kept canids with the writing "*Canis aureus*". However, their attribution to *Canis anther*, rather than to *Canis aureus*, is not to be excluded (see the discussion section).

	T	df	p-value
Canis familiaris versus *Canis aureus*	8.1	38.5	5.1e-09
Canis familiaris versus *Vulpes vulpes*	16.1	42.3	1.1e-12
Canis familiaris versus *Vulpes rueppellii*	22.7	13.5	2.6e-11
Canis aureus versus *Vulpes vulpes*	907	16.8	1.4e-07
Canis aureus versus *Vulpes rueppellii*	18.0	6.9	1.8e-06
Vulpes rueppellii versus *Vulpes vulpes*	9.8	6.7	1.1e-04

Table 3. Results of the post-hoc tests on size.

	F	R2	p-value	adjusted p-value
Canis familiaris versus *Canis aureus*	10.87	0.19	<0.001	0.006
Canis familiaris versus *Vulpes vulpes*	26.38	0.36	<0.001	0.006
Canis familiaris versus *Vulpes rueppellii*	22.92	0.36	<0.001	0.006
Canis aureus versus *Vulpes vulpes*	11.630	0.41	<0.001	0.006
Canis aureus versus *Vulpes rueppellii*	23.00	0.68	0.003	0.018
Vulpes rueppellii versus *Vulpes vulpes*	11.73	0.49	0.002	0.012

Table 4. Results of the post-hoc tests on form. The results indicate there is a difference in dispersion among species (p = 0.019; df=3; F=3.746), and this difference is clearly due to the dog (Table 5). All wild canid species have a similar dispersion.

	Observed p-value	Permuted p-value
Canis familiaris versus *Canis aureus*	0.034	0.027
Canis familiaris versus *Vulpes vulpes*	0.045	0.035
Canis familiaris versus *Vulpes rueppellii*	0.072	0.068
Canis aureus versus *Vulpes vulpes*	0.75	0.77
Canis aureus versus *Vulpes rueppellii*	0.57	0.59
Vulpes rueppellii versus *Vulpes vulpes*	0.37	

Table 5. Results of the pairwise comparison of the dispersion test.

In summary, our analyses detected a significant difference in size and shape between the red fox and a Rueppel's fox, between the red fox and jackals, and finally between Rueppel's fox and jackals. Comparisons with dogs require to check the overlap on the PCA: statistical differences between dogs and jackals are more likely to be due to a high disparity in dog cranial form, but there is clearly a difference in the form of foxes and dogs.

Hierarchical Cluster Analysis (HCA)

The dendrogram we obtained (Figure 2) is an appropriate summary of the data, since correlation between the

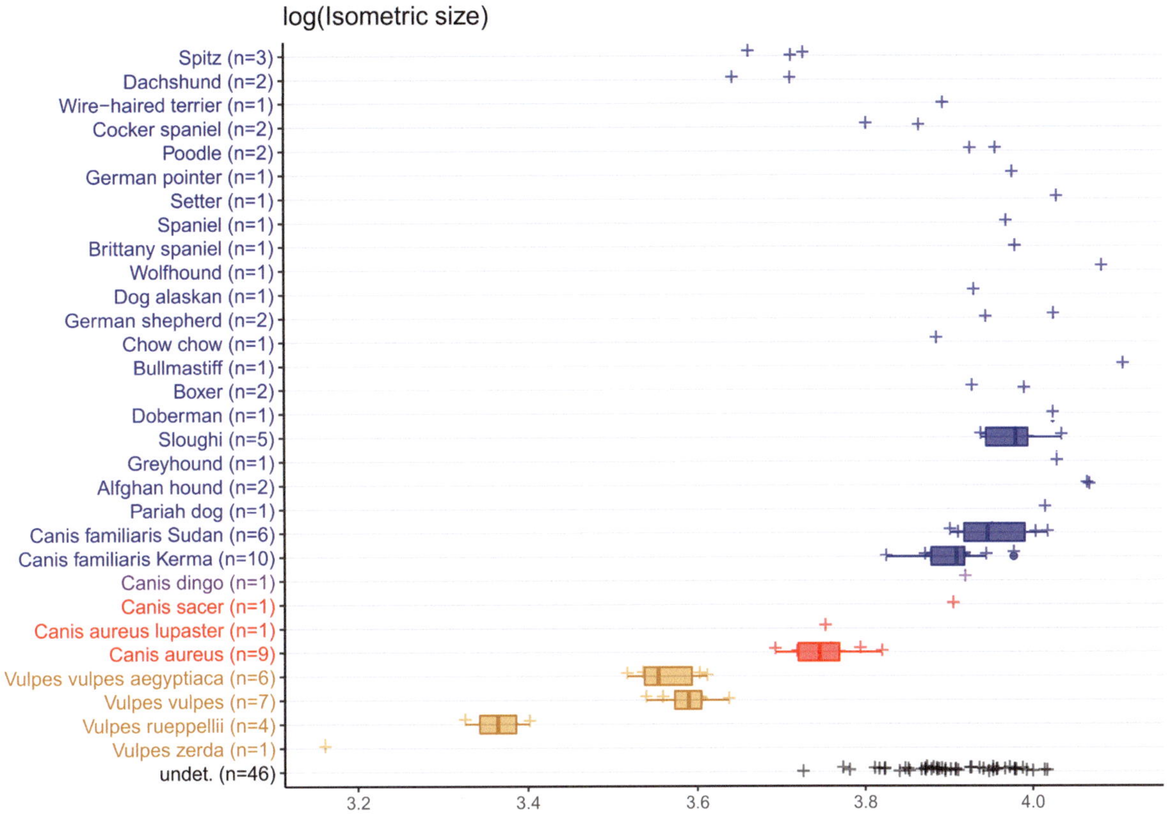

Figure 1. Size of the canids used in this study. There is also a highly significant difference in cranial shape between species ($p<0.001$; df=3; F=19.105) and all species have significantly different cranial forms (Table 4).

original distances and the cophenetic distances (the intergroup dissimilarity at which the two observations are first combined into a single cluster) is high (0.73). There are two main groups in the classification. We kept a decomposition into 6 groups to refine the description of current and ancient canids. Modern individuals from *Vulpes* and *Canis* genera are clustered in two distinct groups, while modern jackals (*Canis aureus*) and dogs are clustered together, all jackals being all grouped along the same branch (Figure 2). The HCA group with modern jackals also contains two ancient canids that were identified previously as "*Canis lupaster typicus*" or "*Canis lupaster*" by Louis Lortet and Claude Gaillard (1909: fig. 194-1957; Table 6) and three other ancient crania. One cranium with writings that mention "*Canis aureus*" was grouped with the modern dogs.

Vulpes zerda is only represented by one individual, which may explain why it clustered with *Vulpes rueppellii.*

The first two axes of the PCA represent almost 70% of the variability, and the first three axes 80% (Figure 3). Axis 1 is more related to size and axis 2 to the elongation of the face (Figure 3). The dogs are placed in three groups (Figures 2, 4). Group E contains dogs with an elongated crania and thin snouts (greyhound type), that have been previously identified as *Canis sacer* or *Canis doederleini* by Lortet and Gaillard (Table 6). Group D contains a few rare dogs with globular crania that project themselves into the morphological space and contains dachshund and small Pomeranian lulus. Finally, group F contains medium dogs. The "stray" dogs, pariahs, and the dingo (a feral Australian dog) project themselves mainly into this third group, even if some look more like greyhounds. This highlights the importance of interpreting the results in terms of morphological variation, rather than trying to categorize dog 'types'.

All the dogs of ancient Egypt are positioned at the level of the spitz/dachshund group and are young (about 6-7 months; the permanent teeth are all present; the canine is in place, and the basisphenoidal suture is still open). The shape of the cranium therefore corresponds approximately to the shape of the adult. The bulky shape of the cranium is likely not only due to the young age of the animals.

Supervised Analyses

All the three foxes identified as "*Vulpes* sp." were identified as *Vulpes vulpes* by the three supervised method (and by the HCA as well). Cross-validation on linear discriminant

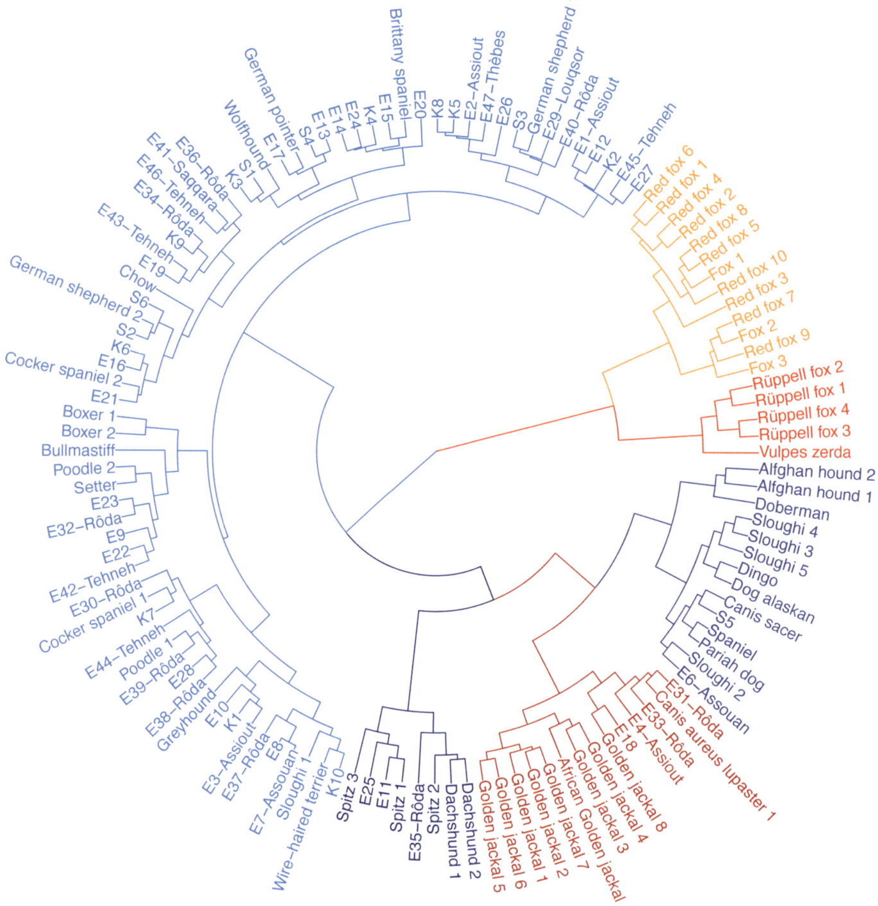

Figure 2. Plotfan representing the results of the classification.

analyses and neural networks lead to a success rate of 100%. With regard to the PLS-DA, the success rate is 98.4%, because one red fox was identified as a jackal. The (modern) canid identified as a *Canis doederleini* (Lortet and Gaillard considered it as a kind of jackal) was identified as a dog by all supervised analyses.

Identifications obtained for archaeological specimens are listed in Table 6. The LDA, PLS-DA and neural networks all identified the same three ancient jackals, including specimens that have been previously identified as "*Canis lupaster typicus*" and "*Canis lupaster*" by Louis Lortet and Claude Gaillard (1909: fig. 194-1957; Table 6). The LDA identified five supplemental ancient jackals, including specimens that have been previously identified as "*Canis sacer*" or stray dogs by Lortet and Gaillard (Table 6). One cranium with writing that mentions "*Canis aureus*" was identified as a dog by all supervised analyses.

Discussion

Wild and domestic canids can be distinguished quite easily using cranial size and shape, with both unsupervised and supervised multivariate methods. Supervised methods lead to high success rates, yet may not be entirely reliable as our cross-validation was performed on the same specimens that were used to establish the decision rules. Consequently, it is essential to enrich the database and check the identification using a larger assemblage of remains. Moreover, the reference dataset misses more well-identified European and African jackals, fennecs (*Vulpes zerda*; Wacher *et al.*, 2015), and other wild canids whose presence in ancient Egypt remains to be explored. For instance, it is not impossible that the black-backed jackal, *Canis mesomelas,* or the pale fox, *Vulpes pallida,* were present in Egypt in the past. Indeed, their current distribution is to the east (Ethiopia) and south (Hoffmann, 2014) for the black-backed jackal and towards central Africa for the pale fox (Sillero-Zubiri & Wacher, 2012).

The multivariate approach provides, however, a new classification of the ancient Egyptian canids from the Musée des Confluences and is complementary to the descriptions provided by previous authors. The correspondence between the determination provided by our different methods and Lortet and Gaillard's classification is given in Table 6. In most cases, supervised and non-supervised

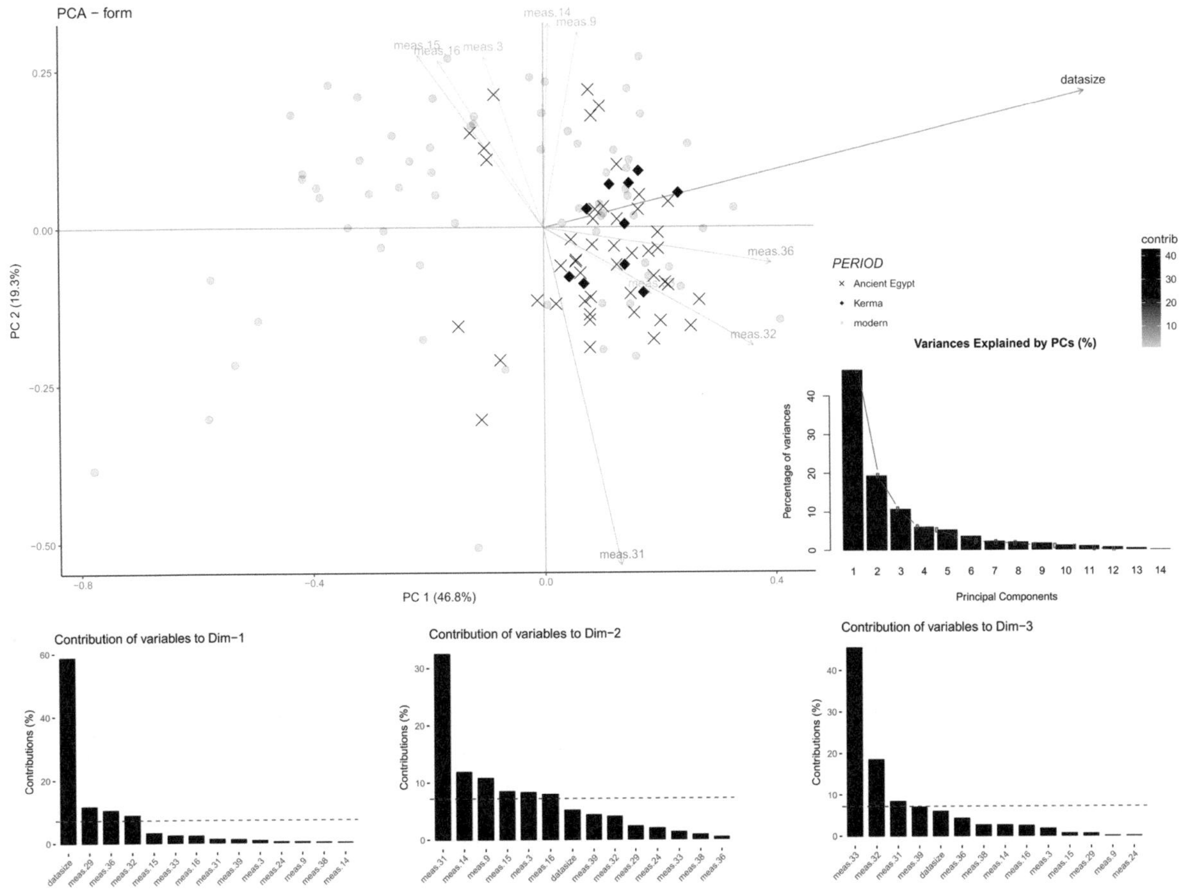

Figure 3. Contribution of the shape variables and size to PCA axes.

analyses lead to the same identifications. We find some communalities with the classification provided by Lortet and Gaillard. However, some species names are no longer appropriate (such as *Canis doederleini* or *Canis sacer*) because canid taxonomy has changed since the last century. This complicates the interpretation of the results. We did not, however, identify any red foxes, in contrast to Lortet and Gaillard, but none of our determinations contradicts those of the previous authors.

The case of the jackals remains difficult. Depending on the zooarchaeologist, some jackals could be identified as dogs and inversely, as testified by some contradictory information written on the skulls or labels. Multivariate metrics appear to provide a more reliable tool for species identification. However, some intermediate shapes also could not be reliably classified by our methods. This is related to the relatively few anatomical differences, the high genetic proximity and possible hybridization between jackals and dogs (Galov *et al.*, 2015). Lortet and Gaillard distinguished two jackals (*Canis aureus* and *Canis lupaster*), but because of changes in taxonomy, *Canis lupaster* is no longer used. Moreover, genetic analyses recently demonstrated that jackals from Europe and northern Africa are not the same, resulting in a difference between *Canis aureus* (Europe) and *Canis anthus* (northern Africa). Since Egypt is at the crossroads between the Near East and North Africa, it is possible that the two species are present in the country and that they were already there several hundred years ago. This point remains unclear and has led to confusion among researchers (Koepfli *et al.*, 2015). As a consequence, the identification of ancient jackals is still unclear and would need further investigation. A few canids that had been identified as jackals are projected into the morphological variability of the dogs in our analyses. On the contrary, we also identified jackals among the dogs identified by Lortet and Gaillard.

Like Lortet and Gaillard in 1903, we identified mostly dogs in the collection of the *Musée des Confluences*. Because of the continuity in shapes, Lortet and Gaillard were limited in their understanding of these big assemblages, and further refused to separate the different breeds of dogs because of the numerous intermediate shapes. Simple qualitative descriptions are limited and consequently multivariate statistical methods may

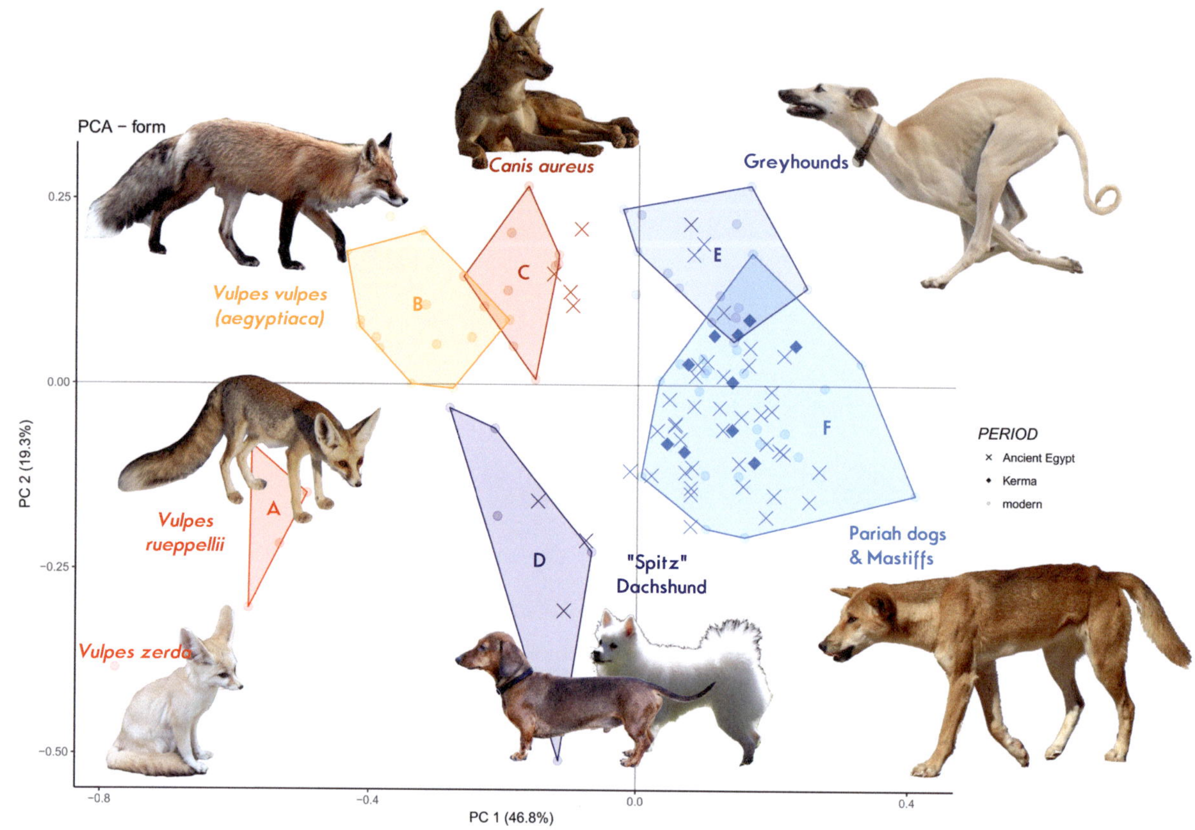

Figure 4. Results of the classification projected on the first two axes of the PCA.

provide more objective tools for classification. Although it remains difficult to separate distinct groups within the continuum of the great diversity of cranial shapes, we identified several morphotypes of dogs that more or less match the three "breeds" of dogs first described by Lortet and Gaillard. We identified mainly intermediate dogs (such as the dogs from Kerma or dingoes), including some with a larger cranium and a shorter snout (mastiff-like dogs). These dogs seem to correspond to the feral dogs and the Egyptian dogs (which size is intermediate and shows a shorter, wider and rounder cranium) described by Lortet and Gaillard. One of our groups contains dolichocephalic dogs with more elongated crania (such as the modern greyhound), which may correspond to the "tesem" of Lortet and Gaillard. As previously claimed by these authors, all dogs look like pariahs from southern regions such as dingoes, greyhounds, or Tibet Dogs. However, a few dogs are clearly different from these two groups. They have a very small and rounded cranium, as observed in modern loulou or dachshunds. Lortet and Gaillard described a spitz-like or Egyptian loulou based on a single mummified cranium from Asyut. Yet, this breed is bigger than the Pomeranian and looks like a Tibetan Dog, which does not match the morphology of these small dogs with rounded crania. Lortet and Gaillard do not mention the presence of bassets in mummified dogs, but their presence at that time is attested by representations on the graves. For instance, a "low-slung, curly-tailed hound" dog, named Ankhu, is represented beneath his owner's chair in the Tomb of Djehutyhotep (12th Dynasty, 1850 BC, Deir al-Bersha) on a limestone fragment of a painted relief (Russmann *et al.*, 2001). Short-limbed dogs are also represented on wall reliefs from the tombs of Khnumhotep and Tehuihotep (12th Dynasty; Brewer *et al.*, 2001: 36, 450). This is worth keeping in mind for future studies.

Conclusion and Perspectives

We here propose a simple metric method that we hope will make morphometric analyses available to every researcher working on Egyptian animal remains. We show that even with few measurements – commonly taken in the field – the description of big canid assemblages can be improved. We chose to focus on the cranium, but long bones could also provide species-specific information. Geometric morphometrics, which offers good visualisation and a more accurate distinction of groups could provide further resolution (Drake & Klingenberg, 2010; Drake *et al.*, 2017; Jojić *et al.*, 2017; Rezic *et al.*, 2017; Schmitt & Wallace, 2014)

	HCA	LDA	PLSDA	NN	previous identification
Fox1 (modern) (50000343)	*V. vulpes*	*V. vulpes*	*V. vulpes*	*V. vulpes*	
Fox2 (modern) (50000341)		*V. vulpes*	*V. vulpes*	*V. vulpes*	
Fox3 (modern) (50000344)		*V. vulpes*	*V. vulpes*	*V. vulpes*	
Canis aureus lupaster (modern) (50000332)	*C. aureus*	*C. aureus*	*C. aureus*	*C. aureus*	*C. lupaster typicus* (Lortet & Gaillard, 1909: fig. 194-195)
E18 (90010311)					
E31-Rôda (50000345)		*C. aureus*	*C. aureus*	*C. aureus*	*Canis lupaster* (Lortet & Gaillard, 1909: fig. 196-197)
E33-Rôda (90010309)		*C. aureus*			
E4-Assiout (51000161)		*C. aureus*	*C. aureus*	*C. aureus*	
E11 (51000028)	Dachshunds – Spitz				
E25 (51000014)					
E35-Rôda (51000030)		*C. aureus*			
Canis sacer (modern) (50000331)	Greyhounds	*C. aureus*			"*Canis sacer – Canis aureus* – Egypt"
E6-Assouan (51000004)					*Canis doederleini* (Lortet & Gaillard, 1909: fig. 202-203)
E1-Assiout (90001303)	Mastiffs / Parias	*C. aureus*			stray dog from Assiut, black coat (Lortet & Gaillard, 1909: fig. 206-207)
E2-Assiout (51000005)					Tesem
E3-Assiout (90010313)					
E7-Assouan (51000008)					
E8 (51000003)					"*Canis aureus*"
E9 (51000021)					
E10 (51000031)		*C. aureus*			
E12 (51000242)					
E13 (90010305)					Egyptian spitz (Lortet & Gaillard, 1909: fig. 204-205)
E14 (51000164)					
E15 (51000165)					
E16 (90010307)					
E17 (90010310)					
E19 (51000163)					
E20 (51000029)					
E21 (90010306)					
E22 (51000018)					
E23 (510000016)					
E24 (51000015)					Egyptian dog (Lortet & Gaillard, 1905: fig. 8)
E26-Tehneh (51000017)					
E27 (51000024)					
E28-Tehneh (51000026)					
E29-Louqsor (51000010)					*Canis sacer* (Lortet & Gaillard, 1909: fig. 199-200)
E30 (51000013)					"*Canis aureus*"
E32-Rôda (90010308)					"*spitz ou dogue*"
E34-Rôda (51000027)					
E36-Rôda (51000160)					
E37-Rôda (51000011)					
E38-Rôda (51000009)					stray dog from Rôda (Lortet & Gaillard, 1905: fig. 5)
E39-Rôda (51000006)					
E40-Rôda (51000007)		*C. aureus*			stray dog from Rôda (Lortet & Gaillard, 1905: fig. 4)
E41-Saqqara (51000025)					

	HCA	LDA	PLSDA	NN	previous identification
E42-Tehneh (51000019)					
E43-Tehneh (51000020)					
E44-Tehneh (51000023)					
E45-Tehneh (51000022)					
E46-Tehneh (90010304)					
E47-Thèbes (51000012)					
K1 (K T83)					
K2 (K212/2)					
K3 (K133/4)					
K4 (K T89)					
K5 (K212/1)					
K6 (K151/1)					
K7 (K T67)					
K8 (K1251/)					
K9 (K249/1)					

Table 6. Classification of the extant and ancient individuals whom species is uncertain.

and merits to be tested on archaeological assemblages of dog remains. However, although this method requires 3D data acquisition and processing, it would be ideal in combination with medical imaging of complete wrapped mummies. Our preliminary database should be extended to other Egyptian canids and reinforces the necessity of an online reference database of ancient and modern Egyptian canids. Applying the method to the thousands of canids mummies from other archaeological sites should enable to better understand selection practices with regards to species and morphotypes, and particularly the possible presence of "spitz"-like dogs.

Acknowledgements

This research was carried out within the framework of the MAHES Project (Momies Animales et Humaines EgyptienneS) supported by the Agence Nationale de la Recherche from the "*Investissement d'Avenir*" program ANR-11-LABX-0032-01 LabEx ARCHIMEDE. We are grateful to Anthony Herrel for proofreading the English of this article. We thank the *Centre de conservation et d'étude des collections* of *Musée des Confluences* in Lyon, especially Didier Berthet, the *Muséum national d'Histoire naturelle* in Paris, in particular Géraldine Veron, and the Natural museum in Geneva, especially Jacqueline Studer and Louis Chaix, for providing access to their collections. We thank the reviewers for their constructive feedback on an earlier version of this manuscript.

Bibliography

Baber, T.T. 2019. Early Travellers and the Animal 'Mummy Pits' of Egypt: Exploration and Exploitation of the Animal Catacombs in the Early Age of Travel. In: Porcier, S., S. Ikram & S. Pasquali. Eds. Creatures of Earth, Water, and Sky. Essays on Animals in Ancient Egypt and Nubia. – Leiden, Sidestone Press: 67-86.

Brassard, C. 2017. Le chien en Egypte ancienne : approche archéozoologique et apports de la craniologie. Application a une série de chiens momifiés (El-Deir) et comparaison avec des chiens actuels et anciens (Kerma). – Lyon, école nationale vétérinaire de Lyon (Mémoire de Master 2, VetAgro Sup).

Brassard, C. 2018. Le chien en Egypte ancienne : approche archéozoologique et apports de la craniologie. – Paleobios 20 : 1-213.

Brassard, C. & C. Callou. 2020. Sex Determination of Archaeological Dogs Using the Skull: Evaluation of Morphological and Metric Traits on Various Modern Breeds. – Journal of Archaeological Science Reports 31: 102294.

Brewer, D., A. Phillips & T. Clark. 2001. Dogs in Antiquity: Anubis to Cerberus: The origins of the Domestic Dog. – Warminster, Aris & Phillips.

Dunand, F., R. Lichtenberg, C. Callou & F. Letellier-Willemin. 2017. El-Deir nécropoles IV, Les chiens momifiés d'El-Deir. – Paris, Editions Cybèle.

Dunand, F., R. Lichtenberg & C. Callou. 2015. Dogs at El-Deir. In: Ikram, S., J. Kaiser & R. Walker. 2015. Egyptian Bioarchaeology. Humans, Animals, and the Environment. – Leiden, Sidestone Press: 169-176.

Drake, A.G. & C.P. Klingenberg. 2010. Large-Scale Diversification of Skull Shape in Domestic Dogs: Disparity and Modularity. – The American Naturalist 175: 289-301.
Drake, A.G., M. Coquerelle, P.A. Kosintsev, O.P. Bachura, M.Sablin, A.V. Gusev, L.S. Fleming & R.J. Losey. 2017. Three-Dimensional Geometric Morphometric Analysis of Fossil Canid Mandibles and Skulls. – Scientific Reports 7: 9508.
Galov, A., E. Fabbri, R. Caniglia, H. Arbanasić, S. Lapalombella, T. Florijančić, I. Bošković, M. Galaverni & E. Randi. 2015. First Evidence of Hybridization Between Golden Jackal (*Canis aureus*) and Domestic Dog (*Canis familiaris*) as Revealed by Genetic Markers. – Royal Society Open Science 2: 150450.
Hartley, M. 2017. Paws in the Sand. The Emergence and Development of the Use of Canids in the Funerary Practice of the ancient Egyptians (CA. 5000 BC – 395 AD). Volume 1. – Unpublished PhD thesis, Macquarie University Sydney.
Herdman, A. 1889-1890. Note on Some Mummy Cats &c, from Egypt. – Proceedings and Transactions of the Liverpool Biological Society IV: 95-96.
Hoffmann, M. 2014. *Canis mesomelas*. - The IUCN Red List of Threatened Species 2014: e.T3755A46122476.
Ikram, S. 2013. Man's Best Friend For Eternity: Dog And Human Burials In Ancient Egypt. – Anthropozoologica 48: 299-307.
Ikram, S., P. Nicholson, L. Bertini & D. Hurley. 2013. Killing Man's Best Friend? – Archaeologcal Review from Cambridge 28.2: 48-66.
Ikram, S., L. Bertini, D. Hurley & S. Vann. 2020. The Faunal Material. In: Nicholson P., S. Ikram & S. Mills. The Dog Catacombs at Saqqara. – Leuven/London, Peeters/ British Museum Publications on Egypt and Sudan.
Jojić, V., J. Porobić & D. Ćirović. 2017. Cranial Variability of the Serbian Red Fox. – Zoologischer Anzeiger 267: 41-48.
Kitagawa, C. 2013. Tomb of the Dogs in Gebel Asyut al-Gharbi (Middle Egypt, Late to Ptolemaic/Roman Period): Preliminary Results on the Canid Remains. In: Cupere, B.D., V. Linseele & S. Hamilton-Dyer. Eds. Archaeozoology of the Near East X: Proceedings of the Tenth International Symposium on the Archaeozoology of South-Western Asia and Adjacent Areas. – Leuven, Peeters (Ancient Near Eastern Studies, Supp. 44): 343-356.
Kitagawa, C. 2016. The Tomb of the Dogs at Asyut: Faunal Remains and Other Selected Objects. The Asyut Project 9. – Wiesbaden, Harrassowitz.
Kitagawa, C. 2019. Mummies from the City of Canine Deities: An Analysis of Canid Osteofaunal Remains from the Tomb of the Dogs on Gebel Asyut al-Gharbi, Asyut, Middle Egypt. In: Porcier S, S. Ikram & S. Pasquali. Eds. 2018. Creatures of Earth, Water, and Sky. Essays on Animals in Ancient Egypt and Nubia. – Leiden, Sidestone Press: 193-210.
Koepfli, K.-P., J. Pollinger, R. Godinho, J. Robinson, A. Lea, S.Hendricks, R.M. Schweizer, O. Thalmann, P. Silva, Z. Fan, A.A. Yurchenko, P. Dobrynin, A. Makunin, J.A. Cahill, B. Shapiro, F. Álvares, J.C. Brito, E. Geffen, J.A. Leonard, K.M Helgen., W.E. Johnson, S.J. O'Brien, B. Van Valkenburgh & R.K. Wayne. 2015. Genome-Wide Evidence Reveals that African and Eurasian Golden Jackals Are Distinct Species. – Current Biology 25: 2158-2165.
Lortet, L. & C. Gaillard. 1903. La Faune momifiée de l'ancienne Egypte I. – Lyon, Henri Georg (Archives du Muséum d'Histoire naturelle de Lyon 8).
Lortet, L. & C. Gaillard. 1907. La Faune momifiée de l'ancienne Egypte II. – Lyon, Henri Georg (Archives du Muséum d'Histoire naturelle de Lyon 9).
Lortet, L. & C. Gaillard. 1909. La Faune momifiée de l'ancienne Egypte III. – Lyon, Henri Georg (Archives du Muséum d'Histoire naturelle de Lyon 10).
McKnight, L.M., S.D. Atherton-Woolham & J.E. Adams. 2015. Imaging of Ancient Egyptian Animal Mummies. – RadioGraphics 35: 2108-2120.
Mosimann, J.E. 1970. Size Allometry: Size and Shape Variables with Characterizations of the Lognormal and Generalized Gamma Distributions. – Journal of the American Statistical Association 65: 930-948.
Nicholson, P.T., J. Harrison, S. Ikram, E. Early & Y. Qin. 2013. Geoarchaeological and Environmental Work at the Sacred Animal Necropolis, North Saqqara, Egypt. – Studia Quaternaria 30: 83-89.
Nicholson, P.T., S. Ikram & S. Mills. 2015. The Catacombs of Anubis at North Saqqara. – Antiquity 89: 645-661.
Porcier, S. & D. Berthet. 2014. Quand momies animales et Museum d'histoire naturelle font bon ménage : la prestigieuse collection du musée des Confluences à Lyon. – Archéothéma 37: 41-45.
Porcier, S., P. Richardin, G. Louarn, S. Ikram & D. Berthet. 2019. Datations par le carbone 14 de 63 momies animales du musée des Confluences à Lyon (France). In: Porcier, S., S. Ikram & S. Pasquali. Eds. Creatures of Earth, Water, and Sky. Essays on Animals in Ancient Egypt and Nubia. – Leiden, Sidestone Press: 283-292.
Reyment, R.A. 1982. Multivariate Morphometrics. In: Krishnaiah, P.R. & L.N. Kanal. Eds. Handbook of Statistics. – Amsterdam, Elsevier: 721-745.
Rezic, A., I. Bošković, P. Lubinu, M. Piria, T. Florijančić, M. Scandura & N. Šprem. 2017. Dimorphism in the Skull Form of Golden Jackals (*Canis aureus Linnaeus*, 1758) in the Western Balkans: A Geometric Morphometric Approach. – Pakistan Journal of Zoology 49: 989-997.
Richardin, P., S. Porcier, G. Louarn, S. Ikram & D. Berthet. 2017. Cats, Crocodiles, Cattle, and More: Initial Steps

Toward Establishing a Chronology of Ancient Egyptian Animal Mummies. – Radiocarbon 59/2: 595-607.

Russmann, E. R. & H.J. Thomas Garnet. 2001. Eternal Egypt: Masterworks of Ancient Art from the British Museum. – London, British Museum Press.

Schmitt, E. & S. Wallace. 2014. Shape Change and Variation in the Cranial Morphology of Wild Canids *(Canis lupus, Canis latrans, Canis rufus)* Compared to Domestic Dogs (*Canis familiaris*). Using Geometric Morphometrics. – International Journal of Osteoarchaeology 24: 42-50.

Sillero-Zubiri, C. & T. Wacher. 2012. *Vulpes pallida*. - The IUCN Red List of Threatened Species 2012: e.T23052A16813736.

von den Driesch, A. 1976. A Guide to the Measurements of Animal Bones from Archaeological Sites. – Cambridge, MA, Peabody Museum (Peabody Museum Bulletin 1).

Wacher, T., K. Bauman & F. Cuzin. 2015. *Vulpes zerda*. - The IUCN Red List of Threatened Species 2015: e.T41588A46173447.

Supplementary Data

Origin of the specimens and measurements taken in this study ID: names given in the present study; Inv.Num: inventory number; Coll.: Museum of origin (CCEC: *Musée des Confluences* in Lyon, France; MNHN: Museum of natural History, Paris, France; MG: Museum of Geneva, Switzerland); VD: measurement according to von den Driesch's nomenclature.

ID	Inv. Num.	Coll.	VD3	VD9	VD14	VD15	VD16	VD24	VD29	VD31	VD32	VD33	VD36	VD38	VD39
Dingo	50000255	CCEC	180	110	33.9	63.6	20.5	62.4	57.2	30.9	45.2	30.6	35.1	58.3	49.2
African Golden jackal	50000319	CCEC	137.5	84	26.7	54.8	18.9	51.5	52.2	30.5	42.4	29	25	46.7	44.7
Canis aureus lupaster 1	50000332	CCEC	141.3	92.8	25.9	58.4	18.4	50.6	50.5	28	36.1	28.3	26.6	46.9	46.3
Canis sacer	50000331	CCEC	171	100.7	31.4	67.2	19.5	62.5	53.6	32.4	46.7	32.2	31.8	57.9	52.5
Golden jackal 1	CG 1972-385	MNHN	148.1	92.9	27.6	58.6	16.7	50.6	48.0	27.8	38.1	28.4	28.2	44.9	41.1
Golden jackal 2	CG 1973-126	MNHN	149.8	95.5	30.6	60.7	17.9	54.5	51.6	29.1	42.1	28.5	27.0	47.0	44.4
Golden jackal 3	CG 1973-129	MNHN	134.8	85.1	27.4	55.4	17.0	41.8	48.7	29.3	36.8	24.9	23.7	44.1	41.5
Golden jackal 4	CG 1973-128	MNHN	136.7	84.8	28.3	55.7	16.0	49.9	50.5	30.0	37.9	25.1	25.7	44.6	41.9
Golden jackal 5	CG 1972-382	MNHN	150.6	100.3	30.2	61.4	18.1	50.3	51.7	26.5	42.4	26.0	26.0	45.3	42.0
Golden jackal 6	CG 1969-476	MNHN	144.5	93.4	27.6	56.0	18.0	49.0	46.4	26.9	39.1	26.8	23.7	43.0	40.3
Golden jackal 7	eric	MNHN	154	103.3	30.2	65.0	18.7	55.7	53.5	32.0	40.7	30.1	27.0	46.1	43.3
Golden jackal 8	50000321	CCEC	141.7	89.1	27.8	57.5	17.6	49.8	50	28.2	33.6	21.7	27	49	48.7
Rüppell fox 1	CG 1930-73	MNHN	91.9	50.4	16.2	37.9	9.6	35.1	39.4	23.3	25.4	19.6	13.8	31.8	31.2
Rüppell fox 2	CG 1982-1009	MNHN	95.5	51.0	17.0	36.6	10.4	35.4	39.2	20.7	26.6	20.8	15.1	33.4	31.9
Rüppell fox 3	1922-266	MNHN	98.5	53.5	17.7	41.0	12.4	37.4	40.5	20.7	24.7	19.7	16.2	32.9	31.7
Rüppell fox 4	1922-264	MNHN	97.9	55.6	18.5	39.1	12.3	37.4	41.0	23.1	26.8	20.0	17.1	33.1	30.8
Fox 1	50000345	CCEC	125.7	74.9	20.3	51.1	13.5	44.7	44.7	21	33.9	26.2	22.5	40.8	39.2
Fox 2	50000341	CCEC	124.7	75.7	24.1	52.8	14.1	42.9	43.7	22.8	31.3	25.5	22.2	42.3	41.5
Fox 3	50000344	CCEC	130.5	75.5	23.9	51.9	13.8	44.2	43.9	18	31.9	24.8	21.2	43.9	42.3
Red fox 1	50000286	CCEC	133.3	79.2	21.8	50	15.6	47.3	43.5	22.4	37	29.4	24.7	40.1	38.6
Red fox 2	CG 1963-1362	MNHN	124.1	75.6	19.8	51.3	15.0	44.8	46.1	23.9	37.8	26.3	21.4	39.8	37.1
Red fox 3	CG 1903-151	MNHN	123.4	75.7	18.1	50.8	16.4	45.7	46.3	22.7	29.7	25.4	20.0	35.6	33.9
Red fox 4	CG 1999-120	MNHN	116.2	68.8	20.7	46.6	13.6	42.1	44.5	22.0	34.0	25.4	19.9	36.8	34.9
Red fox 10	50000292	CCEC	132	77.6	23	53.2	14.9	44.6	45.5	20	33.4	25.9	22.7	39.1	37.5
Red fox 5	50000294	CCEC	118	69	19.2	47.2	13.6	40.3	42.2	19.7	31	23.3	20.4	38.5	37.6
Red fox 6	50000298	CCEC	126.7	75.6	20.6	50.1	15.5	42.4	44.2	21.5	37.6	27.6	22.9	41.6	40.2
Red fox 7	50000323	CCEC	117.8	69.6	21.3	47.6	13.6	40.5	43.1	20.4	29	23	20.2	41.4	40.5
Red fox 8	50000324	CCEC	118.3	70	20.3	46.8	14.6	42.5	44.6	21.4	31.5	22.9	20.6	39.1	37
Red fox 9	50000293	CCEC	129.20	75.40	24.20	47.70	14.50	44.00	44.60	21.20	28.60	23.30	18.90	40.20	38.60
Vulpes zerda	MNHN-1936-20	MNHN	72.3	40.1	13.1	27.6	9.8	33.3	36.2	18.6	20.1	15.1	12.1	29.1	28.5

A) Modern wild canids.

ID	Inv. Num.	Coll.	VD3	VD9	VD14	VD15	VD16	VD24	VD29	VD31	VD32	VD33	VD36	VD38	VD39
Alfghan hound 1	MNHN-1248	MNHN	208	137.9	36.9	75.3	23.1	58.9	61.3	36	52.7	42.4	39.1	71.2	59.6
Alfghan hound 2	MNHN-1285	MNHN	202	127.3	37.3	76	20.8	63.8	57.4	39.2	62.9	43.9	36.8	71.6	58.2
Boxer 1	MNHN-1254	MNHN	150	92.6	28	52.9	20.6	70.3	61	36.8	52	32.8	38.1	61.3	56.3
Boxer 2	MNHN-1214	MNHN	151.2	96.5	31.2	59.6	22.3	68.1	58.5	43.4	51.4	35.4	40.5	64.5	67.4
Brittany spaniel	MNHN-1208	MNHN	167	105	34.1	65	19.4	62	59.2	40.4	56.5	38.1	36.9	57.8	56.6
Bullmastiff	MNHN-1304	MNHN	180	106.2	35.3	63.7	23.1	78.8	62.4	41.6	75.4	50.1	47.2	67.4	63.5
Chow	MNHN-1298	MNHN	167	101.4	28.8	55.7	17.8	62.4	55.2	29.6	48.8	29.3	38.1	58.2	55.6
Cocker spaniel 1	MNHN-1219	MNHN	139	95.1	25.7	56.2	16.4	53.1	52.3	36.3	44.9	28.2	32.1	49.1	46.1
Cocker spaniel 2	MNHN-1250	MNHN	150	101.3	29	59.4	18.8	56.8	53.1	34.4	44.7	30.7	34.7	53.7	51
Dachshund 1	MNHN-1320	MNHN	119.3	80	24.3	46.3	15	46	49	26.3	31.9	21.6	25.8	46.5	46.5
Dachshund 2	MNHN-1212	MNHN	125.5	87.4	24.3	52.1	16.6	49.2	51.8	31.4	34.9	21.2	28.7	47.8	47.8
Doberman	MNHN-1316	MNHN	198	137.1	36.4	77.6	19.4	48.7	64.5	37	54.1	37.8	38.7	63.9	60
Dog alaskan	1902-688	MNHN	172	107.8	33.7	64.8	20.6	62.7	53.5	33.8	45.7	30.4	35.9	61	53.6
German pointer	MNHN-1297	MNHN	170	105.5	31.8	62	20.5	67.5	58.4	41	56.2	38.3	37.1	55.2	53.9
German shepherd 1	MNHN-1244	MNHN	189	120.4	36.2	69.1	22.8	66.1	58.1	35.2	55.3	37.6	39.3	64	57.6
German shepherd 2	MNHN-1215	MNHN	170	114.5	29.4	60.4	19.9	63.4	57	34.9	51.7	34.1	35.4	60.5	56.6
Greyhound	50000325	CCEC	178	118.5	33.5	69.7	18.6	67.7	59.3	43.8	64.6	42.8	34	62.3	59
Pariah dog	50000314	CCEC	179	116.4	35.3	69.5	21.2	68.2	59.8	39.3	54.1	35.6	37.6	61.5	59.8
Poodle 1	50000261	CCEC	155.1	99.5	30.6	61.8	18.5	58.9	57.2	40.3	50.3	34.5	34.7	59.2	55.8
Poodle 2	MNHN-1218	MNHN	167	103.8	28.6	60.5	19.4	64.5	59.8	39.6	53.2	36.8	37.7	61.4	53.3
Setter	MNHN-1301	MNHN	178	108.4	32.7	64.2	22.4	67	59.8	42.2	54.9	40.2	41.6	65.7	60.1
Sloughi 1	MNHN-1987-258	MNHN	190	126.2	30.1	69.1	18.4	63.2	53.8	38.3	52.6	38.5	34	61.3	58.2
Sloughi 2	CG 1962-1515	MNHN	181	119.2	33.6	70.7	20.6	65	58.3	38.1	56.3	37.9	34.7	59.2	52.9
Sloughi 3	1901-564	MNHN	176	116.3	34.4	68.8	20.7	61.4	59.2	35.5	45.5	32.2	32.8	56.3	53.2
Sloughi 4	1901-546	MNHN	179	108.3	36.8	69.8	21.1	63	58.6	33.7	44.3	31.1	33.5	55.1	51.5
Sloughi 5	1902-665	MNHN	186	123.5	39.4	73.3	21.3	69.1	62.7	42.1	51.7	36.3	36.2	59.7	57.3
Spitz 1	MNHN-1226	MNHN	127	82.3	24.8	47.1	14.8	47.8	51.6	33.7	42.1	28.2	27.9	47.2	46.9
Spitz 2	MNHN-1308	MNHN	127.4	82.7	23.2	51.4	14.6	48.4	52.4	32.2	35.2	23.3	27.1	52.1	52.1
Spitz 3	MNHN-1224	MNHN	107.7	68.9	21	42.3	12.6	50.2	50.7	36.4	39.4	28.1	24.9	47.2	47
Wire-haired terrier	MNHN-1205	MNHN	158	112.4	30.2	63.5	17	54.8	53.8	34.3	49.5	35.8	33.4	54.8	50.5
Spaniel	50000328	CCEC	174	108.4	34.6	65.4	20.6	63	57	35.3	52	35.4	36.1	59.1	57.9
S1	S3602.35	MG	174	111.7	34.9	65.2	21.1	64.5	56.2	39.1	62.9	42.1	39.8	62.2	55.5
S2	S3602.66	MG	159	108.8	28.6	59.6	18.8	59.8	52.9	35.1	48.1	34.1	34.8	56.9	53.2
S3	S3602.52	MG	186	109.1	35.9	67.5	22.2	64.7	59.6	37.5	54.2	38	37.9	59.6	54.3
S4	S3602.69	MG	166	110.6	32.4	64.3	19.6	61	58	37.9	53.8	36.7	37.4	54.2	50.5
S5	S3602.8	MG	177	117	32.9	65.1	18.6	63.4	55.5	33.6	46.6	34.5	34.1	60.4	54.7
S6	S3602.5	MG	164	111.5	30.1	62.5	18.1	61	52.4	35.4	51.4	32.3	34.1	57.2	52.8
Wolfhound	50000253	CCEC	190	122.7	39.2	71.9	20.7	69.9	58.9	43	62.6	47.4	42.2	63.7	58

B) Modern dogs.

ID	Inv. Num.	Coll.	VD3	VD9	VD14	VD15	VD16	VD24	VD29	VD31	VD32	VD33	VD36	VD38	VD39
E1-Assiout	90010303	CCEC	161	100.3	27.7	63.6	19.7	61.1	55.6	34.9	52.4	36.7	33.4	54.4	51.8
E2-Assiout	51000005	CCEC	167	107.6	32.4	66.4	20.6	62.6	54.5	33.8	52.8	37.5	35.8	57.8	51.3
E3-Assiout	90010313	CCEC	162	101.6	29.3	58.7	16.3	58	53.9	39.3	56.8	39.1	34.8	54.2	48.7
E4-Assiout	51000161	CCEC	154.1	96.6	31.7	60.1	20.3	57.4	53.8	27.4	41.3	30.4	31.4	51.5	48.9
E6-Assouan	51000004	CCEC	186	120.5	32.8	73.7	22.2	65.2	56.2	34.5	52.4	34	34.5	59.2	51.9
E7-Assouan	51000008	CCEC	162	108.2	27.2	61	15.9	59.5	54.9	37.6	49.1	34.4	32.5	55.1	54.5
E8	51000003	CCEC	160	107.9	27	60.6	15.9	59.2	56	37.6	48.2	34.6	32.5	54.8	54
E9	51000021	CCEC	162	107.8	28.5	65	19.5	67.5	58.7	42.1	58.2	39.1	36.7	61.5	60.2
E10	51000031	CCEC	153.8	99.5	26.8	58.2	15.9	59.1	51.9	34.5	57.1	35.9	33.8	51.8	50.9
E11	51000028	CCEC	130.9	87.4	26	50.8	16.3	54.7	54.8	36.7	41.8	26	29.6	50.6	50.5
E12	51000242	CCEC	151.8	93.2	28	59.7	19.2	58.6	55.9	34.4	50.9	34.1	33	53.4	47.7
E13	90010305	CCEC	171	110	34.2	61.8	19.5	63.7	60.2	40.5	55.8	37.8	38.7	57	51.5
E14	51000164	CCEC	154.2	92.8	30.7	59	18.1	62	52.4	35.7	55.3	36.3	34.7	57.5	53.2
E15	51000165	CCEC	159	104.8	33.8	66.3	19.7	61.8	57.8	36.9	52	35.6	37.3	58.8	55.1
E16	90010307	CCEC	149.5	91.3	29.2	57.4	17.6	56.2	52.6	34.8	47.7	32.7	33.9	51.6	50.2
E17	90010310	CCEC	162	101.8	31.2	55.9	18.7	58.9	55.6	38	51.7	37.3	33.6	50.2	49.3
E18	90010311	CCEC	154.1	100.2	31.4	61.4	19.8	55.7	54	33	39.5	27.2	30.8	52.5	52.2
E19	51000163	CCEC	165	101.8	30.9	66.4	19.7	63.1	56.9	38.8	51.2	38.9	38.2	58.2	52.4
E20	51000029	CCEC	165	103.7	34.5	58.9	19.8	59.8	56.1	38.9	60.1	36.9	35.2	62.6	60.5
E21	900010306	CCEC	158	96.9	29.8	63	18.2	61.6	51.8	36.2	46.3	32.3	33.3	53.2	51.4
E22	51000018	CCEC	173	108.3	29.8	67.9	19.8	67.2	58.1	40	62.9	44.3	39.9	63.3	57.8
E23	51000016	CCEC	168	105.5	32.6	66	17.5	66.2	57.4	38.7	56.7	38.5	39.1	61.2	57.4
E24	51000015	CCEC	150.3	93	33	60	17	57.2	51.7	33.8	52.9	33.9	34.5	53.9	51
E25	51000014	CCEC	126.2	81	21.5	54	13.7	51	52.2	36.7	37.8	25.5	30	48.8	48.6
E26-Tehneh	51000017	CCEC	162	107.4	35.9	66.3	24.5	67	56.9	39.3	56.2	41.1	40.2	59	55
E27	51000024	CCEC	146.3	92.1	26.9	58.4	20.8	59.7	53.9	34.8	46.6	33	33.9	52.7	50.7
E28-Tehneh	51000026	CCEC	140.9	91.4	26.2	55	16.9	57.4	49.3	35.5	47.5	34.4	32.3	48.3	47.6
E29-Louqsor	51000010	CCEC	171	107.3	33.2	68.2	22.2	56.8	54	33.9	50.5	35	35.1	54.9	51
E30	51000013	CCEC	148.7	97	22.3	58.2	17.8	56.6	52.6	34	46.8	32.2	30	51.4	49
E31-Rôda	50000345	CCEC	150.3	94.9	28.2	61.8	19.1	55.2	54.2	28.3	40.8	29.5	29.3	49	47.2
E32-Rôda	90010308	CCEC	163	102.9	32.2	66.4	17.6	69.2	58.5	39.5	60.9	41.5	40.1	62.4	59.9
E33-Rôda	90010309	CCEC	151.9	96.5	27.1	61.1	20	54.3	52	31.5	39.6	28.1	31.3	48.6	48.1
E34-Rôda	51000027	CCEC	171	109.3	32.2	67.2	18.6	64.7	53.9	36.8	54.1	35.4	37.5	55.9	53.1
E35-Rôda	51000030	CCEC	128.4	71	27.8	57.6	18.5	53.4	53.8	36.7	38	25.5	29.9	49.5	49.3
E36-Rôda	51000160	CCEC	158	105.4	30.2	61.6	16.1	62	53.9	31	49.6	34.9	34.4	55.8	50.6
E37-Rôda	51000011	CCEC	154.1	103.3	28.5	62.1	16.7	60.1	54	36.2	49.1	32.4	31.9	52.1	49.6
E38-Rôda	51000009	CCEC	143.1	94.8	26.4	56.8	17	55.5	52.1	38.2	50.7	33.8	31.1	52.7	51.9
E39-Rôda	51000006	CCEC	157	98.6	28.7	62	17.6	59.3	54.9	41	50.8	32.6	32.4	53.9	52.7
E40-Rôda	51000007	CCEC	172	104.5	34.4	68.9	21.7	63	54.4	31.9	48.4	38.1	37.1	58.7	50.7
E41-Saqqara	51000025	CCEC	170	107.7	29.9	64.1	16.8	62.1	57.1	35.5	54.2	38.4	38.7	57.8	51.3
E42-Tehneh	51000019	CCEC	166	101.6	29.4	63.3	19.6	64.9	54.8	37.3	58.3	38.9	39.2	61.4	55.5
E43-Tehneh	51000020	CCEC	171	105.2	29.1	64.3	19.2	62.8	55.3	34.7	48.3	36.6	37.1	55.3	51.2
E44-Tehneh	51000023	CCEC	148.3	100.5	28	61.3	18.2	60.7	53.9	40.2	42.1	33.6	32.5	56	55.3
E45-Tehneh	51000022	CCEC	155	99.8	30.1	58.4	21.1	57.7	54.6	35.1	49.7	34.3	35.1	52.2	49.9
E46-Tehneh	90010304	CCEC	177	109.6	33.3	64.8	19	69.7	58.9	37.3	55.6	40.3	38.7	61.4	53.4

ID	Inv. Num.	Coll.	VD3	VD9	VD14	VD15	VD16	VD24	VD29	VD31	VD32	VD33	VD36	VD38	VD39
E47-Thèbes	51000012	CCEC	161	107.2	31.3	61.4	21.2	62	50.4	35.1	56.8	36.1	34.9	53.2	50.2
K1	KT83	MG	153.1	99	28.4	57.9	17.3	56.9	51.5	36.4	54.9	41.1	33.1	49.5	46.8
K2	K212.2	MG	153.3	101.9	30.1	60.4	20.1	59.3	54.1	35.1	51.8	33.9	32.4	48.8	45.8
K3	K133.4	MG	172	111.7	35	65.8	19.8	62.6	54.9	37.3	60	37.4	39.2	56	51.3
K4	KT89	MG	159	100	30.5	59	18.4	57.4	53.8	36.4	56.8	33.9	34.2	54.5	51.5
K5	K212.1	MG	161.7	108.1	31.3	62.8	20.3	58.5	50.9	35.3	51.5	35.9	33.3	55.3	50.7
K6	K151.1	MG	144.5	92	27.7	56.4	16.9	53.1	48.4	34.2	44.9	31.9	32.4	54.3	50.5
K7	KT67	MG	152.1	100.1	25.5	60.3	19.4	57.9	51.8	37.7	47.8	31.6	34.7	52.7	51.2
K8	K125.1	MG	167	108	34.4	63.1	20.9	59.8	52.2	35.8	55	38	35	54.9	50.6
K9	K249.1	MG	166	106.6	32.2	65.6	18.6	60	51.1	35.3	51.3	35.5	36.8	52.4	49.8
K10	K233.1	MG	165	110.5	30.5	64.2	17.9	57.5	52.3	36.4	54.7	37.1	32.8	55	51.9

C) Ancient canids.

Newcomers in the Bestiary

A Review of the Presence of *Lycaon pictus in* Late Predynastic and Early Dynastic Environment and Iconography

Axelle Brémont

Introduction: Weighing the "*Biocénose*" Against the "*Iconocénose*"

One of the best instances of possible collaboration between bioarchaeologists and "traditional" Egyptologists is the comparison between the three spheres of animal presence defined by Djindjian (2012). What this author calls "*zoocénose*", but which would usually (and perhaps best) be called "*biocénose*", can be defined as the whole fauna dwelling in a given environment at a given time. The "*taphocénose*" is, in turn, restricted to those taxa purposefully exploited by humans. And what he calls "*iconocénose*" is another type of selection operated by man into the available "pool" of animal species: the ones that are depicted in graphic manifestations.

In order to allow for meaningful comparisons and avoid any kind of misinterpretation, it is necessary to assert "*biocénose*" through geological records rather than only faunal remains recovered from human sites, whenever they are available, so as not to blur the distinction with the "*taphocénose*". Another necessary guideline is not to discard what are presumed to be intrusive remains from publications of a site – as has sometimes been done in the past or in very short publications – as those, especially when penecontemporaneous to the studied assemblage, also yield information on taxa present in the vicinity. Finally, it is essential to refrain from using iconographic occurrences as a way of proving the existence of a specific species in the regional "*biocénose*".

This tendency, especially common in past Egyptological studies, of making insufficient use of rigorous zooarchaeological data, can create misinterpretations and especially mistake social and cultural changes for an undistorted reflection of biological or climatic evolution. Before zooarchaeological studies began to increasingly be conducted on Egyptian sites in the 1980s and 1990s, Egyptologists commonly assumed, from a handful of depictions, that the aardvark could be encountered in ancient Egypt (*e.g.* Keimer, 1944). This is especially relevant regarding images from the Predynastic period, as a number of forgeries and suspect artifacts have been ascribed to this period on the sole basis of crudeness of depiction and lack of conformity to the Dynastic canon.

On the contrary, what is precisely of interest to a cultural history of animals is the fact that "*biocénose*", "*taphocénose*" and "*zoocénose*" never quite perfectly overlap, a phenomenon that François Djindjian has underlined for Paleolithic European societies: "*why do they keep on representing bisons and horses when they only eat reindeer?*"

In: Ikram, S., J. Kaiser & S. Porcier. Eds. 2021. *The Ancient Egyptians and the Natural World. Flora, Fauna, and Science.* – Leiden, Sidestone Press: 61-70.

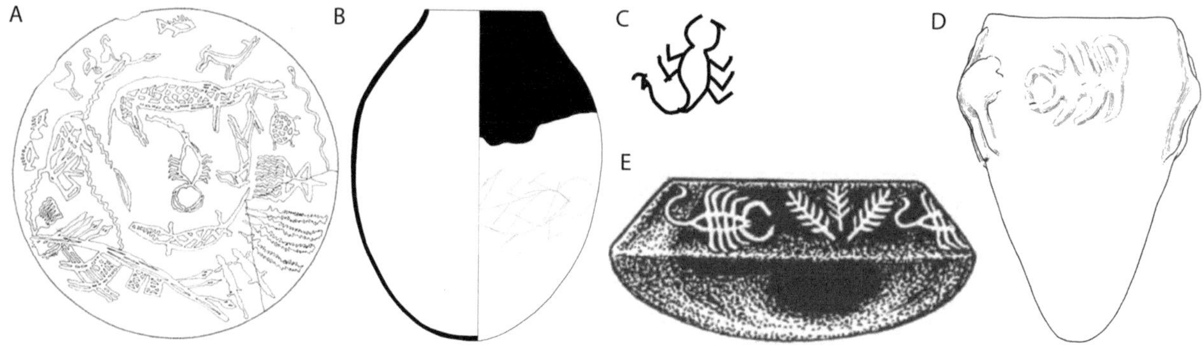

Figure 1. Scorpion depictions pre-dating Naqada IIC. A) C-Ware bowl, Abydos t. U-264, Naqada IC-IIA (Cairo Museum CG2076). After Hartmann (2008); B) Graffito on B-Ware ovoid jar, Dakka cem. 103, Naqada IIA-IIB (Boston MFA 19.1555). Drawing by the author; C) Graffito on pottery, Naqada t. 472 (current location lost), Naqada IIA-IIB. After Petrie (1896: LI.36); D) Modeled vase, Naqada t. 1787, Naqada IIA (Ashmolean Museum 1895.233). After Payne (2000: fig. 32.604); E) C-Ware carinated bowl, Petrie Museum UC 15326 (unprovenanced). After Graff (2009: cat.no. 128).

(Djindjian, 2012: 313, author's translation). Indeed, if we focus on what is known to us from the archaeological record, it appears that not all species known to the Egyptians are present in their iconographic bestiaries; perhaps more surprisingly, the reverse is also true, as not all species depicted in art seem to have actually dwelled in fourth-millennium Egypt.

Newcomers in the Bestiary: Those which have Always been Around...

Around the Naqada IIC period, a clear shift is perceptible towards what one might call a novel "Early Dynastic bestiary", with many new species suddenly becoming the subject of representation, which had not previously attracted much or even any attention in iconography, while others, such as donkeys or dogs, progressively fade from the imagery.

Among such "newcomers" is the scorpion, of which only a handful of depictions are securely dated to before Naqada IIC[1] (Figure 1), although its continued presence in fourth-millennium Egypt can hardly be doubted (even if they are puzzlingly rendered with a very inconsistent number of legs). After this point, however, scorpions suddenly multiply (Hendrickx, 1998) and especially, of course, during the Naqada III period, in relation to the so-called "king(s) Scorpion" from Abydos and Hierakonpolis.

A similar case could be made for frogs, whose geographical and ecological distribution has obviously not varied in ancient times, but which only enter the depicted bestiary around the later stages of Naqada culture.

... And Those which Never Have Been

Another, very contrasted case, is provided by baboon depictions. The bulk of their dated representations does not occur prior to the 6th Dynasty, and unfortunately the very numerous figurines enclosed in "early temples" votive assemblages can still not be dated very confidently (Bussmann, 2010). However, a few depictions are attested at least as early as Naqada IIIA1, including a minute bead from Tarkhan (t. 1552),[2] two faience figurines from Tell Farkha dated to the Naqada IIIB period (Cialowicz *et al.*, 2012: 207), and evidence in writing at least from the later 1st Dynasty, especially with several ivory labels one of them dating from the reign of Semerkhet (Petrie, 1900: pl. XII.26).

However, while baboons appear on this material as well as many faience figurines from the late Old Kingdom and several 4th to 6th Dynasty reliefs (listed in Vandier d'Abbadie, 1964), their material remains have never been found in contexts this old, be it in man-made or geological assemblages, apart from one very specific instance in Hierakonpolis that will be discussed in detail below. This lack of osteological remains is puzzling especially

1 Petrie Museum UC 15326 could date to Naqada I considering the chronological range of carinated vases C04.3Ba according to Hartmann (2016); Boston MFA 19.1555 is typologically dated to Naqada IIA – IIB according to the same author; Ashmolean Museum 1895.323 from Naqada t. 1787 is dated by Hendrickx (1989) to Naqada IIA

2 Two statuettes bearing the names of 1st Dynasty royalty (Berlin ÄM 22607) and another one inscribed with the name of Merneith from a private collection (Wiese, 2001: 35), are unfortunately without known provenance and it is unsure whether they do date to the Early dynastic period (Patch *et al.*, 2011: 236, note 70). The inscription on the second one is considered by some authors to be a modern addition (Vernus, 2009)

for settlements, as primate bones are very distinctive and one could expect them to sometimes venture near human settlements in search for food, especially since they would have dwelt near the river (J. Lesur, personal communication).

Indeed, primates in the New Kingdom are definitely imported, and it appears to be the case already from earlier periods on, as monkeys also feature amongst the Puntite products brought back by Sahure (El-Awady, 2009: pl. 5). In the Old Kingdom reliefs already mentioned, baboons never appear in a natural setting but exclusively in anthropic contexts and as tamed animals, as opposed to species depicted to characterize a biotope in the traditional "fowling in the marshes" and "hunting in the desert" scenes. Vandier d'Abbadie (1964) already took for granted the fact that baboons had never been native to the Egyptian Nile Valley and would exclusively have been imported from further south.

This opinion is supported by the only primate remains recovered so far from early contexts: the *Papio anubis* baboons (as well as green monkeys) burials from the Hierakonpolis elite – some even say royal – cemetery HK6, dating to Naqada IC-IIA. Van Neer *et al.* (2017: 397) strongly suggest that these would have been imported from further south, meaning that they could not be found even as far south as Hierakonpolis, much like the two young African elephants buried at the same site. Indeed, many of the species attested in the cemetery, if not completely exotic, were probably rarely encountered by the inhabitants of the valley, since they were either only found on the margins, such as the leopard, or had already dramatically decreased in population at that time, such as the hartebeest (Lesur, 2013: 44; Van Neer *et al.*, 2004: 111).

Was *Lycaon pictus* Present in Egypt in Predynastic Times?

The two case studies developed above should serve as a theoretical framework for studying the disparities between "*biocénose*" and "*iconocénose*". Let us now look into a third such "newcomer in the bestiary": the species of canids which appears exclusively on seven large Early Dynastic slate palettes (complete list in Hendrickx, 2006: 740), the most famous of them being that from the Hierakonpolis temple "Main Deposit" (Ashmolean Museum E3924).

It has been first proposed by Henry Fischer (1958), following a suggestion by Anthony Arkell, that this animal was an African wild dog (*Lycaon pictus*), a wild canid with spotted pelt now restricted, since the 19th century, at least to the Sudanese savannah belt (Manlius, 1996: 108, quoting Murray, 1866), if not to the extreme southern fringe of the Sahara (D'Huy & Le Quellec, 2009: 89).

This identification has created a general consensus in subsequent literature, including discussions by Krzysztof Cialowicz (1991), John Baines (1993) and Stan Hendrickx (2006). Indeed, the characteristic large, rounded ears speak in its favor and exclude the possibility that this might be another wild canid such as jackal or hyena. The bushy tail is consistent with this identification as well, though it could pertain to other animals as well (see *infra*). These representations thus generally conform to the anatomy of the animal, so much so that it has been assumed from there that the lycaon was familiar to the Egyptians, and that it could be encountered at least up until the Thinite period in the semi-desert margins of the Nile Valley. Some authors have even suggested it might have been used as a hunting auxiliary alongside the *ṯsm* sighthound (*e.g.* Baines, 1993; Bodenheimer, 1960; Boessneck, 1988; Hendrickx, 2006; Keller, 1909; Osborn & Osbornova, 1998).

However, despite quite undisputable iconographic evidence, some doubt can be cast on the actual presence of African wild dogs in Egypt through the Early Dynastic period, especially as this animal is completely absent from every zooarchaeological record throughout the Egyptian territory. This entails even purely geological ones as studied from the Western Desert (Peters & Pöllath, 2004; Pöllath, 2009). Of course, the animal's distribution could have varied according to climate modifications, but it might be useful to note that this species is actually less sensitive to aridification than other savannah animals such as elephant or giraffe: "*ses besoins en eau étant très réduits, il peut même atteindre des zones désertiques*" (Lopez, 1995: 11). Had it ever been present in Egypt in the fourth millennium, there would therefore have been no reason for it to disappear in the subsequent periods. There is nonetheless no trace of this animal ever appearing in faunal remains or iconography in the Old Kingdom, except for a few misidentified specimens that are now considered hyenas (Lopez, 1995: 11, *contra* Keller, 1909).

Despite Gautier (1993: 262-264) considering the African wild dog's absence from human-made assemblages consistent with its not being consumed, many carnivores are usually attested on Predynastic settlements and middens, especially hyenas and jackals, as they come to scavenge on human leftovers or are killed when roaming around settlements. *Hyaena hyaena* was spotted in the Naqada/al-Khattara area as well as in Lower Egypt while *Canis aureus*[3] is attested in the Maadi remains. At Hierakonpolis HK29A, where the ratio of carnivores is abnormally high and varied compared to other sites (including *e.g. Vulpes zerda, Vulpes rueppelli*), the lycaon is nowhere mentioned (Linseele *et al.*, 2009: 124-125). This is all the more surprising that the authors suppose the

3 Recent genetic work has brought the presence of true *Canis aureus* into question (Rueness *et al.*, 2011; Gaubert *et al.*, 2012; Koepfli *et al.*, 2015); for discussion see Ikram and Bertini (2021) and Kitagawa (2016). However, for the purposes of this paper, the identifications in the publications will be used.

Figure 2. Figures of earlier date previously proposed to represent *Lycaon pictus*. A, left) C-Ware necked bottle, unknown provenance (Petrie Museum UC 15332). Courtesy of Stan Hendrickx (personal communication), and the Petrie Museum of Archaeology; B, right) Ivory plaque (inlay?), Hierakonpolis Main Deposit (Petrie Museum UC 14864). Courtesy of the Petrie Museum of Archaeology.

many bones of carnivores bearing cut marks may be proof of skinning and pelt-processing rather than consumption as food, and one would indeed expect the lycaon to be especially sought after in this context.

While possible confusions with other canids might partially account for such a lack (Steder, 2013: 45), the African wild dog has indeed been spotted at other sites: it is especially listed as "rare", but nevertheless present, in Neolithic contexts from Wadi Shaw (nowadays northern Sudan) (Van Neer & Uerpmann, 1989: tab. 3). It is however not recorded at Wadi Halfa for the Late Paleolithic period (Gautier & Van Neer, 1989: tab. 6.19). Only a modern comparison specimen in the British Museum that reportedly comes from Gebel Elba in the southeastern corner of Egypt (Setzer, 1956), but in any case, even if this provenance were confirmed, the area is known to be a "biological hotspot" that may have been home to residual populations. Conversely, one isolated mention of lycaons in the Kharga area at the end of the 19th century has convincingly been demonstrated by Manlius (1996) to be, most probably, a misidentification of hyenas.

Of course, some taxa that were, and still are, definitely present in Egypt in ancient times, such as ostriches, Nubian asses or ibexes, have never or extremely seldom[4] been detected in man-made bone assemblages from the Valley, but it should be emphasized that they do occur in geological zoological records, especially the well-studied ones in the Western Desert (Peters & Pöllath 2004, Pöllath 2009).

Encounter in the Flesh, Encounter through Picture

Moreover, compared to these other species with countless iconographic occurrences, the clues for a possible presence of *Lycaon pictus* in Predynastic Egypt are limited to a handful of attestations in the iconography of the later phases. Some have suggested that it might have been represented earlier on, but the Petrie Museum C-Ware UC 15332 is best qualified as hyena (Ikram, 2001; Navajas, 2005) while another previously cited example, on the Hierakonpolis ivory plaque held in the Petrie Museum (UC 14864) should most probably be re-identified as a leopard, as is obvious from the pattern of the spotted hide and the length and slenderness of the curvy tail (Figure 2).

Stan Hendrickx has suggested that animal tails hung from the belt of human figures on the "Hunters' Palette" and on some earlier material such as the painted vase Abydos t. U-239 (Dreyer *et al.*, 1998: Abb. 12) or the Was-Ha-Waset rock art panel WHW86 (Darnell, 2009: 88) should be considered lycaon tails (Hendrickx, 2006: 739-742; 2013: 252). However, one could argue that they might as well be hyena, fox, or jackal tails, which are similarly bushy. Depictions as schematic as the ones occurring on C-Ware could even represent the tails of basically any furry animal. In any case, those depictions could not be considered compelling proof of the presence of this animal in Egypt, as lycaon skins could very well have been imported from further south.[5]

It is therefore safe to state that the occurrence of African wild dogs is restricted to these seven decorated palettes. They occur in an extremely similar pose and morphology and these artifacts may indeed even have been copied from each other or from a common model. The Brussels fragment in particular (MRAH E6196) is completely identical to the

4 An example of an ostrich tibiotarsi has been identified by S. Ikram at Kom Ombo in an Old Kingdom settlement level that was excavated during a wastewater rescue mission (S. Ikram, personal communication).

5 W. van Neer (personal communication) provides a similar hypothesis for a single giraffe phalanx found at Adaima.

ones on the small Hierakonpolis palette, while the ears of the smaller individuals treated in relief in the center of the same palette are depicted in a stylized manner identical with the Metropolitan Museum fragment (MET 28.9.8) and the "Michailidis palette" (Figure 3).

Furthermore, Henry Fischer (1958: 81) already noticed that all lycaons on the Hierakonpolis palette consistently display five toes on their hind paws, like most other, probably more familiar canids (*Canis familiaris*, *Canis aureus*, *Vulpes vulpes*...), whereas African wild dogs actually have four, both on hind and forelegs. The Michailidis palette even presents only three. Moreover:

> "*it is true that one of the most striking features that allies Lycaon to the hyenas, the absence of a dewclaw on the forefoot, is not consistently observed on the palettes, assuming that this extra digit would be represented like the others*" (Fischer 1958: 81). (Figure 4)

All these arguments would seem to indicate that Predynastic artists actually only had limited familiarity with African wild dogs. This situation is reminiscent of that of other species that seem to have attracted specific attention in Late Predynastic times, although completely exogenous to Egypt. This is for example the case for fallow deer, which unmistakably and indisputably appear on several media from the Naqada IIC period onwards (Coptos colossus "Coptos 1", Ashmolean Museum 1894.105d; Sayala mace handle, Cairo Museum, now lost; "Hunters' Palette" British Museum E20790). After reviewing all available evidence, Chiori Kitagawa (2008) concluded that deer had never been indigenous to Egypt, despite being occasionally represented on a handful of artifacts throughout the Predynastic and Dynastic period. The conclusion was supported by the complete absence of cervid bones in any archaeological assemblage, man-made and geological alike. In this particular case, borrowings from Mesopotamian iconography are the most likely reason to explain the

Figure 3A (top row). Comparison of the treatment of *Lycaon pictus* on the Hierakonpolis palette (left, courtesy of the Ashmolean Museum, Oxford) and on the Brussels fragment MRAH E6196 (right, after Fischer, 1958: Fig. 6); 3B (bottom row). Comparison of the treatment of the ears of *Lycaon pictus* on the Hierakonpolis palette (left), the MET fragment 28.9.8 (center) and the "Michailidis palette" (right). Courtesy of The Metropolitan Museum of Art, Gift of Edward S. Harkness, 1928.

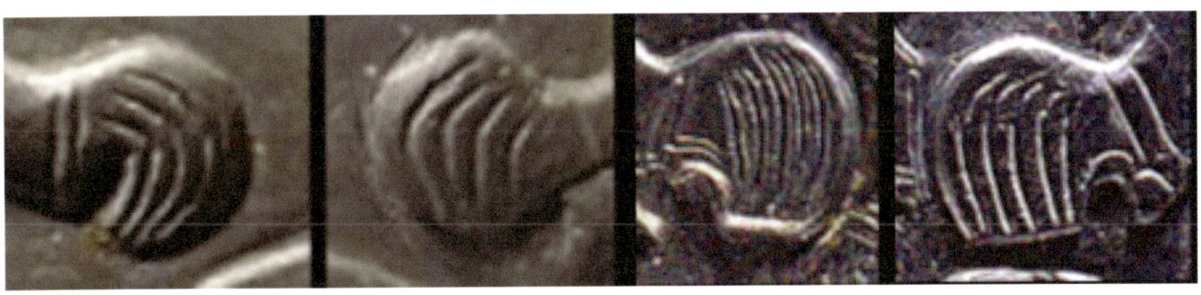

Figure 4. Hierakonpolis palette, details of all hind legs of *Lycaon pictus* (recto and verso). Courtesy of the Ashmolean Museum, Oxford.

Figure 5.
Meir, Tomb of Ukhhotep (t. B2), south wall, central and eastern parts (detail). After Blackman (1915: pl. VII-VIII).

few cases of deer depictions, as happens at the very same period with the serpopard motif and several other mythical animals first appearing on Uruk material.

An Early Taste for Exotic Wonders

How should we then interpret these scarce African wild dog depictions? Are they fading memories of what the Egyptian fauna looked like before the desert dried up again, just like the elephant pictures which linger on until they become less and less recognizable for almost a millennium after the animal actually disappeared from Egypt (Brémont, 2020; Schott, 1971)? Or are they rather demonstrations of an early taste for exotic species, as we know for a fact was very significant in other periods of Egyptian history, from Sahure's Punt expedition in the 5th Dynasty to the so-called "Botanical Garden" of Thoutmosis III in Karnak (Beaux, 1990) to the countless scenes of foreign "tributes"?

One particularly interesting example for our purpose is the Middle Kingdom hunting scene from Ukhhotep's tomb in Meir (t. B2; Figure 5). Several species are there depicted set apart from the hunting scene itself; their layout conveys the impression of a descriptive catalogue rather than a narrative rendering, as opposed to the gazelles and hare on the left part, which are engaged in action, assaulted by dogs and shot by arrows.

Among these species are the fallow deer and baboon once again, as well as a lion, a giraffe and a probable lycaon (although its ears are somewhat less rounded than expected). All of these are definitely exotic in Egypt at that time, and indeed it looks almost as if the artist, or the tomb owner, was really keen on having them represented in the tomb, yet also decided to acknowledge the fact that he had actually never been able to hunt them himself by spatially dissociating them from the actual hunting scene. The precise rendering of the animals' morphological characteristics should not surprise us, considering examples of perfectly naturalistically depicted species that are far from indigenous to Egypt, such as the Deir al-Bahari secretary bird (Taterka, 2019).

From this detour by Middle and New Kingdom iconography, it becomes clearer than ever that animals which are never encountered as such in the *biocénose* may well be the subject of detailed naturalistic representation, be it because some individuals are occasionally imported from other regions or because people in Egypt were familiar with their image rather than with the animal itself. In both cases, interest in depicting them is triggered by their exotic feel. In this, the sudden and brief appearance of the African wild dog in the Protodynastic bestiary is best compared with the situation of the baboon as developed *supra* rather than with that of the scorpion or the frog.

Conclusion: An Exotic Bestiary for the Naqada IIC-IIIA Period

The lycaon and baboon are not alone in this situation. Apart from the Near Eastern fallow deer already discussed, it might be worth noting that the only records of lions previously mentioned in the *taphocénose* for the Predynastic period have recently been re-identified as probable leopards (Van Neer, 2013: 298).

As for the storks that make an appearance in the iconography in the same period, they have been positively identified to depict saddle-billed storks, due to the large, characteristic caruncle on the beak of all known detailed examples (*e.g.* Davis comb handle, Metropolitan Museum inv.no. 30.8.224). Much like the other species mentioned in this article, their distribution is now restricted to the southern half of Sudan. Janak (2011: 149) notes, after (Keimer, 1930) that:

> *"the best (...) depictions of the saddle-billed stork come from the earliest periods of Egyptian history. During the second phase of the Old Kingdom, the sign became schematized with (...) inaccuracies (sharp but short bill, shorter neck, shorter legs, different posture, black wattle, white head etc.)".*

If it had lived in Egypt before, it had decidedly regressed south at this period (Houlihan, 1986: 25). While storks

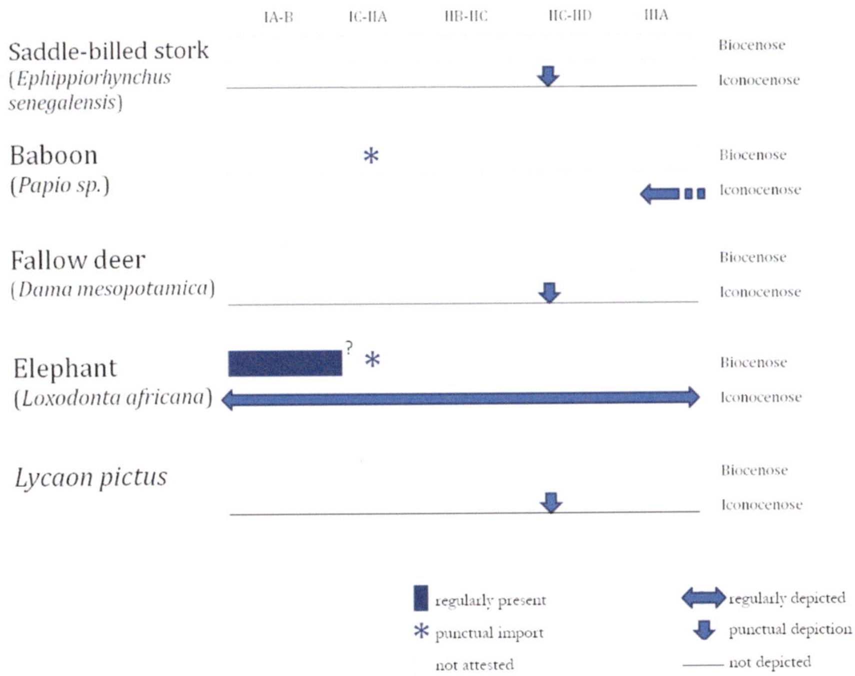

Figure 6. Summary of available osteological and iconographic evidence for the above-mentioned species throughout the Predynastic period.

have been reported from many Predynastic sites, the saddle-billed stork is never evoked; it should however be more easily distinguishable as the tallest in this category of birds.

If we sum up all of the evidence (Figure 6), most of these "newcomers in the bestiary" at the dawn of the Early Dynastic period seem to be species of foreign origin, mostly African fauna, some of them Near Eastern. Although an absence of proof is not to be taken as a proof of absence, not one of these species appearing in the late Predynastic bestiary has ever been spotted in zooarchaeological records, be they geological or man-made. Even according to the most optimistic recent estimations (Kuper & Kröpelin, 2006; Marriner *et al.*, 2012; Peters & Pöllath, 2004: 43), the Sahelian savannah belt could not have gone up more than 800 km, which barely takes it to the current latitude of Edfu. This is also corroborated by the fact that even as far south as in Hierakonpolis, elephants or baboons apparently had to be acquired from distant locations (Van Neer, 2004: 111-112).

The fact that those species enter the Naqada bestiary at the very end of the period rather than early on would seem to confirm that they are not the memory of a "ghost" fauna that was present when the Egyptian climate was still more humid and savannah-like – but rather a clue as to Egyptian interest into exotic species imported from further south.[6]

6 Due to space constraints, the presence of elephants in rock art will not be addressed here.

We know that the spreading of Naqadan material culture both north- and southward is increasingly important starting from Naqada IIB onward and especially so for the Naqada IID- IIIA period. This would seem to agree with a possible exposure of Upper and Middle Egyptian populations to more exotic faunas, owing to more sustained contacts with regions further north and further south. This is true of plant remains as well: the first interest in acclimatizing vine crops is dated to the turn of the Naqada IIIA-IIIB period (Dreyer *et al.*, 1996: 49-57; McGovern & Hartung, 1997: 10; Serpico & White, 1996; Tallet, 1998: 13), while endeavors to transplant the date palm, not originally indigenous to the Nile Valley, may have started at the same period (Farout, 2018, Tallet, in preparation).

The spreading of borrowed fantastic animal motifs at the exact same period may confirm this idea, and indeed Egyptian image makers may very well have considered serpopards or griffins actual animals indigenous to the Mesopotamian environment – how would they know, if the proof of their existence was only conveyed to them by images?

All species are represented on media more codified and standardized than prior iconography, which appear to have been copied from one occurrence to the other rather than individual, original creations. The study of the permanence of elephant depictions, as well as of other similar cases in other areas (*e.g.* giraffes in China, Liscomb, 2016) seems to confirm that the memory of rare animals' morphology may persist for centuries without major deformations if contact with ancient images in maintained.

In any case, it is only thanks to continued interaction between bioarchaeological and historical approached that we will be able the assert the exact extent of the Egyptian "*biocénose*" and its relationship to what human societies selected from it as worth eating and as worth representing.

Acknowledgements

I am deeply indebted to Joséphine Lesur for her help and thorough reading on early drafts of this paper. Any remaining mistake or misinterpretation of zooarchaeological data is my own.

Bibliography

el-Awady, T. 2009. Abusir XVI. Sahure The Pyramid Causeway. History and Decoration Program in the Old Kingdom. – Prague, Czech Institute of Egyptology.

Baines, J. 1993. Symbolic Roles of Canine Figures on Early Monuments. – Archéo-Nil 3: 57-74.

Beaux, N. 1990. Le cabinet de curiosités de Thoutmosis III. – Plantes et animaux du Jardin botanique de Karnak. – Leuven, Peeters.

Blackman, A.M. 1915. The Rock Tombs of Meir. Part II. The Tomb-Chapel of Senbi's Son Ukh-Hotp (B, No. 2). – London, Egypt Exploration Fund.

Bodenheimer, F.S. 1960. Animal and Man in Bible Lands. – Leiden, Brill.

Boessneck, J. 1988. Die Tierwelt des Alten Ägypten: untersucht anhand kulturgeschichtlicher und zoologischer Quellen. – Munich, C.H. Beck.

Brémont, A. 2020. Un éléphant, ça trompe énormément. Animal fantastique ou perte de référent de *Loxodonta africana* dans l'Egypte prédynastique (ca. 4000/2500 av. JC). In: Paillet, E., M. Sepulveda, E. Robert, P. Paillet & N. Mélard. Éds. Caractérisation, continuités et discontinuités des manifestations graphiques des sociétés préhistoriques. Proceedings of the XVIII UISPP World Congress (4-9 June 2018, Paris, France). Volume 3. – Oxford, Archaeopress: 87-104.

Bussmann, R. 2010. Die Provinztempel Ägyptens von der 0. bis zur 11. Dynastie. Archäologie und Geschichte einer gesellschaftlichen Institution zwischen Residenz und Provinz. – Leiden/Boston, Brill.

Cialowicz, K. 1991. Les palettes égyptiennes aux motifs zoomorphes et sans décoration : Etudes de l'art prédynastique. – Krakow, Uniwersytet Jagiellonski.

Darnell, J.C. 2009. Iconographic Attraction, Iconographic Syntax and the Tableaux of Royal Ritual Power in the Pre- and Proto-Dynastic Rock Inscriptions of the Theban Western Desert. – Archéo-Nil 19: 83-107.

Djindjian, F. 2012. Fonctions, significations et symbolismes des représentations animalières paléolithiques. In: Clottes, J. Ed. L'art pléistocène dans le monde. Actes du congrès IFRAO, Tarascon-sur-Ariège, septembre 2010. – Tarascon-sur-Ariège, Société préhistorique Ariège-Pyrénées: 312-320.

Dreyer, G., E.V. Engel, E.C. Köhler, U. Hartung, T. Hikade & F. Pumpenmeier. 1996. Umm el-Qaab. 7./8. Vorbericht. – Mitteilungen des Deutschen Instituts für Ägyptische Altertumskunde in Kairo 52: 49-57.

Dreyer, G., E.C. Köhler, U. Hartung, T. Hikade, F. Pumpenmeier & V. Müller. 1998. Umm el Qaab. Nachuntersuchungen im frühzeitlichen Königsfriedhof. 9./10. Vorbericht. – Mitteilungen des Deutschen Instituts für Ägyptische Altertumskunde in Kairo 54: 77-167.

Farout, D. 2018. La tête de massue du roi Scorpion : un géant sylviculteur ? – Egypte, Afrique & Orient 89: 31-40.

Fischer, H. 1958. A Fragment of Late Predynastic Egyptian Relief from the Eastern Delta. – Artibus Asiae 21: 64-88.

Gaubert, P., C. Bloch, S. Benyacoub, A. Abdelhamid, P. Pagani, C.A.M.S. Djagoun, Arnaud Couloux & Sylvain Dufour. 2012. Reviving the African Wolf *Canis lupus lupaster* in North and West Africa: A Mitochondrial Lineage Ranging More than 6,000 km Wide. – PLoS ONE 7(8): e42740. https://doi.org/10.1371/journal.pone.0042740

Gautier, A. 1993. The Middle Palaeolithic Archaeofaunas from Bir Tarfawi (Western Desert, Egypt). In: Wendorf, F., R. Schild & A.E. Close. Egypt during the Last Interglacial. The Middle Palaeolithic at Bi'r Tarfawi and Bi'r Sahara East. – New York, Springer US: 121-143.

Gautier, A. & W. Van Neer. 1989. Animal Remains from the Late Palaeolithic Sequence at Wadi Kubbaniya. In: Wendorf, F. & R. Schild. The Prehistory of Wadi Kubbaniya. Stratigraphy, Palaeoeconomy, and Environment. Vol. 2. – Dallas, Southern Methodist University Press: 130-162.

Graff, G. 2009. Les peintures sur vases de Nagada I – Nagada II. Nouvelle approche sémiologique de l'iconographie prédynastique. – Leuven, Leuven University Press.

Hartmann, R. 2008. Zwei Fragmente der White Cross-Lined Ware aus dem Friedhof U in Abydos zu Gefässen aus dem ägyptischen Museum Kairo. In: Engel, E.M., V. Müller & U. Hartung. Eds. Zeichen aus dem Sand. Streiflichter aus Ägyptens Geschichte zu Ehren von Günter Dreyer. – Wiesbaden, Harrassowitz: 163-182.

Hartmann, R. 2016. Umm el-Qaab IV. Die Keramik der älteren und mittleren Naqadakultur aus dem prädynastischen Friedhof U in Abydos (Umm el-Qaab). – Wiesbaden, Harrassowitz.

Heinrich, D. & H. Ratjen. 1978. Vergleichende Untersuchungen an den Metapodien von Füchsen und

Hunden. – Schriften aus der Archäologisch-Zoologischen Arbeitsgruppe Schleswig-Kiel 4: 3-50.

Hendrickx, S. 1989. De Grafvelden der Naqada-cultuur in Zuid-Egypte, met bijzondere aandacht voor het Naqada III Grafveld te Elkab. Interne chronologie en sociale differentiatie. Vol. II. Tabellen en bibliografie. – Unpublished PhD thesis, Katholieke Universiteit Leuven.

Hendrickx, S. 1998. Peaux d'animaux comme symboles prédynastiques. A propos de quelques représentations sur les vases White Cross-Lined. – Chronique d'Egypte 73, 146: 203-230.

Hendrickx, S. 2006. The Dog, the *Lycaon pictus* and Order over Chaos in Predynastic Egypt. In: Kroeper, K., M. Chlodnicki & M. Kobusiewicz. Eds. Archaeology of Early Northeastern Africa. In Memory of Lech Krzyzaniak. – Poznan, Poznan Archaeological Museum: 724-746.

d'Huy, J. & J.L. Le Quellec. 2009. Du Sahara au Nil : la faible représentation d'animaux dangereux dans l'art rupestre du désert Libyque pourrait être liée à la crainte de leur animation. – Cahiers de l'Association des Amis de l'Art Rupestre Saharien 13: 85-98.

Hendrickx, S. 2013. Hunting and Social Complexity in Predynastic Egypt. – Bulletin des Séances 57.2-4: 237-263.

Houlihan, P.F. & S.M. Goodman. 1986. The Birds of Ancient Egypt. – Warminster, Aris & Phillips.

Ikram, S. 2001. The Iconography of the Hyena in Ancient Egypt. – Mitteilungen des Deutschen Instituts für Ägyptische Altertumskunde in Kairo 57: 127-140.

Ikram, S. & L. Bertini (with contributions by D. Hurley and S. Vann). 2021. "The Fauna". In: Nicholson, P. Ed. The Catacombs of Anubis at North Saqqara: An Archaeological Perspective. – Leuven/London, Peeters/ British Museum Publications in Egypt and Sudan.

Janak, J. 2011. A Question of Size. A Remark on Early Attestations of the ba Hieroglyph. – Studien zur Altägyptischen Kultur 40: 143-153.

Keimer, L. 1930. Quelques hiéroglyphes représentant des oiseaux. – Annales du Service des Antiquités de l'Égypte 30: 1-30.

Keimer, L. 1944. L'oryctérope dans l'Egypte ancienne. In: Keimer, L. Zoologica I-III. – Le Caire, Imprimerie de l'Institut Français d'Archéologie Orientale: 1-20.

Keller, O. 1909. Die antike Tierwelt. – Liepzig, W. Engelmann.

Kitagawa, C. 2008. The Status of Fallow Deer in Ancient Egypt: Autochthonous or Introduced? In: Vila, E., L. Gourichon, A.M. Choyke & h. Buitenhuis. Eds. Archaeozoology of the Near East VIII. – Lyon, Publications de la Maison de l'Orient et de la Méditerranée: 541-552.

Kitagawa, C. 2016. The Tomb of the Dogs at Asyut. Faunal Remains and Other Selected Objects. – Wiesbaden, Harrassowitz (The Asyut Project 9).

Koepfli, K.-P., J. Pollinger, R. Godinho, J. Robinson, A. Lea, S. Hendricks, R.M. Schweizer, O. Thalmann, P. Silva, Z. Fan, A.A. Yurchenko, P. Dobrynin, A. Makunin, J.A. Cahill, B. Shapiro, F. Álvares, J.C. Brito, E. Geffen, J.A. Leonard, K.M. Helgen, W.E. Johnson, S.J. O'Brien, B. Van Valkenburgh & R.K. Wayne. 2015. Genome-Wide Evidence Reveals that African and Eurasian Golden Jackals are Distinct Species. – Current Biology 25.16: 2158-2165, ISSN 0960-9822, https://doi.org/10.1016/j.cub.2015.06.060.

Kuper, R. & S. Kröpelin. 2006. Climate-Controlled Holocene Occupation in the Sahara: Motor of Africa's evolution. – Science, New Series 313, 5788: 803-807.

Lesur, J. 2013. Des animaux et des hommes en Egypte au Néolithique et Prédynastique. – Archéo-Nil 23 : 33-54

Linseele, V., W. Van Neer & R. Friedman. 2009. Special Animals from a Special Place? The Fauna from HK29A at Predynastic Hierakonpolis. – Journal of the American Research Center in Egypt 45: 105-136.

Liscomb, K. 2016. How the Giraffe Became a Qilin: Intercultural Signification in Ming Dynasty Arts. In: Silbergeld, J. & E.Y. Wang. Eds. The Zoomorphic Imagination in Chinese Art and Culture. – Honolulu, University of Hawai Press: 341-378.

Lopez, A. 1995. Le Lycaon (*Lycaon pictus*: Mammifères, Canidae) dans l'ancienne Egypte. – Nilus 4: 10-14.

Manlius, N. 1996. Le Lycaon présent en Egypte au XIXème siècle ? – Bulletin de la société linéenne de Bordeaux 24, 2: 107-109.

Marriner, N., C. Flaux, D. Kaniewski & C. Morhange. 2012. ITCZ and ENSO-Like Pacing of Nile Delta Hydro-Geomorphology during the Holocene. – Quaternary Science Reviews 45: 73-84.

McGovern, P.E., U. Hartung, V.R. Badler, D.L. Glusker & L.J. Exner. 1997. The Beginnings of Winemaking and Viticulture in the Ancient Near East and Egypt. – Expedition 39, 1: 3-21.

Murray, M. 1866. The Geographical Distribution of Mammals. – London, Day & sons.

Navajas Jiménez, A.I. 2005. The Petrie Museum's Beaker UC15332 or the First Representation of a *Hyaena hyaena* in Egypt. – Göttinger Miszellen 206: 69-77.

Osborn, D. & J. Osbornova. 1998. The Mammals of Ancient Egypt. – Warminster, Aris & Phillips.

Patch, D.C. 2011. Dawn of Egyptian Art. – New Haven, Yale University Press.

Payne, J.C. 2000. Catalogue of the Predynastic Egyptian Collection in the Ashmolean Museum. – Leuven, Peeters.

Peters, P. & N. Pöllath. 2004. Holocene Faunas of the Eastern Sahara: Zoogeographical and Palaeoecological Aspects. In: Lauwerier, R. & I. Plug. The Future from the Past. Archaeozoology in Wildlife Conservation and Heritage Management. – Oxford, Oxbow Books: 34-50.

Pöllath, N. 2009. The Prehistoric Gamebag: The Archaeozoological Record from Sites in the Western Desert of Egypt. In: Riemer, H., F. Förster, M. Herb & N. Pöllath. Eds. Desert Animals in the Eastern Sahara: Status, Economic Significance, and Cultural Reflection in Antiquity; Proceedings of an Interdisciplinary ACACIA Workshop Held at the University of Cologne, December 14-15, 2007. – Cologne, Heinrich Barth Institut: 309-340.

Rueness, E.K., M.G. Asmyhr, C. Sillero-Zubiri, D.W. Macdonald, A. Bekele, A. Atickem & N.C. Stenseth. 2011. The Cryptic African Wolf: *Canis aureus lupaster* is Not a Golden Jackal and is Not Endemic to Egypt. – PloS one, 6(1), e16385. https://doi.org/10.1371/journal.pone.0016385

Schott, E. 1971. Ein Stempelsiegel in Form eines Elefanten. – Mitteilungen des Deutschen Instituts für Ägyptische Altertumskunde in Kairo 27: 99-110.

Serpico, M. & R. White. 1996. A Report on the Analysis of the Contents of a Cache of Jars from the Tomb of Djer. In: Spencer, A. Ed. Aspects of Early Egypt. – London, The British Museum Press: 128-139.

Setzer, H.W. 1956. Mammals of the Anglo-Egyptian Sudan. – Proceedings of the United States National Museum 106, 3377: 447-587.

Steder, D. 2013. Hyäne oder Hyänehund? Untersuchungen anhand der Zooarchäologie sowie der Darstellungen und Textzeugen aus dem alten Ägypten. – Rahden, Verlag Marie Leidorf.

Tallet, P. In press. Entre prise en main et mise en valeur agricole : Les deux faces de la politique égyptienne dans l'ouest du Delta au début de l'histoire égyptienne. In: Yoyotte, M. Ed. Riverine Harbours. – Cairo, Institut Français d'Archéologie Orientale.

Taterka, F. 2019. The Secretary Bird of Deir el-Bahari; One More Piece to the Puzzle of the Location of the Land of Punt. – Revue d'Egyptologie 69: 231-249.

Van Neer, W., V. Linseele & R. Friedman. 2004. Animal Burials and Food Offerings at the Elite Cemetery HK6 of Hierakonpolis. In: Hendrickx, S., R. Friedman, K. Cialowicz & M. Chlodnicki. Eds. Egypt at its Origins. Studies in Memory of Barbara Adams. – Leuven, Peeters: 67-130.

Van Neer, W., V. Linseele & R. Friedman. 2017. More Animal Burials from the Predynastic Elite Cemetery of Hierakonpolis (Upper Egypt): The 2008 Season. In: Mashkour, M. & M. Beech. Eds. Archaeozoology of the Near East 9. Vol. 2. – Oxford. Oxbow Books: 388-402.

Vandier d'Abbadie, J. 1964. Les singes familiers dans l'ancienne Egypte (peinture et bas-reliefs). I. L'Ancien Empire. – Revue d'Egyptologie 16: 146-177.

Van Neer, W. & H.P. Uerpmann. 1989. Palaeoecological Significance of the Holocene Faunal Remains of the B.O.S.-Missions. In: Kuper, R. Ed. Forschungen zur Umweltgeschichte der Ostsahara. – Cologne, Heinrich Barth Institut: 309-341.

Vernus, P. 2009. Dictionnaire amoureux de l'Egypte antique. – Paris, Plon.

Wiese, A. 2001. Antikenmuseum Basel und Sammlung Ludwig: Die Ägyptische Abteilung. – Mainz am Rhein, Philipp von Zabern.

Dévots et animaux sacrés

Alain Charron

Dévots et animaux sacrés[1]

Nous passons beaucoup de temps à étudier les momies animales ou les relations des animaux avec les cultes. Un domaine est cependant moins observé, c'est celui du rôle des Égyptiens, notables, hommes du peuple ou des clergés qui nous ont laissé des témoignages innombrables, malheureusement souvent peu explicites.

Au préalable, il faut revenir sur un problème de formulation. Assez fréquemment, des études mentionnent les pèlerins qui se seraient rendus dans les nécropoles. Pourtant, l'emploi de ce mot n'est sans doute pas le plus judicieux, surtout pour la période pharaonique, ce que faisait déjà remarquer Jean Yoyotte :

> « ... *la masse formidable des souvenirs privés trouvés dans les temples officiels n'atteste aucunement...une pratique intense du pèlerinage dans l'Égypte ancienne* » (Yoyotte, 1960 : 23 ; voir aussi Bernand, 1988 : 50).

Le pèlerinage vers Abydos est fréquemment mentionné mais les Égyptiens étaient loin de tous se rendre dans la cité d'Osiris (Volokhine, 1998 : 71-76). L'Égyptien des petites classes sociales voyageait peu et ce sont surtout

> « *des fonctionnaires en mission, des militaires qui ont profité de leur présence sur ou près d'un lieu sacré pour rendre hommage aux dieux, et pour pérenniser sur pierre leur présence en ces lieux* » (Volokhine, 1998: 84 ; voir auss Malaise, 1987: 80-81).

Le pèlerinage suppose un voyage nécessaire, voire obligatoire, pour aller en un lieu sacré où l'on peut rendre hommage à la divinité de son choix. Cela a sensiblement changé à l'époque ptolémaïque et des déplacements, surtout à Saqqarah pour se rendre au Sérapéum, sont bien attestés, même si la majorité des visiteurs est issue de la province ou des nomes limitrophes (Wilcken, 1927 : 50-52). Il ne faut donc sans doute pas réfuter totalement cette idée (Volokhine, 1998 : 52-53), mais la moduler. Il est préférable d'employer le terme de dévot (Devauchelle, 2017 : 97), un homme pieux attaché aux pratiques religieuses.

1 Tous mes remerciements vont à Salima Ikram et son équipe pour leur aimable invitation et la magnifique organisation de ce colloque.

In: Ikram, S., J. Kaiser & S. Porcier. Eds. 2021. *The Ancient Egyptians and the Natural World. Flora, Fauna, and Science.* – Leiden, Sidestone Press: 71-82.

Origine des dévots

Les textes qui nous sont parvenus indiquent généralement que c'est la population locale, celle du nome, qui honorait le dieu de la cité principale (Malaise, 1987 : 60-62). Ainsi,

> « *Parmi les centaines de stèles qui étaient encastrées au monument funéraire des taureaux Apis à Memphis, la plupart sont signées de gens de Memphis* » (Yoyotte, 1960 : 23 ; voir aussi Volokhine, 1998 : 68).

Quand aucune origine n'est donnée pour le dédicant d'une stèle du Sérapéum, il doit s'agir de membres du clergé de Ptah et sans doute de prêtres serviteurs d'autres temples de l'ancienne capitale.

De nombreuses personnes qui travaillaient aux enterrements des taureaux ou qui suivaient les cérémonies des funérailles pouvaient quand même venir d'autres villes (Vercoutter, 1962 : 128), certains religieux portant en même temps des titres provinciaux et memphites.

Jean Vercoutter, dans l'étude de stèles du Sérapéum, a trouvé des membres de clergés assez proches de Memphis comme Létopolis. Selon lui, d'autres prêtres venaient du Delta (Nebesheh, Behbet al-Haggar et Senet (dans le nome métélite) et de Haute Égypte (Antinoé, al-Kab, la région d'Héracléopolis et Hebenou (près de l'actuelle Minieh). Memphis était la grande ville du pays et les dévotions envers Apis dépassaient de loin celles réservées aux autres animaux.

Sur des stèles dévolues au taureau Mnévis sont figurés des prêtres de Rê ou des fonctionnaires dépendant du domaine sacré, montrant ainsi que c'est ce clergé qui réglait les funérailles de l'animal. Les stèles ont été également offertes par des gens de conditions modestes au Nouvel Empire (Porcier, 2009 : 63-68). Des monuments plus tardifs montrent que des hauts fonctionnaires ont pris part aux dévotions en l'honneur du taureau comme le vizir Psammétique-séneb à la XXXe dynastie (Perdu, 2006 : 41-52).

De même, parmi les stèles découvertes dans le Bucheum d'Ermant, où les souverains sont seuls cités, figure quand même un monument daté de l'an 1 d'Auguste laissé par Kalasiris, un prêtre de Montou d'Ermant (Fairman, 1934 : 11-13 ; Valbelle, 1992 : 9).[2]

À Mendès, un prêtre de l'époque ptolémaïque,

> « *ayant dressé son image dans le temple du Bélier, seigneur de la ville et maître de fécondité, peut compter sur les pèlerins pour prier en sa faveur : Ô vous qui naviguez d'amont ou d'aval pour venir voir les grands béliers sacrés, priez le dieu en faveur de cette mienne statue* » (Yoyotte, 1960 : 20).

La navigation demandée par le prêtre, terme que l'on retrouve sur de nombreux monuments, ne fait peut-être pas référence à un long voyage, mais simplement aux déplacements dans le nome.

Les stèles du Nouvel Empire trouvées à Assiout en l'honneur d'Oupouaout ont été déposées par des personnages dont les titres montrent des prêtres, des artisans, des militaires... originaires de la région (DuQuesne, 2009 : 56). L'absence de précisions concernant les dédicants vient certainement du fait que ceux-ci n'avaient pas besoin d'insister sur leur région d'origine, c'était la même que celle de la divinité honorée.

Il est intéressant de voir en parallèle les graffitis démotiques laissés dans le temple de Philae par des serviteurs habitant certainement dans une région proche du temple :

> « *On se serait attendu à trouver les noms de pèlerins, mais, en fait, la masse des noms appartient à des personnes qui faisaient partie du personnel même du temple, ou qui y furent du moins temporairement employées* » (Yoyotte, 1960 : 55 ; voir aussi Griffith, 1937 : 11).

Le temple a pourtant attiré des pèlerins venant parfois de loin dans le monde hellénistique à l'époque lagide puis sous la domination romaine, certains ayant laissé des graffitis en grec (Rutherford, 1998 : 229-256 ; Volokhine, 1998 : 91-92).

L'étude des inscriptions laissées dans le grand temple de Karnak montre que la majorité des lapicides lettrés exerçait des fonctions en ce lieu (Traunecker, 1979 : 24).[3] Le graffiti peut sans doute remplacer l'offrande d'une statuette ou d'une momie dans l'esprit du dévot de passage dans le sanctuaire et son nom pouvait être lu dans le temple ou la nécropole (Yoyotte, 1960 : 59). Ainsi,

> « *...on a pu se demander, à Basse Époque, si le droit de laisser un proscynème sur les murs des temples n'était pas acheté par des dons et si les meilleurs des graffitis n'étaient pas l'œuvre rétribuée des graveurs attachés à l'édifice* » (Yoyotte, 1960 :62).

---

2 Des Kalasiris sont présents sur plusieurs documents liés au taureau Bouchis, une stèle de l'époque d'Antonin le Pieux nous apprend que l'animal était né dans leur étable (Mond & Myers, 1934 : 15)

3 Le même constat concerne les graffitis laissés sur la troisième terrasse du temple de Deir al-Bahari en hommage à Amenhotep fils de Hapou et à Imhotep (Bernand, 1988 : 54).

Classes sociales des dévots

A priori, n'importe qui était susceptible de faire l'offrande d'une momie ou d'un reliquaire. Les exemples connus ne relèvent pas d'une religion « populaire » contrairement à ce qui a pu être écrit, mais plutôt d'une piété personnelle (Dunand, 2018 : 151), car tous les actes que les dévots pouvaient accomplir étaient encadrés par les clergés.

Ce sont même des lettrés qui sont à la base de l'intérêt pour les animaux dans la religion. Le plus ancien témoignage, conservé au musée du Caire, concerne les honneurs qu'une chatte a connus grâce au prince Thoutmosis, fils d'Amenhotep III. Elle a eu droit à un sarcophage en calcaire, découvert à Memphis, dont les scènes et les inscriptions sont fortement inspirées par les monuments funéraires des hommes. Ce prince, également grand prêtre de Ptah, a surtout conduit le premier enterrement d'un taureau Apis. Sur des bas-reliefs de la chapelle funéraire de l'animal, le roi, accompagné du prince, offre l'encens au taureau (Maystre, 1992 : 132).

Le prince et grand prêtre de Ptah, Khâemouaset, a développé les hommages rendus au taureau Apis à l'époque de Ramsès II. Deux sépultures isolées ont livré un très riche matériel et on lui doit l'ouverture des Petits Souterrains.

Le musée de Munich conserve un fragment de jarre ayant contenu le corps d'un ibis dont la provenance est inconnue. L'enterrement de cet oiseau à la fin de la XIXe dynastie, relaté sur la céramique, est dû à un scribe (Spiegelberg, 1928 : 14-17). Il aurait trouvé l'animal mort noyé dans un lieu nommé « le canal de Ramsès Ier ». Le nom d'un sanctuaire, Per-Ousir n'est pas révélateur d'un lieu, même si cela peut évoquer Abydos où une nécropole d'ibis plus tardive a été trouvée.

Un pectoral en bois (Morenz, 1963 : 42-47) de provenance inconnue et en collection particulière quand il a été publié, présente un homme, le scribe royal Amenmose, faisant une libation et une fumigation devant un ibis momifié qui, en tant que netjery ("divin "), a subi les rites et est devenu un Osiris. Entre eux se trouvent des offrandes et des textes qui renvoient au rite de l'ouverture de la bouche. L'objet, daté du Nouvel Empire, aurait été appliqué sur une momie, ce qu'attesteraient les restes de lin encore conservés au verso.

Les souverains ont été fortement impliqués à la Basse Époque, les épitaphes royales officielles du Sérapéum datées de la XXVIe dynastie nous le montrent bien, comme cette stèle dédiée par Amasis au taureau mort en l'an 23 et conservée au musée du Louvre (Charron & Farout, 2008). Son successeur, le futur Psammétique III (Vercoutter, 1962 : 37-43) ne s'est pas contenté de contrôler et d'aider depuis son palais, il a participé activement aux rites de l'enterrement comme n'importe quel dévot.

Plus tard, à partir de la XXXe dynastie et surtout durant l'époque gréco-romaine, quand le phénomène des animaux en lien avec les cultes a pris une ampleur considérable, beaucoup d'Égyptiens ont été concernés, quelle que soit leur classe sociale. Les Lagides accompagnent cette ferveur par des résolutions et des dons en faveur des animaux sacrés. Ptolémée Ier se serait contenté de prêter cinquante talents d'argent pour les funérailles d'Apis, mais son fils aurait chargé son ministre des finances, Apollonios, de donner pour cent talents de myrrhe pour l'enterrement d'une vache Isis Hesat d'Aphroditopolis. Le décret de Canope daté de Ptolémée III décrit l'attachement des souverains envers Apis, Mnévis et les autres animaux sacrés dignes de considération. Il en est de même sur le décret de Memphis de l'an 1 de Ptolémée V Épiphane ainsi que sur un autre de l'an 23 (Charron, 1998 : 193-195). En revanche, les exemples sont très rares rapportant une visite royale à ces animaux. L'exemple le plus fameux est une stèle de Mendès qui relate la visite de Ptolémée II au bélier/bouc Banebded (De Meulenaere & MacKay, 1976 : 174-177 ; Pasquali, 2019).

Mais il s'agit là d'un exemple assez isolé. Il est en effet possible que Ptolémée VI Philomètor soit venu voir le taureau Bouchis à Thèbes en l'an 24 (Preys, 2017 : 354), mais il est moins assuré que la grande Cléopâtre ait assisté à l'installation d'un autre de ces taureaux en 51 avant J.-C. (Chauveau, 1998 : 26). Les rois et reines ne se rendent que très rarement auprès des animaux sacrés, et surtout pour leur intronisation, cela montre que cet attachement est essentiellement politique, orchestré par les clergés, et ne présente aucune régularité.

De nombreux personnages des classes plus favorisées laissent, eux aussi, des témoignages de leur dévotion. Un des plus beaux exemples est la statue guérisseuse de Djed-Her le sauveur d'Athribis, sur laquelle ce dernier a fait graver les bonnes actions qu'il a accomplies pour le dieu de sa ville l'Horus Khenty-Khety, notamment en assurant la momification des faucons (Jelínková-Reymond, 1956 : 96-112).

Les membres du clergé sont ceux qui ont laissé le plus de traces de leur implication. On leur doit l'essentiel des graffitis du Sérapéum de Saqqarah par exemple (Malek & Ray, 2017 : 52),

Il y a également des personnes de conditions plus modestes. Parmi celles-ci figurent le personnel des temples en charge des animaux, notamment des élevages ou de la momification, ainsi que tous les corps de métiers en lien avec l'enterrement des animaux sacrés : soldats, pleureuses, ouvriers de la tombe et d'autres, dont le lien direct n'est pas inscrit sur les divers monuments. Elles ont généralement laissé peu d'inscriptions, comme par exemple, les stèles d'Assiout déjà mentionnées (DuQuesne, 2009 : 62-63) et d'autres du Sérapéum (Devauchelle, 2016 : 138-139, 144-145 ; Porcier, 2016 : 150-151) datées du Nouvel Empire. Plus tardivement, on en trouve à Saqqarah, toujours au Sérapéum (Malek & Ray, 2017 : 43), dans la nécropole des mères d'Apis (Smith *et al.*, 2011 : 242-247) ou dans les galeries de babouins (Ray, 2011 : 367). Les plus pauvres ont certainement plutôt procédé aux offrandes de momies, plus simples pour des personnes ne sachant pas écrire, et certainement moins onéreuses.

Cependant, les enterrements des animaux n'étaient pas laissés au hasard, ce sont les temples et les membres du clergé qui géraient les cérémonies (Preisigke & Spiegelberg, 1914 : 4).

Les associations religieuses

Cela est d'autant plus vrai que le public n'avait pas libre-accès aux nécropoles (Kessler, 1985 : 579). L'exemple le plus célèbre est celui du Sérapéum qui restait clos et n'était ouvert que pour placer la dépouille de l'Apis qui venait de mourir. En dehors des ouvriers qui préparaient la nouvelle chambre funéraire, seuls les prêtres pouvaient se rendre dans les galeries (Vercoutter, 1962 : 124).

Le dévot pouvait montrer un attachement supplémentaire à la divinité en devenant membre d'une confrérie chargée notamment de conduire les animaux dans les nécropoles (Dils, 1995 ; Quaegebeur, 1984). C'est ce que laissent penser les inscriptions relevées sur les ostraca de d'Ombos connus sous le nom de Prinz-Joachim-Ostraka. Une partie des desservants ne porte en effet aucun titre administratif (Preisigke & Spiegelberg, 1914 : 31). Cependant, nous ne connaissons pas les conditions d'admission dans la plupart de ces confréries, mais il semble qu'elles aient concerné des prêtres et du personnel du temple et que les successions se soient souvent faites de père en fils.

Elles apparaissent tardivement, à l'époque saïte, mais les attestations et règlements connus datent seulement de l'époque gréco-romaine. Ces associations (Cenival, 1969 : 5-19 ; 1972 ; 1977 : 1-49 ; Muszynski, 1977 : 145-174) étaient encouragées par le pouvoir car elles s'occupaient en premier lieu du culte royal.

Parmi ces associations, certaines étaient spécialisées dans le transport des « dieux » vers leur dernière demeure. Parmi les engagements figurant dans les règlements de certaines confréries ayant en charge des animaux, une clause spécifique concernait l'enterrement de ces derniers. Tous les membres devaient y participer sous peine de se voir infliger une amende.

Ce sont de véritables processions qui étaient organisées lors de la mise en place des momies animales dans la nécropole. Le papyrus Lille 29, daté de l'an 24 de Ptolémée III, précise, à propos de l'association d'Horus de Behedet, afin d'appliquer le règlement dans le village de Sebek Pisaï :

> « *Nous devons traîner le faucon, les jours que les membres choisiront pour traîner le dieu durant la période susdite, marchant tous ensemble derrière le « supérieur du faucon » et le reste des « supérieurs » de la « maison »* » (Cenival, 1969 : 10).

Les membres de l'association, dirigés par ce supérieur du faucon (Cenival, 1972 :162-164), se réunissaient dans le temple.

Différents papyrus (P. Hambourg I et Caire 30606, 31179 et 30605), correspondant à quatre années d'exercice datées des Ptolémée VI et VIII, nous apprennent que l'association du dieu Sobek de Tebtynis, dirigée par un stratège, avait pour mission d'apporter les momies de crocodiles jusqu'à la nécropole :

> « *Nous traînerons les dieux de Sebek, et nous les ferons parvenir à leur tombe. Celui d'entre nous qui ne viendra pas avec nous pour traîner les dieux de Sebek et celui qui ne les fera pas parvenir à leur tombe, son amende sera de 25 deben, et la malédiction des dieux de Sebek le poursuivra* » (P. Hambourg I, Cenival, 1972 : 61).

Cette société, sans doute assez importante (Cenival, 1972 : 150-151, 172-173), se réunissait dans le temple, le dromos, les sanctuaires d'autres divinités ou encore dans le cimetière du crocodile du village de Sebek Tebtynis (Cenival, 1972 : 177-178).

A d'Ombos, ce sont des ostraca datés entre 79 et 53 avant J.-C., sous le règne de Ptolémée Aulète, qui nous assurent que les ibis de Thot et les faucons d'Haroeris étaient conduits d'une manière identique jusqu'à leur sépulture (Preisigke & Spiegelberg, 1914). À la tête de la confrérie figure un militaire, un stratège, administrateur de la vie économique du temple (Cenival, 1972 : 159-162).

A Kom Ombo, des porteurs des dieux-crocodiles sont attestés par des graffitis d'époque romaine (Quaegebeur, 1984 : 161-176). Ils se retrouvaient en un lieu de la cour du temple. Jan Quaegebeur a remarqué que les porteurs de crocodiles, les « dieux de Sobek », avaient reçu des noms montrant leur attachement à la divinité (Quaegebeur, 1984 : 168, note 67 ; Cenival, 1972 : 241, 244-245, 256-258), notamment plusieurs Padisobek ou Sobekemheb.

À Drah Aboul Nagga, des tombes du début du Nouvel Empire, remployées vers 145 avant J.-C. afin de recueillir des momies d'ibis et de faucons, ont été fouillées au début du XXe siècle (Kessler, 1989 : 159-164 ; Northampton *et al.*, 1908 : 19-23). Diverses inscriptions démotiques d'époque ptolémaïque, assez stéréotypées, y ont été relevées. Généralement, nous avons juste le nom d'un personnage et celui de son père et rarement un titre, prêtre en chef (lesonis) de Thot (n°11), pastophore (n°28), scribe (n°12, 14, 15). Quelques vœux sont exprimés, comme celui-ci : « *la bonne réputation de Phibis (?) (et) Imuthes (?) reste ici, devant Osiris-Ibis, Osiris-Faucon et les dieux du lieu de repos de l'ibis dans l'éternité* » (n°7). Il peut être également demandé que les " dieux " de la nécropole donnent vie à celui qui a dédié l'inscription (n°8). Ce sont certainement des personnes ayant pris part à l'enterrement des oiseaux qui ont laissé ces marques, un même personnage, Psen-Min, apparaissant dans cinq inscriptions (Northampton *et al.*, 1908 : 23).

Les relations du dévot avec les animaux

Les relations dépendaient du type d'animal concerné. Il y avait une différence, bien entendu, entre Apis ou Mnévis et l'une de ces nombreuses bêtes sacralisées par la momification. Mais dans tous les cas, le dévot était laissé à l'écart des opérations qui ne relevaient que des temples.

Des hommes ont pu, par dévotion, faire une donation de terrain dont les revenus devaient permettre l'entretien des animaux hérauts d'un dieu ou des simples représentants de la divinité. Il fallait nourrir les bêtes et c'est ainsi que dans un texte de l'ancienne collection Miramar, un scribe d'Abydos a pu écrire cette phrase :

> « *j'ai donné de la nourriture à l'ibis, au faucon, au chat et au chacal* » (Bergmann, 1879 : 7-8, pl. VI ; voir aussi Te Velde, 1980 : 78).

En ce qui concerne les animaux sacrés, le lien était certainement plus régulier car la bête était visible de tous comme le mentionne Strabon à propos d'Apis.[4] Celui-ci pouvait être approché quand il s'ébattait dans l'enclos qui lui était réservé à Memphis. Il était également visible lors de fêtes. L'animal avait sans doute une valeur particulière pour les croyants car il pouvait alors être le réceptacle de la force divine. Le dévot n'avait pas accès au dieu, dont la statue était cachée au plus profond du temple, mais à une manifestation de la divinité, un intercesseur.

Les liens se faisaient également, même indirectement, quand des personnages, religieux ou non, grands ou humbles, traitaient le corps du taureau Apis, taillaient son sarcophage, aidaient au creusement de la tombe, participaient aux funérailles en traînant le sarcophage... Parmi les dévots (Vercoutter, 1962 : 37-43), on connaît le rôle de divers hommes importants comme le général Amasis qui vécut à l'époque de Darius. Il était chargé de la garde militaire qui accompagnait la procession funéraire et responsable des émissaires envoyés dans le pays pour récupérer les dons faits par les nomes et villes afin d'assurer la momification d'Apis. Il a fait que l'on apporte

> « *...de l'argent, de l'or et du tissu royal, de la résine odoriférante, toutes espèces de pierres précieuses et toutes sortes de bonnes choses* » (Vercoutter, 1962 : 59-64).

Les dévots pouvaient plus simplement suivre les pratiques religieuses. Le jour même de la mort du bovidé, ils prenaient le deuil (Vercoutter, 1962 : 125-127). Cela implique d'ailleurs que la population concernée était dans l'ancienne capitale ou alors à proximité, pour être prévenue dans les temps. Ils étaient volontaires pour des veilles et un double jeûne, total pendant quatre jours et partiel pendant le reste des soixante-dix jours, ne consommant que du pain, de l'eau et des herbes. Prostrations, dénudation du corps et lamentations accompagnaient la préparation des funérailles.

Certains dévots ont même participé financièrement aux funérailles de quelques animaux sacrés. Diodore de Sicile rapporte qu'à l'époque de Ptolémée, juste après la mort d'Alexandre, un Apis mourut et que

> « *celui qui avait soin de lui non seulement dépensa pour ses funérailles toute la somme, qui était tout à fait considérable, réservée à cet effet, mais encore emprunta à Ptolémée cinquante talents d'argent. Et même de nos jours, certains de ceux qui élèvent ces animaux n'ont pas dépensé moins de cent talents pour leurs funérailles* » (Diodore, I, Vernière, 1993 : 84).

Ces propos sont corroborés par quelques monuments. Le texte d'un fragment de statue naophore du musée de Saint-Pétersbourg, datée des IVe-IIIe siècles avant J.-C., indique que son propriétaire, Ankh-Psammétique, a apporté, sur ses deniers personnels, du matériel dans la Ouabet où était traité le corps d'un Mnévis, (Jansen-Winkeln, 2001 : 102 ; Porcier, 2009 : 70). Le bélier/bouc de Mendès a pu bénéficier, lui aussi, de dons octroyés par des particuliers. Le diocète Harkhebi, qui vécut à l'époque de Ptolémée II, a laissé une inscription sur une statue dans laquelle il vante son engagement financier pour l'inhumation de l'animal (Klotz, 2009 : 288-289 et 296-297 ; Pasquali, 2019 : 278).[5]

Strabon relate la visite effectuée au crocodile sacré représentant Soukhos dans la ville d'Arsinoè (Strabon, XVII, 1, 38 ; Charvet, 1997 : 151). Il rapporte que le reptile était élevé dans un lac et que des visiteurs pouvaient lui apporter différents mets et boissons que des prêtres étaient chargés de lui faire engloutir. Le sénateur Lucius Memmius avait fait la même expérience en 112 avant J.-C. Celui-ci, en mission à Alexandrie, était allé faire du « tourisme » dans le Fayoum et n'avait pas manqué

4 « *En avant du sècos se trouve une cour...C'est dans cette cour qu'on lâche Apis à une certaine heure, principalement pour le montrer aux étrangers...* » (Strabon, XVII, 1, 31 ; Charvet *et al.*, 1997 : 135).

5 L'engagement des particuliers a été très importants dans la construction et la restauration de lieux de culte à l'époque ptolémaïque (Thiers, 2006).

cette attraction (Hohlwein, 1940 : 274). Il est probable que les dévots du Fayoum soient venus rencontrer l'animal et aient agi de la même façon. Une statue conservée au musée du Petit-Palais à Paris (Bernand, 1975 : 39-41) a été dédiée par un personnage nommé Apollonios, fils d'Apollonios, originaire de Talésis, une bourgade certainement proche de la capitale Arsinoé. L'inscription portée sur le socle mentionne l'épiphanie d'un crocodile sacré le 21 juin 60 avant J.-C. sous le règne de Ptolémée Aulète.

Strabon relate la visite à « l'oiseau éthiopien » de Philae (Strabon, XVII, 1 49 ; Charvet, 1997 : 183), un rapace de grande taille qui était sur le point de mourir de maladie. Cela indique que cet animal sacré était accessible aux visiteurs du site. Les dévots avaient sans doute également accès au faucon d'Edfou (Alliot, 1954 : 667). Lors des fêtes du couronnement de l'oiseau qui commençaient le Ier tybi, le peuple participait aux réjouissances sans que l'on sache exactement ce qu'il pouvait voir, peut-être l'apparition du rapace choisi par la statue du dieu.

En ce qui concerne les animaux sacralisés, le lien devait certainement être établi lors d'une fête consacrée à la divinité. C'est ce que peut nous laisser croire un commentaire d'Hérodote sur la navigation de dévots de Bastet vers Bubastis (Hérodote, II ; Legrand, 1982 : 60). D'après lui, soixante-dix myriades d'Égyptiens se rendaient à ces fêtes (Volokhine, 1998 : 69), un nombre considérable que la cité aurait eu sans doute bien des difficultés à accueillir. Le but d'Hérodote est de montrer la notoriété de la fête et suggérer un nombre très important de visiteurs désirant honorer la déesse. C'est peut-être à l'occasion de ces fêtes que les fameux bronzes, qui font la joie des visiteurs de musées, étaient offerts à la divinité.

Grâce aux Prinz-Joachim Ostraca, nous apprenons que dans le nome d'Ombos, les enterrements des rapaces étaient organisés habituellement une fois par an environ, mais il est arrivé que deux ans se passent, voire trois, sans enterrement alors que des animaux continuaient à être momifiés comme le montre l'ostraca 13 daté de l'an 16 de Ptolémée XII Aulète. Un de ces documents indique qu'une cérémonie a eu lieu à l'occasion des mystères d'Osiris pratiqués dans le temple d'Ombos (Preisigke & Spiegelberg, 1914 : 28-34). Les associations religieuses ont d'ailleurs notamment pour but « *une participation massive aux offrandes, lors des fêtes* » (Cenival, 1969 : 13).

Il est certain que les anciens Égyptiens croyaient que ces intermédiaires les rapprochaient du divin (Meeks, 2018 : 197). N'ayant pas accès au saint des saints dans les temples c'était une façon, notamment pour tous ceux qui n'avaient pas la possibilité de laisser des témoignages écrits ou plus consistants comme des statues, de laisser une trace de leur piété et de ce qu'ils attendaient de la divinité.

Les incubations

Une autre manière de rencontrer le dieu était le songe. Le dévot cherchait des réponses à ses soucis personnels en passant la nuit dans un temple avec l'espoir que la divinité lui apparaisse et s'adresse à lui. Cette pratique se développe à l'époque ptolémaïque (Volokhine, 1998 : 32-33). Hor de Sébennytos l'a employée en invoquant Osorapis et Osormnévis deux jours de suite dans la chapelle funéraire de l'ibis sacré à Saqqarah. Il a obtenu une réponse le troisième, le 10 mars 155 avant J.-C. (Ray, 1976 : 130-136 et texte 13 : 55-56). Il a également eu un songe alors qu'il était dans la ouâbet d'Osiris-Mnévis à Héliopolis.

Un petit monument en calcaire peint de la fin du IIIe siècle avant J.-C. présente le taureau Apis. Il a été découvert près de l'Anubeion et est conservé au musée du Caire (Charron, 1998 : 195). Il montre qu'un Crétois pouvait interpréter les songes envoyés par la divinité aux personnes qui passaient la nuit dans l'enceinte de son temple.

Les oracles

La question de l'oracle et des animaux est plus difficile. Pour Pline l'Ancien, Apis

> « *a deux temples, qu'on appelle thalames, et qui servent d'augures aux Égyptiens ; s'il entre dans l'un, c'est un bon présage ; dans l'autre, il annonce des malheurs. Il donne des réponses aux particuliers en prenant des aliments de la main de ceux qui le consultent. Il se détourna de la main de Germanicus, qui mourut peu après.* » (Pline l'Ancien, VIII, 71 ; Ernout, 1952 : VIII, 71: 87-88 ; voir aussi Solin, 33 ; Agnant, 1847).

De même, Eudoxe de Cnide, le grand astronome et mathématicien du IVe siècle avant J.-C., se serait vu annoncer sa fin prochaine parce que le taureau lui aurait léché son manteau (Meeks & Favard-Meeks, 1993 : 202). Une histoire identique aurait touché un Ptolémée qui serait mort après qu'un crocodile sacré eut refusé de le voir (Plutarque, 23 ; Bouffartigue, 2012; Hani, 1976 : 431).

Ce genre d'oracle a-t-il vraiment existé ? Les sources égyptiennes ne nous en disent rien, seules des sources extérieures viennent nous renseigner. Cela ressemble d'ailleurs plus à de la superstition populaire, un système qui n'aurait pas vraiment été régi par les clergés. Les Égyptiens consultaient en fait le dieu figuré sous forme animale comme Thot-ibis « *qui écoute les prières et qui parle* » (Quaegebeur, 1975 : 19-24) dans le petit temple de Kasr al-Agouz. L'oracle qui aurait été rendu concernerait plutôt une statue que l'animal vivant (Valbelle, 1992 : 14-15).

Les oracles pouvaient être rendus pas la voix des prêtres.

« *Les questionneurs ne pouvaient ignorer que la réponse passait par le truchement des hommes et, cependant, leur adhésion au système n'était pas remise en cause. Non par naïveté, mais parce que, à travers l'image du dieu, même manipulée par des hommes semblables à eux, ils estimaient qu'ils s'adressaient à une puissance plus qu'humaine et avaient ainsi la possibilité d'établir une communication directe avec la divinité* » (Dunand & Zivie-Coche, 1991 : 126).

Les offrandes

L'aspect le plus visible de l'honneur rendu par des dévots est l'offrande. Le principe du troc est bien connu et, grâce aux inscriptions qui nous sont parvenues, nous savons ce que le dédicant attendait de la divinité en retour.

Quelques animaux sacrés ont bénéficié de stèles offertes par des personnes ayant participé aux funérailles. En revanche, d'autres n'ont eu qu'une stèle au nom du roi comme dans le cas du taureau Bouchis[6]. Ce droit, octroyé par le pouvoir ou le clergé ne nous apporte cependant pas de réponses sur la manière dont le monument était réalisé, payé et placé dans la nécropole.

Il était également possible d'offrir une statuette, ce sont par exemple les petits bronzes d'Apis que l'on trouve dans les collections égyptiennes du monde entier.

Pour les animaux sacralisés, nous ignorons également quel était le processus permettant à quelqu'un d'acquérir les différents objets à déposer devant la divinité, un bronze, statuette ou reliquaire, type d'offrande originaire du Delta (Charron, 2012 : 295-296), une momie que celle-ci soit « nue « comme certains crocodiles de Kom Ombo ou recouverte de tissus, voire de parties stuquées, un cercueil en terre cuite ou en bois reprenant ou non la forme animale du dieu.

Car c'est bien la divinité, sous sa forme animale qui est honorée sur les reliquaires et non une bête quelconque. L'animal du dieu est pris à l'intérieur. Un décor représentant la divinité pouvait être appliqué au moyen de tissus sur certaines momies.

Ce type d'artefact pouvait avoir une valeur religieuse assez identique à celle d'une statue, que ce soit une idole ou une offrande, c'est ce que peut nous laisser croire une fresque située dans une niche de la première cour du temple de Théadelphie. La momie de crocodile, couchée sur un brancard et portée en procession par des prêtres, incarnait l'image cultuelle de Pneferôs (Breccia, 1926 : 105, pl. 64,3 ; Crawford, 1971 : 95).

En revanche, la momie d'un animal sacralisé devait coûter moins cher pour un dédicant et permettait de répondre à une ample demande.

L'objet devait être choisi, payé et remis à un représentant du clergé et nous ne pouvons que supposer avec Dimitri Meeks que :

« *Le fidèle pourra contempler de près et nourrir ces multiples reflets du dieu qu'il est venu honorer, puis ayant payé pour une momie, déjà sans doute toute prête, il pourra, avant qu'elle n'aille rejoindre ses semblables entassées dans de vastes catacombes, toucher à sa façon un bref instant au divin* » (Meeks, 1986 : 190).

Un certain nombre de reliquaires portent une inscription, ce qui montre que soit les donateurs savaient écrire, soit qu'il y avait une personne susceptible de les aider au moment de l'achat de la pièce. Le texte donne le nom du dieu suivi d'une formule assez stéréotypée : puisse-t-il donner la vie, la prospérité, la santé, une longue existence...à X (Yoyotte, 1960 : 61).

Cette pratique ne perdure guère et nous n'avons quasiment pas d'inscriptions sur les emballages de momies comme ceux en terre cuite des ibis de Saqqarah alors que leur nombre est plus important que jamais.

Il y a heureusement des exceptions comme un sarcophage en calcaire polychrome d'un ibis sacré conservé au musée du Louvre dont la cuve prend l'aspect du corps de l'animal. Celle-ci contient, protégée par une couche de mortier, la momie de l'un de ces oiseaux. Sous le bec de l'ibis est sculptée une stèle dans laquelle un personnage nommé Petepâgem s'est placé, avec ses parents, sous la protection de « Thot qui donne la vie ».

Les demandes

A la Basse Époque, suite aux évolutions conduites depuis le Nouvel Empire, notamment à travers les fêtes religieuses, les pratiques oraculaires, le culte des rois divinisés ou les dieux qui écoutent les prières, les emplacements dévolus à la piété personnelle se sont développés (Kruchten, 1997 : 31-33), doublant en quelque sorte le temple réservé à la religion traditionnelle et mettant en valeur certaines statues de la divinité et les images de procession. Les dévots souhaitaient pouvoir atteindre la divinité et faire connaître leurs suppliques, mais ils n'attendaient pas de choses incroyables.

Ils voyaient dans les animaux sacralisés un autre moyen d'accéder au dieu qu'ils n'avaient pas la possibilité de voir dans son temple (Malaise, 1987 : 71), un intermédiaire, support de la puissance divine, auquel ils confiaient leurs demandes. La divinité paraissait ainsi plus proche, mais, ne livrant qu'une de ses apparences, elle n'était pourtant pas plus compréhensible.

6 Concernant ce dernier, il y a bien eu quelques petites stèles en calcaire portant des noms, mais elles n'apportent aucun renseignement sur la participation de la population (Mond & Myers, 1934 : 15).

Dans le cas du taureau de Memphis, les inscriptions portées sur les stèles peuvent donner des détails intéressants. Le petit monument laissé par le général Amasis, la stèle Louvre IM 4017 précise ce que le personnage attend des efforts fournis pour Apis :

> « *...récompense-le en proportion de ce qu'il a fait pour toi. Prolonge ses années, rend stable son nom éternellement, accorde pour ta part que cette stèle soit établie fermement dans la nécropole afin que l'on se souvienne de son nom éternellement* ». (Vercoutter, 1962 : 59-64).

La stèle du Louvre IM 4053 provenant elle aussi du Sérapéum montre le fils du roi Amasis et de Nakhtbasteterou, le prince Pasenenkhonsou en adoration devant le taureau Apis. Les demandes sont assez simples,

> « *que le nom du fils royal demeure... une vie parfaite (?), un enterrement excellent (?) dans la nécropole...* » (Devauchelle, 2011 : 145).

Les numéros 23 et 25 des Prinz-Joachim-ostraka présentent des invocations conventionnelles. Sur le premier, un certain Petechonsis, fils de Peteharoëris, demande à ce que son nom reste ici, devant les dieux des lieux de l'embaumement avec ses enfants pour l'éternité[7]. Dans le second, c'est simplement que le beau nom de Kêmi-Hor demeure pour l'éternité. On peut comprendre que ces lieux sont les tombes des ibis et des faucons de d'Ombos. Ces hommes avaient pris une part à l'entretien ou aux funérailles des oiseaux et laissaient leur nom près d'eux.

C'est pour une raison identique que des noms ont été inscrits sur les parois des tombes d'ibis et de faucons de Drah Aboul Nagga (Northampton *et al.*, 1908 : 19-23).

Il en est de même sur un certain nombre de petits bronzes qui portent une dédicace (Kessler & Nur el-Din, 2015 : 151-152). Le dévot a laissé son nom et souvent celui de son père sur l'objet qui reproduit la divinité.

L'emplacement dévolu aux offrandes

Les entrées peu importantes des nécropoles, les galeries pas toujours aisément praticables et l'empilement des momies, comme dans les nécropoles d'ibis de Saqqarah, incitent à penser que le public n'avait pas accès aux catacombes (Kessler, 1985 : 579). C'est encore plus vrai quand ce sont des puits de mastabas qui étaient remplis comme à Saqqarah-Nord ou à Abou Rawash. Seuls ceux qui aidaient aux inhumations avaient le droit d'y pénétrer. Les graffitis parfois découverts à l'intérieur des nécropoles devaient avoir pour origine le personnel, comme à Saqqarah dans les galeries de babouins et de faucons (Ray, 1978 : 29-30). Les textes laissés dans les galeries d'ibis et de babouins de Tounah al-Gebel mentionnent des administrateurs, des scribes, du personnel du temple, des prêtres et les porteurs de momies (Kessler, & Nur el-Din, 2015 :150).

Le Sérapéum n'était ouvert que pour creuser la tombe et organiser les funérailles d'un Apis. Mais les travaux entrepris pour les bovidés sacrés pouvaient être très longs, trois ans comme cela se voit pour l'Isis Ihyroudj, mère de l'Apis mort en l'an 3 de Nectanébo II, et pour laquelle les travaux du caveau ont commencé en juin 375 et se sont achevés en juin 372, après son inhumation en janvier 372 (Devauchelle, 2017 : 995). Les stèles découvertes dans les souterrains du Sérapéum, soit prises dans les murs de fermeture des chambres funéraires, soit à l'entrée, ont été laissées par les participants, notamment celles en démotique par les équipes de carriers.

Une publication récente des graffitis du Sérapéum (Malek & Ray, 2017 : 51), datés ente le VIe siècle et la fin de l'époque ptolémaïque, le montre bien. Ils ont été laissés par des prêtres de Ptah. On y trouve également des prêtres d'une forme de ce grand dieu : Ptah *nb pḥty*, qui pourrait être particulièrement honoré par les travailleurs intervenant dans les souterrains. On trouve également des serviteurs d'Onouris dont le culte est attesté dans cette zone. Alors que les titres en rapport avec Apis sont peu nombreux, un prêtre lecteur en chef d'Apis, titre qui pourrait avoir été créé en imitant le clergé de Ptah, un père divin et trois « *s3 st* » d'Apis sur une même inscription, il n'y a aucune référence à Osiris-Apis (Malek & Ray, 2017 : 53). En revanche plusieurs personnages ont des noms construits avec celui d'Apis : Ankh-hapi ou Hapi par exemple.

Un graffiti démotique relevé dans la galerie des babouins de Saqqarah (Ray, 2011 : 134-137), bien que non daté, a pu être attribué à la mère du taureau Apis mort en l'an 3 de Nectanébo Ier. En effet, il y est mentionné la bénédiction de Ptah et d'Apis-Osiris en faveur de personnes ayant travaillé dans les catacombes pour Isis Taimen (Devauchelle, 2017 : 98).

Des inscriptions ont également été laissées en dehors des souterrains, notamment sur les montants des portes d'entrée. Au Sérapéum, une stèle datée de l'époque d'Amasis dont il ne reste qu'une partie du cintre (Devauchelle, 2011 : 145),[8] ce qui interdit de connaître le bénéficiaire, a été trouvée en avant des débris du grand pylône. Ce serait une marque de dévotion privée déposée dans le temple funéraire d'Apis. Cela implique que tous les témoignages dont nous disposons, notamment les bronzes anonymes, n'ont pas toujours la nécropole pour origine,

7 De telles demandes étaient habituelles dans les sanctuaires (Bernand, 1988 : 56) à propos de Bès honoré par un particulier à Abydos à l'époque romaine.

8 Stèle Louvre IM 4074.

mais des annexes. Une autre stèle (Devauchelle, 2011 : 146)[9] indique que le même souverain a accordé le droit au dédicant, malheureusement anonyme, d'inscrire son nom dans le « pavillon divin », un bâtiment du Sérapéum.

D'autres offrandes ont été également laissées à l'extérieur des nécropoles, quelque part dans le sanctuaire et sans doute dans des chapelles : il s'agit des petits bronzes figurant la divinité, comme ceux du taureau Apis, d'autres étant même des reliquaires pouvant contenir un animal ou au moins un morceau, comme les chats de Bubastis. Ce sont eux qui portent parfois des inscriptions donnant les noms de dévots.

En conclusion, il est possible de dire que les inhumations d'animaux, surtout si elles correspondent à des enterrements en nombres importants, sont le fait des temples et que ce sont des membres du clergé qui en assuraient l'organisation et le suivi. Tout devait être prêt pour l'arrivée des dévots qui venaient sans doute depuis tout le nome, assister à la fête de la divinité. Le système était contrôlé par les temples et les recettes dues à la vente des momies comme des bronzes représentaient certainement des ressources non négligeables pour les clergés[10], même si ce n'était certainement pas la raison première.

Ces quelques exemples ne permettent pas de répondre à toutes les questions concernant les dévots et leur attitude envers les animaux. Le sujet, rarement abordé pour lui-même, mérite d'être davantage étudié. Cela nous permettrait sans doute de mieux appréhender le phénomène des animaux en lien avec la religion.

Bibliographie

Agnant, A. Trad. 1847. Solin. Polyhistor. - Paris, Pancoucke. http://remacle.org/bloodwolf/erudits/solin/deux.htm#XIII, vu le 25/05/2021.

Alliot, M. 1954. Le culte d'Horus à Edfou au temps des Ptolémées. – Le Caire, Institut d'archéologie orientale (Bibliothèque d'Étude 20, 2) : 667.

Bernand, É. 1975. Recueil des inscriptions grecques du Fayoum I, La « Méris » d'Hérakleidès. – Leyde, Brill.

Bernand, É. 1988. Pèlerins dans l'Égypte grecque et romaine. In: Mactoux, M. & E. Genys. Eds. Mélanges Pierre Lévêque I, religion. – Besançon, Université de Besançon (Annales littéraires de l'université de Besançon) : 49-63.

Bergmann, E. von. 1879. Hieroglyphische Inschriften. – Vienne, Von Faesy & Frick.

Bouffartigue, J. Trad. 2012. Plutarque, L'intelligence des animaux, 23. – Paris, Les Belles Lettres.

Breccia, E. 1926. Teadelfia e il tempio di Pneferôs. – Bergame, Officine dell'Istituto Italiano d'Arti Grafiche (Monuments de l'Égypte gréco-romaine I).

Cenival, de, F. 1969. Les associations dans les temples égyptiens d'après les données fournies par les papyrus démotiques, Religions en Égypte hellénistique et romaine, Colloque de Strasbourg, 16-18 mai 1967. – Paris, Presses universitaires de France : 5-19.

Cenival, de, F. 1972. Les associations religieuses en Égypte d'après les documents démotiques. – Le Caire, Institut d'archéologie orientale (Bibliothèque d'Étude 46).

Cenival, de, F. 1977. Deux papyrus inédits de Lille avec une révision du P. dém. Lille 31. – Enchoria 7 : 1-49.

Charron, A. 1998. Enseigne d'un interprète des songes crétois. In : Empereur, J.-Y. Ed. Catalogue La gloire d'Alexandrie, Paris, musée du Petit Palais, 7 mai – 26 juillet 1998. – Paris, Paris musées : 195.

Charron, A. 1998. Les Ptolémées et les animaux sacrés. In : Empereur, J.-Y. Ed. Catalogue La gloire d'Alexandrie, Paris, musée du Petit Palais, 7 mai-26 juillet 1998. – Paris, Paris musées : 192-198.

Charron, A. & D. Farout. 2008. Le premier sarcophage en pierre du taureau Apis. – Égypte Afrique & Orient 48 : 39-50.

Charron, A. 2012. Les bronzes « reliquaires » d'animaux à la Basse Époque. In : Zivie-Coche Ch. & I. Guermeur. Eds. « Parcourir l'éternité ». Hommages à Jean Yoyotte I. – Turnhout, Brepols (Bibliothèque de l'École des Hautes Études, sciences religieuses 156): 281-304.

Charron, A. & C. Barbotin. Eds. 2017. Catalogue Savoir et pouvoir à l'époque de Ramsès II, Khâemouaset, le prince archéologue, Arles, Musée départemental Arles antique, 8 octobre 2016 – 22 janvier 2017. – Gand/Arles, Snoeck/Musée départemental Arles antique.

Charvet, P. Trad. 1997. Strabon, XVII, 1. Géographie. In : Yoyotte, J., P. Charvet & S. Gompertz, 1997. Strabon. Le voyage en Égypte. – Paris, Nil.

Chauveau, M. 1998. Cléopâtre au-delà du mythe. – Paris, Liana Levi.

Crawford, D.J. 1971. Kerkeosiris. An Egyptian Village in the Ptolemaic Period. – Cambridge, Cambridge University Press.

De Meulenaere, H. & P. MacKay. 1976. Mendes II. – Warminster, Aris & Phillips.

Devauchelle, D. 2011. La XXVIe dynastie au Sérapéum de Memphis. In : Devauchelle D. Ed. La XXVIe dynastie, continuités et ruptures, promenade saïte avec Jean Yoyotte, Université Charles-de-Gaulle, Lille, Actes du colloque international, 26-27 novembre 2004. – Paris, Cybèle.

9 Stèle Louvre 4113.

10 « *... en période de lourde pression fiscale, les prêtres, si infimes qu'aient pu être leurs fonctions et les revenus y afférant, recevaient une petite portion du bénéfice global. Ils avaient donc tout intérêt à remuer les sistres et à nourrir les chats sacrés, à les enterrer en grande pompe* » (Cenival, 1977 : 30).

Devauchelle, D. 2017. Les enterrements d'Apis au temps des Nectanébos. In : Jurman, C., B. Bader & D. A. Aston. Eds. A True Scribe of Abydos. Essays on First Millenium Egypt in Honour of Anthony Leahy. – Louvain, Peeters (Orientalia Lovaniensa Analecta 265) : 95-117.

Devauchelle, D. 2016. Les stèles du Nouvel Empire au Sérapéum de Memphis. In : Charron A. & C. Barbotin. Eds. Savoir et pouvoir à l'époque de Ramsès II, Khâemouaset, le prince archéologue, Arles, Musée départemental Arles antique, 8 octobre 2016- 22 janvier 2017. – Gand/Arles, Snoeck/Musée départemental Arles antique : 138-139.

Dils, P. 1995. Les ṯ3j (n3) ntr.w ou θεαγοί. Fonction religieuse et place dans la vie civile. -Bulletin de l'Institut Français d'Archéologie Orientale 95 : 153-171.

Dunand, F. 2018. Corps vivants des dieux. – Paris, Picard.

DuQuesne, T. 2009. The Salakhana Trove. Votive Stelae and other Objects from Asyut. – London, Darengo Publications (Oxfordshire Communications in Egyptology 7).

Ernout, A. Trad. 1952. Pline l'Ancien. Histoire naturelle. – Paris, Les Belles Lettres.

Fairman, H.W. 1934. The Hieroglyphic Inscriptions. In: Mond R. & O.H. Myers. Eds. The Bucheum II. The Inscriptions. – London, Humphrey Milford (Memoir of the Egypt Exploration Found 41).

Griffith, F.L. 1937. Catalogue of the Demotic Graffiti of the Dodecaschoenus. I. Text. – Oxford, Oxford University Press.

Hani, J. 1976. La religion égyptienne dans la pensée de Plutarque. – Paris, Les Belles Lettres.

Hohlwein, N. 1940. Déplacements et tourisme dans l'Égypte romaine. – Chronique d'Égypte 29 : 153-278.

Jansen-Winkeln, K. 2001. Die Biographie eines Priesters aus Heliopolis. – Studien zur Altägyptischen Kultur 29 : 97-110.

Jelínková-Reymond, E. 1956. Les inscriptions de la statue guérisseuse de Djed-Ḥer-le-Sauveur. – Le Caire, Institut d'archéologie orientale (Bibliothèque d'étude 23).

Kessler, D. 1985. Tierkult, In : Helck, W. & E. Otto. Eds. Lexikon der Ägyptologie VI. Stele – Zypresse. – Wiesbaden, Harrassowitz : 571-587.

Kessler, D. 1989. Die heiligen Tiere und der König I, Beiträge zu Organisation, Kult und Theologie der spätzeitlichen Tierfriedhöfe. – Wiesbaden, Harrassowitz (Ägypten und Altes Testament 16).

Kessler D. & A. Nur el-Din, 2015. Tuna al-Gebel. In : Ikram S. Ed. Divine Creatures. Animal Mummies in Ancient Egypt. – Cairo, The American University in Cairo Press : 120-163.

Klotz, D. 2009. The Statue of the *dioikêtês* Harchebi/ Archibios. Nelson-Atkins Museum of Art 47-12. – Bulletin de l'Institut Français d'Archéologie Orientale 109 : 281-310.

Kruchten, J.-M. 1997. Profane et sacré dans le temple égyptien. – Bulletin de la Société d'Egyptologie, Genève 21 : 23-37.

Legrand, Ph.-E. Trad. 1982. Hérodote. Histoires II. – Paris, Les Belles lettres.

Malaise, M. 1987. Pèlerinages et pèlerins dans l'Égypte ancienne. In Chélini J. & H. Branthomme. Eds. Histoire des pèlerinages non chrétiens. – Paris, Hachette.

Malek, J. & J.D. Ray, 2017. The Graffiti from the Memphite Serapeum. – Oxford, Perdita Press.

Maystre, C. 1992. Les grands prêtres de Ptah de Memphis. – Fribourg/Göttingen, Universitätsverlag/ Vandenkoeck & Ruprecht (Orbis Biblicus et Orientalis 113).

Meeks, D. 1986. Zoomorphie et image des dieux dans l'Égypte ancienne. – Le temps de la réflexion 7 : 171-191.

Meeks D. & Ch. Favard-Meeks. 1993. La vie quotidienne des dieux égyptiens. – Paris, Hachette.

Meeks, D. 2018. Les Égyptiens et leurs mythes. Appréhender un polythéisme. – Paris, Hazan (Louvre Éditions).

Mond R. & O.H. Myers 1934. The Bucheum I. The History and Archaeology of the Site. – London, Humphrey Milford.

Morenz, S. 1963. Ein neues Dokument der Tierbestatung. – Zeitschrift für Ägyptische Sprache und Altertumskunde 88 : 42-47.

Muszynski, M. 1977. Les « associations religieuses » en Égypte d'après les sources hiéroglyphiques, démotiques et grecques. – Orientalia Lovaniensia Periodica 8 : 145-174.

Northampton, W.G.S.S.C., W. Spiegelberg & P.E. Newberry. 1908. Report on Some Excavations in the Theban Necropolis during the Winter of 1898-9. – London, A. Constable and Co.

Pasquali, S. 2019. Sur la manière dont un Égyptien a raconté l'installation de l'animal sacré de Mendès. In: Porcier, S., S. Ikram & S. Pasquali. Eds. Creatures of Earth, Water and Sky. Essays on Animals in Ancient Egypt and Nubia. – Leyde, Sidestone Press : 273-282.

Perdu, O. 2006. Psammétique Séneb – un vizir d'Héliopolis avant la conquête d'Alexandre. – Égypte, Afrique & Orient 42 : 41-52.

Porcier, S. 2009. Mnévis. Recherches sur le culte du taureau sacré d'Héliopolis. – Inédite PhD thèse de l'université Montpellier III – Paul Valéry.

Porcier S. 2016. Stèle déposée en l'honneur d'un taureau Apis et du prêtre Pyiay. In: Charron A. & C. Barbotin. Eds. Savoir et pouvoir à l'époque de Ramsès II, Khâemouaset, le prince archéologue, Arles, Musée départemental Arles antique, 8 octobre 2016-22 janvier 2017. – Gand/Arles, Snoeck/Musée départemental Arles antique : 150-151.

Preisigke F. & W. Spiegelberg. 1914. Die Prinz-Joachim-Ostraka. Grieschiche und demotische Beisetzungsurjunden für Ibis- und Falkenmumien aus Ombos. – Strasbourg, Karl J. Trübner.

Preys, R. 2017. Le cas "Philométor" dans les temples égyptiens. – Bulletin de l'Institut Français d'Archéologie Orientale 117 : 329-356.

Quaegebeur, J. 1975. Teëphibis, dieu oraculaire ?. – Enchoria 5 : 19-24.

Quaegebeur, J. 1984. La désignation « *porteur(s) des dieux* » et le culte des dieux-crocodiles dans les textes des époques tardives, Mélanges Adolphe Gutbub. – Montpellier, Université de Montpellier (Publication de la recherche) : 161-176.

Ray, J.D. 1976. The Archive of Ḥor. – London, Egypt Exploration Society (Egypt Exploration Society Memoir 2).

Ray, J.D. 1978. The Non-Literary Material from North Saqqâra: A Short Progress Report. Internationales Demotisten – Colloquium, Berlin, 1977. – Enchoria 8 : 29- 30.

Ray, J.D. 2011. Texts from the Baboon and Falcon Galleries. Demotic, Hieroglyphic and Greek Inscriptions from the Sacred Animal Necropolis, North Saqqara. – London, Egypt Exploration Society (Texts from Excavations 15).

Rutherford, I. 1998. Island of the Extremity: Space, Language and Power in the Pilgrimage Traditions of Philae. In: Frankfurter, D. Ed. Pilgrimage and Holy Space in Late Antique Egypt. – Leiden/Boston/Cologne, Brill (Religions in the Graeco-roman world 134): 229-256.

Smith, H.S., C.A.R. Andrews & S. Davies 2011. The Sacred Animal Necropolis at North Saqqara. The Mother of Apis Inscriptions II. Commentaries and Plates. – London, Egypt Exploration Society (Texts from Excavations 14).

Spiegelberg, W. 1928. Neue Urkunden zum ägyptischen Tierkultus. – Munich, R. Oldenbourg.

Te Velde, H. 1980. A Few Remarks upon the Religious Significance of Animals in Ancient Egyt. – Numen 27, 1: 76-82.

Thiers, C. 2006. Égyptiens et Grecs au service des cultes indigènes. Un aspect de l'évergétisme en Égypte lagide. In : Molin M. Ed. Les régulations sociales dans l'Antiquité. – Rennes, Presses universitaires de Rennes : 275-301.

Traunecker, C. 1979. Manifestations de piété personnelle à Karnak. – Bulletin de la société française d'égyptologie 85 : 22-31.

Valbelle, D. 1992. Les métamorphoses d'une hypostase divine en Égypte. – Revue de l'histoire des religions 209, 1 : 3-21.

Vercoutter, J. 1962. Textes biographiques du Sérapéum de Memphis. – Paris, Honoré Champion.

Vernière, Y. Trad. 1993. Diodore de Sicile. – Paris, Les Belles Lettres (Bibliothèque historique, I).

Volokhine, Y. 1998. Les déplacements pieux en Égypte pharaonique : sites et pratiques cultuelles. In : Frankfurter, D. Ed. Pilgrimage and Holy Space in Late Antique Egypt. – Leiden/Boston/Cologne, Brill (Religions in the Graeco-roman world 134): 51-97.

Wilcken, U. 1927. Urkunden der Ptolemäerzeit. – Berlin/ Leipzig, W. de Gruyter.

Yoyotte, J. 1960. Les pèlerinages dans l'Égypte ancienne. – Paris, Seuil (Les pèlerinages, Sources orientales 3): 17-74.

Tuberculosis at Tell-el Amarna

A Theoretical Exercise in the Economic and Social Effects of Chronic, Terminal Disease in Ancient Egypt

Gretchen R. Dabbs

Introduction

Tuberculosis is a leading cause of death world-wide. In 2018, an estimated 10 million individuals contracted tuberculosis and 1.5 million individuals died from the disease. It is, to date, the most deadly infectious disease known to man (World Health Organization 2020).[1] In 1996 Murry and Lopez predicted tuberculosis would become the fourth leading cause of death world-wide by 2020. However, there is perhaps a little good news. In 2016 tuberculosis was ranked tenth on the list of causes of death, down from sixth in 2000 (World Health Organization 2018). With the emergence of COVID-19, it is possible tuberculosis will fall off the list entirely.

Tuberculosis (TB) is a bacterial infection caused by several species of genus *Mycobacterium.* Symptoms of active infection include lethargy, chronic cough, cough with bloody sputum, difficulty breathing, weakness, loss of appetite, wastage of the physical body, fever, chills, irritability, night sweats, chest pains, female amenorrhea, and male impotence. Often tuberculosis is called a disease of contrasts, as the individual will suffer alternating phases of certain symptoms, such as fever with hot flashes alternating with pallor versus periods of euphoria and increased appetite during general malaise and decreased appetite (Roberts & Buikstra, 2003: 20).

Not everyone exposed to tuberculosis will develop the full blown disease or even, necessarily, symptoms. In modern clinical situations, fewer than 10% of individuals infected with tuberculosis bacilli will develop symptoms (O'Reilly & Daborn, 1995: 10) and a multitude of biological, economic, and environmental conditions can affect the probability of infection and/or development of the disease. These things include, but are certainly not limited to, the prevalence of open cases of pulmonary tuberculosis in a community, density of the bacteria in expelled sputum, density of bacteria in the air around the individual, number of people present, duration of contact with the infected individual, dietary resources and diversity of the individual, age at exposure, portal into the body (respiratory or gastrointestinal), level of immune response, social conditions of life, local cultural factors (example: food taboos or alcohol consumption/prohibition), size and nature of inoculum, environment, and attempts at treatment (summarized in Roberts & Buikstra, 2003: 44-45; Zakrzewski *et al.*, 2016: 159, 161-162, 167-168). Without treatment 25% of patients with the active disease die within two years, 50% will die

1 This paper was originally written in 2018 and revised in mid-2020. This footnote serves as an acknowledgement that the current COVID-19 pandemic may, ultimately, prove the preceding sentence false.

In: Ikram, S., J. Kaiser & S. Porcier. Eds. 2021. *The Ancient Egyptians and the Natural World. Flora, Fauna, and Science.* – Leiden, Sidestone Press: 83-98.

within five years, and the remaining individuals undergo spontaneous remission of the disease (Enarson & Rouillan, 1998: 40, 46), although underlying complications related to the disease can ultimately shorten lifespans. Of the individuals who develop the disease, bony changes are rare as well, particularly in the pre-antibiotic era (Steyn *et al.*, 2013), only 3-5% of untreated individuals developing bony response (Roberts, 2015: S118). Skeletal tuberculosis is a prime example of the Osteological Paradox, where what you see in the skeletal assemblage is not directly reflective of the prevalence of the disease within the living population (Wood *et al.*, 1992).

Skeletal tuberculosis lesions tend to develop in areas of high oxygen tension, as the *Mycobacterium* thrives in such an environment (Roberts & Buikstra, 2003: 91). The most common area of skeletal lesion development is in the vertebral column, where it differentially affects the anterior body and the infection develops between the bodies and the anterior longitudinal ligament. Vertebral involvement is present in approximately 40% of skeletal tuberculosis cases (Aufderheide & Rodríguez-Martín, 1998: 121) and these lesions are most commonly observed in the thoracic and lumbar spine, with cervical and sacral involvement being very rare (Resnick & Niwayama, 1995: 2463). Once the vertebral body is destroyed due to expansion of infectious bolus, the vertebral column can collapse, resulting in kyphosis of the spine, or Pott's Disease (Resnick & Niwayama, 1995: 2463). An estimated 50% of individuals with Pott's Disease die within five years, which may be due to complications of paralysis and/or respiratory dysfunction resulting from changed thoracic shape (Aufderheide & Rodríguez-Martín, 1998: 123). Tuberculosis can also manifest in the appendicular joints, and is commonly found in the hip (20% of skeletal cases), knee (16% of skeletal cases), and sacroiliac joint (2% of cases) (Aufderheide & Rodríguez-Martín, 1998: 139). Tuberculosis with non-vertebral joint involvement is commonly associated with disease progression during childhood (Aufderheide & Rodríguez-Martín, 1998: 139).

Tuberculosis was recognised in mummified tissues from Egypt early in Egyptology and there is now soft tissue, skeletal, and molecular evidence that demonstrates tuberculosis was present in Egypt from at least as early as the pre-Dynastic period and it continued to affect the peoples of Egypt through its entire history into the modern era. Elliott-Smith and Ruffer first described tuberculosis in human remains based on the macroscopic examination of the mummy of the 21st Dynasty priest Nesperhan (1910). Additional sources of information concerning the presence of tuberculosis in ancient Egypt could be the written record, although no currently known documents or medical papyri describe the symptoms or treatments of any similar condition (Cave, 1939: 144; S. Schøidt, personal communication) and artistic representations of the disease, of which there are many, ranging from pre-Dynastic statuettes with kyphotic spines to two-dimensional representations in the New Kingdom Theban tomb of Ipuy. It should be acknowledged that while these representations depict physical conditions known to affect individuals with tuberculosis infections, there are other conditions with similar manifestations.

Molecular analysis of human remains from multiple sites in Egypt through time has demonstrated that tuberculosis was present in Egypt as early as the pre-Dynastic period and it was present in individuals with no skeletal markers for tuberculosis, which Zink and colleagues used to suggest high tuberculosis infection rates (Zink *et al.*, 2003a: 247). Dabernat and Crubezy (2010: 719) also argued for tuberculosis as an endemic condition as early as the pre-Dynastic Period, based on skeletal evidence of tuberculosis in subadult skeletons from Adaima, Upper Egypt.

Amarna: The Site

Tell el-Amarna is the archaeological site in Middle Egypt that preserves the remains of the short-lived capital city (Akhetaten) of Akhenaten and Nefertiti during the Amarna Period (ca. 1353-1332 BC) (Figure 1). Akhetaten was built, occupied, and abandoned in approximately 15-20 years during the apogee of the New Kingdom, one of Egypt's most prosperous times. The city itself sits nestled along the Nile River separated from a crescent-shaped arc of desert cliffs by an expanse of seemingly lightly used desert landscape. These cliffs delineate the boundaries of Akhetaten, as described on the multiple boundary stelae carved into them. They also house the rock cut tombs of the high officials, although there is scant evidence most of these tombs were ever used (Kemp, 2012: 251). Excavations have been ongoing periodically at Amarna for over a century and have yielded an impressive array of information concerning life in the ancient city. However, for much of this time, the location of the cemeteries remained a mystery and therefore our understanding of the lives of the non-elites who populated the city was not as complete as it could otherwise be. In the early part of this millennium, Helen Fenwick of the University of Hull surveyed the desert areas surrounding Amarna as part of her Desert Hinterlands Survey project, identifying four potential cemeteries during the process (Fenwick, 2003: 11). Anna Stevens has suggested these four cemeteries represent 'public' cemeteries and are different in nature to previously identified cemeteries at the Stone and Workmen's Villages, which she calls 'community' cemeteries (Stevens, 2017). To date, all of these public cemeteries have undergone some amount of excavation, but the skeletal analysis lags behind. The South Tombs Cemetery, the largest of the identified non-elite cemeteries, was excavated from 2006-2013 and the skeletal material

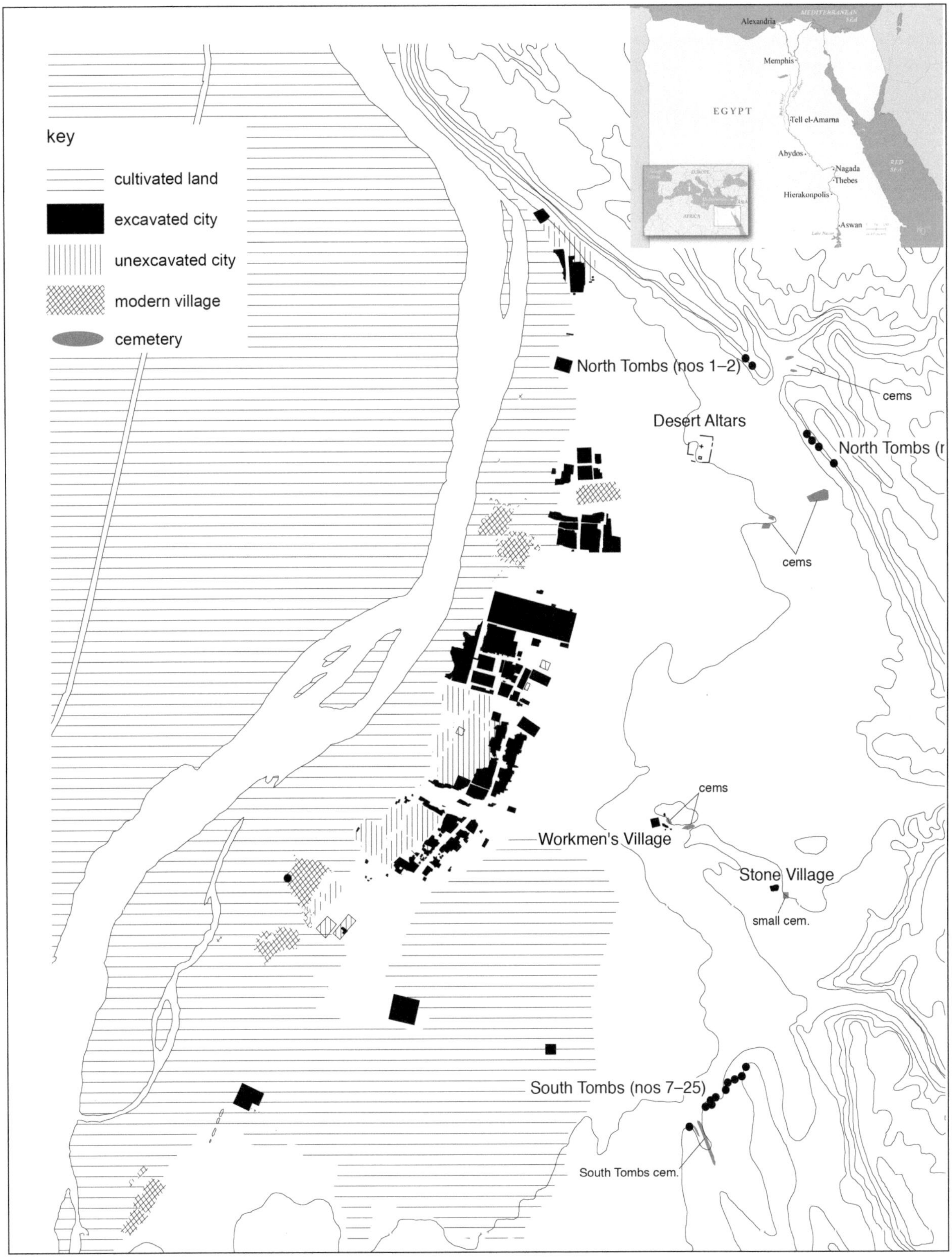

Figure 1. Map showing Tell el-Amarna within Egypt (inset) and the city as currently known. Courtesy of the Amarna Project.

analysed from 2006-2018. This is the only cemetery for which both excavations of specific sample areas and the bioarchaeological analysis of the skeletal remains recovered in those areas are complete and it is the focus of the current discussion.

The South Tombs Cemetery (STC) is a large cemetery situated within a dry wadi bisecting the cliff face into which the 19 tombs for officials known as the South Tombs were cut. There are an estimated 4,000-5,000 individuals buried within the STC (Stevens, 2017: 10). It reflects a typical bioarchaeological assemblage, with representation of individuals in all phases of life from fetuses to old age (Dabbs *et al.*, 2015: 34). Archaeologically, the STC is a difficult site, as much of it was robbed, likely beginning in antiquity, resulting in information loss, destruction of the site, and disassociation of the archaeological and skeletal remains from their original position (Kemp *et al.*, 2013: 67). In total, the skeletal material excavated from the STC represents a minimum of 440 individuals, although only 429 of those individuals could be assigned individual numbers by virtue of their completeness and/or direct association with *in situ* remains in an identified grave pit. Variation in burial treatments and grave goods, where present, suggests the individuals in the STC likely represent a broad slice of society save the high elites of Akhetaten and they may have been buried in this cemetery due to either association with one of the high officials who had a rock-cut tomb in the southern cliff face, or possibly because they lived in the southern part of the city (modern name Main City), or both (Stevens, 2017: 17). Skeletal analysis suggests the lives of these individuals were difficult, with heavy workloads, disease, and nutritional deficiency being reflected often (Dabbs *et al.*, 2015: 39-40).

This article presents a differential diagnosis of skeletal lesions observed in one individual. The lesions observed are consistent with tuberculosis and establish the presence of TB within the ancient population. Zakrzewski and colleagues (2016: 162) have suggested Akhetaten may not have been a significant disease reservoir, as it did not have the time depth for infectious diseases to become established. However, given the demonstrable presence of tuberculosis in the population and our growing understanding of the harsh living conditions at Akhetaten, it is becoming more likely that many of the residents of Akhetaten succumbed to disease before skeletal tuberculosis lesions could develop. The presence of a single likely case of TB within the sample should not be seen as a unique, but relatively useless, piece of skeletal trivia. Using cemetery size estimates, population estimates, and data on the propensity for tuberculosis to manifest on the skeleton to estimate potential population-wide tuberculosis infection rates, I will use the evidence of TB within the sample to discuss the broader impacts of such a disease and potential implications of a wide-spread, chronic, terminal disease on the economic and social systems of Akhetaten.

Materials and Methods

Individual 277

Individual 277 was excavated from the South Tombs Cemetery in 2012. The body was interred singly in a grave that was robbed at some point, likely in antiquity. While the majority of the thorax was displaced, most of the remains were recovered from within the confines of the burial pit. The cranium was lost in the postmortem period, but the remainder of the skeleton is nearly complete, including a mandible (Figure 2). The body was originally wrapped in textile and a mat made of plant material, a common burial treatment at the South Tombs Cemetery. The *in situ* arms, right leg, and lower left leg show the individual was buried in the extended, supine position typical of the South Tombs Cemetery, with the body oriented with the head at the southeast end of the grave pit.

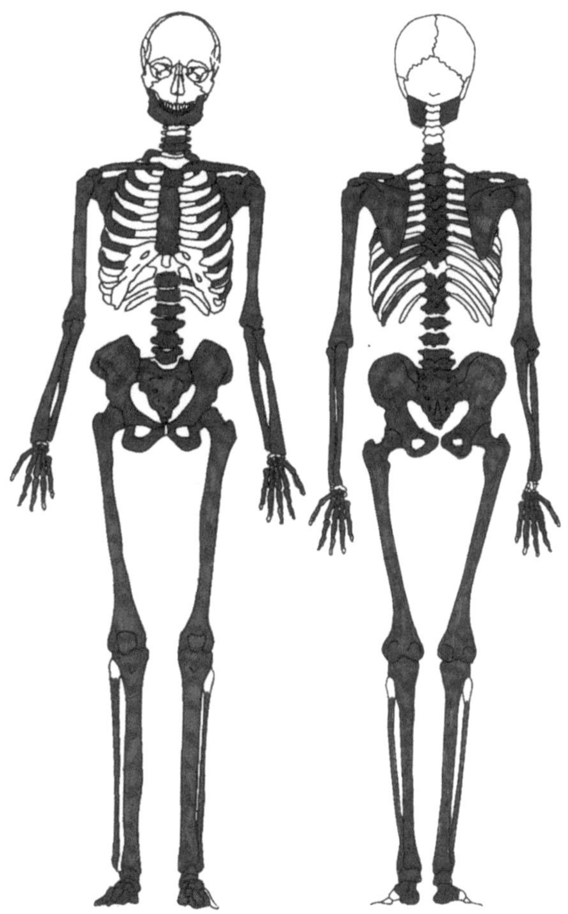

Figure 2. Representation of material present from Ind 277 (in black). Diagram by G. Dabbs.

Individual 277 is an adult probable female, estimated to be 40-44 years old at the time of death based on the auricular surface morphology (Lovejoy *et al.*, 1985). Designation as a probable female is based on the morphology of the os coxae and metric analysis of the postcranial remains using published Amarna-specific metric standards (Dabbs, 2010; 2020). Stature, calculated using the length of the right tibia (without eminences) (Raxter *et al.*, 2008), is consistent with the STC average for females and sits at 152 cm. The remains are well preserved, and with the exception of a few missing elements (see Figure 2), degree of preservation is not considered an inhibitory issue in the skeletal analysis. Overall, the skeleton of this individual exhibits lower levels of biological stress than is typical for someone of her age at Akhetaten, but still shows the characteristics of heavy workloads, trauma, and disease. The mandibular canines each show one linear enamel hypoplasia. The elbows and left wrist exhibit minor lipping around the articular surface margins. The left second rib has a healed fracture near the sternal end. There are lesions of active periosteal new bone formation (radii and fibulae) and hypertrophic osteoarthropathy (tibiae). The twelfth thoracic and fifth lumbar vertebrae have compression fractures on the anterior superior bodies that resulted in a loss of height of the anterior margin, which may reflect compensatory modifications to the disease process described below.

The vertebral column of Individual 277 shows extensive destruction of the vertebral bodies by lytic lesions with the resultant bony modifications necessary to maintain spinal functionality. The fifth thoracic vertebra has a large lytic lesion on the anterior body and the body itself is slightly shortened on the left side (Figure 3). The sixth thoracic has reactive bone growth on the lateral body margins. The seventh, eighth, and ninth thoracic vertebrae are fused together in the posterior aspects including the spinous processes and articular facets (Figure 4). The bodies of these vertebrae are largely destroyed without any evidence of healing (Figure 5). The tenth thoracic was not recovered, but the exposure of trabecular bone along the inferior margin of the ninth thoracic vertebra lamina suggests it was also fused to the others. The eleventh and twelfth thoracic vertebrae are fused together by ossified ligaments on the left side of the bodies (Figure 6). The eleventh thoracic has two lytic lesions on the anterior superior body margin (Figure 7). Given the morphology of the recovered elements, it is likely the vertebral column was kyphotic either above the destroyed 7th thoracic vertebra or between the 10th and 11th thoracic vertebrae, potentially in both areas, representing a substantial alteration of the thorax shape. The first lumbar vertebra has two large lytic lesions on the anterior surface of the body with ossified ligaments on the left inferior body margin (Figures 8, 9). The third lumbar vertebra also has a large lytic lesion resulting in the loss of the left superior

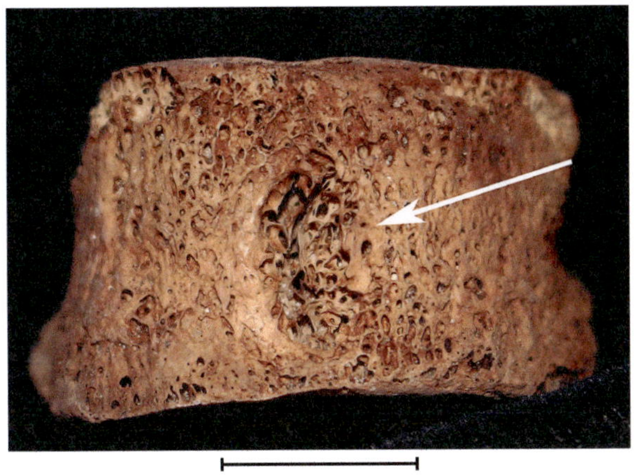

Figure 3. Lytic lesion (white arrow) on anterior surface of fifth thoracic vertebra. Photograph by G. Dabbs.

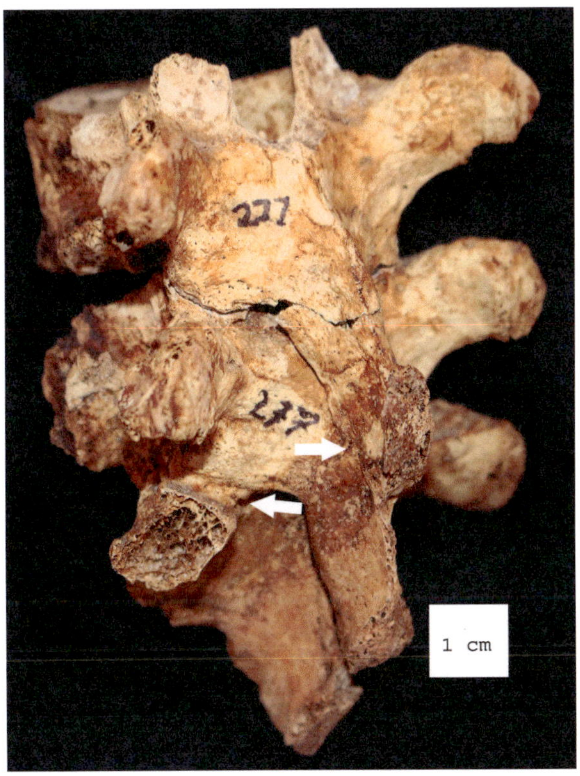

Figure 4. Posterior view of thoracic seven through nine showing fused posterior aspect (white arrows). Photograph by G. Dabbs.

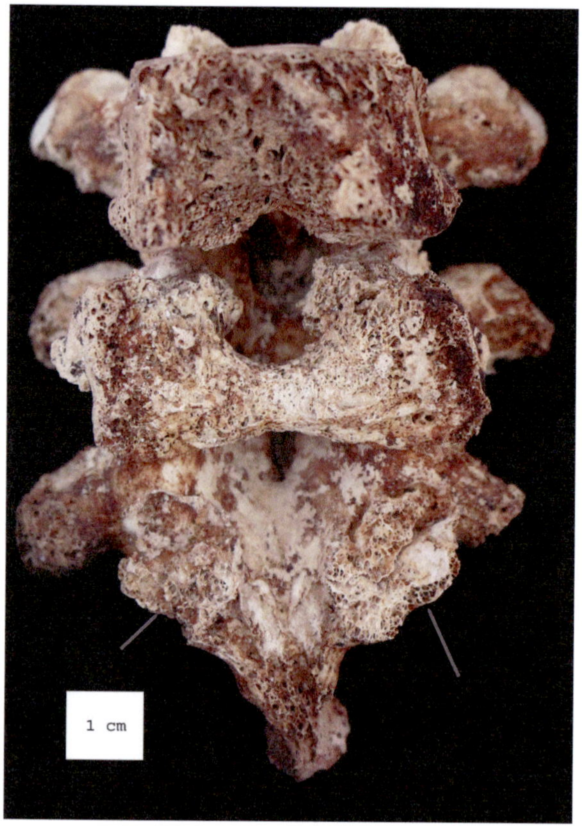

Figure 5. Anterior view of thoracic seven through nine showing loss of bodies, and taphonomic damage along inferior ninth (gray arrows). Photograph by G. Dabbs.

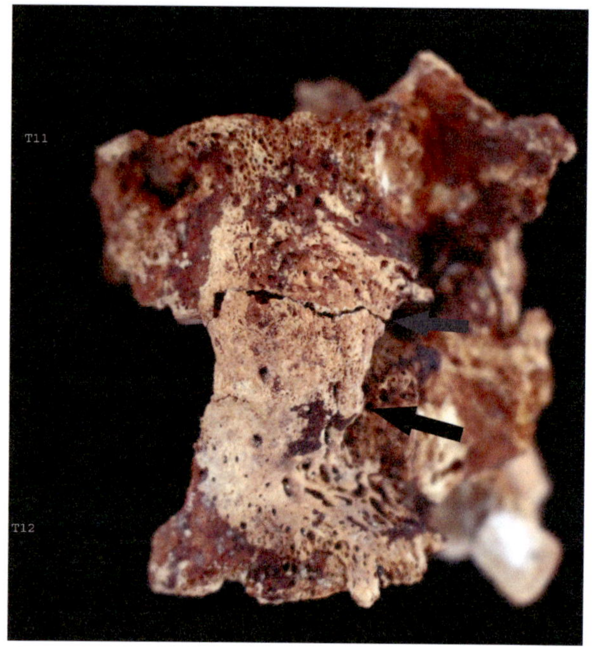

Figure 6. Ligamentous bridging (black arrow) on the lateral surface between thoracic eleven and twelve (gray arrow indicates postmortem breakage of ossified ligament). Photograph by G. Dabbs.

Figure 7. Lytic lesions on the right anterior side of eleventh thoracic vertebra. Photograph by G. Dabbs.

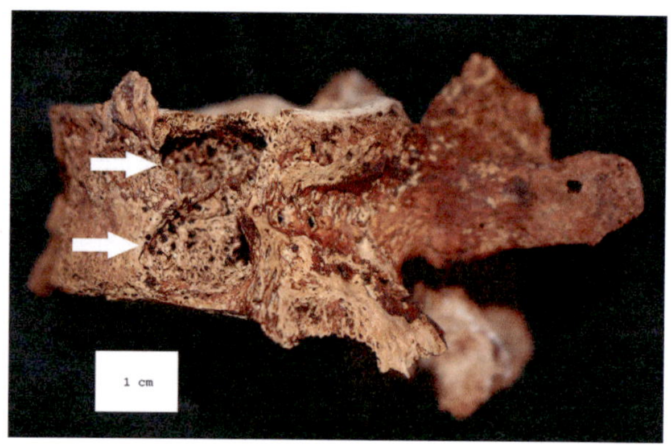

Figure 8. Lytic lesions (white arrows) on the left lateral side of the first lumbar vertebra. Photograph by G. Dabbs.

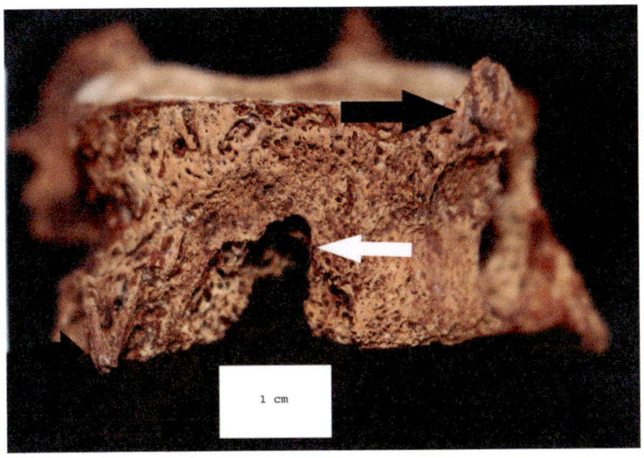

Figure 9. Lytic lesion (white arrow) on the anterior body of the first lumbar vertebra; ossification of ligaments (black arrow). Photograph by G. Dabbs.

Figure 10. Destruction of left anterior body of third lumbar vertebra (white arrow) and ossification of ligaments (black arrow). Photograph by G. Dabbs.

half of the anterior body (Figure 10). There are small osteophytes and ligamentous ossifications surrounding this lesion. The fourth lumbar vertebra has beginning ossification of ligaments on both sides of the body.

Results (Differential Diagnosis)

Individual skeletal lesions can be difficult to interpret, especially when the lesions observed are consistent with more than one disease process. When a single lesion type may manifest from several different disease processes, a systematic, thorough differential diagnosis must be performed (Buikstra, 1976a: 316). Several bacterial, parasitic, and fungal infections exhibit similar lesion morphology, including actinomycosis, brucellosis, coccidioidomycosis, echinococcosis, histiocytosis, histoplasmosis, sarcoidosis, tuberculosis, and suppurative osteomyelitis. Additionally, malignant tumors, osteitis deformans (Paget's Disease), rheumatoid arthritis, ankylosing spondylitis, Scheuermann's disease, and traumatic injury can also cause morphologically similar lesions. Each of these conditions was considered as a possible causative agent for the lesions observed in Individual 277. After careful consideration of the presence of pathognomonic lesions, the distribution of lesions, and other factors as listed individually, tuberculosis is the condition that most closely fits the lesions observed and is therefore the most likely causative agent. However, molecular testing would be required to confirm this differential diagnosis. Table 11 outlines the major diagnostic criteria for each condition, along with the observed characteristics for the lesions on Individual 277.

Additional conditions were eliminated from consideration as a result of epidemiological factors, such as global distribution of the underlying agent. For example, Coccidioidomycosis, which produces lesions others have identified as similar enough to tuberculosis lesions to include in differential diagnoses for other sites (Buikstra, 1976b: 360), is only found in the Western Hemisphere (Hector & Laniado-Laborin, 2005: 15) and is therefore excluded as a potential causative agent based on distribution.

Discussion

How Prevalent was Tuberculosis at Amarna?

If the identification of tuberculosis above is accepted, tuberculosis at Amarna is a classic example of the Osteological Paradox (Wood *et al.*, 1992). There are, broadly speaking, three types of people buried at the STC with respect to TB: 1) Individuals with tuberculosis that have skeletal lesions indicating disease infection; 2) Individuals with tuberculosis without skeletal lesions indicating the disease, and 3) Individuals without tuberculosis, who then also do not have the skeletal lesions representing the disease. The paradox is, of course, that the latter two categories are morphologically indistinguishable from one another. Molecular analysis could potentially separate these individuals, but due to government restrictions, this is not currently likely.

It is perhaps possible to use clinical data on rates of tuberculosis infection, rates of skeletal lesion development, and estimates of cemetery and city size as the basis for a theoretical exercise to discuss the potential for wider distribution of tuberculosis throughout the ancient city of Akhetaten and further how that distribution may have affected the overall population of the city in terms of economics, social structure, and overall well-being.

Based on the number of individuals thus far excavated from the STC (n=429), the single representation of skeletal tuberculosis is unsurprising, given that tuberculosis only

Possible Causative Agent	**Common Diagnostic Criteria**	**Ind 277 manifestation**
	Bacterial Infections	
Actinomycosis	Osseous involvement rare	Osseous involvement
	Neural arch lesions are as common as body lesions	Neural arches are not affected
	Spheroid defects with associated periosteal reaction (Rothschild *et al.*, 2006: 25)	Not observed
Brucellosis	Skeletal involvement rare (~10%)	Skeletal involvement observed
	Presents both lytic and proliferative lesions simultaneously	Lytic lesions only; no proliferative lesions
	Final stages include hypertrophic new bone formation	Hypertrophic bone formation observed on tibiae
	Does not cause vertebral collapse	Vertebral collapse observed
	Usually only involves lumbar vertebrae	Thoracic and lumbar involved
	No periosteal proliferation on long bones	Periosteal lesions present on several long bone diaphyses
Bacterial Osteomyelitis	Involves posterior portions of the vertebrae	No posterior vertebral involvement
	Generally only involves a single vertebra	Multiple vertebrae involved
	Most common in individuals less than 12 years old	Ind 277 estimated to be 40-44 years old
Tuberculosis (Roberts, 2015)	Destructive lesions of the thoracic/ lumber vertebrae, knee, hip, or other joint with little to no bone growth (pathognomonic)	Destructive lesions of both thoracic and lumbar vertebrae present, with no associated bone growth
	New bone formation on the visceral ribs	Not observed
	Calcified pleura or granulomatous lung nodules	Not observed
	Destructive lesions underlying skin lesions of lupus vulgaris	n/a – skull not present for analysis
	Bone formation on long bones	Bone formation observed on radii, tibiae, and fibulae
	Tuberculous dactylitis of the short bones of the hands and feet	Not observed, but limited recovery of short bones of hands and feet
	Bone changes on the endocranium or pelvic bones	Endocranium not available; no changes on the pelvic bones
	Fungal Infections	
Histoplasmosis	Skeletal lesions rare, with vertebral lesions being more uncommon	Vertebral lesions present
	Most commonly affects the skull, hands, feet, and radius	Hands, feet, and radius not affected (cranium not present)
Fungal infections (general)	Involvement of posterior vertebral elements common	No posterior vertebral involvement.
	Parasitic Infection	
Echinococcosis	Osseous involvement is rare (1-2%) (Schneppenheim & Jerosch 2003, 107)	Osseous involvement
	Osseous involvement limited to a single element	Multiple vertebral bodies affected
	Pathological fractures occur at late stages	Vertebral compression fractures present, but on vertebrae without lytic lesions
	Posterior elements of the vertebral column (lamina) and associated ribs are often involved	Anterior aspects only affected; no rib involvement
	Internal Dysfunction	
Histiocytosis	Generally only affects children	Adult individual observed
Malignant Tumor	High mortality of young adults	Ind 277 is 40-44 years old
	Often affects two non-contiguous vertebrae	Multiple contiguous thoracic and lumbar vertebrae affected
	Neural arches and ribs often affected	Neural arches and ribs not affected
Osteitis Deformans (Paget's Disease)	Simultaneous resorption and proliferation of bone	No generalised proliferation of bone, isolated to vertebral responses to changing mechanical stress
	Vertebral column only fifth most commonly affected region of the skeleton	Vertebral bodies only area affected by lytic lesions
	Kyphosis due to vertebral collapse	Kyphosis due to vertebral destruction
	All portions of the vertebrae are affected when vertebral manifestations are present	Only anterior vertebrae affected
Rheumatoid Arthritis	Bilaterally symmetrical	Lesions concentrated on the left side of the body
	Generally affects bones of the hands and feet	Bones of the hands and feet non-pathological
Ankylosing Spondylitis	Bony ankyloses of sacroiliac joint and ligaments	No bony ankyloses of sacroiliac joint
	Fusion of anterior vertebral bodies	Some present, but not systematic and not in the typical candlewax manifestation
Sarcoidosis	Most often affects bones of the hands and, less commonly, the feet	Hands and feet not affected
	Vertebral pathology very rare	Vertebral pathology present
	Multifocal vertebral lesions with paraspinal masses	No masses observed
Scheuermann's Disease	Age of onset is usually 12-18 years (±5 years)	Ind 277 estimated to be 40-44 years old
	Lytic lesions are rectangular	Lytic lesions are round to oval shaped

Possible Causative Agent	Common Diagnostic Criteria	Ind 277 manifestation
	External Causes	
Vertebral fracture	Generally only one vertebra	Multiple vertebrae involved
	Less destruction of vertebral body	Complete destruction of vertebral bodies
	Kyphosis not angular	Kyphosis angular
	Callus may be present	No callus present
Traumatic Arthritis	Affects any region of the spine	Isolated changes to the thoracic and lumbar vertebrae
	Associated with traumatic fracturing, often	Vertebral compression fractures present, but not on vertebrae associated with lytic lesions

Table 1. A summary of the differential diagnosis of Individual 277, diagnostic criteria for individual conditions as noted in Aufderheide & Rodríguez-Martín (1998), unless otherwise specified.

manifests skeletal lesions in 3-5% of untreated people (Roberts, 2013: S118), although slightly more examples might be expected.

Estimation of Akhetaten's population size is fraught with difficulties, including unknown household composition and the assumption of house density in unexcavated areas. Early estimates of Akhetaten's population size vary widely, with Kemp (2012: 272) estimating the population as 20,000-50,000 and Janssen (1983: 288) estimating double that at 50,000-100,000. For the purposes of this paper a conservative estimate of 30,000 individuals will be used. The total of individuals buried in the cemeteries is estimated to be 10,000-13,000 (Stevens, 2017: 10). If the ratio of individuals with evidence of tuberculosis in the STC (1:429; 0.21%) is applied to these population estimates, it then suggests 21-27 skeletal individuals from the cemeteries would exhibit lesions of tuberculosis if the entirety of all known cemeteries were to be excavated (an impossibility) and no unknown burial grounds exist. Extrapolating that to the estimated living population yields and estimate of upwards of 63 individuals exhibiting skeletal lesions. This, however, ignores the point that skeletal manifestation of tuberculosis is but an extremely small percentage of individuals who actually have the disease. Zink and colleagues (2003a: 246) demonstrated 14.0% of individuals from different sites in Upper Egypt with no evidence of skeletal tuberculosis tested molecularly positive for tuberculosis bacilli. These samples spanned from the Early Pre-Dynastic through the Late Period. Although the sample sizes are relatively small, (n=50 representing entirety of ancient Egyptian population), this value provides a starting point for discussing how widespread tuberculosis may have been in ancient Egypt and is preferable to data from clinical literature, because it reflects similar living conditions and life experiences, particularly the absence of modern medical technology. In a different study of Middle and New Kingdom elites from Theban tombs, Zink and colleagues' (2003b: 362) molecularly identified tuberculosis in 17% of sampled individuals with amplifiable aDNA (n=48) that did not exhibit skeletal lesions indicative of the disease. However, since the Amarna STC is distinctly not of elites (Kemp *et al.*, 2013: 65), this exercise will utilise the more conservative 14.0% value from the more widely distributed study (Zink *et al.*, 2003a: 246).

Thus, if we assume one person in the STC demonstrates skeletal lesions and 14.0% of those excavated individuals who do not exhibit lesions actually were infected with tuberculosis (428*0.14=60 individuals), we can assume a 14.2% (61/429) overall infection rate for the STC. If we then apply that rate to the city as a whole, using the estimated Akhetaten population size (30,000) yields an estimate of 4,260 individuals with possible tuberculosis. In modern developing countries, only about 10% of those individuals with the tubercle bacillus develop clinical disease (O'Reilly & Daborn, 1995: 10). It is likely that rates of clinical disease would have been higher in the past, particularly at sites such as Akhetaten, where numerous risk factors for developing the condition intersect, such as dry dusty conditions, which have been demonstrated as conditions favorable to development of infection (Roberts & Buikstra, 2003: 62-64), poor nutrition (Dabbs *et al.*, 2015: 37-38; Kemp *et al.*, 2013: 71-72), and compact, close living quarters with high population density. However, in the vein of conservative estimations, we can then calculate that of the 4,260 individuals with possible tuberculosis infection, perhaps as many as 426 individuals (1.42% of estimated city population) in the ancient city expressed clinical symptoms.

Kemp (2012: 272) has estimated the number of houses within the main part of the city at Amarna to be approximately 3,040, with an average of 10 individuals per household. If these values are correct, this analysis suggests an average of one in every seven households may have been infected with tuberculosis. It is highly unlikely tuberculosis infections would be spread evenly across the entire city. There would, of course, be areas where infection rates are higher, exacerbated by conditions such as close living quarters, occupational respiratory hazards, exposure to infected individuals, poor/poorer nutrition, depressed immune response, sanitation issues, etc. There would also be areas where no tuberculosis infections existed.

Physical manifestations of active tuberculosis include coughing, coughing up blood, difficulty breathing, weakness, lethargy, loss of appetite and weight, chills, night sweats, irritability, fever, amenorrhea, and male impotence (Loddenkemper *et al.*, 2016: 2; Roberts & Buikstra, 2003: 20). Some of these symptoms are so severe as to incapacitate an individual near the end of life. Records from Deir el-Medina, the village of workmen dedicated to rock-tomb cutting in the Valley of the Kings, show that royal tomb workers could and did miss work due to illness without fear of loss of wages (Austin, 2015: 85; Eyre, 1987: 178). It is not clear if this benefit is extended to the general population. However, it can be expected that significant missed work, including complete withdrawal due to incapacitation, would have resulted in a loss of compensation. A modern study of TB patients in India reports the average person missed 83 working days due to tuberculosis (Rajeswari *et al.*, 1999: 873), but this should be considered an absolute minimum, as these patients were seeking treatment and their participation in the study ended when TB was no longer detectible. It is possible the period of missed work in the ancient world, where no treatment is known by modern scholars, would have resulted in more days of lost labour than in modern times. While it may be difficult to assess economic impact in a society that lacked currency, modern analyses demonstrate tuberculosis is a sizeable economic burden. Funds are spent on treatments, whether successful or not. Worker hours are lost, both during illness and after death. Individuals who die from tuberculosis are no longer part of an active economy throughout their naturally projected life, thus reducing consumerism. Further, the deaths of young individuals deplete the quality of the future workforce and can have substantial economic costs (Kirigia *et al.*, 2016: 10). In the aforementioned study of modern individuals with tuberculosis in India, the direct costs of tuberculosis, which involve treatment, medication, and transportation costs for treatment are only one third of the economic cost involved in dealing with TB. The indirect costs, which include loss of wages due to missed work, decreased ability to work due to illness, and disability necessitating occupational changes, including retirement, constitute almost twice the economic burden of treating tuberculosis, and this study focused only on the infected individual, discounting the costs involved in caretaking (Rajeswari *et al.*, 1999: 873). Modern African countries experience an average loss of 1.37% of the gross domestic product annually due to non-medically related (i.e., indirect) costs of tuberculosis (Kirigia *et al.*, 2016: 10), demonstrating significant economic burden attributable to the disease.

The Impact of Chronic Disease Beyond the Infected Individual

In addition to individuals infected with tuberculosis, we must also consider those who cared for the ailing, as tuberculosis is likely to have been a chronic, terminal disease in the pre-antibiotic era, especially when environmental and social conditions were primed for depressed immune systems, as they were at Akhetaten. Modern clinical literature has devoted significant effort to understanding the effects of acting as a familial caregiver for individuals with chronic, often terminal diseases, such as cancer (variety of types) and dementia/Alzheimer's disease. Further, research on familial caregiving for the elderly and individuals with chronic, non-terminal diseases such as diabetes, highlights the stress of terminal conditions and notes the degree of strain on family caregivers is higher for those caring for terminal patients than those caring for elderly or individuals with chronic, but not terminal conditions (Kim & Schulz, 2008: 483). Consistently, across cultures and geography, family caregivers for individuals with chronic, potentially terminal diseases, reflect higher allostatic loads than control samples, with mental and physical health both being affected. Over 200 psychological and physical problems have been reported associated with caregiving (Stenberg *et al.*, 2010: 1015), including physical pain, sleep issues, indigestion, loss of appetite, higher blood pressure, depression, anxiety, and a host of other minor conditions when considered individually, but things that would be additive to increase the caregiver burden, a multidimentional, biophychosocial reaction resulting from an imbalance of care demands relative to a caregiver's personal time, social roles, physical and emotional states, financial resources, and formal care resources given the other multiple roles they fulfill (Given *et al.*, 2001: 3).

Psychological consequences of acting as a caregiver most commonly include increases in anxiety and depression. In a United States based study, 57% of female caregivers of spouses (wives) and 29% of male caregivers of spouses (husbands) experienced clinical depression, versus the 8-9% observed in control samples (Haley *et al.*, 2000: 13). Other works support this finding (*cf.* Clark, 2002; Covinsky *et al.*, 2003; Rabins *et al.*, 1990; Stommel *et al.*, 1990). Depression is so common among caregivers, "caregiver depression" has been defined specifically as a mood disturbance resulting from the stress of providing care, which may be manifest as feelings of loneliness, isolation, fearfulness, and easily being bothered (Fortinsky *et al.*, 2002: 157; Harris *et al.*, 2001: 226). Further, clinical symptoms of depression are maintained after the death of the ill individual, often for several years (Bodnar & Kiecolt-Glaser, 1994: 378; Haley, 2003: 25).

Familial caregivers typically report lower levels of physical health, including self-rated health (Haley *et al.*, 2000: 15). Physical symptoms often reported/observed include conditions such as elevated blood pressure, altered lipid profiles, impaired immune functioning, and *greater vulnerability to infectious illness* (Haley & Bailey, 1999: 326;

emphasis added). Further, caretakers report more pain (headaches, back pain, leg pain, muscle tension), sleep issues (disturbed sleep, poor sleep quality, restless sleep, lack of sleep), fatigue (loss of energy, physical exhaustion), and other symptoms than non-caregiver controls (Stenberg *et al.*, 2010: 1015). The totality of the reactions to these stressors in caregivers has been demonstrated to increase caregiver mortality (Schulz & Beach, 1999: 2218).

Family caregivers of chronically, terminally ill patients experience high levels of chronic stress, which has been shown to increase the risk of infection (Haley & Bailey, 1999: 326; Kiecolt-Glaser *et al.*, 1991: 358), slow wound healing (Kiecolt-Glaser *et al.*, 1995: 1195-1196), and speed chromosome aging via telomere shortening (Damjanovic *et al.*, 2007: 4249). Clearly, familial caregiving is an activity directly detrimental to overall health of the caregiver. From an economic standpoint, familial caregivers often dedicate large portions of their daily waking hours to caring for the ill, with reports of an average of 10 hours per week for cancer patients in the early palliative stage (Haley, 2003: 26) to more than 40 hours per week (Bevans & Sternberg, 2012: 399) reported for other conditions.

If we then consider that the infected individual and their caretaker are both no longer contributing to the household economy at previous levels, the severity of a tuberculosis infection on the household becomes apparent. The infected individual no longer positively contributes to the household economy and is an economic non-contributor, likely s/he exerts a substantial negative burden due to demands of care and expenditure for treatment, but there is no solid evidence for treatment in the Egyptian literature. Thus, it is discounted for this discussion. The caregiver's ability to contribute will vary depending on patient needs, but near the terminal phase, contribution to the household economy is likely to be very limited, if existent at all. If a quarter of the caregiver's time is used caring for the infected individual, the net decline in household economy is 12.5%. If all of the caregiver's time is utilised, the net decline is 20%, assuming a household size of ten individuals (Kemp, 2012: 272) and all other household members are contributing at a stable rate.

If we take a step back from the details and examine the larger picture of tuberculosis and its effects at Akhetaten, this work has shown that even though skeletal evidence of tuberculosis is not common among the excavated sample from the STC, there is significant potential for the disease to have had major social and economic consequences within the ancient city. If the estimates are accurate and the ancient city fits into the assumptions made herein, upwards of one in seven houses potentially had an individual with active tuberculosis infection. Further, each of these houses may have had a familial caregiver to that individual, resulting in up to 20% of the household being economically negative, and perhaps substantially so. A quick calculation suggests that if 14.2% of the households lose 20% of their economic output, the overall economic system declines about 2.8%, a dramatic reduction for populations living on the edge, as the bioarchaeological analysis of Amarna suggests they were. To put this in perspective, modern Indian tuberculosis patients report losing approximately 26% of their annual family-level income to the indirect (non-medically related) costs of tuberculosis infection (Rajeswari *et al.*, 1999: 873). Additionally, at least in the case of familial caregivers, the stress of caregiving may have reduced immune function, making it more likely s/he would contract tuberculosis. Contact with tuberculotic patients, along with individual immune response, is a major contributing factor to contracting tuberculosis (Roberts & Buikstra, 2003). This perpetuation of the cycle of infection would likely have affected some households more so than others, resulting in potentially expanding economic inequality within the Akhetaten population.

Alternate Explanations

Of course, all theoretical exercises require alternate explanations, and in this case there are many. Individual 277 may have been a recent transplant to the city and died shortly after arrival, or on her way to the city. Thus, she would not represent the population well at all. The disease identified as tuberculosis may not, in fact, be tuberculosis. Only molecular analysis can identify with certainty, and as discussed above, this is not a current possibility at Amarna. Tuberculotic individuals could be buried in another, as of yet unexcavated, cemetery at Amarna or within an excavated cemetery, but not included in the current excavated sample. Tuberculotic individuals could be buried within a different area of the STC, or transferred to burial areas external to Akhetaten entirely.

Assumptions and Limitations

In the foregoing economic assessment of the potential impact of tuberculosis on the ancient city of Akhetaten, certain assumptions had to be made to progress in the exercise. Where possible, these assumptions are based in published literature from/about ancient Egypt. However, certain aspects of life in ancient Egyptian cities are not generally discoverable, due to the rampant illiteracy, which has left a dearth of written documentation about the state of non-elite households. Insofar as possible, where assumptions had to be made without evidence, I tried to be as conservative as possible to avoid unintentionally and artificially inflating the projected rates of tuberculosis and its effect on the ancient city.

One glaring assumption made is that all members of the household were equal economic contributors. This assumption ignores the presence of non-working children, elderly, and potentially disabled individuals.

There is evidence (written, artistic, and/or archaeological) from ancient Egypt to support the presence of all of these individuals within a household. However, for the purposes of this exercise, it is not possible to parse out the economic contribution of each individual member of a household and therefore refine the analysis further.

In this same vein, the assumption is made that caregiving is household based and second that the ill individual and the caregiver are both economic contributors to the household. Certainly, children contract and die from tuberculosis. If children are not direct economic contributors, the consequences on family economic status would be less than projected. Also, caregivers, although generally likely to be adults, may not have been direct contributors to the household economy. Elderly individuals or otherwise disabled individuals could serve as caregivers and in doing so, their contribution to care does not cost the household economy anything, as they were potentially not contributing toward it. Further, if community caregiving was practiced, the individual household level impacts would be reduced.

This exercise also assumes a one to one caregiver to patient relationship, assuming each household only had one ill individual. Given that exposure to ill individuals is one of the primary risk factors for contracting tuberculosis, it is likely tuberculosis clustered in some houses and/or areas of the city. However, other than acknowledging that this assumption likely does not apply to all cases, it is impossible to derive any more specific ratio on which to base the calculations.

Finally, building on Wood and colleagues' (1992) idea of hidden heterogeneity of risk and differences in individual frailty, there are very likely differences in the lengths of time ill individuals require assistance before death. There is also variation to be seen in the degree of care required, with some individuals requiring complete care for long periods of time, while other individuals would require little care and only for short periods of time. In this at least, the use of modern clinical data, while not perfect, does provide some foundation on which to build this analysis, focusing on average values for the number of workdays missed and the amount of time caregivers devote to palliative care for ill individuals.

Further research and future finds may provide more information that will allow for the refinement of some of the assumptions made within this exercise. It is my sincere hope that other researchers will take up this flag and directly address some of the broader aspects of health and disease in ancient populations and explicitly consider the broader implications of the biological disease and illness we so often focus on in bioarchaeology.

Conclusion

Zink and colleagues (2003a: 247) postulated that tuberculosis may have been widespread in ancient Egypt, with systemic spread of tuberculosis occurring just prior to death. Such a wave of disease would not be reflected as physical skeletal lesions, as the body would not have had time to respond to the infection before death, but instead can be hypothesised once the presence of tuberculosis is established at a site, either macroscopically or molecularly. Conditions at Akhetaten may have been primed for rapid spread of the disease and quick death. The environment at ancient Akhetaten was likely dry, dusty and hot, much as it is today (Meskell, 2002: 23). A West African study has shown that tuberculosis infection rates peak in the dry season (Tomkins, 1993 cited in Roberts & Buikstra, 2003: 63). Further, skeletal analysis has suggested high levels of chronic stress in the form of dietary deficiency, disease, and heavy workloads (Dabbs *et al.*, 2015: 37-38; Kemp *et al.*, 2013: 71-72) as evidenced by short statures, high frequency of linear enamel hypoplasias, degenerative joint disease, and work related traumatic injury which may have led to compromised immune response.

This paper has presented the skeletal evidence for likely tuberculosis at the ancient Egyptian capital city of Akhetaten. While there is only one skeletally manifest case, it is possible to envision a reality wherein a substantial portion of the population was affected, either directly or indirectly, by the disease. It is projected that approximately one in seven households would have housed someone with tuberculosis, and that it is likely these individuals would have required care, causing substantial psychological and physical stress to the familial caregivers, something that would have reduced their immune response and potentially opened them up to a tuberculosis infection themselves. While it is not possible to assess this effect with clinical precision, it is suggested herein that the presence of tuberculosis could have been a major health issue of the ancient city population and its health and economic consequences could have been severe at a household level, which would have then manifested further into the community through Egypt's redistributive economic system.

Funding

The STC excavations were supported by grant funding from the King Fahd Center for Middle Eastern Studies at the University of Arkansas, National Geographic, the British Academy, the Amarna Trust, and the Institute for Bioarchaeology. This particular project received no specific grant funding. No funding agency had any influence on the scope of the work, data collection protocols, or interpretation of results.

Acknowledgements

The author would like to thank Sofie Schøidt for sharing her medical papyrus expertise. Jerry Rose, Anna Stevens, and Christine Sellars offered useful comments on earlier drafts of this manuscript. Anna Stevens, as always, encouraged me to think bigger and broader about the lives of the Akhetaten residents. Anonymous reviewers provided helpful critique. All of them have made this a better piece than it would have otherwise been. I extend my gratitude to the editors for their efforts in putting together this volume. The research conducted at Amarna is done so under permissions granted by the Ministry of Tourism and Antiquities and I thank them for their continued support.

Bibliography

Aufderheide, A.C. & C. Rodríguez-Martín. 1998. The Cambridge Encyclopedia of Human Paleopathology – Cambridge, Cambridge University Press.

Austin, A. 2015. Accounting for Sick Days: A Scalar Approach to Health and Disease at Deir el-Medina. – Journal of Near Eastern Studies 74 (1): 75-85.

Bevans, M. & E.M. Stenberg. 2012. Caregiving Burden, Stress, and Health Effects Among Family Caregivers of Adult Cancer Patients. – Journal of the American Medical Association 307 (4): 398-403.

Bodnar, J.C. & J.K. Kiecolt-Glaser. 1994. Caregiver Depression after Bereavement: Chronic Stress isn't Over when it's Over. – Psychology and Aging 9: 372-380.

Buikstra, J.E. 1976a. Differential Diagnosis: An Epidemiological Model. – Yearbook of Physical Anthropology 20: 316-328.

Buikstra, J.E. 1976b. The Caribou Eskimo: General and Specific Disease. – American Journal of Physical Anthropology 45: 351-368.

Cave, A.J.E. 1939. The Evidence for the Incidence of Tuberculosis in Ancient Egypt. – British Journal of Tuberculosis 33 (3): 142-152.

Clark, P.C. 2002. Effects of Individual and Family Hardiness on Caregiver Depression and Fatigue. – Research in Nursing and Health 25: 37-48.

Covinsky, K.E., R. Newcomer, P. Fox, J. Wood, L. Sands, K. Dane & K. Yaffe. 2003. Patient and Caregiver Characteristics Associated with Depression in Caregivers of Patients with Dementia. – Journal of General Internal Medicine 18: 1006-1014.

Dabbs, G.R. 2010. Sex Determination Using the Scapula in New Kingdom Skeletons from Tell El-Amarna. – Homo: Journal of Comparative Human Biology 61 (6): 413-420.

Dabbs, G.R., J.C. Rose & M. Zabecki. 2015. The Bioarchaeology of Akhetaten: Unexpected Results from a Capital City. In: Ikram, S., J. Kaiser, & R. Walker. Eds. Egyptian Bioarchaeology: Humans, Animals, and the Environment. – Leiden, Sidestone Press: 43-52.

Dabbs, G.R. 2020. Estimating the Sex of Ancient Egyptian Skeletal Remains: Methods from Tell el-Amarna. – Bioarchaeology of the Near East 14: 1-16.

Dabernat, H. & E. Crubezy. 2010. Multiple Bone Tuberculosis in a Child From Predynastic Upper Egypt (3200BC). – International Journal of Osteoarchaeology 20: 719-730.

Damjanovic, A.K., Y. Yang, R. Glaser, J.K. Kiecolt-Glaser, H. Nguyen, B. Laskowski, Y. Zou & D.Q. Beversdorf. 2007. Accelerated Telomere Erosion is Associated with a Declining Immune Function of Caregivers of Alzheimer's Disease Patients. – The Journal of Immunology 179: 4249-4254.

Enarson, D.A. & A. Rouillon. 1998. The Epidemiological Basis of Tuberculosis Control. In: Davies, P.D.O. Ed. Clinical Tuberculosis. – London, Chapman and Hall Medical: 35-52.

Eyre, C. 1987. Work and the Organisation of Work in the New Kingdom. In: M.A. Powell, Ed. Labor in the Ancient Near East. – University Park, PA, Eisenbrauns: 167-221.

Fenwick, H. 2003. Desert Survey. In: Kemp, B.J. Ed. Tell el-Amarna, 2002-03. – Journal of Egyptian Archaeology 89: 10-21.

Fortinsky, R., K. Kercher & C. Burant. 2002. Measurement and Correlates of Family Caregiver Self-Efficacy for Managing Dementia. – Aging and Mental Health 6 (2): 153-160.

Given, B., S. Kozachik, C. Collins, D. Devoss & C.W. Given. 2001. Caregiver Role Strain. In: Mass, M., K. Buckwalter, M. Hardy, T. Tripp-Reimer & M. Titler. Eds. Nursing Care of Older Adults Diagnosis: Outcome and Interventions. – St. Louis, MO, Mosby: 679-695.

Haley, W.E., L.A. LaMonde, B. Han, S. Narramore & R. Schonwetter. 2000. Family Caregiving in Hospice: Effects on Psychological and Health Functioning among Spousal Caregivers of Hospice Patients with Lung Cancer or Dementia. – The Hospice Journal 15 (4): 1-18.

Haley, W.E. 2003. Family Caregivers of Elderly Patients with Cancer: Understanding and Minimizing the Burden of Care. – Journal of Supportive Oncology 1 (supplement 2): 25-29.

Haley, W.E. & S. Bailey. 1999. Research on Family Caregiving in Alzheimer's Disease: Implications for Practice and Policy. In: Vellas, B. & J.L. Fitten. Eds. Research and Practice in Alzheimer's Disease. Vol. 2. – Paris, Carma: 321-332.

Harris, J., H. Godfrey, F. Partridge & R. Knight. 2001. Caregiver Depression following Traumatic Brain Injury: A Consequence of Adverse Effects on Family Members? – Brain Injury 15 (3): 223-238.

Hector, R.A. & R. Laniado-Laborin. 2005. Coccidioidomycosis – A Fungal Disease of the Americas. – PLoS Medicine 2: 15-18.

Janssen, J.J. 1983. El-Amarna as a Residential City. – Bibliotheca Orientalis 3-4: 274-288.

Kemp, B.J. 2012. The City of Akhenaten and Nefertiti: Amarna and Its People. – London, Thames & Hudson.

Kemp, B.J., A.K. Stevens, G.R. Dabbs, M. Zabecki & J.C. Rose. 2013. Life, Death, and Beyond in Akhenaten's Egypt: Excavating the South Tombs Cemetery at Amarna. – Antiquity 87 (335): 64-78.

Kenyan/British Medical Research Councils. 1989. Tuberculosis in Kenya 1984: A Third National Survey and a Comparison with Earlier Surveys in 1964 and 1974. – Tubercle 70: 5-20.

Kiecolt-Glaser J.K., J.R. Dura, C.E. Speicher, O.J. Trask & R. Glaser. 1991. Spousal Caregivers of Dementia Victims: Longitudinal Changes in Immunity and Health. – Psychosomatic Medicine 53: 345-362.

Kiecolt-Glaser, J.K., P.T. Marucha, W.B. Malarkey, A.M. Mercado & R. Glaser. 1995. Slowing of Wound Healing by Psychological Stress. – Lancet 346: 1194-1196.

Kim, Y. & R. Schulz. 2008. Family Caregivers' Strains: Comparative Analysis of Cancer Caregiving with Dementia, Diabetes, and Frail Elderly Caregiving. – Journal of Aging Health 20: 483-503.

Kirigia, J.M. & R.D.K. Muthuri. 2016. Productivity Losses Associated with Tuberculosis Deaths in the World Health Organization African Region. – Infectious Diseases of Poverty 5: 43.

Loddenkemper, R., M. Lipman & A. Zumla. 2016. Clinical Aspects of Adult Tuberculosis. – Cold Springs Harbor Perspectives in Medicine 6: a017848.

Meskell, L. 2002. Private Life in New Kingdom Egypt. – Princeton, Princeton University Press.

Murry, C.J.L. & A.D. Lopez. 1996. The Global Burden of Disease: A Comprehensive Assessment of Mortality and Disability from Diseases, Injuries, and Risk Factors in 1990 and Projected to 2020. – Cambridge, World Health Organization.

Lovejoy, C.O., R.S. Meindl, T.R. Pryzbeck & R.P Mensforth. 1985. Chronological Metamorphosis of the Auricular Surface of the Ilium: A New Method for the Determination of Adult Skeletal Age at Death. – American Journal of Physical Anthropology 68: 15-28.

O'Reilly, L.M. & C.J. Daborn. 1995. The Epidemiology of *Mycobacterium bovis* Infections in Animals and Man: A Review. – Tubercle and Lung Disease 76 (supplement 1): 1-46.

Rabins, P.V., M.D. Fitting, J. Eastham & J. Fitting. 1990. The Emotional Impact of Caring for the Chronically Ill. – Psychosomatics 31: 331-336.

Rajeswari, R., R. Balasurbramanian, M. Muniyandi, S. Geetharamani, X. Thresa & P. Venkatesan. 1999. Socio-Economic Impact of Tuberculosis on Patients and Family in India. – The International Journal of Tuberculosis and Lung Disease 3 (10): 869-877.

Raxter, M.H., C.B. Ruff, A. Azab, M. Erfan, M. Soliman & A. El-Sawaf. 2008. Stature Estimation in Ancient Egyptians: A New Technique Based on Anatomical Reconstruction of Stature. - American Journal of Physical Anthropology 136 (2): 147-155.

Resnick, D. & G. Niwayama. 1995. Osteomyelitis, Septic Arthritis, and Soft Tissue Infection: Organisms. In: Resnick, D. Ed. Diagnosis of Bone and Joint Disorders. – Edinburgh, Saunders: 2448-2558.

Roberts, C.A. 2015. Old World Tuberculosis: Evidence from Human Remains with a Review of Current Research and Future Prospects. – Tuberculosis 95: S117-S121.

Roberts, C.A. & J.E. Buikstra. 2003. The Bioarchaeology of Tuberculosis: A Global Perspective on a Re-Emerging Disease. – Gainsville, University of Florida Press.

Rothschild, B., V. Naples, & L. Barbian. 2006. Bone Manifestations of Actinomycosis. – Annals of Diagnostic Pathology 10: 24-27.

Schneppemheim, M. & J. Jerosch. 2003. Echinococcosis Granlosus/Cysticus of the Tibia. – Archives of Orthopaedic Trauma Surgery 123: 107-111.

Schulz, R. & S.R. Beach. 1999. Caregiving as a Risk Factor for Mortality: The Caregiver Health Effects Study. – Journal of the American Medical Association 282: 2215-2219.

Smith, G.E., M.A. Ruffer & K. Sudhoff. 1910. Pott'sche Krankheit an einer ägyptischen Mumie aus der Zeit der 21. Dynastie (um 1000 v. Chr.). Vol. 3. – Töpelmann, Giessen.

Stenberg, U., C.M. Ruland & C. Miaskowski. 2010. Review of the Literature on the Effects of Caring for a Patient with Cancer. – Psycho-Oncology 19: 1013-1025.

Stevens, A.K. 2017. Death and the City: The Cemeteries of Amarna in their Urban Context. – Cambridge Archaeological Journal 28 (1): 103-126.

Steyn, M., Y. Scholtz, D. Botha & S. Pretorius. 2013. The Changing Face of Tuberculosis: Trends in Tuberculosis-Associated Skeletal Changes. – Tuberculosis 93: 467-474.

Stommel, M., C.W. Given & B. Given. 1990. Depression as an Overriding Variable Explaining Caregiver Burdens. – Journal of Aging Health 2 (1): 81-102.

Tomkins, A. 1993. Environment, Season and Infection. In: Ulijaszek, S.J. & S.S. Strickland. Eds. Seasonality of Human Ecology 35th Symposium Volume of the Society for the Study of Human Biology. – Cambridge, Cambridge University Press: 123-134.

Wood, J.W., G.R. Milner, H.C. Harpending, K.M. Weiss, M.N. Cohen, L.E. Eisenberg, D.L. Hutchinson, R. Jankauskas, R. Cesnys, G. Cesnys & M.A. Katzenberg. 1992. The Osteological Paradox: Problems of Inferring Prehistoric Health from Skeletal Samples. – Current Anthropology 33 (4): 343-370.

World Health Organization. 2020. World Tuberculosis Day 2020. https://www.who.int/campaigns/world-tb-day/2020. Accessed 27.08.2020.

World Health Organization. 2018. The Top 10 Causes of Death. http://www.who.int/news-room/fact-sheets/detail/the-top-10-causes-of-death. Accessed 10.11.2018.

Zakrzewski, S., A. Shortland & J. Rowland. 2016. Science in the Study of Ancient Egypt. – London, Routledge.

Zink, A.R., W. Grabner, U. Reischl, H. Wolf & A.G. Nerlich. 2003a. Molecular study on Human Tuberculosis in three geographically distinct and time delineated Populations from ancient Egypt. – Epidemiology and Infection 130: 239-249.

Zink, A.R., C. Sola, U. Reischl, W. Grabner, N. Rastogi, H. Wolf & A.G. Nerlich. 2003b. Characterization of *Mycobacterium tuberculosis* Complex DNAs from Egyptian Mummies by Spoligotyping. – Journal of Clinical Microbiology 41 (1): 359-367.

Burial Practies in the West Delta: Cases from Kom Aziza

Shereen El-Morsi & Aya M. Salem

Introduction

Kom Aziza is located in the western part of the Delta some 9 km north of the city of Abu Homs and 18 km southeast of Kafr al-Dawar (Figure 1). It is part of a larger network of ancient sites, including Kom Hashim, Kom al-Dibaa, Kom Palm and Kom Takala. In modern times the archaeological site of Kom Aziza has been divided in two (north and south) by the Seif Canal and an asphalt road that run parallel in a north-south direction. The main modern village and the mosque are now located on the south side of the road and canal, and the archaeological tell site on the north (Figure 2).

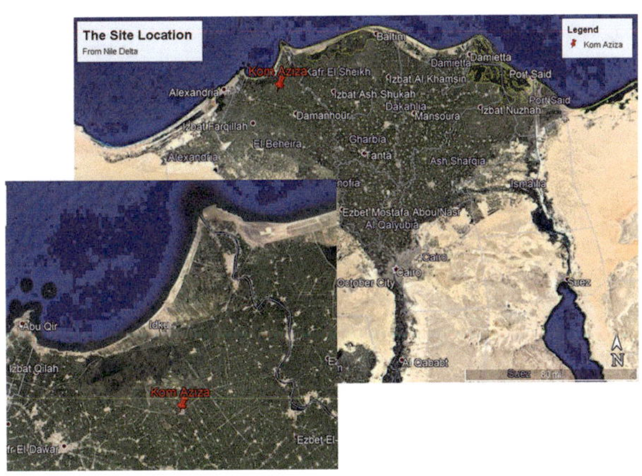

Figure 1. The location of Kom Aziza in the Delta. Courtesy Google Earth.

Figure 2. The site of Kom Aziza. Courtesy Google Earth.

In: Ikram, S., J. Kaiser & S. Porcier. Eds. 2021. *The Ancient Egyptians and the Natural World. Flora, Fauna, and Science.* – Leiden, Sidestone Press: 99-104.

History of Excavation

The site was first excavated in 1976 by the Egyptian Antiquities Organization (now the Ministry of Tourism and Antiquities [MoTA]) prior to the construction of a primary school The excavations were carried out under the direction of Ahmed El-Nashar in the southern part of the area, where no archeological remains were found. Subsequently, in 1988 the MoTA excavated in the north, under the direction of Mr. Abdel Fatah, General Manager of Department of Antiquities of the Western Delta. Excavations showed that the area was used as a cemetery in the Roman era. Subsequently, in 2004 and 2005, Penny Wilson worked here as part of the Delta Survey, mapping and drill-coring. Her results showed considerable Roman use of the site. The present excavation was carried out in the cemetery in the north in 2015[1] (Figure 2).

The Burials

This paper provides an overview of the burials of 55 individual. The bodies were found in a variety of graves. Pit graves, consisting of simple pits cut into the earth, were most common (Figure 3). These were followed by rectangular pits cut into the earth, surrounded completely or partially by a mudbrick wall (Figure 4). The next most common were rectangular mudbrick constructions surrounding the burial (Figure 5) and after them, those with fired brick constructions.The most unusual and unique grave type is the Amphora grave, found first during the 1988 excavations (Figure 6). Here, individual pottery coffins are surrounded by 12 amphorae on each side, standing on their toes, and covered with mud plaster, thus creating an elongated pyramid over each coffin.

Coffin Types

Four types of coffins were found at Kom Aziza: pottery jars, amphorae, anthropoid pottery coffins, and mud plaster coffins. The first three types are generally found

Figure 3 Simple pit burials (burial 18, 19, on top of burial 20). Photo by S. El-Morsi.

Figure 4. Mudbrick enforced tomb (burial 19). Photo by A. Salem.

1 The team members are: Ibrahim Sobhy, Ramadan Mohamed Kamel, Walid Abd El-Bary and Ahmed Mohamed Naeem.

Figure 5. Mudbrick tomb. Photo by I. Sobhy.

Figure 6. Example of an amphora tomb. Photo by A. Naem.

in very good condition, but the last type is often destroyed due to the wet environment of the Delta. Thirty of the fifty-five Kom Aziza burials were encoffined in a supine position. Of these, twenty were in bi-partite pottery jar coffins sealed with mud plaster, and three in anthropoid pottery coffins; these were found in a group, with the head at the west (Figures 7-10). If the individual was too tall for the bi-partite coffins, it was extended by the addition

Figure 7. Pottery jar coffin with burial 34 after it was opened. Photo by A. Salem.

Figure 8. Anthropoid coffins with burials 13, 14, &15. Photo by S. El-Morsi.

Figure 9. Pottery jar coffin with burial 34, facing north. Photo by A. Salem.

Figure 10. Sealed amphora coffin with burial 9. Photo by R. Mohamed.

of an amphora. Only five amphora coffins were noted. These are like the pottery coffins, but simpler. These too accommodated supine bodies of young individuals.

The last type of coffin found at Kom Aziza is a mud plaster coffin, of which two examples were found, both poorly preserved. They were made of plaster, constructed around the dead body.

Methods

Each burial was treated individually. Determination of sex was based on discriminatory characteristics of the skull and pelvis (Ublaker, 1994: 16-21). Estimation of age was based on several methods: degree of tooth wear (Lovejoy, 1984: 47-56); morphology of the auricular surface (Meindle, 1985: 15-28), age-related modification of the pubic symphysis (Suchey, 1990: 227-238), while estimation of sub-adult age was based on tooth development and eruption (Alqahtani, 2010), and bone size (Maresh, 1970). After excavation, recording, and preliminary analysis on site, the bones were taken to the storeroom for cleaning and future analysis.

Results

The tombs yielded five infants, thee child, thirteen young adult, thirty-one middle adults and three old adults (Figure 11). In terms of sex, 29 male and 19 female

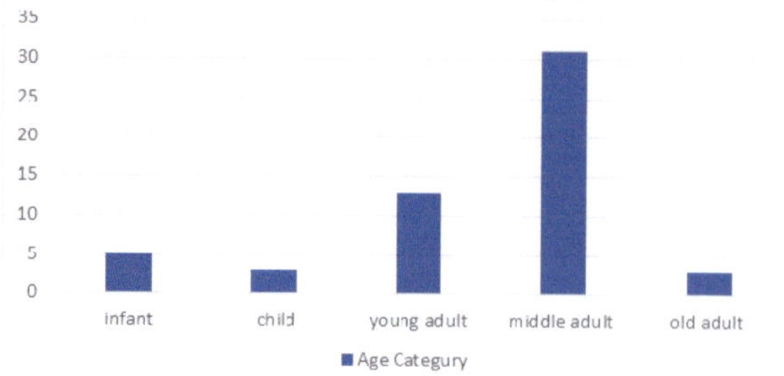

Figure 11. Graph showing the primary sex distribution of burials. By A. Salem.

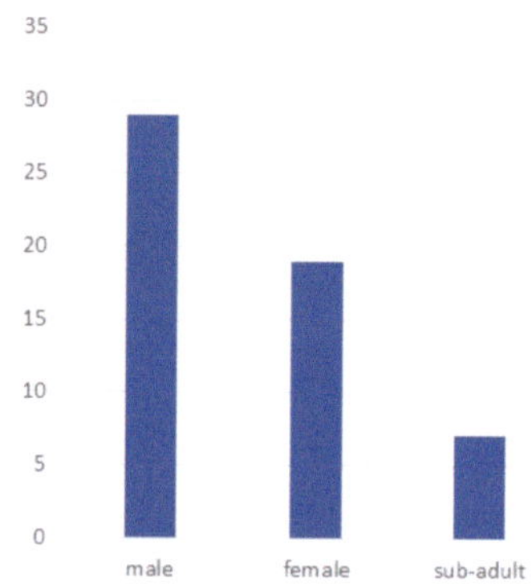

Figure 12. Graph showing the age distribution of burials. By A. Salem.

skeletons were identified. Seven individuals were too young to be sexed (Figure 12). Clearly the cemetery was used for a broad range of population, though one that was not wealthy, as attested by the absence of grave goods.

Because of the generally poor level of preservation at Kom Aziza, we do not have much direct evidence for mummification. There was very little in the way of textile remains, though in some instances we found some traces of black material in the abdomen or throat region (possiblly resin/oil impregnated textiles) or mud packing. However, the body position of some individuals suggest that they had been tightly wrapped before being put in the coffin. For example Burial 10 skeleton 17 (Figure 13), a supine burial of a child aged about 11-13 years, lying with the right hand on the femur and the left hand on the pelvis, had the clavicles squeezed together, indicating tight bandaging of the body before placing it in a pottery coffin. The body was also packed with mud in the abdominal area and perhaps as a means of re-shaping the body as is found in Late Period mummies (Ikram & Dodson, 1998: 128-29). This phenomenon has also been observed at the delta site of Quesna, as well as at many other Egyptian sites.

The part of the cemetery at Kom Aziza that was explored seemed to have been used by a variety of

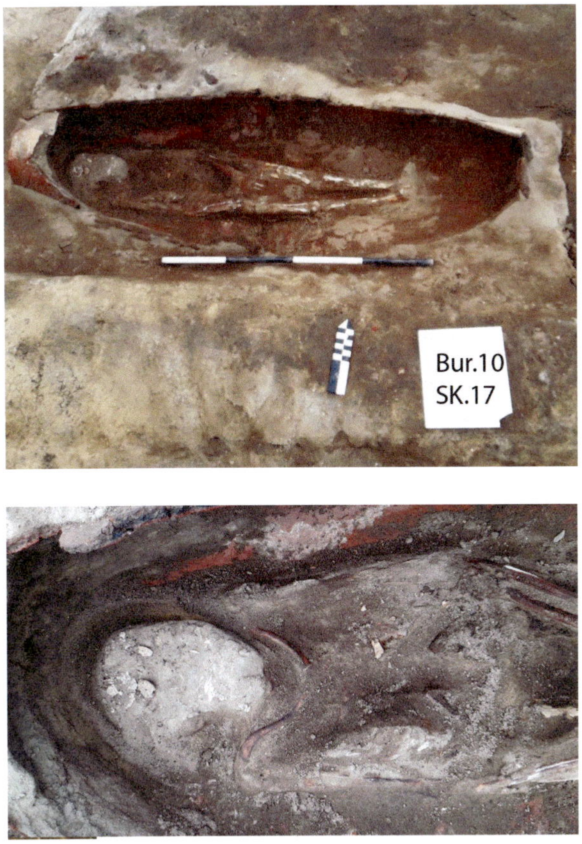

Figure 13. A child burial (burial 9) within an amphora coffin (with detail). Photographs by W. Abd El-Bary.

individuals of all ages and sexes. Although they all died at different ages, no pathologies were noted during the preliminary examination – future work in the laboratory might alter this. It is probable that the different burial types (pits as opposed to constructed tombs, different types of coffins) hint at differences in status, as well as, possiblly, diachronic change, but more work needs to be carried out to increase the sample size and to further analyse the remains.

Acknowledgement

A word of appreciation to the Ministry of Antiquities for encouraging our participation in the project, and the team of the Kom Aziza site, in particular Ibrahim Sobhy, Ramadan Mohamed, Walid Abd El-Bary and Ahmed Naem. We would also like to thank Penny Wilson for her help and support.

Bibliography

Aufderheide, A. 2003. The Scientific Study of Mummies. – Cambridge, Cambridge University Press.

Gunnasdottir, H.L.S. 2007. A Short History of Linen. Online at: https://www.timeless-creations.ca/Linen.pdf

Ikram, S. & A. Dodson. 1998. The Mummy in Ancient Egypt: Equipping the Dead for Eternity. – London, Thames & Hudson.

Kaiser, J. 2011. Human Osteology Training Manual. – Ancient Egypt Research Associates in-house publication.

Lovejoy, O. 1984. Dental Wear in Libben Population: Its Functional Pattern and Role. – American Journal of Physical Anthropology 68.1: 47-56.

Maksoud, G.A. 2011. A Review on the Materials used During the Mummification Process in Ancient Egypt. – Mediterranean Archaeology and Archaeometry 11.2: 129-150.

Meindle, L.A. 1985. Chronological Metamorphosis of the Auricular Surface of the Ilium: A New Method for the Auricular surface of the illum. – American Journal of Physical Anthropology 68.1 : 15-28.

Shoukri, S.A. 1988. Kom Aziza project de report preliminaire de la campagne de fouilles de 1988. – Unpublished report.

Brooks, S. & J.M. Suchey. 1990. Skeletal Age Determination Based on the Os Pubis: A Comparison of the Acsádi-di-Nemeskéri and Suchey-Brooks Methods. – Human Evolution 5.3: 227-238.

Buikstra, J.E. & D. Ublaker. 1994. Standards for Data Collection from Human Skeletal Remains. – Fayetteville, AK, Arkansas Archaeological Survey.

Venit, M.S. 2002. Monumental Tombs of Ancient Alexandria: The Theater of the Dead. – Cambridge, Cambridge University Press.

A Structure-from-Motion Pipeline for Bone Morphology 3D Analysis

Margaret Farmer & Angelique Corthals

Introduction

Digital 3D imaging techniques have recently emerged as powerful tools in physical anthropology and archaeology (Bevan *et al.*, 2014; Errickson *et al.*, 2017; Jurda & Urbanova, 2016; Morgan *et al.*, 2019). Virtual reconstructions are particularly important when the materials cannot be moved off-site for analysis.

On an excavation site, where thousands of individual pieces of material are being sorted, imaging and 3D rendering is often cumbersome and time consuming. But it is also necessary for circumstances in which none of the material, especially human remains, can be transported into a lab for further analysis. Therefore, the only record will be the images taken at great cost of time and effort to the team. We have therefore sought an imaging system that would not only streamline the acquisition of digital pictures, but also render them in 3D for further studies. Ultimately, we are seeking a computing solution to perform a 'cradle-to-grave' operation, a series of tools (or toolbox) that would operate all functions from acquisition to analyses and even hypothesis generation.

The test site we are using to develop such a tool is the Theban Tomb Complex 29, the excavation of which is (TTC 29), directed by Drs. Laurent Bavay and Dimitri Laboury, from the Universities of Brussels and Liege (Belgium), respectively. The site has been under investigation by this team since 2008, and has a complicated history of multiple explorations, tomb robbing, and environmental damage (most notably, earthquakes and landslides).

There are several techniques available for capturing 3D data, including laser scanning, CT scanning, and structured light scanning, but photogrammetry is one of the most versatile and inexpensive methods currently available (Aliberti *et al.*, 2018; Aragón *et al.*, 2018; Fernández-Hernandez *et al.*, 2015; Galeazzi, 2016; Jalandoni *et al.*, 2018; Magnani *et al.*, 2016). Photogrammetry requires the building of point clouds. Photogrammetric point clouds are simplified structures that represent the surface detail and shape of an object in space. In the case of skeletal remains, these point clouds can replace the traditional morphometric (craniometric) landmarks that are measured by hand and fraught with errors. The only input required for generating a point cloud is image data, such as digital photographs, which can be collected with equipment as simple as a mobile device.

Whatever the device, however, the time necessary to acquire and process images remains a major bottleneck in documenting larger numbers of artifacts and/or remains. Time constraints sometimes lead to the lack of photographic record of items considered less informative or less important at the time of triage. Our philosophy is that no artifact or remains should be excavated or removed without images. Therefore, to speed up the

In: Ikram, S., J. Kaiser & S. Porcier. Eds. 2021. *The Ancient Egyptians and the Natural World. Flora, Fauna, and Science.* – Leiden, Sidestone Press: 105-112.

process, it is necessary to build an informatics structure called a 'Pipeline' (*i.e.* a software that leads data through a series of processes to aggregate and standardize the data), where images are acquired and then cleaned up and processed, not by the handler, but by the computer/device itself, thereby saving a tremendous amount of time in the field. A pipeline is the first tool we are developing as part of the greater toolbox mentioned above.

Today's most common photogrammetry pipelines utilize Structure from Motion (SfM), the process of assembling a 3D virtual structure by triangulating a series of 2D images, taken of a subject from different viewpoints. Through SfM algorithms, common features are extracted and matched among the images, the geometry is verified, and a sparse cloud of 3D points is reconstructed (Snavely *et al.*, 2008). To render the human remains catalogued at TTC 29, the open-source software COLMAP was used to generate, then export, point cloud data to other software for further analysis (Schönberger & Frahm, 2016; Schönberger *et al.*, 2016). COLMAP was chosen for this project because of its accessible Graphical User Interface (GUI) and cross-platform compatibility. COLMAP has also been shown to generate point clouds with higher accuracy than other open-source SfM pipelines when rendering predetermined ground-point synthetic datasets (Bianco *et al.*, 2018).

Traditionally, Geographic Information Systems (GIS) are used for analyzing geospatial data. However, GIS software can also automate the segmentation of osteological data, treating the bone surface as a landscape at a smaller scale. By automating the analysis of the point cloud generated with SfM software , the need for selecting landmarks or doing any type of selection by hand is removed. Support vector machines and convolutional neural networks will be explored as possible supervised and unsupervised machine learning tools.

Bones that have been damaged antemortem, perimortem, or postmortem have unique surface features that are difficult to describe. Traditionally, the morphology of bone samples have been quantified with geometric morphometrics (GMM), based on selected landmarks. While there is support in the literature for the repeatability of results using GMM, any method requiring physical measurements between landmarks inevitably introduces bias (Hirst *et al.*, 2018.). However, a sparse point cloud rendering of an osteological feature has the potential to be used as classification criteria, just as point clouds of terrestrial data are used in Geographic Information Systems (GIS) analysis. Because different photogrammetry software packages can export point clouds in many different formats and at different resolutions, it is crucial to maintain an accurate scale for all models.

Within the last decade, there has been a proliferation of new open-source photogrammetry software, making 3D data collection protocols more accessible than ever (Hassett, 2018; Knapitsch *et al.*, 2017; Siebke *et al.*, 2018; Sung & Lin, 2017). However, there are few cohesive methods for researchers to upload and share information with other teams via the Internet. 3D data may be saved in different formats, which can only be analyzed with specific programs, and database construction can be software-specific. These limitations hinder the ability of researchers to update their findings, which is pertinent for teams who may not have direct access to the materials.

We propose a pipeline with a streamlined protocol for analyzing point cloud data, by incorporating existing open-source software and algorithms. The initial steps of the process are: 1) data acquisition; 2) upload of the information to a database hosted on a remote server; 3) use of the image data to generate a 3D point cloud with SfM; and 4) application of machine-learning algorithms utilized in GIS software to classify distinguishing surface features on the point cloud. Additional features of the pipeline include collecting, cleaning, analyzing, and making predictions about new data.

Materials and Methods

Image data was obtained from samples in TTC 29, Tomb C3. The tomb contained commingled remains from multiple individuals. Bones of pathological or taphonomical interest – those damaged antemortem, perimortem, or postmortem – were chosen for analysis. The photographic setup consisted of a white turntable and a paper green-screen background. The ambient light was filtered, non-directional sunlight. Two separate image sets were acquired for each sample: one with a Nikon D7200 DSLR (1920 x 1080 pixel frames, 30 fps for video) and the other with an iPhone SE (1920 x 1080 pixel frames at 30 fps for video, 2016 model). Both cameras were mounted on a tripod approximately 20 cm from the subject. The sample was placed in the middle of the turntable with a metric scale.

The camera was set up for three rotations: one at 45° above the horizontal; one at the horizontal, and one 20° below the horizontal. The sample was placed in the center of the turntable and rotated 360°, captured with video, for each vertical angle. The sample was then turned over and the process was repeated to cover any undocumented surfaces. The metric scale was only included while filming one side of the subject to avoid mismatching by the software in the processing step.

After documentation, the videos were uploaded to a PC hard drive (Intel Core i7-6700HQ CPU, 2.60 GHz, 16.0 GB RAM). VLC Media Player was used to extract individual frames as .jpg files in batches (every 10-20 frames) (VLC). Files were named with unique identification keys. A JavaScript application was designed to compress files and convert the image data to JSON format in order to upload

Action	Parameters	Purpose
Statistical Outlier Removal (SOR) (CloudCompare)	mean distance estimation: 6 points, standard deviation multiplier threshold: 1.00	removes isolated points and outliers
Label Connected Components (CloudCompare)	octree level: 8 min. points per component: 10	segmentation and simplification of point cloud
Multiscale Model to Model Cloud Comparison (M3C2) (CloudCompare)	octree depth 8, minimum 10 points per component; use DSLR cloud normals for projection diameter and depth	calculates cloud-to-cloud distance
Iterative Closest Point (ICP) (CloudCompare)	RMS difference: $1x10^{-5}$ max thread count: 4 reference: DSLR model	fine registration of two point clouds
Screened Poisson Surface Reconstruction (Meshlab)	reconstruction depth: 8 minimum number of samples: 1.5 points interpolation weight: 4	generate polygon mesh from point cloud
Detect markers (cross non-coded) (Metashape)	tolerance: 10 pixels maximum residual (pix): 5	automatically detects metric scale in scene
Principal Component Analysis (PCA) (CloudCompare)	reference: DSLR model	automatic scaling of point clouds to same size

Table 1. Algorithms and features used in the SfM pipeline.

them to MongoDB, an open-source, cloud-based, noSQL document database. The database will be linked to existing data repositories through an application programming interface (API) that can connect different software, handle any file type, and mine for new data.

The structure-from-motion pipeline was completed with COLMAP software using the default settings. Image sets consisted of ~100-200 .jpg files for an individual sample. The top and bottom views were automatically merged by the software. The point cloud was then exported as a polygon file (.ply). Models were also rendered in the commercial software Agisoft Metashape in order to calibrate the scale for the model (high accuracy with adaptive camera model fitting) (Agisoft Metashape, 2019). In Metashape, a subset of images was selected to detect the features of the metric scale. The patterns on the scale served as ground control points automatically detected by the software and then assigned distance values of 0.010 +/- 0.001 m (projections >20 and pixel error <0.5) between crosses. Estimated camera positions were calculated and xyz coordinates were preserved when the point clouds were exported in .ply (polygon file) format. COLMAP extensions for geo-referencing were not used because of the need for pre-determined ground point coordinates in a text file, which is unavailable in an image set with no geographical data in the EXIF tags (a type of image metadata).

Semi-automated processing was completed in the open source software CloudCompare (CloudCompare, 2019). The point cloud was "cleaned" in CloudCompare by applying the Statistical Outlier Removal (SOR) filter (the mean distance between each point and its neighbors was calculated, and, assuming a Gaussian distribution, all points that fell outside the interval defined by the mean and standard deviation of the global distances were eliminated). SOR was followed by the Label Connected Components segmentation algorithm (the point cloud was split into segments based on a set of point-to-point distance parameters [Table 1]) (Figure 1, 2) (Girardeau-Montaut *et al.*, 2005). For the models exported from COLMAP to CloudCompare with no coordinate data, principal dimension (deduced from Principal Component Analysis) was used to automatically scale the clouds (one from the iPhone, the other from the DSLR) to the same size using the DSLR cloud as the reference (Figure 3). The point clouds exported from Metashape were already scaled to a coordinate system. The Iterative Closest Point algorithm (ICP) was used to align and superimpose pairs of point clouds, and the cloud-to-cloud distance was calculated with the Multiscale Model to Model Cloud Comparison (M3C2) algorithm (Figure 4) (Lague *et al.*, 2013). After the difference in point clouds from two different cameras and two different software was quantified, the most detailed model was chosen for further analysis (see Table 1 for summary of algorithms and features used). To visualize the general shape of the model, the .ply file was exported to the open source software Meshlab (Cignoni *et al.*, 2008;. Screened Poisson surface reconstruction was applied to generate a simple mesh of the point cloud in Meshlab (Kazhdan & Hoppe, 2013). This mesh can be viewed and shared in a browser window via threejs.org (Figure 5) (threejs, 2019).

Results

After the data acquisition step of the pipeline, the JavaScript application successfully uploaded a series of individual images to MongoDB, which automatically generated a record for each image from the metadata. After the images had been processed into point clouds, the Metashape models were more robust than the COLMAP models, with almost twice as many points per cloud. However, processing time for the Metashape clouds

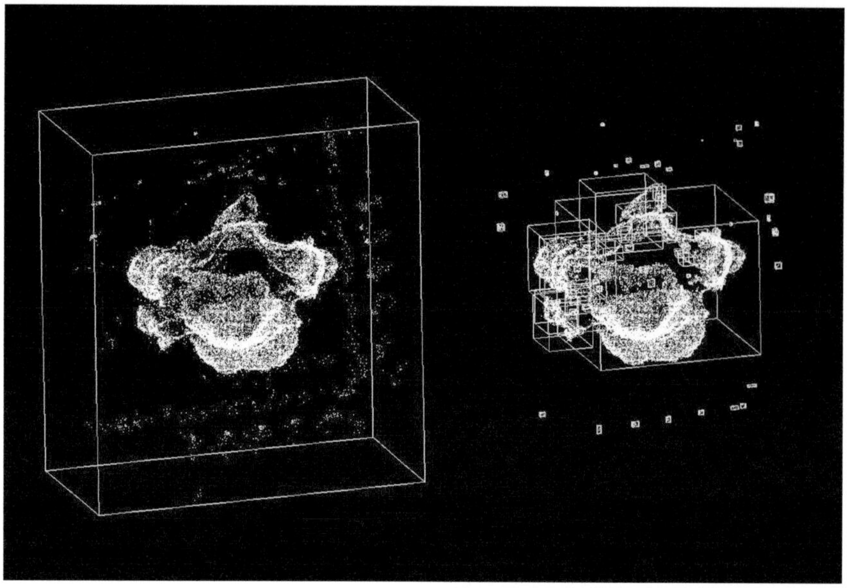

Figure 1. Point cloud cleaning step in CloudCompare. The Statistical Outlier Removal filter, then The Label Connected Components algorithm, was used to simplify the structure of the point cloud.

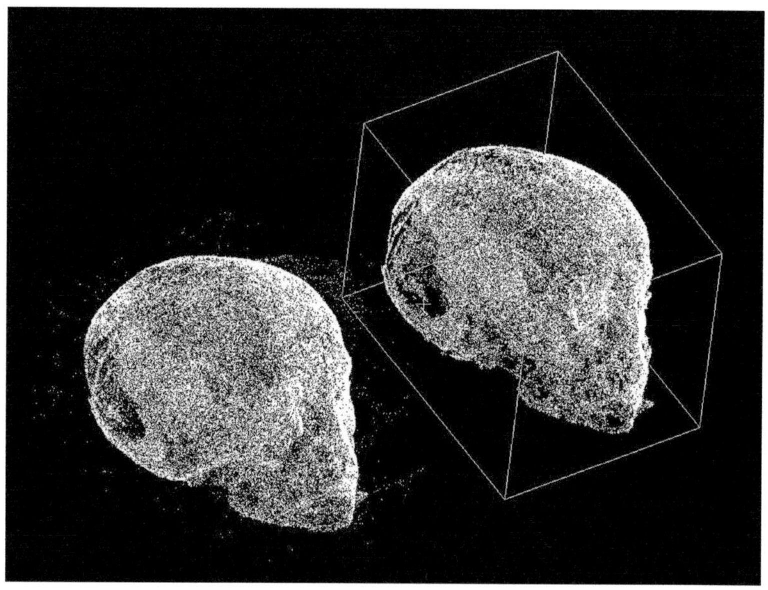

Figure 2. Point cloud of mummified juvenile head before and after outlier removal and segmentation step in CloudCompare. The raw point cloud (left) contains background noise. The cleaned point cloud (right) consists of extracted key points.

was much longer than that of the COLMAP clouds (up to 5 hours *vs.* 1 hour). Point clouds rendered with the iPhone video frames had, on average, 5-10% fewer points after the simplification step (removing outliers, segmentation) than the point clouds with the DSLR video frames. Metashape DSLR/iPhone clouds did not require scaling because they were exported with a coordinate system, rendering them the same size. However, the COLMAP models required scaling with PCA (using the Metashape DSLR cloud as reference). Simplifying the point clouds by automatically removing statistical outliers effectively cleared most of the background noise for all models. The M3C2 algorithm calculated a mean distance of ≤0.11 mm (s ≤ 1.0 mm) between DSLR and iPhone point clouds rendered of the samples in Metashape. The mean distance between the COLMAP models was ≤0.010 mm (s ≤ 0.87 mm). The Metashape DSLR models were more sensitive to the Label Connected Components algorithm with segmentation of up to 150 sub-sections, while the Metashape iPhone clouds segmented into less than 50. The results for the COLMAP models followed the opposite pattern, with ~25 segments for the DSLR and ~40 for the iPhone. In general, iPhone point clouds contained more noise, lower resolution (and thus fewer points), and different pixel RGB values than the DSLR point clouds.

Discussion

We have established a pipeline for image processing that minimizes the time required for data acquisition, by exporting still frames from videos. The point clouds

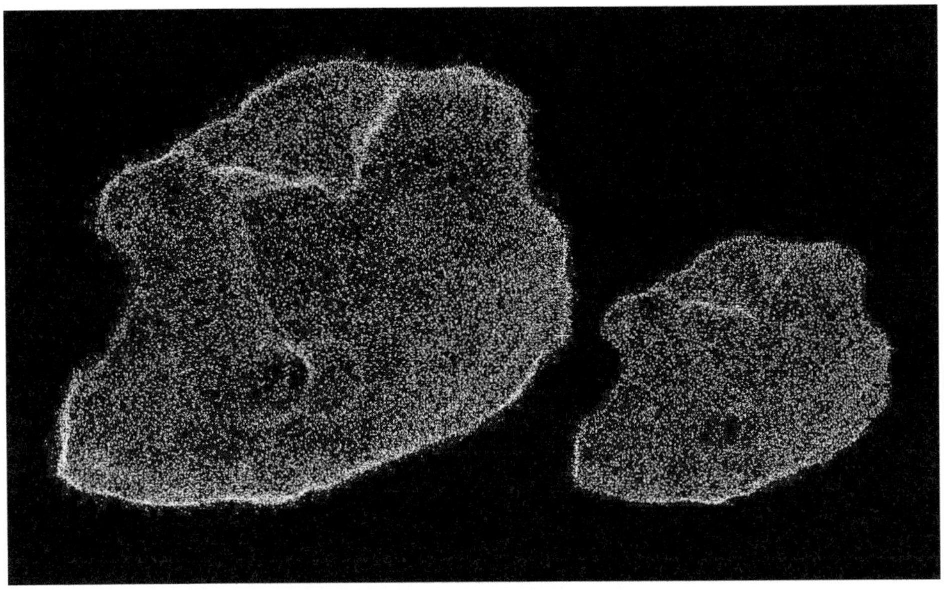

Figure 3. Size discrepancy in COLMAP DSLR (left) and iPhone (right) point clouds due to resolution (CloudCompare). When a coordinate system is not assigned, the images with lower resolution lead to smaller point clouds and additional scaling is required.

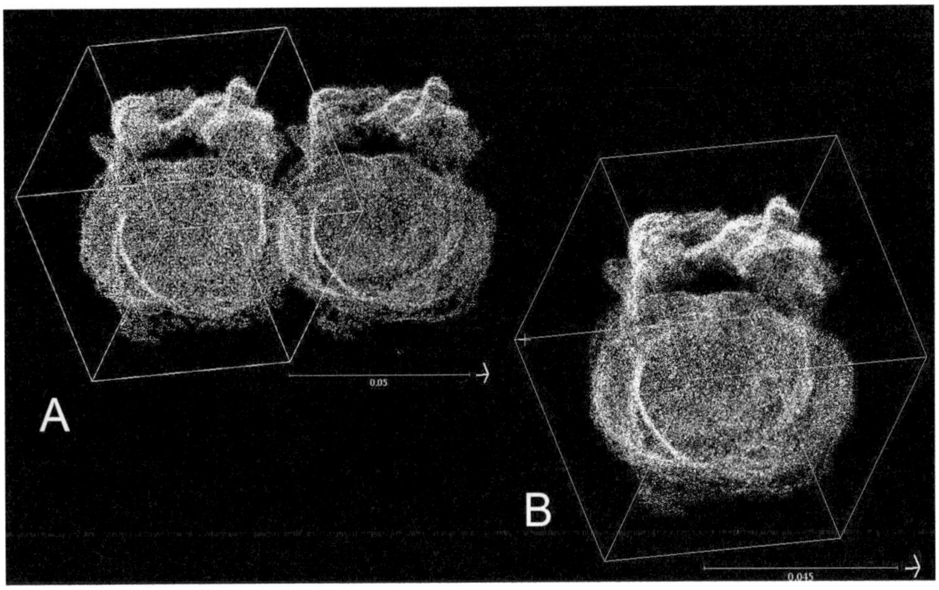

Figure 4. Aligning Metashape point clouds in CloudCompare. A) The iPhone point cloud (left) and DSLR point cloud (right) before alignment; B) After alignment with the Iterative Closest Point algorithm. The scale is in meters.

constructed from iPhone images were smaller and more sparse than the DSLR images, due to the digital zoom of the camera lowering the resolution of the images. However, the phone is a 2016 model and current models have better cameras. Moreover, the distance between point clouds once they were scaled was relatively small (≤ 0.11 mm). It should be noted that the measurements and camera coordinates are estimated from a calibration scale, which makes absolute error difficult to quantify after scaling. Automatic marker detection is not currently available in most open-source software. However, AliceVision's Meshroom is a newer open-source software with algorithms for detecting concentric circles, which will be explored for potential use in the pipeline (Meshroom, 2018). It may also be beneficial to deploy scale-invariant algorithms so the machine learning analysis is conducted on specific areas of the bone surface, not the overall size of the sample.

In a trade-off between quality and speed, the COLMAP models have fewer points than the Metashape models. However, there was little difference in the key points extracted from each cloud during the simplification step. The quality of the models in COLMAP can also be modified with additional scripts and executables through the command line. COLMAP is a suitable software for the pipeline without the additional scripts, but we are investigating if it is necessary to use the command line to construct better 3D point clouds. In the future we will analyze the results of combining images from different cameras with different focal lengths and lighting

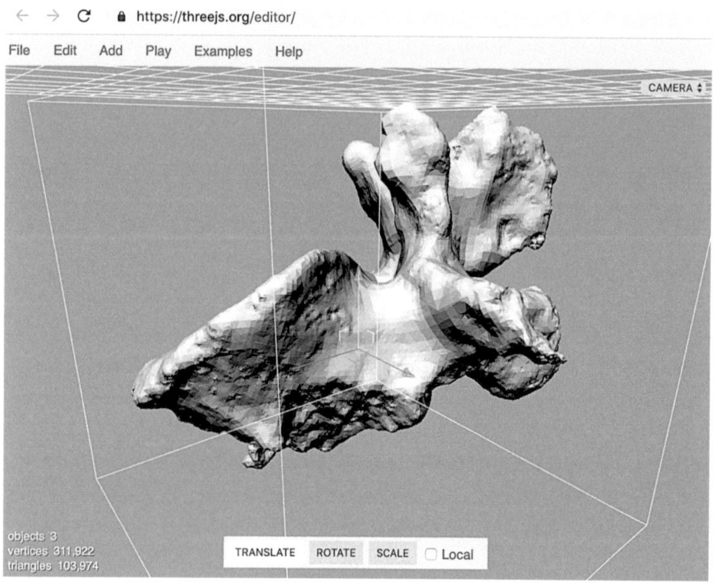

Figure 5. A simplified triangular mesh from Meshlab (.obj format) uploaded to threejs.org. The open-source JavaScript library contains algorithms for manipulating 3D data and publishing 3D models to the web.

conditions for any significant difference in quality or shape of the point cloud. The SfM analysis in the API will use images extracted from the Internet by the toolkit to render new 3D point clouds and extract key features to compare with samples in the database.

The CloudCompare algorithm for isolating connected components, which is based on a classical image processing algorithm, was an effective tool for fast, preliminary segmentation of point clouds. The prototype of the application for uploading the image files to MongoDB is currently stable enough to load and delete image files through a browser window. The next step will be to make the application deployable and integrate it with the DANTe API.

There are a few web-based SfM pipelines available, but due to the heavy computational load, it is common to encounter connection timeouts or poor alignment for sparse clouds. We will continue to monitor the progress of the technology in order to incorporate web-based SfM into the pipeline, eliminating the need to download software. The code for the machine learning algorithms used in this project will also be integrated into the API.

The next step for the SfM pipeline will incorporate additional machine learning. The key points extracted from the sparse clouds should be stable, distinctive, and identifiable with specific criteria. Supervised learning requires the computer to analyze a training set, isolate criteria that describe how the structures differ from each other, and then use these criteria to classify new samples (the testing set). Support Vector Machine-based classification, frequently used in GIS, will be examined as a possible method for the pipeline (Zhang, Lin & Ning, 2013).The sensitivity step of the pipeline will measure the rate of true similarities between the feature queried and the data in the repositories, while the specificity step will measure the sensitivity rate against the rest of the repository. In other words, statistical software (such as R) can be incorporated into the pipeline in order to test how accurately the supervised learning algorithms recognized the similarity (or dissimilarity) between the point cloud model being tested and point cloud models that have already been identified (R Core Team, 2013).

Unsupervised learning algorithms will parse the data based on patterns the computer detects without any input from a training set and without prior connections. Some proposed models for unsupervised point cloud analysis include the minimum description length principle algorithm, which clusters data based on patterns of variation, and the convolutional neural network, which uses a system of layers to reduce the dimensionality of image data and output a defining characteristic (Zhang *et al.*, 2013; Zhi *et al.*, 2017).

The Datascape Analysis and Navigation Tool (DANTe) is a publicly available web-based toolkit which allows users to access multiple resources simultaneously and compartmentalize the information in a user-friendly web browser interface. The API is made up of commands, functions, and protocols for the user to write software or interact with external sources, such as databanks, online journals, and cloud drives. Search criteria can be generated from the GIS algorithmic data of the point cloud and metadata to query other public data repositories (for example, ResearchGate, Dryad Digital Repository, Google Scholar, and Digitised Diseases, among many others), without the need for keyword searching, and return the results all in one window. Accessing data across disciplines enhances the usability and value of separate biological/anthropological collections which may currently be confined to their own disciplines (Corthals *et al.*, 2016). In general, the fields of archaeology and anthropology are also moving towards integrating a global dataset and more collaborations (White *et al.*, 2018). The point cloud will then be available for visual comparison with other information returned in the search. Information will be kept up-to-date and available to the scientific community. The code for Dante is currently under construction and will be freely available through Github (https://github.com/aspcorthals/DANTe) when it is ready to share.

Conclusion

The data collection and simplification stage of the SfM pipeline was successfully implemented with open-source software, with the exception of automatic detection of the metric scale in Metashape. The simplification and segmentation of the point clouds was semi-automated with algorithms applied in CloudCompare and MeshLab that may be incorporated into the API. By comparing models rendered under the same conditions but with images from two different cameras, we showed that smartphone video capture can be as accurate for key point extraction as high-resolution digital cameras, and as mobile devices they have the potential to directly upload images to the database. The application created to upload image files to MongoDB will be incorporated into the generalized API, as will the machine learning algorithms for segmentation and analysis. Future work will include improving automated measurement accuracy for point clouds, virtual reconstruction of fragmented bones, and streamlining the SfM algorithms with the DANTe pipeline.

Bibliography

AgiSoft Metashape Professional (Version 1.5.2) [Software]. 2019. Retrieved from http://www.agisoft.com/downloads/installer/

Aliberti, L., M. Ángel & A. Rodríguez. 2018. Digital Photogrammetry for the Geometrical Analysis of the Umbrella-Shaped Dome in Baia (Naples). – The International Archives of the Photogrammetry, Remote Sensing and Spatial Information Sciences XLII-2.2: 23-28.

Aragón, E., S. Munar, J. Rodríguez & K. Yamafune. 2018. Underwater Photogrammetric Monitoring Techniques for Mid-Depth Shipwrecks. – Journal of Cultural Heritage 34: 255-60.

Bevan, A., X. Li, M. Martinón-Torres, S. Green, Y. Xia, K. Zhao, Z. Zhao, S. Ma, W. Cao & T. Rehren. 2014. Computer Vision. Archaeological Classification and China's Terracotta Warriors. – Journal of Archaeological Science 49(1): 249-254.

Bianco, S., G. Ciocca & D. Marelli. 2018. Evaluating the Performance of Structure from Motion Pipelines. – Journal of Imaging 4 (8): 98 (https://doi.org/10.3390/jimaging4080098).

Cignoni, P., M. Callieri, M. Corsini, M. Dellepiane, F. Ganovelli & G. Ranzuglia. 2008. MeshLab: An Open-Source Mesh Processing Tool. – Sixth Eurographics Italian Chapter Conference: 129-136.

CloudCompare (version 2.10.2) [Software]. 2019. Retrieved from http://www.cloudcompare.org

Corthals, A., J. Almeida & L. Davalos, 2016. DANTe – Datascape Analysis and Navigation Tools: a Genelab Data Mining, Aggregation and Machine-learning Toolbox for the Next Generation of Space Biological Research – A Proposal in Response to NASA Research Announcement NNH16ZTT001N-GL Appendix A: GeneLab Innovation Awards for Translational Systems Biology and Informatics Research Using the GeneLab Data System; A Systems Biology Informatics Research Using the GeneLab Data System. – NASA proposal 16-2016GL_FP-0020.

Errickson, D., I. Grueso, S.J. Griffith, J.M. Setchell, T.J.U. Thompson, C.E.L. Thompson & R.L. Gowland. 2017. Towards a Best Practice for the Use of Active Non-Contact Surface Scanning to Record Human Skeletal Remains from Archaeological Contexts. – International Journal of Osteoarchaeology 27: 650-661.

Fernández-Hernandez, J., D. González-Aguilera, P. Rodríguez-Gonzálvez & J. Mancera-Taboada. 2015. Image-Based Modeling from Unmanned Aerial Vehicle (UAV) Photogrammetry: An Effective, Low-Cost Tool for Archaeological Applications. – Archaeometry 57.1: 128-145.

Galeazzi, F. 2016. Towards the Definition of Best 3D Practices in Archaeology: Assessing 3D Documentation Techniques for Intra-site Data Recording. – Journal of Cultural Heritage 17: 159-169.

Girardeau-Montaut, D., M. Roux, R. Marc & G. Thibault. 2005. Change Detection on Points Cloud Data Acquired With a Ground Laser Scanner. WG III/3-4 V/3 – Proceedings of the ISPRS Workshop Laser Scanning 2005, XXXVI (January): 30-35.

Hassett, B. 2018. Which Bone to Pick: Creation, Curation, and Dissemination of Online 3D Digital Bioarchaeological Data. – Archaeologies 14.2: 231-249.

Hirst, C.S., S. White & S.E. Smith. 2018. Standardisation in 3D Geometric Morphometrics: Ethics, Ownership, and Methods. – Archaeologies 14 (2): 272-298.

Jalandoni, A., I. Domingo & P.S.C. Taçon. 2018. Testing the Value of Low-Cost Structure-from-Motion (SfM) Photogrammetry for Metric and Visual Analysis of Rock Art. – Journal of Archaeological Science: Reports 17: 605-616.

Jurda, M. & P. Urbanova. 2016. Three-Dimensional Documentation of Dolní Věstonice Skeletal Remains: Can Photogrammetry Substitute Laser Scanning? – Anthropologie 52 (2): 109-118.

Kazhdan, M. & H. Hoppe. 2013. Screened Poisson Surface Reconstruction. – ACM Transactions on Graphics 32 (3): 29: 1-13.

Knapitsch, A., J. Park, Q. Zhou & V. Koltun. 2017. Tanks and Temples: Benchmarking Large-scale Scene Reconstruction. – ACM Transactions on Graphics (TOG) 36.4: 1-13.

Lague, D., N. Brodu & J. Leroux. 2013. Accurate 3D Comparison of Complex Topography with Terrestrial Laser Scanner: Application to the Rangitikei Canyon (N-Z). – ISPRS Journal of Photogrammetry and Remote Sensing 82: 10-26.

Magnani, M., M. Douglass & S.T. Porter. 2016. Closing the Seams: Resolving Frequently Encountered Issues in Photogrammetric Modelling. – Antiquity 90.354: 1654-1669.

Meshroom: A 3D Reconstruction Software. [Software]. AliceVision. 2018. Retrieved from https://github.com/alicevision/meshroom.

Morgan, B., A.L.J. Ford & M.J. Smith. 2019. Standard Methods for Creating Digital Skeletal Models Using Structure-from-Motion Photogrammetry. – American Journal of Physical Anthropology 169.1: 152-160.

R Core Team. 2013. R: A Language and Environment for Statistical Computing. – Vienna, R Foundation for Statistical Computing.

Rusu, R.B. & S. Cousins. 2011. 3D is Here: Point Cloud Library (PCL). 2011 IEEE International Conference on Robotics and Automation, Shanghai: 1-4.

Schönberger, J. L. & J.M. Frahm. 2016. Structure-from-Motion Revisited. – Proceedings of the IEEE Computer Society Conference on Computer Vision and Pattern Recognition: 4104-4113.

Schönberger, J. L., E. Zheng, M. Pollefeys & J.M. Frahm. 2016. Pixelwise View Selection for Unstructured Multi-View Stereo. Retrieved from https://github.com/colmap/colmap.

Siebke, I., L. Campana, M. Ramstein, A. Furtwängler, A. Hafner & S. Lösch. 2018. The Application of Different 3D-Scan-Systems and Photogrammetry at an Excavation: A Neolithic Dolmen from Switzerland. – Digital Applications in Archaeology and Cultural Heritage 10: E00078.

Snavely, N., S.M. Seitz & R. Szeliski. 2008. Modeling the World from Internet Photo Collections. – International Journal of Computer Vision 80.2: 189-210.

Sung, B. & C. Lin. 2017. A Fast 3D Scene Reconstructing Method Using Continuous Video. – EURASIP Journal on Image and Video Processing 2017.1: 1-14.

threejs [JavaScript library]. 2019. Retrieved from https://github.com/mrdoob/three.js

VLC 3.0.6. [Software]. VideoLAN. 2019. Retrieved from www.videolan.org.

White, S., C. Hirst & S.E. Smith. 2018. The Suitability of 3D Data: 3D Digitisation of Human Remains. – Archaeologies 14 (2): 250-271.

Zhang, J., X. Lin & X. Ning. 2013. SVM-Based Classification of Segmented Airborne LiDAR Point Clouds in Urban Areas. – Remote Sensing 5 (8): 3749-3775.

Zhang, Q., X. Song, X. Shao, H. Zhao & R. Shibasaki. 2013. Unsupervised 3D Category Discovery and Point Labeling from a Large Urban Environment. Proceedings. IEEE International Conference on Robotics and Automation, May 2013. – Karlsruhe, Institute of Electrical and Electronics Engineers: 2685-2692.

Zhi, S., Y. Liu, X. Li & Y. Guo. 2017. Toward real-time 3D Object Recognition: A Lightweight Volumetric CNN Framework Using Multitask Learning. – Computers & Graphics 71: 199-207.

Lions and Science and Whorls, Oh My!

Karen Polinger Foster

Introduction

Egyptian art and thought are exceptionally rich in lions. From the ravening beasts on slate palettes to the subdued mascots in royal scenes to the conniving creatures of satirical papyri, lions play a multitude of crucial roles. The interaction theme of this conference afforded a welcome opportunity to synthesize the latest scientific findings and fresh iconographic insights, with particular attention to the hair whorls often shown as shoulder emblems. The results indicate that many long-held assumptions should now be replaced or modified, leading to a new understanding of lions in ancient Egypt.

Out of Africa

Of the world's large, terrestrial mammals, lions are second only to humans in their erstwhile global presence throughout Africa, Asia, Europe, and the Americas. A recent study suggests that their diaspora began about two million years ago in the lush wetlands of south-central Africa (Moore *et al.*, 2015). At some point during arid Early Pleistocene cycles, certain of those lion populations stayed put, while others evolved to exploit the drier, grassland habitats that were increasingly rife with herbivore prey species, moving northward over the eons across the savannahs of Africa. The DNA of modern Serengeti lions, for instance, reflects an admixture of three major waves of expansion, the latest during the Pleistocene/Holocene transition (14,000 to 7000 years ago) (Antunes *et al.*, 2008).

Although this wetland/grassland scenario may be overly conjectural, lions' evolutionary success and diversification certainly took them far and wide. By about 500,000 years ago, if not before, they were established in the distant reaches of Europe. Among many other finds of lion fossils, a toe bone turned up, rather fittingly, in excavations beneath the bronze lions of London's Trafalgar Square (Jackson, 2010: 9-10). In the opposite direction, lions traversed the Bering Strait on the land bridge of the last Ice Age and headed south into the Americas, following the animals grazing on the steppes emerging as the glaciers melted.

Early human contact with lions may be gauged from discoveries of contemporaneous stone tools and lion bones. Thanks to cave art, we glean an idea of how similar Paleolithic European lions were to modern ones. The oldest drawings, made about 30,000 years ago, come from the Chauvet cave in southeastern France (Chauvet *et al.*, 1996). The body language, dark tail tuft, and individuated black whisker spots (to this day, each lion still has its own pattern) attest to the artists' careful, first-hand observation. A major reason for lions' notable ascendance is their group-living and other social behaviors, unique among felids (Estes *et al.*, 1991: 370-377; see also below). The cave artists captured these too. Prides of gimlet-eyed lions stalk their prey, or slink in twos and threes along the suggestively undulating walls.

In: Ikram, S., J. Kaiser & S. Porcier. Eds. 2021. *The Ancient Egyptians and the Natural World. Flora, Fauna, and Science.* – Leiden, Sidestone Press: 113-122.

Of special interest for the present inquiry is the matter of their manes. Cave lions, some of which are clearly male, do not seem to have manes. This apparently missing element has led to the theory that a division between maneless and maned lions arced across southeastern Europe (Yamaguchi *et al.*, 2004: 338). Yet close inspection of several Chauvet lions shows dark vertical striations behind the ears and extending a short distance down the neck, precisely where a mane would be, albeit a thin one.

The Lion's Mane

The mane is the pre-eminent feature of the species' sexual dimorphism, again unique among felids. In addition, what distinguishes the lion from animals with similarly dramatic male/female pelage and plumage differences is the mane's great variability, from individual to individual, even over the course of a lifetime, as well as from region to region. The manes of adult lions run the gamut from scant, sparse, and short to extremely thick, enveloping, and long. Regardless of amplitude, hair color can range from blond to rust to very dark brown/black, with many nuances in contiguous or zonal shading. Mane growth appears responsive to such ecological factors as ambient temperature (West & Packer, 2002; Yamaguchi *et al.*, 2004: 330).

Why do lions have manes, and what does their phenotype plasticity signify? As we shall see, the answers illuminate much about ancient lion art. An early hypothesis was that the mane protects the lion from the jaws and claws of a rival male, so the denser the mane, the better for the lion (Schaller, 1972: 360). Current research has shed additional light on the subject, demonstrating how integral this sexually selected trait is to the group-living dynamics of lions (Yamaguchi *et al.*, 2004).

The sole felids to do this, lions form prides, which operate in defined territories and can include as many as nine males and eighteen females, together with cubs, yearlings, and subadults (Estes, 1991; Jackson, 2010; Schaller, 1972). Once they are a few years old, males are usually evicted from the pride. Sometimes traveling in coalitions of two or three, they seek to join another pride by forcing out its resident males. The newcomer lions promptly kill nursing cubs, causing their mothers, and synchronously the pride's other lionesses, to enter estrous. Mating ensues. Both males and females thus have multiple opportunities to assess other males, but on what do they base their evaluations?

In a series of ingenious experiments, scientists had life-size plush lions fabricated, as well as wigs of blond and dark manes in short and long lengths (West & Packer, 2002). A bewigged dummy was placed in the vicinity of lions, and a recording of hyenas gorging at a kill was played to lure them to the spot. Consistently, males were cautious about the dark-maned lions, much preferring to approach the blond ones. Lionesses, on the other hand, showed a marked interest in the dummies with the darkest manes, long or short. The experiments were repeated using taxidermied lions and yak-hair manes, with comparable results (Packer, 2015: 342-345). Observers also noted how the lions reacted when they realized they were being duped, from stalking off to knocking the dummies down and tearing out their stuffing.

Lionesses in hot climates have got it right. A thick, dark mane bespeaks a strong, well-nourished male. But although this type of mane provides good insulation from external heat, it comes with the chief cost of leaving less proximal body area for him to radiate internal heat. Infrared thermographic imaging bears this out (Nagel *et al.*, 2003). A lion able to withstand the stress is a mate who can confer many benefits on his offspring.

Mane variability, prime among other phenotype traits, misled generations of taxonomists into naming anywhere from 8 to 24 lion subspecies (Jackson, 2010: 18-19). Recent phylogeographic studies indicate that there are in fact only two subspecies, *Panthera leo melanochaita*, found primarily in southern and eastern Africa, and *Panthera leo leo*, in the rest of Africa and Eurasia. This has now been formally proposed by the Cat Classification Task Force of the Cat Specialist Group, under the aegis of the International Union for the Conservation of Nature (IUCN) (Kitchener *et al.*, 2017).

The Asian Lion

It is the lion hitherto and in error thought to be a subspecies (*Panthera leo persica*) that is most relevant for Egypt, the Near East, and the Bronze Age Aegean (Blakolmer, 2016; Gransard-Desmond, 2010; Shaplund, 2010). Rather than calling it *Panthera leo leo* or the Northern lion, the inclusive terms now used in scientific discourse, I shall refer to it here as the Asian lion, a term more familiar to art historians and archaeologists. This should be understood as simply a descriptor, with no genetic implications. In ancient art, there are scores of diligent renderings of its distinctive fold of skin along the belly, generally furred in an extension of the mane. The earliest images of this appear on Mesopotamian cylinder seals of the late Uruk period, that is, the final centuries of the fourth millennium BC (*e.g.* Buchanan, 1981: 154).

Another morphological feature is outwardly invisible, and concerns the infraorbital foramen, the cavity through which the lion eye's artery, vein, and nerve pass. A tiny bit of connective bone can produce a divided infraorbital foramen (DIF). Recent inspection of hundreds of felid skulls, including tigers and cave lions, refutes the long-held view that the DIF occurs only in Asian lions (Yamaguchi *et al.*, 2009). Its greater frequency in them, however, invites further inquiry to determine when and where Asian lions began to develop their visible phylogenetic traits.

In this regard, the Levant is critically important, for lions, together with other taxa, made their way from northeast Africa to Eurasia through that region's biogeographic corridor. Current research suggests that early Pleistocene movement occurred continuously, rather than in the discrete pulses previously thought to be the case (Belmaker, 2017). By the beginning of the Holocene, human settlement and ventures, especially agriculture, had come to have a significant impact on the composition and distribution of mammalian species, notably in the southern Levant. Yet lions persisted there, well into Crusader times (Bar-Oz & Weissbrod, 2017: 364). In Western Asia as a whole, dozens of records document lion sightings and killings from 1781 on. The last credible reports from Syria date to 1891; from Iraq, 1914; from Iran, 1942 (Joslin, 1973: 242-249; Schnitzler, 2011).

The last wild vestiges of Asian lions are found exclusively in the Gir Lion Sanctuary in northwestern India (Joslin, 1973; de Planhol, 2004: 715-717; Sharma & Johnsingh, 1996; Singh, 1997). There, initially private and then governmental organizations have created protected areas and a national park, enhanced vegetation and prey animal diversity, and worked with Gir's 14,000 inhabitants to control and monitor human/livestock/lion interaction. Much has been made in the popular press of the increase in lions, from a few dozen at the start of the twentieth century to over 500 in the 2015 census, but their viability and outlook are problematic. Their DNA reveals that Gir lions are so inbred as to be essentially clones of each other, with all the concomitant troubles and serious consequences that one would expect (O'Brien, 2003: 47).

Today's Asian lions, especially at Gir but also in zoos, are victims of a classic population bottleneck, meaning that at a certain point there were too few individuals for genetic variability. This was believed to have happened to them owing to anthropogenic activities in the late nineteenth century, but new genetic research and calculations using various molecular measures have pushed the bottleneck back 3000 years (O'Brien, 2003: 51). It is too soon to know if captive breeding programs, as in the new "Land of the Lions" exhibit at the London Zoo, or lion relocation to other reserves, as proposed in India, will remedy the situation (Evans, 2018: 194-212).

All this has made it challenging to reconstruct their demographic and evolutionary history. Taking bone and tissue samples from museum specimens has proven to be a productive way around the problem (Barnett *et al.*, 2014). Molecular and statistical analyses of this venerable mitochondrial DNA suggests that the Asian lion branched off several times from African clades over the past 61,500 years, and that the latest splits (from North African lions) took place within the past 21,000 years. As for the lions already in Eurasia, their genetic signature was likely overwritten by the latest arrivals.

Lions in Ancient Egypt?

This brings us to the question of lions in ancient Egypt. From about 8500 to 5300 BC, climatic conditions in northeast Africa were relatively wet, with adequate rainfall to support extensive grasslands, herbivore prey populations, and lions and other predators (Bubenzer *et al.*, 2007). Then, driven principally by progressively weaker summer monsoons, aridification developed. Large-bodied herbivore species mainly retreated to the savannahs further south, setting in motion a series of extinction cascades in Egypt (Yeakel *et al.*, 2014). Lions followed the grazers, for in order to thrive, a lion typically requires a diet of 20 to 30 substantial animals per year, totalling about 2500 kilos (Schaller, 1972: 397, 453). While these numbers are based on the consumption averages for modern Serengeti lions, the figures have plausibly held steady over the millennia, to judge from the fairly constant size of lion molars (Yamaguchi *et al.*, 2004: 335-336).

Archaeological projects in the Wadi Howar region of Nubia have yielded a more detailed picture of the humid/arid transition (Kröpelin & Oehm, 2007; Kuper *et al.*, 2007; Pöllath & Peters, 2007). Through the sixth and into the fifth millennium, the area surrounding this tributary of the Nile comprised a mosaic of biomes – moist, deciduous savannahs, grasslands with thorn bushes, semi-deserts, and deserts. To the west lay the vast Nubian Paleolake, over 5000 km^2 in its heyday. Early fourth-millennium conditions, though drier, could still sustain aquatic species with high ecological demands, as well as elephants and other large mammals. But by the millennium's close, the situation had drastically changed. With the 150 mm annual precipitation isohyet some 200 km to the south, the Wadi Howar found itself surrounded by desert zones of contracted vegetation, and domesticated cattle dominated the faunal assemblage.

Farther north, as arid Egypt stood on the cusp of the dynastic age, few if any wild prides of lions could have managed to maintain themselves. They certainly flourished, however, in art. The most complete compilations of their occurrences remain Hemmer (1962) and Osborn (1998: 113-119). Hemmer proposed that those most often depicted were the subspecies *Panthera leo nubica* (named in 1843), which he pictured as a short-maned lion with an unfurred belly. Osborn distinguished between a long-maned and a short-maned lion, suggesting a subspecies dichotomy, with temporal correlations and distribution for each.

As we have seen, scientific work undertaken since these publications has effectively ruled out all such classification schemas. Yet one continues to find them perpetuated in otherwise rigorous analytical treatments. So, for instance, Yeakel *et al.* (2014) preserved Osborn's dichotomy, listing *Panthera leo* (long-maned) and *Panthera leo* (short-maned) as separate entries in their chart of mammalian extinctions in Egypt over the past 6000 years.

An important related issue is that the relationship between formal Egyptian art and faunal reality, in the sense we understand it, is very complex (Binnberg, 2017; Vernus & Yoyotte, 2005: 38-49). This is largely because tomb, temple, and palace imagery was intended primarily to convey aspects of social, political, or religious ideology (see further below). Egyptian art should not be taken as face-value snapshots, but evaluated in context. Not doing so leads to flawed conclusions, such as those in a recent niche modeling study (Peterson *et al.*, 2014: 555), which accepted that if *"sculptures depict Pharaohs killing male lions on the shores of the Nile,"* then wild lions were there to be killed.

Rock drawings, on the other hand, are more informal, and thus might be expected to reflect a greater degree of direct experience with the animals in question. Indeed, hundreds of creatures were pictured in deftly stylized modes on the cliffs and outcroppings of the Eastern and Western Deserts, Nubia, and the Sinai (Judd, 2009; Suková, 2011). It is striking how few lions there are. Of the 815 rock-art wild and domesticated animals from southeast Egypt, for instance, only two are catlike, neither of them definitively leonine. Even when lions can be confidently identified, the dating of rock art is fraught with difficulty (Edwards, 2006: 55-58; Judd, 2009: 73-86). In short, petroglyphic lions afford unreliable chronological indicators of the presence or absence of *Panthera leo leo*. Nevertheless, the extreme rarity of rock-art lions seems telling.

Lions in Egyptian Imagery

Why, then, do lions abound in Predynastic and pharaonic art and thought? From earliest times, lions provided broad scope for iconography and figurative language to express mastery over inimical forces, real or imagined, be they human enemies, prey animals, or natural elements of chaos (De Wit, 1951; Germond & Livet, 2001: 191-194; Strawn, 2005; Vernus & Yoyotte, 2005: 152-166). The lion was at once an emblem of triumphant, virile kingship and a worthy opponent for the king to slay (Van Essche, 1991). Leonine metaphors could apply as well to ordinary mortals. In the *Tale of the Shipwrecked Sailor*, for example, the narrator proudly asserts that his shipmates' *"hearts were fiercer than those of lions"* (quoted in Simpson, 1972: 51, 53). Cosmologically, an addorsed pair of lions symbolized the eastern and western horizon mountains and, as a corollary, yesterday and tomorrow. In addition, the lion constellation figured prominently in the star-chart ceilings in New Kingdom royal tombs (Wilkinson, 1989). The composite forms – human-headed lion and lion-headed deity – lie beyond our present purview.

Also from earliest times, Egyptians knew the lands beyond the first cataract of the Nile and the bounds of its valley (Creasman & Wilkinson, 2017). Given the logistics, interactions, and biomes involved, we may posit that lions were encountered more regularly to the northeast, rather than to the south (Smith, 2017). For royal hunts, beaters doubtless rounded up lions. This practice must have made it possible for Amenhotep III to claim he slaughtered over one hundred lions in the first decade of his reign. Even if this count is pharaonic hyperbole, enough kings killed enough lions in far-off places to create an oft-repeated trope that united the concept of heroic strength and skill with the idea of dominion and/or influence abroad.

To represent lions in text or image, long-distance travel was not necessarily required. Beginning in the first dynasties, royal menageries included lions, sent to court by foreign leaders or collected in the wild (Foster, 2020: 26-48; Houlihan, 1996: 195-208). Several tomb reliefs depict their transport in wooden cages, pulled along on sledges. Some lions, presumably born in captivity or taken at a young age, are shown sitting or standing quietly by the throne, itself often adorned with leonine motifs (Darnell, 2014: 245-246), or running alongside the royal chariot in battle.

In the *Instructions of Amenemhat*, couched as posthumous advice to his son and successor, the Middle Kingdom pharaoh says he *"curbed lions and carried off crocodiles"* (quoted in Simpson, 1972: 196). The Old Kingdom magician Dedi supposedly knew *"how to make a lion go behind him, its tether on the ground"* (quoted in Simpson, 1972: 22). We might find all this improbable or strictly metaphorical, were it not for such texts as an admonition to a wayward student: *"Lions and horses can be taught, but you..."* (quoted in French in De Wit, 1951: 10). And the numerous Egyptian and Aegean depictions of collared and leashed lions and other felids (Foster, 2008) find outside support in an entry listing a lion leash among the products of a Mesopotamian workshop at the turn of the second millennium (Van De Mieroop, 1987: 42).

However they obtained first-hand knowledge, artists built on the set of lion-specific identifiers developed early on, especially the tail tuft, hind leg saphena vein, and lobes within the ear. Hunting tableaus attest to an acuity of observation of lion predation. Among other authentic behaviors, lions are shown dispatching their prey by suffocation, clamping their jaws around the muzzle (Evans, 2010: 111-115; Kleingütl, 1997: 58-63). In fact, this normally occurs once the quarry has been brought down, but, as in the Old Kingdom tomb of Ptahhotep II, the hapless animal, there a bull, is depicted still standing (Germond & Livet, 2001: Fig. 111). With the deeper meaning involving the close ideological links between lions and bulls (De Wit, 1951: 400-403), this sort of image illustrates well the ways in which Egyptian lion art is simultaneously naturalistic and formulaic.

This may be seen already in Predynastic pictorial works (Germond & Livet, 2001: 39-40). On the so-called Hunters' Palette, for example, three lions threaten ordered society, symbolized by the ranks of well-armed men, who have wounded two of them with arrows (Spencer, 1980: 79). Undaunted, one lion mauls a fallen hunter, but a comrade draws his bow to finish the beast. The palette hails the hero's power and prestige, the celebration signaled by the shrine and its emblem in the corner opposite. A third lion is smaller and as yet unscathed. If it is a cub, the usual identification, why does it have a mature mane? Should we ascribe this to artistic confusion, or to insistence on the mane to ensure reading the animal as lion? Or is it meant to lurk in the distance, a harbinger of future, far-flung triumph, whether fact or fiction?

Likewise, millennia later in the temple vestibule of Abu Simbel, the diminutive, maned lion beside Ramesses II's war chariot vaunts pharaonic superiority over his enemies (Germond & Livet, 2001: Fig. 40). Its coat is spotted, a pattern echoed in the leopard or cheetah pelts used for part of the horses' accouterments. This mélange enhances viewers' admiration for the geographical and cultural range of the king's purported victories, as in the one he claimed at the Battle of Kadesh, monumentally depicted and described on the temple walls. In his military camp, a lion is shown twice, perhaps all the same individual, sitting docilely near the oxen teams and causing them no alarm (Strawn, 2005: Figs. 4.99-100).

Egyptian artists were notably adept at rendering the mane in all its plasticity. Various conventions were used – close-set flame shapes, plain or striated scales, short strokes, long stripes, pointed lappets, and dots, dashes, and curls. The mane often extends to a furred underbelly, which can be handled as a prolongation of the mane's design, or herringbones on a light ground, or hatch marks. As we recall, this pelage feature is a defining characteristic of most Asian lions. Its regular appearance in art lends support to the idea, put forward above, that the Levant and its neighboring regions were the main visual and actual sources for lions in ancient Egypt. We need not affirm continuous, direct exposure abroad or in menageries, for images in temples and tombs, copybooks and sketches surely provided models. On an ostracon found near the tomb of Tutankhamun, for example, someone practiced drawing a pharaoh spearing a lion with a hatched belly (Houlihan, 1996: Fig. 53).

Whorls and Swirls

In addition to their mane patterns, sundry lions in Egyptian art also sport rosettes, pinwheels, stars, or concentric circles, primarily on their shoulders (Figure 1). Sainte Fare Garnot (1937) was the first to call attention to the corpus. A decade later, Kantor (1947) took up the subject in depth, adding Near Eastern and Aegean parallels to her discussion and inspiring a spate of articles (Arkell, 1948; Bate, 1950; van Buren, 1950; Vollgraff-Roes, 1953). From then on, the designs have often figured in studies on stylistic influences, especially in ivories and the decorative arts (*e.g.* Collon, 2005; Feldman, 2006: 95-97; Michel, 1991; Rehak & Younger, 1998: 251; Thomas, 2016a; Younger, 1987). Their solar and stellar associations have been generally accepted (Budde, 2000; Satzinger, 1997; Wilkinson, 1989).

Three theories have emerged as to the origin of these motifs: (1) they derive from an actual aspect of male lion fur, readily seen; (2) they stem from a subtle fur feature observable on some young lions, but not once the mane has developed; and (3) they reference the ribbons occasionally shown adorning captive pharaonic lions, rather than any natural pelage pattern.

In order to shed fresh light on the matter, I examined 55 skins and stuffed specimens of males, females, and cubs in the mammalogy collections of the American Museum of Natural History (AMNH), the Yale Peabody Museum of Natural History (YPM), and the Turkmenistan National Museum (TNM) (Table 1). My findings demonstrate that the motifs are firmly based on localized reversals in the normally backwards sloping hairs of lion and other mammalian pelage. They also indicate that lion art studies have confounded two different types of these reversals.

In my view, the primary inspiration came from the nexus of a male's elbow tuft, which I refer to here as the whorl (Figure 2). This consists of coarse hairs projecting several centimeters from the surface of the skin, most often making a clockwise pinwheel in varying degrees of tightness, with extensions on either side (Figure 3). The

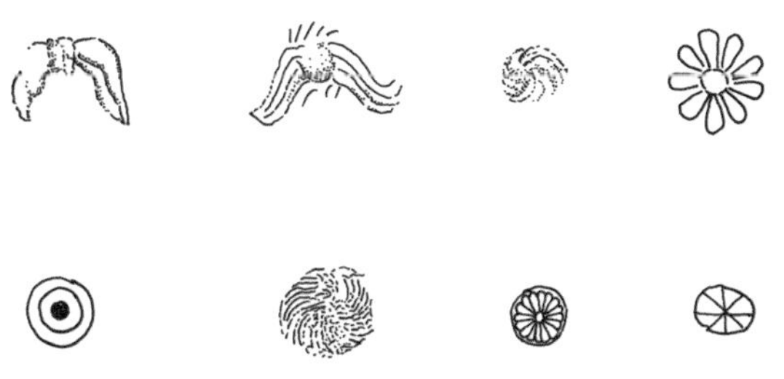

Figure 1. Representative hair whorl motifs, not to scale. Top row, left to right: plaque from Nippur. After Collon (2005, fig. 9); Queen of the Night plaque. After Collon (2005, fig. 12); relief from Nineveh. After Collon (2005, fig. 13); ceremonial axe from Choga Zanbil. After Aruz et. al. (2008: Fig. 148). Bottom row, left to right: wall painting from Deir al-Bahri. After Kantor (1947: fig. 2A); relief from Deir al-Bahri. After Kantor (1947: fig. 2B); relief from Saqqara. After Kantor (1947: fig. 1A); wall painting from Beni Hasan. After Kantor (1947: pl. VIIIC). Drawings by K. Foster.

Museum No.	Right side	Left side	Provenance on tag
AMNH 28151	blonde, dark in center, curved tufts, L 10cm	blonde, dark in center, knotlike, L 11cm	South Africa
AMNH 30244	none	none	British East Africa
AMNH 52076	none	none	Faradje, Congo, 1911
AMNH 52078	none	none	Zaire, 1912
AMNH 52082	none	none	Zaire, 1912
AMNH 80609	dark, knotlike, L 12.5cm	dark, knotlike, L 16.5cm	Capelongo, Angola, 1925
AMNH 83619	dark, knotlike, L 12cm	dark, knotlike with pinwheel center, L 13.3cm	Tsotsoroga Pan, Botswana, 1930
AMNH 85143	none	none	Serengeti, 1928
AMNH 85144	none	none	Serengeti, 1928
AMNH 186959	dark, pinwheel, Ø 8.5cm	dark, knotlike, L 10cm	Serengeti (?), before 1955
AMNH "Africa" exhibit	none	none	Serengeti
AMNH "Africa" exhibit	none	none	Serengeti
AMNH "Asia" exhibit	none	none	unspecified
YPM 2057	none	none	Serengeti
YPM 3217	none	none	unspecified
YPM 3218	light blonde, straight tufts, L 8cm	light blonde, curved tufts, L 11cm	Amorro River, Masai territory, Kenya, 1931
YPM 5251	light blonde to brown, pinwheel, Ø 4.5cm	light blonde to dark, pinwheel, Ø 6cm	Kenya, 1965
YPM 5252	none	none	unspecified
TNM exhibit	light, pinwheel, not measured	none	Turkmenistan (?)

Table 1. Catalogue of Male Specimens Examined.

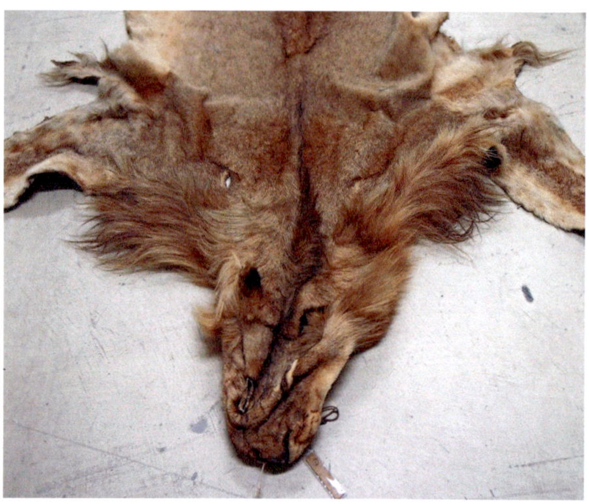

Figure 2. View of lion skin with hair whorls (AMNH 28151). Photograph by K. Foster.

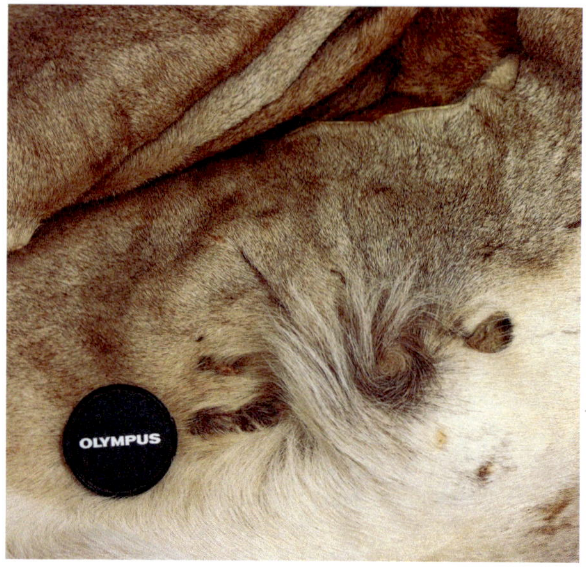

Figure 3. Detail of lion skin with pinwheel hair whorl (YPM 5251, left side). Photograph by K. Foster (the diameter of the lens cap is 4 cm).

whorl's coloration matches that of the individual's mane, or is darker. It occurs on both elbows, but is not identical or mirror image.

I found whorls on about one-third of the males I inspected, which accords with what lion specialists have observed in eastern and southern Africa: "*Some males grow dark prominent elbow tufts at an early age, whereas some males never develop elbow tufts*" (Whitman & Packer, 2006: 42). Asian lions are frequently said to have more pronounced elbow and tail tufts than African ones (Jackson, 2010: 17), but can this be substantiated beyond the Gir near-clones?

I believe that ancient artists saw whorls on lion elbows with sufficient frequency to render them as circular motifs. As for the whorls resembling a short length of tied cloth (Figure 4), it is interesting that this knotlike form does not appear on the lions in Egyptian or Aegean art, but occurs in Mesopotamian lion art from the third millennium on (Figure 1, top row). In another study (Foster, 2019), I have traced its trajectory and meaning in the self-presentation of ancient Near Eastern kings and suggested that its iconographic memory may persist in the heroic lion-knot of Herakles.

The other reversal is colloquially called a cowlick; I refer to it here more formally as the swirl. This forms a small pinwheel of the same hair color and quality as the rest of the coat and lies more or less smoothly on the body. There were no swirls on any of the skins I examined, but the stuffed Turkmenistan lion, a young male, had a swirl high on its right shoulder. First-hand inspection by others of zoo lions confirmed the presence of swirls on several

Figure 4. Detail of lion skin with knotlike hair whorl (AMNH 8060, left side). Photograph by K. Foster (the diameter of the lens cap is 4 cm).

juveniles and a lioness (Arkell, 1948; Bate 1950). According to the superintendent whom Bate consulted at the Dublin zoo, whenever juvenile males have swirls, their manes eventually cover them; the "very very small" one on the lioness was quite inconspicuous.

In my opinion, lion art chiefly renders whorls, not swirls, by reason of their greater visibility and frequency, association with maned males, and suggestive shapes. Accordingly, I would hesitate before reading the pinwheeled Egyptian lions of the horizon as juveniles with swirls, and thus symbols of the youthful sun-god and regeneration (Budde, 2000).

As for the emblematic placement of whorls, most are writ large on the lion's shoulder, framed by the mane; some are where the leg joins the body, just above the elbow; others appear on the back, forehead, or cheek. Migration from the elbow is to be expected, given the norms of Egyptian representation of the living world, which prioritized visual clarity over somatic exactitude.

Where we regularly see swirls is in the rosettes or pinwheels ornamenting many bovine heads in Egyptian and Aegean art (Thomas, 2016a; 2016b). Full investigation of this topic would take us beyond my brief, but it is worth mentioning here that age-old anecdotal reports and early breeding studies (*e.g.* Craft, 1931) correlating a swirl's type and placement on the forehead with that individual's reactivity are now borne out by solid empirical data on cattle, spurring similar work to be initiated on horses and dogs (Grandin *et al.*, 1995; Grandin & Deesing, 1998; Lillebø, 2013; Tomkins & McGreevy, 2010). An explanation may lie in the "*fact that hair patterns in the fetus form at the same time the brain forms*" (Grandin *et al.*, 1995: 117). It seems most probable that ancient observers recognized behavioral keys in the swirls on the livestock under their care.

To my knowledge, no comparable projects have been carried out on lions, nor has anecdotal information been collected systematically. We may, nevertheless, speculate and pose questions for future scientific and ethnographic inquiry. Do lions with hair reversals exhibit particular kinds of behavior or other traits? What genetic linkages might there be? Since only some males grow elbow tufts, are those lions especially robust or superior, or perceived that way? One wonders, for instance, if there are whorls or swirls on the lion skins thought to promote virility for the men who sleep on them in some parts of Africa today (Jackson, 2010: 35). And perhaps of greatest importance for deepening our understanding of lion art, do modern peoples imbue these fur features with symbolic meaning(s)?

Conclusions

As an historian of ancient art, with particular interests in the rendering of fur and feathers (Foster, 2016; 2019; 2020), I have ventured here into scientific realms, where I discovered that much is newly known about lions worldwide, from eons ago to the present. My own examination of 55 lion specimens led me to museum basements and storage drawers. The results of these journeys and findings oblige us to rethink many traditional interpretations and assumptions bearing on the appearance and roles of lions in Egyptian art and thought, reality and imagery. The ramifications take us beyond the Nile Valley to the art of the Near East and Aegean, where likewise lions keep company with kings, hunt and are hunted, and yes, often have whorls on their shoulders. But that is for another day.

Acknowledgments

In the preparation of this article, I am indebted for their patient guidance in lion genetics and other scientific matters to Ross Barnett, Hans Bauer, Constance de Brun, Laurence Frank, Stephen O'Brien, Craig Packer, and Peyton West. At the American Museum of Natural History, Eleanor Hoeger, Sara Ketelsen, and Marisa Surovy kindly facilitated my access to the mammology collections under their care, as did Kristof Zyskowski at the Yale Peabody Museum of Natural History. On the art historical/archaeological side, my thanks for references and comments are due especially to Fritz Blakolmer, John Darnell, and John Younger. Benjamin Foster and Nancy Thomas gave preliminary drafts of the manuscript the benefit of their careful reading. In Cairo, the questions posed by conference participants greatly helped shape and sharpen the final version, as did the comments made by an anonymous reviewer.

Bibliography

Antunes, A. J.L Troyer, M.E. Roelke, J. Pecon-Slattery, C. Packer, C. Winterbach, H. Winterbach, G. Hemson, L. Frank, P. Stander, L. Siefert, M. Driciru, P.J. Funston, K.A. Alexander, K.C. Prager, G. Mills, D. Wildt, M. Bush, S.J. O'Brien & W.E. Johnson. 2008. The Evolutionary Dynamics of the Lion *Panthera leo* Revealed by Host and Viral Population Genomics. – PLOS Genetics 4: 1-11.

Arkell, A.J. 1948. The Shoulder Ornament of Near Eastern Lions. – Journal of Near Eastern Studies 7: 52.

Aruz, J., K. Benzel, J.M. Evans. Eds. 2008. Beyond Babylon: Art, Trade, and Diplomacy in the Second Millennium B.C. – New York, The Metropolitan Museum of Art.

Barnett, R., N. Yamaguchi, B. Shapiro, S.Y.W. Ho, I. Barnes, R. Sabin, L. Werdelin, J. Cuisin & G. Larson. 2014. Revealing the Maternal Demographic History of *Panthera leo* using Ancient DNA and a Spatially Explicit Genealogical Analysis. – BMC Evolutionary Biology 14: 70-81.

Bar-Oz, G. & L. Weissbrod. 2017. The Kaleidoscope of Mammalian Faunas during the Terminal Pleistocene and Holocene in the Southern Levant. In: Enzel, Y. & O. Bar-Yosef. Eds. Quaternary of the Levant. Environments, Climate Change, and Humans. – Cambridge, Cambridge University Press: 363-368.

Bate, D.M.A. 1950. The 'Shoulder Ornament' of Near Eastern Lions. – Journal of Near Eastern Studies 9: 53-54.

Belmaker, M. 2017. Biogeography and Palaeoecology of the Early Pleistocene Large Mammals in the Levant. In: Enzel, Y. & O. Bar-Yosef. Eds. Quaternary of the Levant. Environments, Climate Change, and Humans. – Cambridge, Cambridge University Press: 355-361.

Binnberg, J. 2017. Animal Identification in Iconography. An Interdisciplinary Approach Combining Zoology, Anthropology, and Archaeology. In: O'Sullivan, R., C. Marini & J. Binneberg. Eds. Archaeological Approaches to Breaking Boundaries. Interaction, Integration, and Division. – Oxford, Archaeopress: 279-289.

Blakolmer, F. 2016. Il buono, il brutto, il cattivo. Character, Symbolism and Hierarchy of Animals and Supernatural Creatures in Minoan and Mycenaean Iconography. – Creta Antica 17: 97-182.

Buchanan, B. 1981. Early Near Eastern Seals in the Yale Babylonian Collection. – New Haven, Yale University Press.

Budde, D. 2000. Zur Symbolik der sogenannten Schulterrosette bei Löwendarstellungen. – Zeitschrift für ägyptische Sprache und Altertumskunde 127: 116-135.

Chauvet, J., E.B. Deschamps & C. Hillaire. 1996. Dawn of Art: The Chauvet Cave. The Oldest Known Paintings in the World. – New York, Harry N. Abrams.

Collon, D. 2005. The Queen of the Night. – London, The British Museum Press.

Craft, W.A. 1931. The Inheritance of Hair Whorls. – Proceedings of the Oklahoma Academy of Science for 1931: 7-8.

Creasman, P.P. & R.H. Wilkinson. Eds. 2017. Pharaoh's Land and Beyond. Ancient Egypt and Its Neighbors. – Oxford, Oxford University Press.

Darnell, J.C. 2014. The Stela of the Viceroy Usersatet (Boston MFA 25.632), His Shrine at Qasr Ibrim, and the Festival of Nubian Tribute under Amenhotep II. – Égypte nilotique et méditerranéenne 7: 239-276.

de Planhol, X. 2004. Le paysage animal. L'homme et la grande faune: une zoogéographie historique. – Paris, Fayard.

De Wit, C. 1951 Le rôle et le sens du lion dans l'Égypte ancienne. – Leiden, Brill.

Edwards, D.N. 2006. Drawings on Rocks, the Most Enduring Monuments of Middle Nubia. – Sudan & Nubia 10: 55-63.

Estes, R.D. 1991. The Behavior Guide to African Mammals. Including Hoofed Mammals, Carnivores, Primates. – Berkeley, University of California Press.

Evans, L. 2010. Animal Behaviour in Egyptian Art. Representations of the Natural World in Memphite Tomb Scenes. – Sydney, Australian Centre for Egyptology.

Evans, S. 2018. When the Last Lion Roars. – London, Bloomsbury Publishing.

Feldman, M.H. 2006. Diplomacy by Design. Luxury Arts and an "International Style" in the Ancient Near East, 1400-1200 BCE. – Chicago, University of Chicago Press.

Foster, K.P. 2008. A Taste for the Exotic. In: Hitchcock, L.A., R. Laffineur & J. Crowley. Eds. DAIS. The Aegean Feast. – Leuven, Peeters: 327-338.

Foster, K.P. 2016. Fur and Feathers in Aegean Art. In: Touchais, G., R. Laffineur & R. Rougemont. Eds. PHYSIS. L'environnement naturel et la relation homme-milieu dans le monde égéen protohistorique – Leuven, Peeters: 217-226.

Foster, K.P. 2019. Two New Aegean Memories and Metaphors. Athena as a Swallow, Herakles as a Lion. In: Borgna, E., I. Caloi, F. Carinci & R. Laffineur. Eds. MNEME. Past and Memory in the Aegean Bronze Age. – Leuven, Peeters: 609-618.

Foster, K.P. 2020. Strange and Wonderful. Exotic Flora and Fauna in Image and Imagination. – Oxford, Oxford University Press.

Germond, P. & J. Livet. 2001. An Egyptian Bestiary. Animals in Life and Religion in the Land of the Pharaohs. – London, Thames & Hudson.

Grandin, T. M.J. Deesing, J.J. Struthers & A.M. Swinker. 1995. Cattle with Hair Whorl Patterns above the Eyes are more Behaviourally Agitated During Restraint (Fixation). – Applied Animal Behaviour Science 46: 117-123.

Grandin, T. & M.J. Deesing. Eds. 1998. Genetics and the Behavior of Domestic Animals. – Amsterdam, Elsevier.

Gransard-Desmond, J.-O. 2010. Approche archéologique du lion pour la Syrie du IVe au IIe millénaire. – Akkadica 131: 145-163.

Hemmer, H. 1963. Untersuchungen über das Aussehen des altägyptischen Löwen *Panthera leo* nubica (Blainville, 1843) und seine verwandtschaftlichen Beziehungen zu den benachbarten Löwenuntertarten. – Säugetierkundliche Mitteilungen 12: 117-144.

Houlihan, P.F. 1996. The Animal World of the Pharaohs. – London, Thames & Hudson.

Jackson, D. 2010. Lion. – London, Reaktion Books.

Joslin, P. 1973. The Asiatic Lion. A Study of Ecology and Behaviour. – Edinburgh, University of Edinburgh.

Judd, T. 2009. Rock Art of the Eastern Desert of Egypt. Content, Comparisons, Dating and Significance. – Oxford, Archaeopress.

Kantor, H.J. 1947. The Shoulder Ornament of Near Eastern Lions. – Journal of Near Eastern Studies 6: 250-274.

Kitchener, A.C., C. Breitenmoser-Würsten, E. Eizirik, A. Gentry, L. Werdelin, A. Wilting, N. Yamaguchi, A.V. Abramov, P. Christiansen, C. Driscoll, J.W. Duckworth, W.E. Johnson, S.J. Luo, E. Meijaard, P. O'Donoghue, J. Sanderson, K. Seymour, M. Bruford, C. Groves, M. Hoffmann, K. Nowell, Z. Timmons & S. Tobe. 2017. A Revised Taxonomy of the Felidae. The Final Report of the Cat Classification Task Force of the IUCN/SSC Cat Specialist Group. – Cat News 11 (entire issue).

Kleinsgütl, D. 1997. Feliden in Altägypten. – Vienna, Institut für Afrikanistik und Ägyptologie.

Kröpelin, S. & S. Oehm. 2007. Wadi Howar National Park. In: Bubenzer, O., A. Bolton & F. Darius. Eds. Atlas of Cultural and Environmental Change in Arid Africa. – Cologne, Heinrich Barth Institut: 122-123.

Kuper, R., W. Möhlig, M. Bollig & O. Bubenzer. 2007. Environmental Change and Cultural History in Northeastern and Southwestern Africa. In: Bubenzer, O., A. Bolton & F. Darius. Eds. Atlas of Cultural and Environmental Change in Arid Africa. – Köln, Heinrich Barth Institut: 8-11.

Lillebø, S. 2013. Correlation between Hair Whorls and Different Types of Behavior in Dogs (*Canis familiaris*). – Unpublished MA Thesis, Norwegian University of Life Sciences.

Michel, S. 1991. Haarwirbel und Haarbüschel im skythischen und thrako-getischen Bildrepertoire. – Hamburger Bieträge zur Archäologie 18: 181-215.

Moore, A.E., F.P.D. Woody Cotteril, C. Winterback, H. Winterback, A. Antunes & S. O'Brien. 2015. Genetic Evidence for Contrasting Wetland and Savannah Habitat Specializations in Different Populations of Lions (*Panthera leo*). – Journal of Heredity: 1-3.

Nagel, D., S. Hilsberg, A. Benesch & J. Scholz. 2003. Functional Morphology and Fur Patterns in Recent and fossil *Panthera* species. – Scripta Geologica 126: 227-241.

O'Brien, S.J. 2003. Tears of the Cheetah. And Other Tales from the Genetic Frontier. – New York, Griffin Publishing.

Osborne, D.J. With contributions by J. Osbornova. 1998. The Mammals of Ancient Egypt. – Warminster, Aris & Phillips.

Packer, C. 2015. Lions in the Balance. Man-Eaters, Manes, and Men with Guns. – Chicago, University of Chicago Press.

Peterson, A.T., T. Radocy, E. Hall, J.C. Kerbis Peterhans & G.G. Celesia. 2014. The Potential Distribution of the Vulnerable African lion *Panthera leo* in the Face of Changing Global Climate. – Oryx 48: 555-564.

Pöllath, N. & J. Peters. 2007. Holocene Climate Change, Human Adaptation and Landscape Degradation in Arid Africa as Evidenced by the Faunal Record. In: Bubenzer, O., A. Bolton & F. Darius. Eds. Atlas of Cultural and Environmental Change in Arid Africa. – Köln, Heinrich Barth Institut: 64-67.

Rehak, P. & J.G. Younger. 1998. International Styles in Ivory Carving in the Bronze Age. In: Cline, E. & D. Harris-Cline. Eds. The Aegean and the Orient in the Second Millennium. Proceedings of the 50th Anniversary Symposium, Cincinnati, 18-20 April 1997. – Liège/Austin, Université de Liège/University of Texas at Austin: 230-255.

Sainte Fare Garnot, J. 1937. Le lion dans l'Art égyptien. – Bulletin de l'Institut Français d'Archéologie Orientale 37: 75-91.

Satzinger, H. 1997. The Scarab on the Cat's Forehead. In: National Museum in Warsaw "Pro-Egipt." Eds. Essays in Honour of Prof. Dr. Jadwiga Lipińska. – Warsaw, National Museum: 399-407.

Schaller, G.B. 1972. The Serengeti Lion. A Study of Predator-Prey Relations. – Chicago, University of Chicago Press.

Schnitzler, A.E. 2011. Past and Present Distribution of the North African-Asian Lion Subgroup. A Review. – Mammal Review 41: 220-243.

Shaplund, A. 2010. The Minoan Lion: Presence and Absence on Bronze Age Crete. – World Archaeology 42: 273-289.

Sharma, D. & A.J.T. Johnsingh. 1996. Impacts of Management Practices on Lion and Ungulate Habitats in Gir Protected Area. – Dehradun, Wildlife Institute of India.

Simpson, W.K. Ed. 1972. The Literature of Ancient Egypt. An Anthology of Stories, Instructions, and Poetry. – New Haven, Yale University Press.

Singh, H.S. 1997. Population Dynamics, Group Structure and Natural Dispersal of the Asiatic lion *Panthera leo persica*. – Journal of the Bombay Natural History Society 94: 65-70.

Smith, S.T. 2017. Artisans and Their Products. Interaction in Art and Architecture. In: Creasman, P.P. & R.H. Wilkinson. Eds. Pharaoh's Land and Beyond. Ancient Egypt and Its Neighbors. – Oxford, Oxford University Press: 133-148.

Spencer, A J. 1980. Catalogue of Egyptian Antiquities in the British Museum V. Early Dynastic Objects. – London, The British Museum Press.

Strawn, B.A. 2005. What Is Stronger than a Lion? Leonine Image and Metaphor in the Hebrew Bible and the Ancient Near East. – Fribourg, Vandenhoeck & Ruprecht GmbH & Co KG.

Suková, L. 2011. The Rock Art of Lower Nubia (Czechoslovak Concession), – Prague, Czech Institute of Egyptology.

Thomas, N.R. 2014. A Lion's Eye View of the Greek Bronze Age. In: Touchais, G. R. Laffineur & F. Rougemont. Eds. PHYSIS. L'environnement naturel et la relation homme-milieu dans le monde égéen protohistorique. – Leuven, Peeters: 375-389.

Thomas, N.R. 2016a. 'Hair Stars' and 'Sun Disks' on Bulls and Lions. A Reality Check on Movements of Aegean Symbolic Motifs to Egypt, with Special Reference to the Palace at Malkata. In: Alram-Stern, E., F. Blakolmer, S. Deger-Jalkotzy, R. Laffineur & J. Weilhartner. Eds. METAPHYSIS. Ritual, Myth and Symbolism in the Aegean Bronze Age. – Leuven, Peeters: 129-137.

Thomas, N.R. 2016b. Connecting the Pieces: Egypt, Dendra, and the Elusive 'Keftiu' Cup. In: Driessen, J. Ed. RA-PI-NE-U. Studies on the Mycenaean World offered to Robert Laffineur for his 70th Birthday. – Louvain, UCL Presses Universitaires de Louvain: 327-347.

Tompkins, L. M. & P.D. McGreevy. 2010. Hair Whorls in the Dog (*Canis familiaris*). I. Distribution, Part II. Asymmetries. – The Anatomical Record 293: 338-350, 513-518.

van Buren, E.D. 1950. An Additional Note on the Hair Whirl. – Journal of Near Eastern Studies 9: 54-55.

Van De Mieroop, M. 1987. Crafts in the Early Isin Period. A Study of the Isin Craft Archive from the Reigns of Išbi-Erra and Šū-ilišu. – Leuven, Peeters.

Van Essche, E. 1991. Les félins à la guerre. In: Delvaux, L. & E. Warmenbol. Eds. Les divins chats d'Égypte. Un air subtil, un dangereux parfum. – Leuven, Peeters: 31-48.

Vernus, P. & J. Yoyotte. 2005. Bestiaire des Pharaons. – Paris, Agnès Viénot Ed.: Ed. Perrin.

Vollgraff-Roes, A. 1953. The Lion with Body Markings in Oriental Art. – Journal of Near Eastern Studies 12: 40-49.

West, P.M. & C. Packer. 2002. Sexual Selection, Temperature, and the Lion's Mane. – Science 297: 1339-1343.

Whitman, K.L., & C. Packer. 2006. A Hunter's Guide to Aging Lions in Eastern and Southern Africa. – Huntington Beach, Conservation Force.

Wilkinson, R.H. 1989. A Possible Origin for the 'Shoulder Ornaments' in Egyptian Representations of Lions. – Varia Aegyptiaca 5: 59-71.

Yamaguchi, N., A. Cooper, L. Werdelin & D.W. Macdonald. 2004. Evolution of the Mane and Group-Living in the Lion (*Panthera leo*). A Review. – Journal of Zoology 263: 329-342.

Yamaguchi, N., A.C. Kitchener, C.A. Driscoll & D.W. Macdonald. 2009. Divided Infraorbital Foramen in the Lion (*Panthera leo*). Its Implications for Colonisation History, Population Bottlenecks, and Conservation of the Asian lion (*P. l. persica*). – Contributions to Zoology 78: 77-83.

Yeakel, J.D., M.M. Pires, L. Rudolf, N.J. Dominy, P.L. Koch, P.R. Guimarães Jr. & T. Gross. 2014. Collapse of An Ecological Network in Ancient Egypt. – Proceedings of the National Academy of Sciences of the United States of America 111: 14472-14477.

Younger, J.G. 1987. The End of Mycenaean Art. In: Thomas, E. Ed. Forschungen zur griechischen Vorgeschichte. Das Ende der mykenischen Welt, Akten des internationalen Kolloquiums 7-8 July 1984 in Köln. – Berlin, Berhard Thomashera: 63-72.

Human and Faunal Remains in Egypt: A New Department and a New Approach

Zeinab Hashesh & Ahmed Gabr

Introduction

The Egyptian climate has contributed to the conservation of its bioarchaeological heritage. The majority of excavated sites consist of cemeteries, and thus the gathering, analysis, and curation of the remains from these sites is crucial. However, in many cases this important work has not been given the attention that it deserves. The study of human remains throughout the history of Egyptology has been sporadic at best (Aufderheide, 2003: 10; Baker & Judd, 2012; Ikram, 2015; 2018; Kaiser, 2018) and fraught with ideological bias at worst. Studies on skeletons carried out in the 1880s emphasized political and racial theories, and many of the analyses used skull biometrics. Thus, excavators kept only the skull, pelvis, and pathological cases as seen, for example, in the cases of De Morgan (1897) and Petrie (1885; 1890; 1894; Petrie & Quibell, 1896; 1900; 1901; Petrie *et al.*, 1902-1904; 1907; 1913; 1923; see also Kaiser, 2018: 32). In contrast, later generations, such as Brunton and Reisner (Brunton, 1923: 1; Reisner, 1932) paid more attention to whole bodies in order to better understand health and disease in ancient Egypt, in part as a reflection of the changing nature of the study of human remains worldwide. As time passed, anatomists, medical doctors, and physical anthropologists contributed to the study of human remains, including medical professionals such as Grafton Elliot Smith, Douglas Derry, and Ahmed M. Batrawi (Aufderheide, 2003: 10; Baker & Judd, 2012; Ikram, 2015; 2018; Kaiser, 2018) making it a more main-stream discipline, and one that involved examining and curating entire skeletons/mummies. Now, bioarchaeological remains are more highly valued for the information that they provide about the health of ancient populations, their diet, foodways, and cultural and religious history.

Naturally, large amounts of human remains have been excavated throughout Egypt in the last century and a half. However, until fairly recently they were often not accorded the same protection and were not valued as much as anthropogenically produced artefacts. The Egyptian Law for Antiquities Protection No. 117 of 1983, as amended by laws No. 3, No. 61 and No. 91 of 2010, and according to the first article, point 3, states that *"we should keep our human and faunal remains because the progress of countries is measured by how much they are keeping their heritage."*

Thus, in 1983, the importance of human remains was officially recognized. However, attention paid to the remains and their curation was again sporadic, at best. Finally, in 1994 an anthropology laboratory was established in the Research and Conservation Centre at the Ministry of Antiquities, under the direction of Dr. Samia el-Merghani. The laboratory was able to collect for curation human remains from some archaeological sites, such as Kafr Hassan Dawoud, Tell Samara, Mensheat Ezzat, Saqqara, Marina, Sohag, Luxor, and Aswan. However, with only a small staff and limited resources, it became increasingly

In: Ikram, S., J. Kaiser & S. Porcier. Eds. 2021. *The Ancient Egyptians and the Natural World. Flora, Fauna, and Science.* – Leiden, Sidestone Press: 123-132.

difficult for the anthropology lab to oversee the curation of the vast amounts of human remains excavated by the Ministry all over Egypt.

In 2006 an advanced field school conducted by the Ancient Egypt Research Associates (AERA) in conjunction with the American Research Center in Egypt (ARCE) provided an opportunity for the Ministry of Antiquities archaeologists to be trained in bioarchaeology and how to excavate, record and store human remains. Through 2007, 2008 and 2009 the authors as graduated students participated in a series of further field schools, serving as assistant teachers and finally as supervisors. This was the beginning of establishing human remains as one of the major disciplines in the Egyptian excavations of the Ministry. In addition, in 2007, a conference held in the Sinai, entitled 'Managing Egypt's Cultural Heritage: The First Egyptian Cultural Heritage Organisation Conference Cultural Heritage Management' also emphasized the urgency for curating bioarchaeological remains all over Egypt (Hassan *et al.*, 2009; Ikram, 2009).

After graduating from the AERA fieldschool, the authors started to teach bioarchaeology to young Egyptian archaeologists within the Ministry of Antiquities. Subsequently, more training courses and workshops for both archaeologists and the public were held all over Egypt, in places such as Alexandria, Siwa, Marsa Matruh, North Sinai, Serabit al-Khadem, Cairo, Dakahliyya, Zagazig, Minya, Beni Hasan, Dendara, Abydos, Luxor, and Aswan (Figure 1).

Unfortunately, however, progress toward promoting the importance of human remains research and the attempts at establishing areas for proper storage and curation was halted with the revolution of 25 January 2011, and the situation worsened during the political turmoil that followed. Many archaeological sites suffered different degrees of destruction, including al-Hiba, Saqqara, and Abusir al-Malaq, where vast numbers of burials and skeletal remains were destroyed without documentation. One of the main reasons for the lack of recording was absence of a special department for human and faunal remains that should have been established in order to record and store bioarchaeological remains within the storage magazines and Inspectorates. Finally, in August 2017 such a department was created.

Figure 1. Library of Alexandria Bioarcheology workshop organized by the first author. Photograph by S. Ahmed.

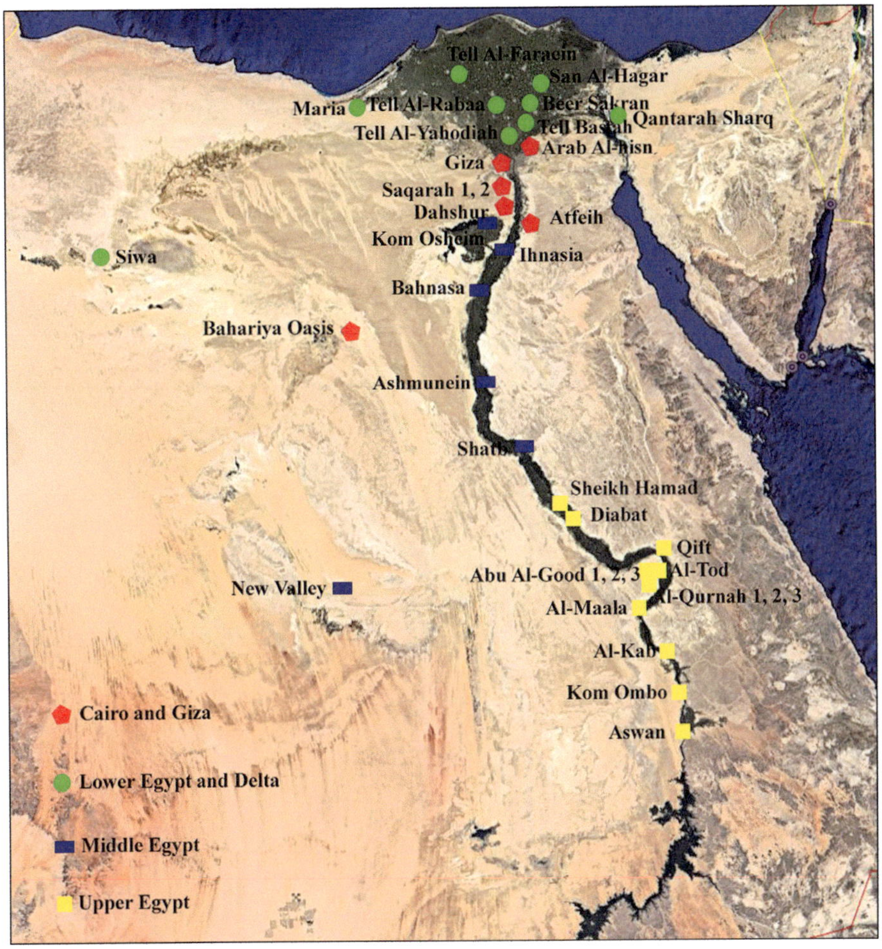

Figure 2. Satellite view of Egypt showing the magazines all over Egypt divided to four main areas; Lower Egypt and Delta, Cairo and Giza, Middle Egypt and upper Egypt. Map adapted from Google by A. Gabr.

The Department for Bioarchaeological Remains

The newly founded Department for human and faunal remains was established under supervision of the head of the Magazines Sector Said Shebl, and it had several goals. These included: 1) Assessment of skeletal materials inside storage magazines; 2) Establishing a processing and recording system; 3) Building a database for human and faunal remains at magazines; 4) Documentation of all skeletal materials and 5) Determination of the future realities and options for recording, storing and maintaining the materials.

Challenges

Even after a full year of preliminary surveys of storage magazines, it is still difficult to estimate the full quantity of human remains that the department will have to deal with. Twenty magazines were surveyed during 2017: in lower Egypt (Buto and Tell Basta Magazines), in Cairo and Giza (Atfih, two in Saqqara, and Arab al-Hisn), Middle Egypt and Oases (Ihnasya al-Medina, al-Kharga oasis, Kom Aushim), in Upper Egypt (Aswan, Kom Ombo, al-Kab, Luxor, two in al-Maala, Qurna, Tod, Deyabat, Qift, and El Sheikh Hamed) (Figure 2). It is not unlikely that the final tally will run hundreds of thousands of bones. In order to deal with the enormous amount of material, the first author divided the Department into three major units: 1) A unit for studying and documentation, which is dedicated to studying and recording what is already inside the magazines; 2) A unit for publishing, which publishes (or republishes) reports and translates Arabic reports into English and 3) A unit for training, which will be responsible for training programs for the archaeologists who work in the magazines and who will be carrying out the documentation process.

One of the major challenges facing the newly minted department was to design and then implement a recording system. In the first instance the authors had to assess the variety of materials to be recorded in order to determine what information should be recorded, and tools were needed to do so successfully, accurately, and economically.

The only way to proceed was to start with visiting magazines. The 34 official ministry magazines are divided into four regions: nine in Lower Egypt, seven in Cairo and Giza, six in Middle Egypt, and 12 in Upper Egypt (Figure 2). Additionally, information was collected from

Figure 3. Human and faunal remains before cleaning, Atfih magazine, room 7. Photograph by A. Gabr.

147 magazines belonging to foreign missions in Egypt, though no internal inspection was carried out at this time.

The storage within the magazines varied. Sometimes the human remains were placed inside wooden or cardboard boxes without sorting, and skeletons were completely mixed-up and not grouped together. In addition, some skulls had been placed on the metal shelving in the magazine where they comingled with stone blocks and organic materials (Figure 3). Labels were minimal, and often missing. It was unusual to find complete information about the excavators, provenance of the material, or the date of its excavation. Our subsequent assessment of some magazines revealed that their contents might range in date from 1898 onward.

After one year of surveys, the authors were able to identify the broader topics and kinds of information that could be retrieved. This allowed us to design a recording system. In parallel with designing the recording system, we established a suitable processing system for starting the magazines survey and evaluation of the human remains proper, carried out after discussions with colleagues all over the world, and guided by best practices for collections management, especially human remains (Antoine & Taylor, 2014).

The following methodology was used: survey sheets were designed that included the main information about the human and faunal remains in each magazineand were used for documentation, a basic cleaning of the materials was carried out, the cleaned remains were photographed, and finally the data were entered into an Excel sheet. The survey sheet included the following fields: magazine name, room, shelf, container number/s, container type, provenance, year and excavator information, general description; identification of whether the bone was human or non-human, presence of related textiles, finds, plant materials, state of preservation, publications related to the material, and any magazine registration information, any other comments, and the recorder's information (name/initials and date) (Figure 4).

Study Case: The Atfih Magazine

The Atfih magazine is situated approximately 90 km south of Cairo near Helwan. It was the first magazine we started with, and it contains skeletal materials from the site of Maadi dated to ca. 3600 BC, which had been excavated by Ibrahim Rizkana from 1930 to 1953 and published in 1990 (Rizkana & Seeher, 1990). These skeletal remains had an eclectic storage history: they had been kept in a Cairo University storeroom since 1948, then moved to the Helwan Magazine where they stayed until 2017, when they were finally moved to Atfih.

Encountering the human remains for the first time in Atfih Magazine was discouraging. They were divided between two different rooms. In Room 1, human and animal skulls were placed directly on shelves, mixed in with other organic materials, including shells, as well as ceramics. A thick layer of dust covered all materials. The other room, Room 7, contained disarticulated human and animal remains in wooden boxes and bags. None had been cleaned after excavation, and no system of recording or inventory existed (Figure 3).

The material was preliminarily analyzed according to standards based on Buikstra and Ubelaker (1994) a minimum number of individuals was established, and individual skeletons reassembled, if possible. Room 7 contained more than 15 boxes.

Human and faunal remains **Survey Sheet** Page......../........

Magazine Name:	inspectorate:	Room:	Shelf:

Container Type:	# of containers:

Provinence:	Year:	Excavator:

Description	

Mummified: Y / N

Animal: Y / N	Species:

Textile: Y / N	fabric:

Related Finds: plants	Objects

preservation Excellent / good / fair / Bad

Published: Y / N

Registered: Y / N		
	If yes:	General / study

comments:	

Recorded By	Initial	Date

Zeinab Hashesh, Ahmed Gabr

Figure 4. The survey sheet form which designed to include the main information about the human and faunal remains in each magazine. Designed by Z. Hashesh and A. Gabr.

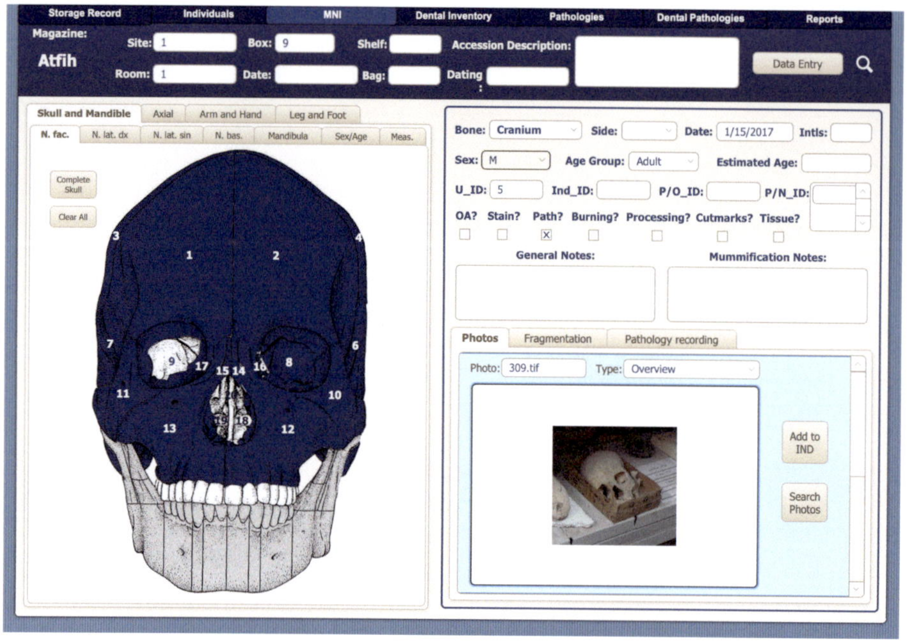

Figure 5. Basic database supported with photos, Atfih Magazine. Originally designed by J. Kaiser for ARCE, and adapted to the survey project.

After surveying the state of the magazine and its contents, we wrote a report that included recommendations for magazine logistics and also for the best methods for storing skeletal materials, in addition to providing the results in a basic database, supported with photos (Figure 5). This database was based on work by Jessica Kaiser (2018), and has since then been refined, and is to be adapted so that it can be used for all Egyptian magazines.

This was the Department's first step. However, with it, one of our long-term aims had been achieved: giving researchers the opportunity to work on these materials. Since the survey, the Atfih materials have been used by a young student, Hajar Fathi, as the basis for her MA thesis at Fayum University's Faculty of Archaeology.

Study Case: The Ihnasya al-Medina Magazine

The Ihnasya Magazine is located at Ihnasya al-Medina, approximately 18 km to the west of Beni Suef. It contains a number of mummified human and animal remains from different periods. The mummified remains come from Bahsamon, Abusir al-Malaq, and al-Moddel, sites that are all close to Beni Suef. The magazine contained seven human mummies, five of which were inside painted coffins, and one mummified crocodile that had been placed on a shelf. All the materials were covered with dust (Figure 6). We carried out a survey collecting data about these materials, then entered it into the database supported with photos. The survey helped us to develop basic guidelines about how skeletal and mummified remains should be moved and stored after excavation (Ikram, 2015: 73) (Figure 7).

Recommendations and guidelines to local employees included the following suggestions:

- Eliminate cotton, sponge, or straw as storing materials;
- Store all bioarchaeological remains in a separate place away from any stone or wooden objects;
- Keep the bioarcheological remains on the middle shelves. If they have to be placed on the lower shelf, it should be at least 20cm above the floor;
- Complete and partial mummified remains should be placed on metal shelves and be covered with white cloth (made from cotton) away from the sun or any artificial light and in a well-ventilated area;
- If using plastic bags, be sure they have perforations and are not completely sealed;
- Every skeleton should be stored in a separate box and labeled with all relevant information;
- In cases where textile remains adhere to the bones; do not remove them. The remains should be handled by a conservator;
- Loose teeth, maxillae, and mandibles should be bagged separately from the cranial vault;
- Do not use aluminum foil. If you have to, use it as a temporarily solution only, not for more than six months and do not place it directly on the bones;
- Do NOT eat, drink or smoke inside the magazine.

We emphasized and explained to the magazine's directors the future steps necessary for the proper storage of this skeletal material, and reminded them of its significance in terms not only of scholarship, but also the Antiquities' laws. This is something that must be emphasized throughout

Figure 6. Mummified remains inside a wooden coffin before cleaning, Ihnasya al-Medina Magazine. Photo by A. Gabr.

ضوابط بشأن استلام البقايا الادمية والحيوانية بالمخازن المتحفية

1- يجب فحص البقايا الادمية والحيوانية التى سيتسلمها المخزن على الطبيعة قبل استلامها وفى حالة وجود اكثر من مصدر للبقايا العظمية يتم فصل كل مصدر على حدى.

2- يكون استلام البقايا الادمية والحيوانية مرفقاً بكشف موضح به مكان الاكتشاف (المصدر) والوصف وصورة ضوئية ملونة لكل هيكل عظمى أو مومياء أو حتى العظام المنفصلة كلا على حدى وذلك سواء كانت هذه البقايا ناتج حفائر ، حرز أو ناتج حفر خلسة على أن يرفق معها تفاصيل القضية وتفاصيل العثور عليها واقرب موقع أثرى منها.

3- يحفظ كل هيكل عظمى أو مومياء فى صندوق خشبى أو بلاستيك منفصل ويدون عليه كافة البيانات

4- يجب ان تغطى المومياوات بقماش ابيض قطن فى مكان جيد التهوية بعيد عن اى اشعة للشمس أول اى مصدر ضوء صناعى مباشر وفى حال استخدام ضوء كهربائى يفضل أن يكون خافت لأن الاضاءة تؤثر على حالة حفظ الاثار العضوية فى العموم .

5- يحذر التعامل مع البقايا الادمية او الحيوانية بدون استخدام جوانتى لليد ، ويجب غسل الايدى جيدا بعد الانتهاء من العمل .

6- كل الادوات المستخدمة سواء كانت أكياس أو صناديق ورقية أو خشبية لابد أن تكون جاف تماما قبل الاستخدام وفى حال استخدام أكياس بلاستيكية يجب عمل أكثر من ثقب بها .

7- يجب وضع البقايا الادمية والحيوانية فى الارفف الوسطى وفى حال انشغال الأرفف الوسطى يتم استخدام الرف السفلى والذى لابد أن يرتفع عن الارض 10 سم .

8- يحذر استخدام القطن والاسفنج داخل الصناديق المستخدمة لحفظ العظام داخل المخزن .

9- التأكد من عدم وجود حشرات وفئران ولذا يمكن استخدام مبيد حشرى و مصيدة فئران داخل المخزن.

10- يحظر دخول أى من العاملين بالمخزن بأى مصدر لطعام او شراب وخاصة فى أماكن حفظ البقايا الأدمية والحيوانية .

11- بعد استلام البقايا الادمية أو الحيوانية يتم مخاطبة ادارة دراسة وتسجيل وتوثيق البقايا الادمية والحيوانية للتنسيق مع مركز بحوث وصيانة الاثار لمعاينتها على الطبيعة و تحديد عما اذا كانت بحاجة لترميم وصيانة من عدمه.

Figure 7. Basic guidelines about the roles of how to receive skeletal remains and the stages of storing. Designed by Z. Hashesh.

Egypt as there has been a trend to rebury such material, even if it has not been studied, due to space constraints.

Conclusion

As previously stated, archaeological sites and magazines in Egypt contain vast amounts of bioarchaeological remains, perhaps the largest collection in the world. These have historically not been curated in an ideal manner, but due to a new initiative this is changing. The Department for Human and Faunal Remains within the Ministry, founded in 2017, has established a protocol for surveying and inventorying the material as well as training staff to identify, catalogue, and to curate and store it properly.

The survey process started in 2017, when 20 out of 34 magazines were surveyed. Two magazines were surveyed in Lower Egypt, three magazines in Cairo and Giza, three magazines in Middle Egypt and 12 magazines in Upper Egypt. In addition, the location and self-reported holdings of 147 magazines belonging to foreign missions were collected in one list for the first time.

One archaeologist from every magazine was selected to be responsible for the human and faunal remains, and those archaeologists will subsequently receive further training through a series of courses. A set of guidelines for magazine directors was designed in order to help them with receiving, handling and storing human remains in all their magazines.

Our expectations about the success of this new department are related to how we can finish the training courses for magazine curators over the next two years, in addition to establishing the magazine database that will be accurately and thoroughly referenced and made available to researchers to give them access to these extremely valuable testaments to the lives of the ancient Egyptians.

Bibliography

Amer, M. 1936. The Excavations of the Egyptian University in the Prehistoric Site at Maadi, Near Cairo. – The Journal of the Royal Anthropological Institute of Great Britain and Ireland 66: 65-69.

Antoine, D. & E. Taylor. 2014. Collection Care: Handling, Storing, and Transporting Human Remains. In: Antoine, D., A. Fletcher & D. Hill. Eds. Regarding the Dead: Human Remains in the British Museum. – London, British Museum Press: 43-48.

Aufderheide, A.C. 2003. The Scientific Study of Mummies. – Cambridge, Cambridge University Press.

Baker, B.J. & M.A. Judd. 2012. Development of Paleopathology in the Nile Valley. In: Buikstra, J.E. & C. Roberts. Eds. The Global History of Paleopathology. Pioneers and Prospects. – Oxford, Oxford University Press: 209-234.

Buikstra, J.E. & D. H. Ubelaker. 1994. Standards for Data Collection from Human Skeletal Remains. – Fayetteville, Arkansas Archaeological Survey Research Series.

De Morgan, J. 1897. Recherches sur les origines de l'Égypte. – Paris, Ernest Leroux.

Hassan, F.A., I. Rizkana & Y. Fayed. 1996. Dental Conditions of the Population of Ma'adi Culture as Affected by the Environment. – Cairo, Geographical Research Series.

Hassan, F.A., L.S. Owen, A. De Trafford, G.J. Tassie & J. Van Wetering. 2009. Managing Egypt's Cultural Heritage: Proceedings of the First Egyptian Cultural Heritage Organisation Conference on Egyptian Cultural Heritage Management. – London, Golden House/Echo Publications.

Ikram, S. 2009. May they Live Forever: The Examination, Care, and Curation of Egyptian Mummies. In: Hassan, F.A., L.S. Owen, A. De Trafford, G.J. Tassie & J. Van Wetering. Eds. Managing Egypt's Cultural Heritage: Proceedings of the First Egyptian Cultural Heritage Organisation Conference on Egyptian Cultural Heritage Management. – London, Golden House/Echo Publications: 258-265.

Ikram, S. 2015 (online). Physical Anthropology and Mummies. In: Shaw, I. & E. Bloxham. Eds. Oxford Handbook of Egyptology. – Oxford, Oxford University Press (online version).

Ikram, S. 2018. An Overview of the History of the Excavation and Treatment of Ancient Human Remains in Egypt. In: O'Donnabhain, B., M.C. Lozada & M. Cecillia. Eds. Archaeological Human Remains: Legacies of Imperialism, Communism and Colonialism. – Cham, Springer: 45-56.

Ikram, S., J.E. Kaiser & R. Walker. 2015. Egyptian Bioarchaeology: Humans, Animals, and the Environment. – Leiden, Sidestone Press.

Kaiser, J.E. 2018. Raising the Dead: The Bioarchaeology of the Saite and Roman Period Wall of the Crow Cemetery at Giza. – Unpublished PhD thesis, University of California.

Petrie, W.M.F. 1890. Kahun, Gurob and Hawara. – London, Kegan Paul.

Petrie, W.M.F. 1894. Tell el Amarna. – London, Methuen & Co.

Petrie, W.M.F. 1900-1901. The Royal Tombs of the First Dynasty. – London, Egypt Exploration Fund/Kegan Paul, Trench, Trübner & Co.

Petrie, W.M.F. 1907. Gizeh and Rifeh. – London, Bernard Quaritch, School of Archaeology, University College.

Petrie, W.M.F., E.R. Ayrton, C.T. Currelly & A.E. Weigall. 1902-1904. Abydos. – London, Egypt Exploration Fund/ Kegan Paul, Trench, Trübner & Co.

Petrie, W.M.F. & J.E. Quibell. 1896. Naqada and Ballas. – London, Bernard Quaritch.

Reisner, G. 1932. A Provincial Cemetery of the Pyramid Age. – Berkeley, University of California Press.

Rizkana, I. & J. Seeher. 1990. The Predynastic Cemeteries of Maadi and Wadi Digla. With an Appendix on the Physical Anthropology by Karl-Georg Beck and Stefan Klug. – Mainz am Rhein, Philipp von Zabern.

Creatures of the Sun, Creatures of the Moon: Animal Mummies from Lisbon's National Archaeological Museum[1]

Salima Ikram[1], Carlos Prates[2], Sandra Sousa[2], and Carlos Oliveira[2]

Introduction

The Lisbon Mummy Project (LMP) was established to work on Lisbon's *Museu Nacional de Arqueologia* (National Archaeological Museum, [MNA]) collection of human and animal mummies. The collection of animal mummies was first imaged in 2007 at IMI-art/Affidea, a Lisbon private medical imaging group. Later, in 2010, with the support of Siemens, the mummies were scanned using Computer Tomography (CT) (Ikram, 2012a; Prates *et al.*, 2011; 2015; also see: https://artsandculture.google.com/exhibit/o-lisbon-mummy-project/qgList1Ii30FLA). The collection is small, but diverse, including humans, ibises, raptors, and crocodiles. This paper will focus on the results of the studies of the animal mummies: Juvenile Crocodile MNA E 128; Small Crocodiles MNA E 129, E 407, E 408; an Ibis mummy MNA E 127; Jar MNA E 304; Falcon Mummy MNA E 128.

Materials and Methods

The MNA has seven animal mummies, none of which is provenanced and it is unclear as to how and when they were donated to the Museum. They all appear to be of the votive type (Ikram, 2015a: 19-15), associated with different deities. Four of these are of crocodiles: one juvenile and three neonates, associated with the solar deity, Sobek (Figure 1). These have tentatively been identified as *Crocodilus niloticus*, pending DNA studies that might ascribe them to *C. succhus* (Hekkala *et al.*, 2011). For the most part, their wrappings have gone, save for a scrap or two (Figures 1, 2), leaving their bodies bare; the animals are stretched out with their limbs lying close to the torso. The animals are blackened due to the application of resinous material some of which survives on the skin (this has not been analysed as yet; Figures 1, 2). It is probably a mixture of resin, oil, and wax, although it is possible that bitumen was also used (Buckley *et al.*, 2004; Clark *et al.*, 2013; Clark *et al.*, 2016; Filcher, 2020; Ikram, 2013; Nissenbaum & Buckley, 2013). Each of the three highly 'resined' neonate crocodile are supported between two twigs (at least one twig is actually a split rib of a palm leaf) that lay along their length and were tied to the body with linen thread or folded pieces of bandages (Figures 1, 2). In one case at least it is clear that the dark resinous material was applied to the animal after it had been secured to the supports. Such supports are commonly used to protect these small,

1 We would like to dedicate this article to the memory of Alvaro Figueiredo, one of the founders of this project, and a dear friend.

In: Ikram, S., J. Kaiser & S. Porcier. Eds. 2021. *The Ancient Egyptians and the Natural World. Flora, Fauna, and Science.* – Leiden, Sidestone Press: 133-140.

Figure 1. The crocodile mummies (juvenile crocodile MNA E 128; neonates: MNA E 129, E 407, E 408). Traces of linen and embalming materials are visible. Photograph courtesy of Museu Nacional de Arqueologia.

Figure 3. A wrapped ibis mummy (MNA E 127). Photograph courtesy of Museu Nacional de Arqueologia.

Figure 4. A ceramic vessel made of Nile silt that contains the remains of an ibis mummy (MNA E 304). Photograph courtesy of Museu Nacional de Arqueologia.

Figure 2. One of the neonate crocodiles supported by the palm fibres and painted with black, resinous material. Photograph courtesy of Museu Nacional de Arqueologia.

fragile mummies from breaking (for example, from the Medelhavsmuseet in Stockholm, numbers 30810, 30721, 10133, and unnumbered ones from the Egyptian Museum, Cairo, that were found in the mouth of CG 29628).

There is one mummy of an ibis-type, a bird that is linked with the god Thoth. It is still enclosed in linen bandages, but parts of the beak, cranium, and feathers show through (Figure 3). Most probably, the surviving bandages were not its ultimate wrappings, but rather ones that were not meant to be seen. Many ibises positioned in this manner have elaborate final wrappings, either in the form of herringbones or coffers (such as those in the Egyptian Museum, Cairo, CG 29696; National Museum of Antiquities, Leiden H.III.EEE I; *Museo Egizio* Turin Suppl. 8188, Suppl. 8189, Suppl. 11013), with the bandaging beneath looking like the Lisbon example. Ibis mummies are perhaps the most common mummified animal to be found throughout Egypt (Ikram, 2012: 44; 2015b: xvi-xx). The positioning of the bird within the wrappings is common and can neither be used for dating nor provenance.

There is one long ceramic vessel made of Nile Silt, that is sealed with a small plate, also of Nile Silt, that is secured with gypsum plaster (Figure 4). Ibises are found interred in jars at sites such as Saqqara (Nicholson, 2015; Nicholson *et al.*, 1999), Abydos (Ikram, 2007; Loat, 1914; Peet & Loat, 1913), and Tuna al-Gebel (Kessler & Nur el-Din, 2015). The style of this particular jar most closely resembles those found at Saqqara, and it is probable that this example originates from there.

The seventh mummy is tightly wrapped, and some of the broad linen bandages are soaked in some sort of dark, shiny, resinous material (see above for its possible components) (Figure 5). It is possible that these bandages were the final wrappings, although bird mummies also had more elaborate outer wrappings (such as Egyptian Museum, Cairo CG 29682, CG 29689), and some even wore masks or had facial details picked out in linen (for example, Egyptian Museum, Cairo CG 29685; Medelhavsmuseet, Stockholm E 3885). The shape of the head suggests that it is a raptor. Raptor mummies are known from all over Egypt (Ikram, 2012b) and are emblems of solar deities, such as Re and Horus.

Figure 5. The wrapped mummy of a raptor (MNA E 128). Photograph courtesy of Museu Nacional de Arqueologia.

The animal mummies were first examined by x-rays (a digital over table from Siemens (Iconos), with digital radiography (DR) in an extended vertebral view/long leg protocol). This was followed by computed radiography (CR), segmental anatomic views, with Kodak imaging plates, KV 60-80; automatic exposure), and MDCT scans (Siemens Multidetector CT, Somatom Sensation 64 slices, Z-sharp technology and an isotropic resolution of 0,33 mm; specially planned protocols used 120 KV, 450 to 700 mAs, 1,5s tube rotation and 0,45 pitch, that were applied selectively to different anatomic areas, particularly in the case of the humans.

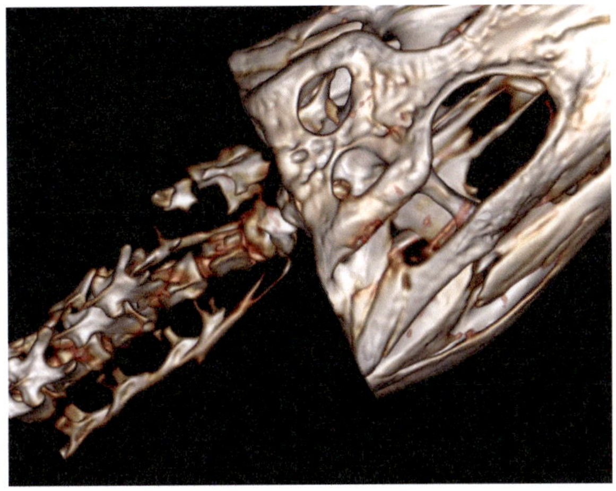

Figure 6. CT 3D reconstruction of the upper view of the juvenile crocodile skull (posterior mandible borders digitally removed) showing the fractured/luxated two upper cervical vertebrae. Photograph courtesy of IMI-art/Affidea.

Results

The imaging of the mummies helped to identify them and provided information about their preparation. It was not really possible to tell if the four crocodiles had been eviscerated, but it seems unlikely from the imaging. One of these might have been killed by having its cranio-vertebral junction severed by a sharp implement similar to a stiletto, as suggested by two luxated first cervical vertebrae (Figure 6) and an adjoining small sharp skin wound, that was not visible in normal x-rays, but was detected by the computer tomography (CT) scans. To accomplish this, the head would have to have been secured, and it is possible that the broken mandible of this mummy (Figure 7) is due to the pressure needed to immobilise the head. Of course, there is always the possibility that the mummies might have suffered broken bones (notably the mandibles) due to post-mortem handling.

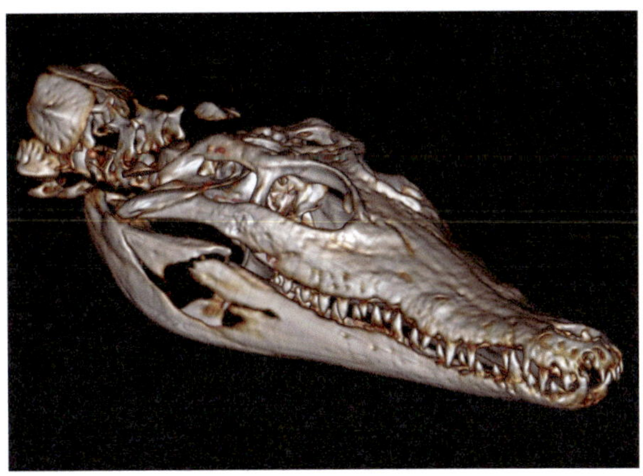

Figure 7. CT 3D reconstruction showing an oblique view of the juvenile crocodile skull showing the broken mandible and posteriorly the luxated cervical vertebrae. Photograph courtesy of IMI-art/Affidea.

Imaging confirmed that the ibis was indeed a mature sacred ibis (*Threskiornis aethiopicus*; Figure 8), with its long, curving beak and limbs. It was wrapped as if it were in the seated position, but pressed down on its feet, its wings neatly folded, and its neck is twisted and compressed with at least four levels of fracture so that its

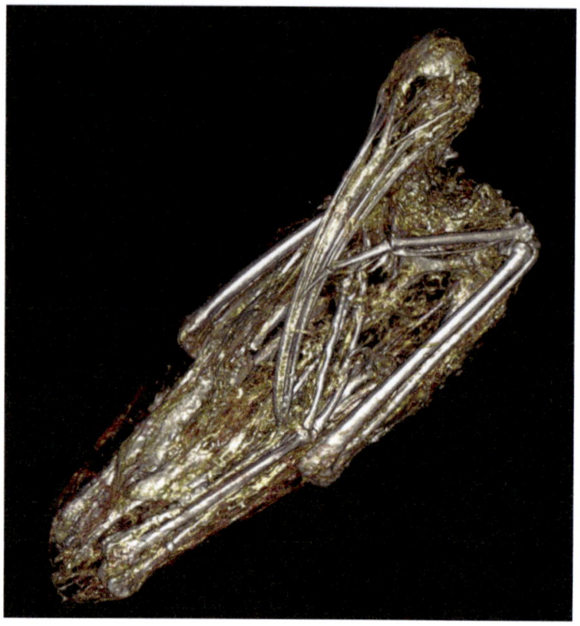

Figure 8. CT 3D reconstructed image of the skeleton of the ibis mummy. Photograph courtesy of IMI-art/Affidea.

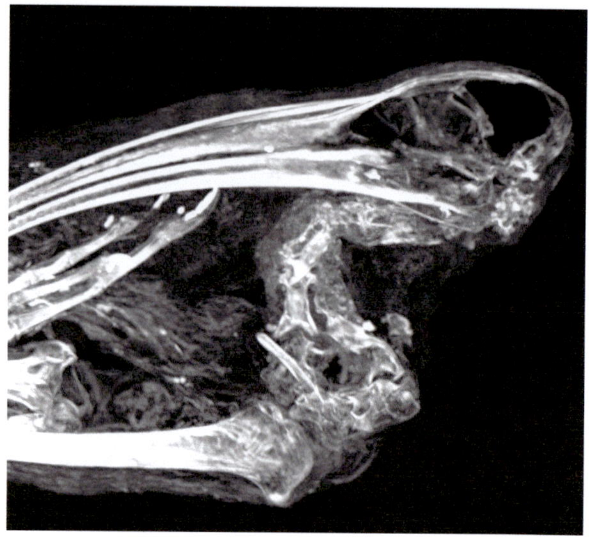

Figure 9. CT MIP (Maximum intensity projection – thick slice view) sagittal oblique view of the compressed ibis neck showing the vertebral trauma. Photograph courtesy of IMI-art/Affidea.

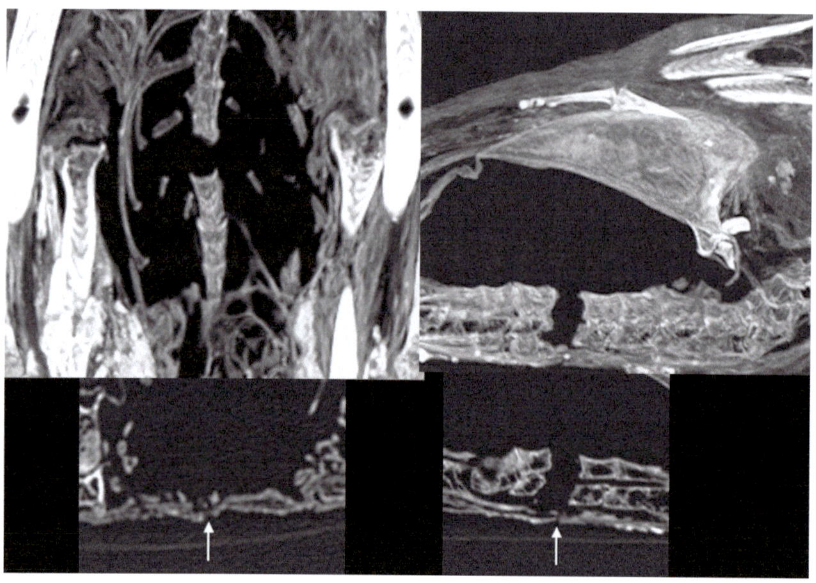

Figure 10. CT MPR (Multiplanar Reconstruction) oblique views (clockwise: coronal, sagittal, sagittal and axial) showing the precise excision of one of the ibis' vertebra, leaving the surrounding ones intact, as well as the emptied thorax. Below detailed views of skin defects (arrows). Photograph courtesy of IMI-art/Affidea.

beak rested on its belly (Figure 9). Having its neck wrung might have been the cause of death.

It was most probably eviscerated abdominally as the area is voided, but the evisceration cut is not clear. A curious feature about this bird is that there appears to be a precise excision of a dorsal vertebra (Figure 10), leaving the flanking vertebrae intact, together with their corresponding lateral ribs, but the reason for this is unclear and it is very probably unrelated to its mode of death. It is possible that the excision was dorsal (no feathers are visible, and there is a small break in the skin. Thus far no parallels have been found in other ibis mummies, but one exists in a human mummy, that of Namenkhetamun (26th Dynasty, ca. 664-525 BC) in the Birmingham Museum and Art Gallery (https://www.birminghammuseums.org.uk/blog/posts/meet-the-mummy-namenkhetamun).

In the past, sealed jars have proved challenging, but with contemporary CT technology one can now view the contents of sealed containers without damaging the artefact. The careful imaging of the sealed pottery jar revealed that the texture of the CT texture of the clay of the vessel body is distinct from that of the lid, indicating

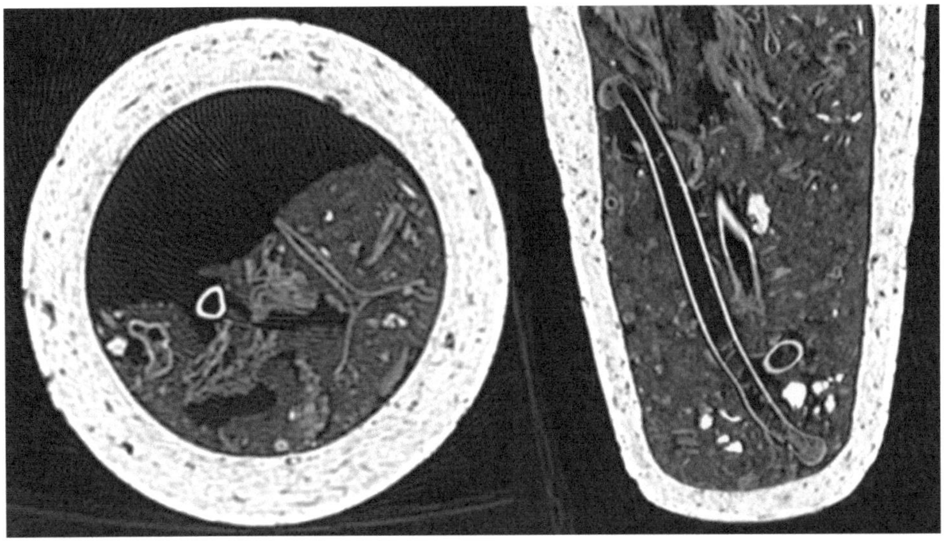

Figure 11. CT MPR (Multiplanar Reconstruction) orthogonal views of the ceramic vessel showing the sternum, humerus and other wing bones, as well as textile, pebbles, and other detritus, including, possibly, sawdust. Photograph courtesy of IMI-art/Affidea.

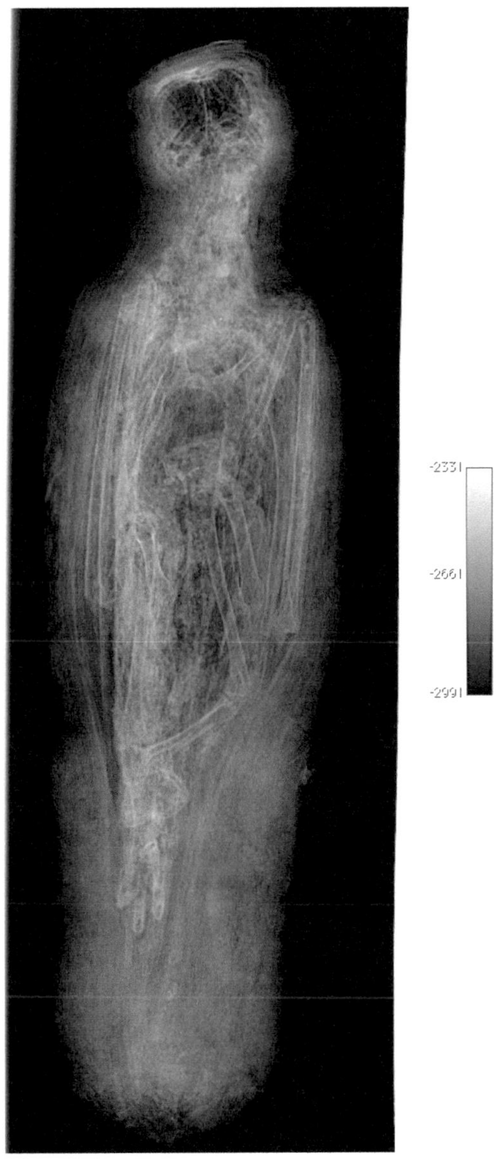

Figure 12 (right). A frontal x-ray showing the raptor mummy, one of its legs slightly crossed over the other. Photograph courtesy of IMI-art/Affidea.

a different model of manufacturing. The contents consist of the partial remains of a sacred ibis (*Threskiornis aethiopicus*) with bones (a set of bones of a wing – some broken – and a sternum), some feathers, mud, pebbles, textiles, and possibly sawdust (Figure 11). It is common to find fragments of animals (or sometimes just mud or other materials) wrapped up and placed in tombs, together with actual, complete, mummified creatures, creating packages/coffins that look as if they contain complete mummies, even if they do not. Whether this was deliberate dissimulation or whether it was a way of piously treating the fragmentary residual remains left over from mummifying creatures is unclear (for an overview of the different interpretations of this practice, see Ikram, 2019, with references).

The well-wrapped bird mummy (total length from skull to end of synsacrum is 20 cm, and with the addition of the extension of the feet, ca. 24 cm; Figure 12) that appeared to be a raptor, was indeed a falcon, a *Falco peregrinus*, based on the skull and the leg proportions (a short tarsometatarsus compared to the tibiotarsus).[2]

The bird was positioned with its legs extended and its wings close to the body. Along its cervical vertebrae there is a dense material apposed, mostly left located, and likely intended to help the bird hold a correct final pose. The vertebrae are irregularly aligned, possibly due to trauma/cause of death (Figure 13). Although it was eviscerated a small oval soft-tissue structure could be seen (Figure 14) in the thoracic cavity above what seems to be a laminar

2 We are grateful to Megan Spitzer of the Smithsonian Institution's Natural History Museum for confirming this identification.

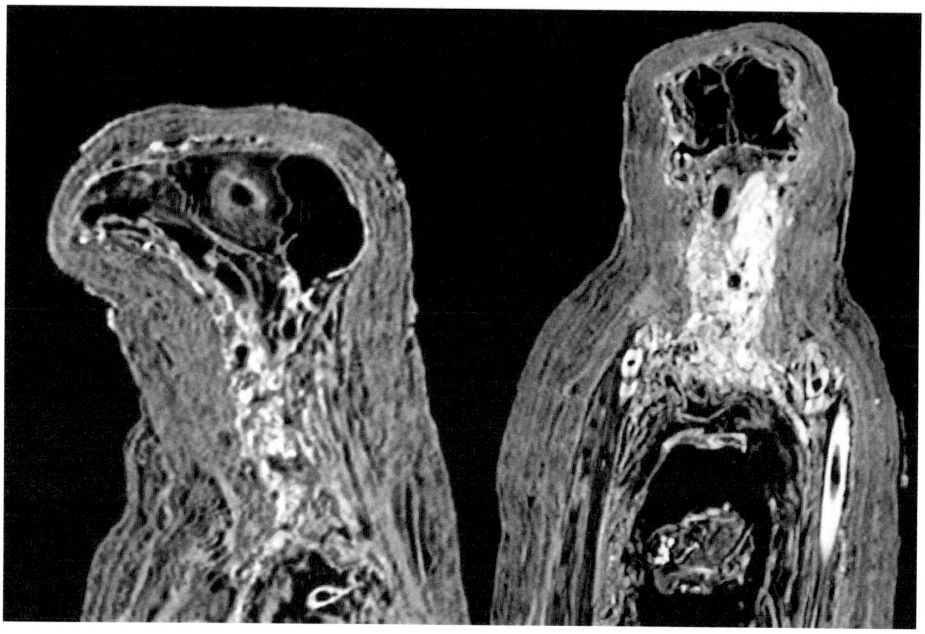

Figure 13. CT MPR (Multiplanar Reconstruction) oblique sagittal view of the raptor head and coronal upper body view, showing the dense neck supporting material. Photograph courtesy of IMI-art/Affidea.

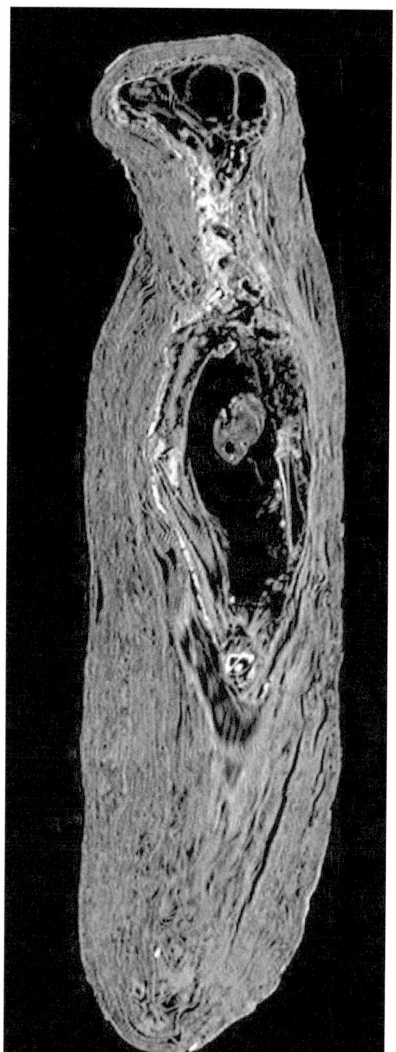

remain of part of an organ. This oval object might be a combination of visceral remains, or, more likely, part of the gizzard, with some of the contents still present, as attested by the presence of some tubular objects that are calcified/very dense, and similar to bone. It is possible that this was re-introduced into the body after evisceration – this is a practice that is seen in other examples of bird mummies (Ikram, 2017; Ikram *et al.*, 2015; Wade *et al.*, 2012).

Discussion

Although animal mummies have been known to naturalists and Egyptologists, interest in these remarkable multi-faceted artefacts has fluctuated through the centuries (Ikram, 2020). Recently, increasing attention has been paid to them (see Ikram, 2019 for a synopsis, and the various conferences, exhibitions, and publications associated with the topic), yet there are still many questions regarding these artefacts that remain unanswered. The very nature of certain groups of animal mummies continues to be a subject of discussion among Egyptologists (Ikram, 2015a; 2019; Kessler, 1986; 1989; 2007; Kessler & Nur el-Din, 2015; Ray, 1976; 2001; Smith & Davies, 2005; Smith *et al.*, 2006; 2011; von den Driesch, 1993; von den Driesch *et al.*, 2005), while the different ways in which the animals were acquired, mummified, and deposited, is still under concentrated study. Trying to establish chronologies for types of mummies (for example Bleiberg

Figure 14. CT MPR (Multiplanar Reconstruction) oblique sagittal view of the raptor body showing in the thoracic cavity a remaining oval soft tissue structure. Photograph courtesy of IMI-art/Affidea.

et al., 2013; Richardin *et al.*, 2017; Wasef *et al.*, 2015), and provenance, based on wrappings and parallels from excavated examples, remain foci of investigation. Thus, it is important to produce as many studies, as holistic and interdisciplinary as possible, to better understand these artefacts. It is only by gathering together a vast amount of information about the diverse types and species of animal mummies and making this accessible to a broad range of scholars that some of the questions that remain unanswered can be addressed. Thus, even when the results of investigations are not startling, they remain important parts of the puzzle that will eventually be able to answer some of the remaining questions concerning animal mummies.

This small group of animal mummies in the LMP is a testament to the interest and curiosity in ancient Egypt and its religious customs and environment evinced by the travellers who acquired them. LMP's investigation of these mummies had yielded a few interesting results. Notable among these is the possible cause of death of one of the crocodiles, the fact that the carefully sealed jar contains a loose amalgam of materials, including part of an ibis mummy, the strange absence of one vertebra in a wrapped ibis mummy, and the presence of part of an internal organ, attesting to either incomplete evisceration, or the deliberate return of viscera to the body of a falcon. Thus, it is very probable that more frequent use of CT scanning on other animal mummies will provide valuable findings, most of which are out of reach of plain radiography. Future work on this collection will hopefully include radiocarbon dating and analyses of the mummification materials. It will be of particular interest to find out if natron was employed for the desiccation of all or any of the species, and to identify the source(s) of the black, resinous material that was used on all of these animals as part of their mummification. Thus, this small group of mummies will contribute to our larger understanding of the ideologies behind the practice of animal mummification, as well as the practicalities of mummification, and thus provide a window into the lives and culture of the ancient Egyptians.

Bibliography

Atherton-Woolham, S. & L. McKnight. 2015. Animals as Votive Offerings in Ancient Egypt. In: McKnight, L. & S. Atherton-Woolham. Eds. Gifts for the Gods: Ancient Egyptian Animal Mummies and the British. – Liverpool, Liverpool University Press: 23-24.

Bleiberg, E., Y. Barbash & L. Bruno. 2013. Soulful Creatures: Animal Mummies in Ancient Egypt. – Brooklyn, Brooklyn Museum.

Buckley, S.A., K.A. Clark & R.P Evershed. 2004. Complex Organic Chemical Balms in Pharaonic Animal Mummies. – Nature 413: 837-841.

Clark, K.A., S. Ikram & R.P. Evershed. 2013. Organic Chemistry of Balms Used in the Preparation of Pharaonic Meat Mummies. – Proceedings of the National Academy of Sciences 110 (51): 20392-20395.

Clark, K.A., S. Ikram & R.P. Evershed. 2016. The Significance of Petroleum Bitumen in Ancient Egyptian Mummies. – Philosophical Transactions of the Royal Society A: Mathematical, Physical and Engineering Sciences 374: 1-15.

Filcher, K. 2020. Ancient Egyptian Coffins and Mystery of 'Black Goo'. – https://blog.britishmuseum.org/ancient-egyptian-coffins-and-mystery-of-black-goo/

Hekkala, E., M.H. Shirley, G. Amato, J.D. Austin, S. Charter, J. Thorbjarnarson, K.A. Vliet, M.L. Houck, R. Desalle & M.J. Blum. 2011. An Ancient Icon Reveals New Mysteries: Mummy DNA Resurrects a Cryptic Species within the Nile Crocodile. – Molecular Ecology 20(20): 4199-4215.

Ikram, S. 2007. Animals in a Ritual Context at Abydos. A Synopsis. In: Hawass, Z. & J. Richards. Eds. The Archaeology and Art of Ancient Egypt: Essays in Honor of David B. O'Connor. – Cairo, Supreme Council of Antiquities: 417-32.

Ikram, S. 2012a. The Lisbon Mummy Project. – KMT: A Modern Journal of Ancient Egypt 23.4: 55-59.

Ikram, S. 2012b. An Eternal Aviary: Bird Mummies from Ancient Egypt. In: Bailleul-LeSuer, R. Ed. Between Heaven and Earth: Birds in Ancient Egypt. – Chicago, Oriental Institute, University of Chicago: 41-48.

Ikram, S. 2013. A Re-Analysis of Part of Prince Amenemhat Q's Eternal Menu. – Journal of the American Research Center in Egypt 48: 119-135.

Ikram, S. 2015a. Divine Creatures: Animal Mummies. In: Ikram, S. Ed. Divine Creatures. Animal Mummies from Ancient Egypt. – Cairo, American University in Cairo: 1-16.

Ikram, S. Ed. 2015b. Divine Creatures. Animal Mummies from Ancient Egypt. – Cairo, American University in Cairo.

Ikram, S. 2017. The Hunter and the Hunted: The Brief Story of an Animal Mummy. In: Verschoor, V., A.J. Stuart & C. Demarée. Eds. Imaging and Imagining the Memphite Necropolis: Liber Amicorum René van Walsem. – Leiden/Leuven, Nederlands Instituut voor Het Nabije Oosten/Peeters: 43-46.

Ikram, S. 2019. Shedding New Light on Old Corpses: Developments in the Field of Animal Mummy Studies. In: Porcier, S., S. Ikram & S. Pasquali. Eds. Creatures of Earth, Water, and Sky. Essays on Animals in Ancient Egypt and Nubia. – Leiden, Sidestone Press: 179-192.

Ikram, S. 2020. Ancient Egyptian Fauna. In: Shaw, I. & E. Bloxham. Eds. Oxford Handbook of Egyptology. –

Oxford, Oxford University Press (Oxford Handbook of Egyptology).
Ikram, S., R. Slabbert, I. Cornelius, A. du Plessis, L.C. Swanepoel & H. Weber. 2015. Fatal Force-Feeding or Gluttonous Gagging? The Death of Kestrel SACHM 2575. – Journal of Archaeological Science 63: 72-77.
Kessler, D. 1986. Tierkult. In: Helck, W. & W. Westendorf. Eds. Lexikon der Ägyptologie VI. Stele – Zypresse. – Wiesbaden, Harrassowitz: 571-87.
Kessler, D. 1989. Die Heiligen Tiere und Der Konig. I. Beiträge zu Organisation, Kult und Theologie der spätzeitlichen Tierfriedhöfe. - Wiesbaden, Harrassowitz (Ägypten und Altes Testament, Band 16).
Kessler, D. 2007 Spitzmaus und Ichneumon im Tierfriedhof von Tuna el-Gebel. – Bulletin of the Egyptian Museum (in Memory of Mohammed Moursi) 4: 71-82.
Kessler, D. & A. Nur el-Din. 2015. Tuna El-Gebel: Millions of Ibises and Other Animals. In: Ikram, S. Ed. Divine Creatures. Animal Mummies from Ancient Egypt. – Cairo, American University in Cairo Press: 120-163.
Loat, W.L.S. 1914. The Ibis Cemetery at Abydos. – Journal of Egyptian Archaeology 1: 40.
Lortet, L.C. & C. Gaillard. 1902. Sur les Oiseaux Momifiés. – Annales du Service des Antiquites d'Egypte 3: 18-21.
Nissenbaum, A. & S. Buckley. 2013. Dead Sea Asphalt in Ancient Egyptian Mummies - Why? – Archaeometry 55.3: 563-68.
Peet, T.E. & W.L.S. Loat. 1913. Cemeteries of Abydos. 3. – London, Egypt Exploration Fund.
Prates, C., S. Sousa, C. Oliveira & S. Ikram. 2011. Prostate Metastatic Bone Cancer in An Egyptian Ptolemaic Mummy, a Proposed Radiological Diagnosis. – International Journal of Paleopathology 1: 98-103.
Prates, C., C. Oliveira, S. Sousa & S. Ikram. 2015. A Kidney's Ingenious Path to Trimillennar pPreservation: Renal Tuberculosis in An Egyptian Mummy? – International Journal of Paleopathology 11: 7-11.
Lisbon Mummy Project: https://artsandculture.google.com/exhibit/o-lisbon-mummy-project/qgList1Ii30FLA
Nicholson, P. 2015. The Sacred Animal Necropolis at North Saqqara, the Cults and Their Catacombs. In: Ikram, S. Ed. Divine Creatures. Animal Mummies from Ancient Egypt. – Cairo, American University in Cairo Press: 44-71.
Nicholson, P., C. Jackson & K. Frazer. 1999. The North Ibis Catacomb at Saqqara: The Tomb of the Birds. In: Leahy, A. & J. Tait. Eds. Studies on Ancient Egypt in Honour of H. S. Smith. – London, Egypt Exploration Society: 209-214.
Ray, J. 2001. Animal Cults. In: Redford, D.B. Ed. The Oxford Encyclopedia of Ancient Egypt. 1. – Oxford, Oxford University Press: 345-348.
Ray, J.D. 1976. The Archive of Hor. – London, Egypt Exploration Society.
Richardin, P., G. Louarn, D. Berthet, S. Ikram & S. Porcier. 2017. Cats, Crocodiles, Cattle, and More: Initial Steps Toward Establishing A Chronology of Ancient Egyptian Animal Mummies. – Radiocarbon 59.2: 595-607.
Smith, H.S., C.A.R. Andrews & S. Davies. 2011. The Sacred Animal Necropolis at North Saqqara: The Mother of Apis. Inscriptions 1-2. – London, Egypt Exploration Society.
Smith, H.S. & S. Davies. 2005. The Sacred Animal Necropolis at North Saqqara: The Falcon Complex and Catacomb, the Archaeological Report. – London, Egypt Exploration Society.
Smith, H.S., S. Davies & K. Frazer. With a contribution by R. Bland. 2006. The Sacred Animal Necropolis at North Saqqara, the Main Temple Complex: The Archaeological Report. – London, Egypt Exploration Society.
von den Driesch, A. 1993. The Keeping and Worshipping of Baboons during the Later Phase in Ancient Egypt. – Sartoniana 6: 15-36.
von den Driesch A., D. Kessler, F. Steinmann, V. Berteaux & J. Peters. 2005. Mummified, Deified and Buried at Hermopolis Magna – The Sacred Birds from Tuna El-Gebel, Middle Egypt. – Ägypten und Levante 15: 203-244.
Wade, A.D., S. Ikram G. Conlogue, R. Beckett, A. Nelson, R. Colten, B. Lawson & D. Tampieri. 2012. Foodstuff Placement in Ibis Mummies and the Role of Viscera in Embalming. – Journal of Archaeological Science 39.5: 1642-1647.
Wasef, S., R. Wood, S. El Merghani, S. Ikram, C. Curtis, B. Holland, E. Willerslev, C.D. Millar & D.M. Lambert. 2015. Radiocarbon Dating of Sacred Ibis Mummies from Ancient Egypt. – Journal of Archaeological Science: Reports 4: 355-361.

Brief Notes about a Mummified Crocodile from the National Archaeological Museum (MANN) of Naples, Italy

Ilaria Incordino & Cinzia Oliva

Introduction

The Egyptian section of the Archaeological Museum of Naples museum holds 2500 objects, ranging in date from 3000 BC-AD 640). They were acquired starting in 1803, from Campania sites, such as Pompei, Ercolano, Stabiae, etc., Egypt, and private collections sold or gifted to the Bourbon royal family. In 2016, after almost ten years of closure, the Neapolitan Egyptian collection was reopened to the public with a complete new exhibition layout: a thematic display, in contrast with the previous one that grouped the objects based on the collectors' donations (Borgia and Picchianti). Among the renewed spaces, Room 22 was dedicated to ancient Egyptian religion and magic, including syncretism and animal cults. The focus of the room is an extraordinary crocodile mummy with its original wrappings, which has been the subject of a conservation analysis and Egyptological study in order to better understand its possible date and provenance.

Description (I. Incordino)

The central case TB1 holds the body of a mummified crocodile (length 2.54 m., width 0.20 m.) of unknown provenance, generally dated to the Late Period (664-332 BC). The animal has been mummified with large bandages. Two baby crocodiles are placed toward the rear of the animal, though this position might not be the original one, and be due to curatorial intervention. The position of the body had been secured through the use of two long palm branches along the two side of the animal, in order to keep the desired form. Moreover, the complete set of original wrapping is still present, with several layers of linen bandages and various textiles and ropes, that still partially cover the crocodile.

Thanks to the careful research of Dr. Mainieri in the archives related to the MANN collection (Mainieri, 2015; In press), it has been possible to reconstruct the arrival of the crocodile at the museum. According to the Arditi's inventory of 1828, we now know the original condition of the mummy:

> *"Palm branch bundle coated with thick canvas and bandages of canvas too. Two more small bundles of straw and canvas flank one of the end*[s]. *It* [is] *supposde to contain an embalmed crocodile long 10 ¼ spans, 1 palm of diameter".*

Clearly, the crocodile mummy was intact at the time of its acquisition by the Naples Museum. In addition, the archives and inventories research has provided new data about 'minor' acquisitions, which also might be related to the crocodile mummy. In fact, some previously unknown documents have revealed a donation of 13 Egyptian antiquities to

In: Ikram, S., J. Kaiser & S. Porcier. Eds. 2021. *The Ancient Egyptians and the Natural World. Flora, Fauna, and Science.* – Leiden, Sidestone Press: 141-146.

Figure 1A (above). The mummified crocodile of Naples (MANN, Inv. No. 2338). Courtesy of MANN. Photograph by I. Incordino. Figure 1B (right). The mummified crocodile of Naples (MANN, Inv. No. 2338) in exhibition. Courtesy of MANN. Photograph of I. Incordino.

the Naples Museum from a Scottish doctor, James Edward Hogg, in 1833[1]. Among those items, we find:

> *"Canvas bundle of dark color with thick bandages of canvas too. Length of 10 spans and maximum diameter of one span. It seems to contain an embalmed crocodile"* (Notamento degli oggetti egizi ricevuti dal dottore inglese Sign. Hogg)

Unfortunately, no information about the provenance of those antiquities has been recorded in this document, nor any reference to a possible dating (Figure 1).

Provenance and Interpretation (I. Incordino)

Since the continuation of research on the acquisitions of the Naples Museum has continued, a travel book published by Hogg in 1835 has been found (Hogg, 1935), in which we can follow his journey made in Egypt in 1832-1833. From his tour we know that he visited some of the most important Egyptian sites following Ibrahim Pasha's campaign. Coming from Jerusalem on the 21st of August 1832, Hogg arrived at Damietta and later at Alexandria, from where he eventually travelled for five months in Egypt, reaching the Second Cataract at Wadi Halfa.

It is not possible to spot any reference to his acquisition of Egyptian antiquities in his book. However, Hogg's acquaintance with William Gell, who resided in Naples, might explain the presence and sale of Hogg's items to the Royal Bourbon Museum of Naples in 1833. Hogg's text does provide a few clues as to where he might have acquired the crocodile. He visited Kom Ombo site on the 12nd of November of 1832, and the book also contains a quick reference to the Fayum region:

> *"Surely the Greek sage* [Herodotus] *must have been fallen in with the nurse of Pharaoh's daughter, and gravely recorded all the 'tales of wonder' she related. His artificial Lake Moeris has proved by recent and actual survey to be a natural reservoir; and his marvelous labyrinth has vanished 'into thin air', and 'like the baseless fabric of a vision', left 'not a wreck behind'. Even the progeny of his tame crocodiles can no longer be found in the Fayum, nor have we yet discovered their sculptures effigies, decorated with the necklaces and bracelets which he has so liberally bestowed upon them"* (Hogg, 1835: 306-307).

1 About 33 Egyptian items are today kept at the British Museum belonging to Dr. Edward Hogg's collection. They are mainly small size objects, such as amulets and scarabs, dated mostly to the New Kingdom (see: Long, 1836: 430-431).

It could be likely, therefore, to suppose an origin of the mummified crocodile of Naples from Kom Ombo area, rather than the Fayum one, though it is not possible to support this idea with any concrete evidence thus far. It is also possible that the crocodile was purchased, together with other objects, at some other location.

The mummy has generally been dated to the Late Period, but actually it has been quite difficult to find strict parallels to the mummy. The main reason lies in the fact that originally the crocodile mummy was completely wrapped, but a few years later its arrive in Naples in 1833, the package was opened to expose the animal for the benefit of the visitors.

The principal evidence for that event is a painting, made in 1875, called *Museo. NekoRanyes-Taketot-Coccodrillo-Hos-Hamis-Tmachmot-Zuleika-Noute-Maj-Giulietta e Rosina*, by Paolo Vietri, which today is kept at the Bank of Naples Collection (Palace Zevallos Stigliano, Naples). This painting (Figure 2) shows the Neapolitan crocodile with its bandages open, hanging from a support, for better visibility. Clearly, between 1833 and 1875 the mummy has been partially exposed for the public. The reasons behind this decision relate to the 19th century trend to exhibit only complete and impressive items to the public in order to gain more attention.

Though the bandages were removed from the mummy, they remain in the museum's collection, which allows for their study in terms of textiles and wrapping style (Figure 3). Among them is an outer shroud formed by different fragments of linen textiles sewn together, with larger weft threads. This was followed by a layer of fabric that still partially adheres to the crocodile's body, which is impregnated with embalming resins, and is secured to the crocodile by several meters of rope. Padded linen that had been folded and impregnated with resinous materials was used to reinforce the tail and emphasize its shape.

Often, complete mummies can be dated according to the style of the wrapping if the bandages pattern is elaborate, as is known from datable Graeco-Roman human mummies, particularly those of the Roman era (Walker & Bierbrier, 1997), but in this case it is not possible to apply those chronological criteria because the bandages are loose, and the original design of the outer wrappings has been lost.

Cleaning and Conservation (C. Oliva)

Initial study of the crocodile mummy showed that the external wrappings have been cut open during the 19th century in order to exhibit the mummy. Subsequently, the crocodile had been moved onto a wooden board to which it was nailed through the surviving textiles.

The first step in the conservation process involved moving the crocodile from the old wooden board to a new, suitable support, made of inert material that was strong enough to allow a safe handling in the future life of the artefacts. To do that, all the nails were removed and a melinex sheet was slipped under the mummy, in order to lift it and move it to the new support. Next, the surface was carefully vacuum cleaned, with the help of a soft brush and the mechanical and gentle action of vulcanized sponges that helped to remove dust and dirt (Figure 4).

One of the main goals of a conservation treatment on mummies is the study and the identification of the different textiles that are present on the artefact and the kind of bandages and mummification that has been used to create the mummy. During the surface cleaning and

Figure 2. Painting of Paolo Vietri. Courtesy of MANN. Photograph by I. Incordino.

Figure 3. Different layer of bandages. Photograph by I. Incordino.

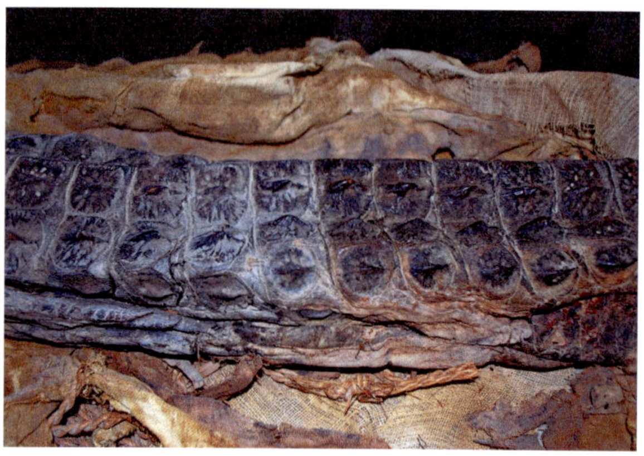

Figure 4. Cleaning of the crocodile mummy. Photograph by C. Oliva.

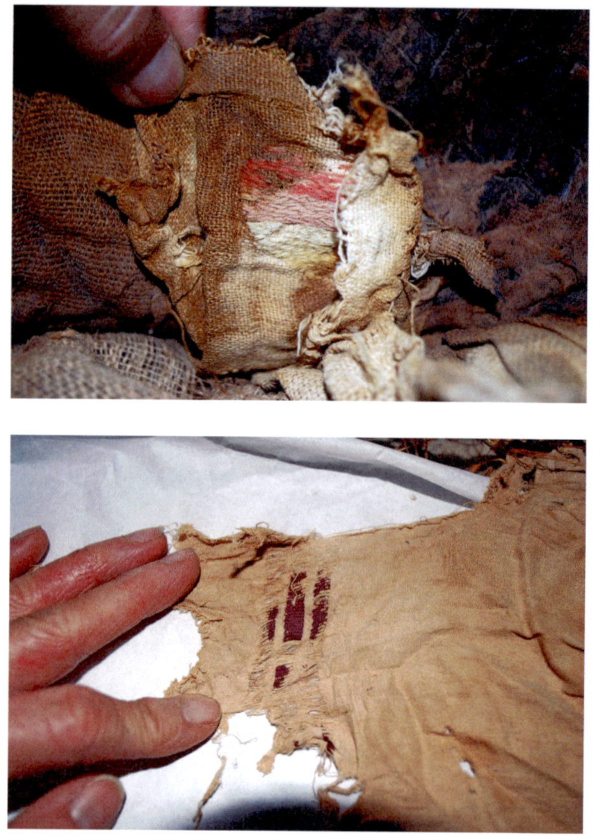

Figure 5A, B. Details of textiles from the mummy bandage. Photographs by C. Oliva.

the consequent study, some fragments of the original bandages were found still adhering to the animal's skin, together with traces of the embalming substances.

Each textile was individually recorded. Important data on these included: types of fiber, torsions, thread count and other technical data, such as the presence of starting and/or finishing borders, fringes, selvedges and traces of any prior use of these clothes. The textiles analysis is still a work in progress, but it was possible to document many interesting textiles, such as a fragment of a tunic with a tapestry weaving decoration in wool (maybe a sleeve decoration?) and another fragment of an embroidery in ivory, pink and red threads on linen (Figure 5).

Those fragments that are sewn together represent a case of material reuse, very frequently found in ancient Egyptian mummy wrappings in both human and animal mummies.

Conclusion

In order to get more information about the mummification process, the species of crocodile involved, the typologies of textiles used for the bandage and a possible dating of this item, several technical analysis are still needed. In particular, a complete CT scan could definitely add more clues about the Naples crocodile mummy. In this respect, an extraction of DNA would provide useful information to define the species of this animal, providing another instance for a world database of Egyptian animal mummies.

Bibliography

Arditi. 1828. Inventario dei monumenti egiziani Arditi (1828). Inventario Antico. – Naples, no publisher.

Pirelli, R., F. Poole, I. Incordino, S. Mainieri, C. Cozzolino R. Di Maria. 2016. Guida alla Collezione egizia del MANN. – Milano, Electa.

Pirelli, R., F. Poole, C. Cozzolino, R. Di Maria, G. Esposito & D. D'Errico. 1989. La collezione egiziana del Museo

Archeologico di Napoli. – Napoli, Museo Archeologico Nazionale di Napoli.
Hogg J.E. 1835. Visit to Alexandria, Damascus and Jerusalem During the Successful Campaign of Ibrahim Pasha. 2 Vols. – London, Saunders and Otley.
Ikram, S. 2015. Divine Creatures. Animal Mummies in Ancient Egypt. – Cairo, American University in Cairo Press.
Long, G. 1836. The British Museum: Egyptian Antiquities. Vol. I. – London, The British Museum.
Mainieri, S. In press. La Collezione egiziana del Museo Archeologico Nazionale di Napoli, voll. I-II, Università di Napoli "L'Orientale". – Naples, Il Torcoliere.
Panceri, P. & F. Gasco. 1876. Viaggio in Egitto dei soci Pancieri e Gasco nell'anno 1873-1874. – Napoli. – Stabilimento tipografico Trani.
Pirelli, R. & S. Mainieri. 2015. Georg Zoëga and the Borgia Collection of Egyptian Antiquities: Cataloguing as a Method. In: Ascani, K., P. Buzi & D. Picchi. Eds. The Forgotten Scholar: Georg Zoëga (1755-1809). At the Dawn of Egyptology and Coptic Studies. – Leiden, Brill (Culture and History of Ancient Near East 74): 151-159.
Raven, M.J. & W.K. Taconis. 2006. Egyptian Mummies. Radiological Atlas of the Collection in the National Museum of Antiquities at Leiden. – Turnhout, Brepols (Papers on Archaeology of the Leiden Museum of Antiquities).
Wilkinson, R.H. 2003. The Complete Gods and Goddess of Ancient Egypt. – London, Thames & Hudson.
Walker, S. & M. Bierbrier. 1997. Ancient Faces: Mummy Portraits from Roman Egypt. – London, The British Museum Press.

Venerunt, Viderunt, Vicerunt: The Roman Conquest and the Non-Elite

Jessica Kaiser

Introduction

Despite the relatively rich historical and archaeological material available for both the Saite and Roman periods of Egyptian history, our understanding of everyday life for the average person in these eras is still very limited. This is to a large extent due not only to the nature of the sources preserved to us, but also to the character of the sources most sought after by modern scholars. Though until recently underutilized in Egyptian archaeology, skeletal remains can offer a wealth of information on health, living conditions and social organization of past populations. This study examines the extent to which sociopolitical changes from the Saite to Roman periods of Egyptian history affected the lives of a non-elite population from the Memphite region in Lower Egypt by investigating patterns of physiological stress in skeletal material from the Wall of the Crow Cemetery in Giza in combination with archaeological and historical data.

The Wall of the Crow Cemetery lies in a low tract of desert on the Giza Plateau, approximately 400 meters southeast of the Sphinx, and immediately south of the massive Old Kingdom wall which has given the cemetery its name. The cemetery overlies a large Old Kingdom urban gallery compound related to the building of the Khafre (2558-2532 BC) and Menkaure (2532-2503 BC)[1] pyramid complexes. The galleries and surrounding areas have been under excavation by Ancient Egypt Research Associates (AERA), directed by Mark Lehner, since 1989, with the overarching goal of finding evidence of the social and economic structures that supported the building and maintenance of the Giza pyramids during the Fourth Dynasty of the Old Kingdom.

Human burials were first encountered at the site in 1998, but it was not until 2000 that large scale cemetery excavations began. Between then and 2009, a total of 348 primary (*in situ*) inhumations and 105 secondary (disturbed) deposits of human bone were excavated, under the supervision of the author. The Wall of the Crow Cemetery was in use primarily during the late 25th Dynasty through the end of the 26th (Saite) Dynasty (ca. 725-525 BC), and then again during the early to mid-Roman period, first to second centuries AD. The focus of this paper is the 226 burials that could be dated, based on pottery or stratigraphy or both, to one of these two phases of cemetery use.

Historical Background

If the Saite use phase of the Wall of the Crow cemetery dates from late 8th century BC (25th Dynasty) through the end of Saite period, there is no denying that the population

1 Dates following Shaw (2000).

In: Ikram, S., J. Kaiser & S. Porcier. Eds. 2021. *The Ancient Egyptians and the Natural World. Flora, Fauna, and Science.* – Leiden, Sidestone Press: 147-162.

burying their dead in the shadow of the wall would have lived in an age of insecurity and instability for Egypt. Prior to the reunification of the country under Psamtik I in 664 BC, the 25th Dynasty Nubian kings ruled the northern parts of the country more as overlords than as traditional kings, and left the administration of the country mainly to the established local rulers (Kitchen, 2009: 387; O'Connor, 1983). Overall, this resulted in political fragmentation; however, the immediate administration of the Memphite region would have fallen under the rulers of Sais, who in general appear to have bestowed a fair amount of stability on their northern kingdom, a stability that would have continued into the Saite era.

In contrast to the relative stability and prosperity of the Saite period, the Roman conquest of Egypt brought with it several changes that would likely have negatively impacted living conditions for the Wall of the Crow cemetery population. No longer independent, under the Romans Egypt was governed as an asset to be exploited and and taxation and corvée labor increased (Capponi, 2010; Lewis, 1983:160). Due to the Roman disdain for the worship of animals, the importance of the Memphite region, which previously had benefitted from government resources spent on the cult of the Apis bull, declined. In addition, upward mobility was greatly restricted for native Egyptians as a result of the segregationist policies implemented by the Romans (Lewis, 1983: 19). The resentment bred by the restrictions and fiscal burdens imposed on the non-Roman population led to several revolts, some lasting for several years. Preserved texts from the period tell of both food and labor shortages, property damage, and the negative social ramifications resulting from men leaving their villages to join the uprisings (Capponi, 2010). Finally, preserved Roman census records tell us of low life expectancy and high infant mortality (Bagnall & Frier, 1994; 2006; Scheidel, 2012). The end of the period under consideration also saw a major disease outbreak in the form of the Antonine plague (ca. 165-180 AD), which had devastating effects on the country and may have reduced the population by as much as ten percent (van Minnen, 2000).

For women, in particular, Roman rule also meant a significant reduction of civil rights. In pharaonic times, Egyptian women enjoyed many priviliges and freedoms denied women of other ancient cultures. They could inherit, buy and sell property, conduct business without the aid of a male guardian, instigate their own divorces and even take men to court (Graves-Brown, 2010: 33). The comparatively strong position of women in Egyptian society persisted through the Ptolemaic period: although women had virtually no legal independence under Greek law, the Ptolemies also retained the native Egyptian legal system and recognized demotic contracts of all kinds as valid. Thus, Egyptian women continued to conduct business and manage properties and inheritance on their own until the end of the Ptolemaic era (Vandorpe, 2002; 2012). With the Roman conquest, however, women's legal independence was severely curtailed. The Romans no longer recognized demotic contracts, meaning that women – to some extent excluding Roman citizens – required the assistance of a male guardian in all legal matters (Vandorpe, 2012). In addition, the Augustan marriage laws severely restricted one of the few avenues of upward mobility available to women, by strongly discouraging intermarriage between native Egyptians and the Hellenized classes (Vandorpe & Waebens, 2010).

Though historical sources available to us may be tainted by the generally bad reputation the 'troublesome' Egyptians enjoyed in Imperial Rome, there is no question that the Roman conquest imposed new hardships on native Egyptians. However, classical sources, just like pharaonic texts before them, often relate to the literate minority. Skeletal remains, in contrast, offer us the opportunity to investigate whether these societal changes had any measurable impact on a subset of the population which would otherwise generally be silent in the historical record. Did the increased taxation and higher demand for corvée labor result in higher workloads for the Roman population compared to the Saite population? Did the increase in the proportion of agricultural product devoted to taxes negatively impact the diet of the Roman period Giza community? Did the decline in both economic and legal independence for women after the Roman conquest impact women in communities even on the lower ends of the status scale? If the policies implemented by the Romans were as far-reaching as textual sources suggest, the decline in living conditions for native Egyptians would likely translate to increased level of physiological indicators of stress in the Roman period skeletal sample as compared to the earlier Saite material.

The Wall of the Crow Cemetery

The majority of the Saite period burials were concentrated around the eastern end of the Wall of the Crow in the northeastern part of the AERA concession. A smaller group was also excavated in the western outskirts of the site (Figure 1). Most of the Roman period burials were also in the northeastern part of the site. In this case, however, the interments were not directly by the Wall of the Crow, but further south and spread out over a larger area. Thus, it appears that the use of the northeastern area as a burial ground started adjacent to the Wall of the Crow in the 25th Dynasty, and subsequently spread southward during the Roman era. The Saite and Roman burial usage of the site was limited to its northern half, and bounded by what is known as "Main Street", a paved Old Kingdom access road transecting the site from west to east, suggesting that the road may have been in use, or was at least visible, during the Late and Roman periods.

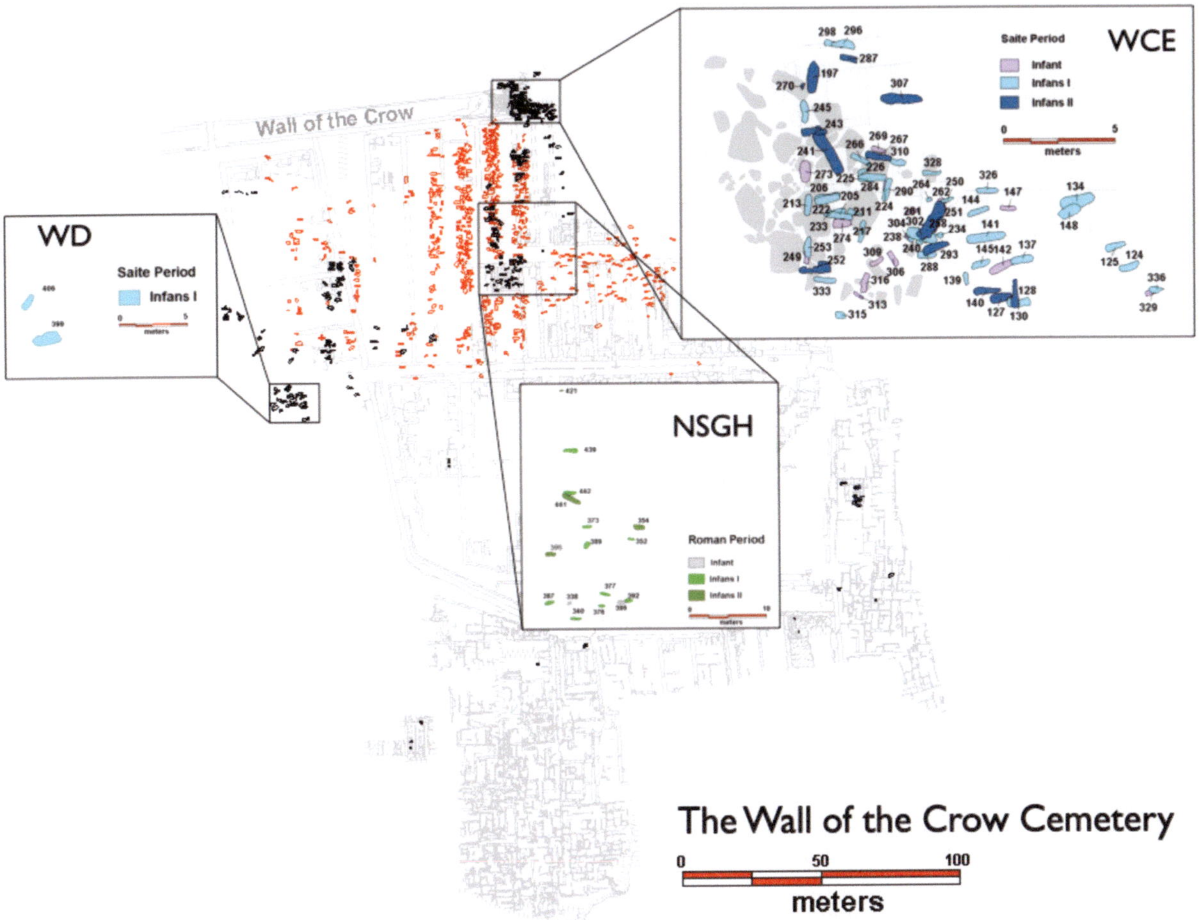

Figure 1. Map of the Wall of the Crow archaeological site, with the areas where Saite (WCE; WD) and Roman (NSGH) period burials were excavated. The insert shows the distribution of subadult burials in each area. Black outlines denote excavated burials across the site; red denotes identified but as yet unexcavated burials.

The Wall of the Crow burials were simple inhumations sunk directly in the sand, with few grave goods. In both the Saite and Roman periods, by far the most common funerary object to be included in the grave was a coffin. These coffins were generally made of painted mud and were without exception very poorly preserved. The majority were anthropomorphic in shape, but rectangular or sub-rectangular tapered coffins were also fairly common, particularly during the Saite period. Generally, the center of the lid of the coffin had collapsed, and the decoration on the coffin lids was not usually preserved. In some cases, however, an inscription could be seen, in a band running down the center of the coffin lid. These inscriptions were all very crude but could be recognized as the standard *htp di nsw* offering formula evoking Osiris. Approximately two thirds (110 of 165; 67%) of the Saite burials and slightly less than half (28 of 63; 44%) of the Roman burials were equipped with coffins. In both periods, coffins were more common among adults than among juveniles and children.

Smaller burial items like amulets and jewelry, in contrast, were more commonly found with children and adolescents than with adults.[2] In the Saite period, in particular, the amulets included in the graves of children often had connotations of fertility or a connection to motherhood and childbirth. Bes amulets, for example, occurred in several graves. This dwarf-like god with leonine features was primarily a household deity, concerned mainly with the protection of women and children, particularly during the vulnerable time of pregnancy and childbirth (Dasen, 1993: 68). Apart from a few bronze pendants depicting Amun-Min, Hathor or inanimate objects, and two lotus flowers made of carnelian and lapis lazuli, all the amulets were made of faience.

2 65% (42 of 73) of the Saite non-adult burials and 70% of the Roman non-adult burials (11 of 16) were equipped with burial items, compared to only 19% (14 of 75) of Saite adults and 28% (11 of 40) of Roman adults.

With few exceptions, all were crudely made. Amulets were most commonly found in the area around the neck of an individual, likely originally suspended and worn as necklaces. Additionally, beads and cowrie shells were often found around the wrists and ankles of the bodies, probably worn as bracelets or anklets.

Very few of the Wall of the Crow burials were equipped with pottery, particularly in the Saite period. Of the 226 datable burials, only 27 had pottery; these included less than a tenth of the Saite burials, and only about one quarter of the Roman burials. To some extent, this discrepancy may be due to the spatial distribution of the burials, since the Saite burials were more densely packed, thus allowing for more relative dating. In both eras, approximately three quarters of the burials with pottery belonged to adults, and only a quarter to children. Storage jars were common in both periods (amphorae in the Roman period and cylindrical storage jars in the Saite period). In addition to the amphorae, the Roman assemblage also contained juglets, a cooking pot, dishes and bowls, and a miniature amphora. The Saite assemblage contained only one dish, as well as two medium-sized jars with post-firing polychrome decoration, pilgrim flasks and an imported Levantine storage jar (Tavares & Laemmel, 2011).

The grave goods were decidedly Egyptian in nature during both the Saite and Roman phases of cemetery use, and surprisingly uniform. There is no evidence of the Hellenistic influences characteristic of funeral receptacles of elite Egyptians of the Roman period and, were it not for the pottery, the graves would have been very difficult to date. Further, the coffins themselves were made mainly of mud, with limited wood and textile reinforcements, and would likely have been difficult to transport over longer distances. With this in mind, it seems likely that the Wall of the Crow population was both native Egyptian and relatively local.

The location of the cemetery on low ground, coupled with the fact that several richly furnished burials were found on the slope above it,[3] further implies that the Wall of the Crow cemetery did not serve the most affluent members of society. Nevertheless, finds in the general vicinity of stelae denoting necropolis areas under choachyte control (Petrie, 1907: 29) suggest that access to the cemetery probably needed permission from these necropolis workers. A choachyte was a type of mortuary priest, who for a fee would be responsible for finding a suitable location for burial and continuing the funerary cult of the deceased. Thus, a grave in the cemetery likely required some kind of remuneration. Further, the inclusion of both coffins and smaller grave goods in many of the graves show that the population burying their dead in the cemetery had sufficient means to divert at least some of their assets toward funerary arrangements. In addition, several of the burials show evidence of post-mortem manipulation and traces of wrapping, and it seems likely that at least a cursory attempt at mummification – which would have come at a cost – was fairly common. Hence, while the Wall of the Crow population certainly can be described as "non-elite," they were not destitute.

Materials and Methods

As noted above, a total of 226 burials from the Wall of the Crow Cemetery could be securely dated. Of these, 165 graves containing 165 individuals dated to the Saite period, and 61 burials containing 63 individuals (two were multiple inhumations) belonged to the Roman phase of cemetery use.

Age and sex assessment followed standard protocols (Buikstra & Ubelaker, 1994). Sex was determined from dimorphic features of the pelvis (Buikstra & Mielke, 1985; Milner, 1992; Phenice, 1969) and cranial morphology (Acsádi & Nemeskéri, 1970). Age was assessed based on dental wear (Brothwell, 1989; Scott, 1979; Smith, 1984; Walker *et al.*, 1991), degenerative changes of the pubic symphysis and auricular surface (Brooks & Suchey, 1990; Lovejoy *et al.*, 1985; Meindl *et al.*, 1985; Todd, 1920; 1921), and cranial suture closure (Meindl & Lovejoy, 1985). For immature remains, age estimation was based primarily on dental development and secondarily on epiphyseal closure (Moorrees *et al.*, 1963a; 1963b; Schaefer *et al.*, 2009: 338-355; Ubelaker, 1999). Adult stature was determined using the formula developed by Raxter *et al.* (2008) for Egyptian skeletal remains.

To investigate whether the sociopolitical impact of the Roman conquest was reflected in health, four non-specific stress markers indicative of generalized physiological stress (linear enamel hypoplasia, *cribra orbitalia*, degenerative joint disease and periosteal new bone formation) were scored closely following the criteria developed for the Global History of Health Project (GHHP) by Richard Steckel and colleagues (2006; 2002). In addition, adult stature was compared between subgroups and temporal samples. Only statistically significant results are presented here; for a fuller treatment of the material, including an examination of the material culture associated with the burials, see Kaiser (2018).

The frequencies of skeletal stress indicators were compared between temporal phases, sex, and age groups through nonparametric statistics. When possible, Pearson's chi-square test was used to test association

3 The area west of the Wall of the Crow cemetery is currently under excavation by the Egyptian Ministry of Antiquities, and the results are as yet unpublished. Several inhumation burials have been found and excavated immediately above the Wall of the Crow cemetery; those seen by the author were invariably more richly furnished, with better quality coffins and more grave goods, than their Wall of the Crow counterparts.

between variables. The chi-square test compares observed counts (O) versus expected counts (E) and measures whether or not the values in the cells are associated to a higher degree than that expected by chance (VanPool & Leonard, 2010: 240-243). Fisher's exact test was used when expected counts were less than one in any cell in the chi-square matrix. The significance level was set as $p < 0.05$.

The only variable measured on a continuous scale was stature. Normal Q-Q plots and the Shapiro-Wilk test were performed on both the Saite and the Roman populations to assess the normality of the sample distributions The test results were not statistically significant, and the Q-Q plots also suggested a normal distribution (Razali & Wah, 2011) prompting the use of the independent samples t-test for the comparison of groups. This test measures whether or not there is a statistically significant difference between the means of two groups with normal distribution of scores. To determine whether the variance within each group was the same, Levene's test for equality of variance was used in conjunction with the t-test (VanPool & VanPool, 2001: 125-127). Finally, effect size was evaluated using Cohen's d statistic, following the guidelines of Cohen (1988) designating .2 or above for small effect, .5 or above for moderate effect and .8 for large effect.

Enamel hypoplasias are defects in dental enamel caused by disruptions during amelogenesis. They are generally believed to most commonly be the result of episodic nonspecific metabolic and nutritional insults (Goodman *et al.*, 1987; White *et al.*, 2012: 455-56). Linear enamel hypoplasias are particularly important in archaeological materials, as they provide a record of stress episodes during childhood, when the teeth were formed, that are permanently preserved (when not worn away) as linear defects in adult dentition.

Cribra orbitalia presents as pitting and porosity in the orbital roof (Waldron, 2009: 137), and has traditionally been linked to iron deficiency anemia (Angel, 1966; Grauer, 1993; Stuart-Macadam, 1985; 1987a; 1987b; 1992). However, recent studies (Fairgrieve & Molto, 2000; Walker *et al.*, 2009; Wapler *et al.*, 2004) have shown that *cribra orbitalia* may have more to do with inflammations in the orbital bone or malnutrition and vitamin deficiencies (particularly of vitamin C) than with anemic responses. Hence, it may be best to view the condition as a general stress indicator rather than a specific nutritional deficiency.

Degenerative Joint Disease (DJD) is by far the most common form of pathological lesion encountered in archaeological skeletal material. Many studies have drawn inferences between DJD and functional stress in the lifestyles of past populations (Cohen & Crane-Kramer, 2007; Goodman & Martin, 2002; Goodman *et al.*, 1984; Larsen *et al.*, 2009; Robb, 1994; Tainter, 1980; Walker & Hollimon, 1989), and some have gone so far as to suggest links between patterns observed in the distribution of osteoarthritis and specific activities (Hershkovitz *et al.*, 1996; Larsen *et al.*, 1995). However, because of the multifactorial nature of the disease, there is no one-to-one relationship between DJD and behavior (Waldron, 2007: 124). Even so, differences between distinct groups that were evaluated based on the same methodology can still be valuable as indicators of variation in activity patterns, as long as the discussion stays general and any specific conclusions are clearly distinguished as speculation (Goodman & Martin, 2002; Ortner, 2003: 550; Waldron, 2007; 2012)

Maternal environmental factors during pregnancy, disease, environmental stress, inbreeding and adverse living conditions during childhood and adolescence can significantly impact genetic growth potential (Kemkes-Grottenthaler, 2005; Mensforth, 1985). Several studies have shown an association between heightened frequencies of chronic infections and decrease in stature on a population level (*e.g.* Lambert, 1993; Lovejoy *et al.*, 1990; Temple, 2007). Short stature in particular is frequently associated with poor nutrition and high disease load during childhood (Bogin, 1988; Goodman, 1991; Kemkes-Grottenthaler, 2005).

Results

Demographic profile

The age distribution of the Giza sample is provided in Table 1. Slightly more than half (115/228, or 50.4%) of the individuals recovered from the Wall of the Crow cemetery had reached adulthood at the time of death, though proportions differed between the Saite and Roman phases. Of the bodies dating to the Saite period (n=165), 45.4% belonged to the young adult group or older, whereas among Roman individuals (n=63) a full 63.5% were 18 or older when they died.

In both phases, individuals under the age of twelve make up a substantial proportion of the sample: 44.2% in the Saite period and 25.4% in the Roman phase. Infants appear to be underrepresented in both phases, with only 12.1% of Saite individuals and a mere 3.2% of Roman individuals belonging to this age group, suggesting that the very youngest were likely buried outside of the cemetery.

Sex assessment was carried out for a total of 127 individuals, 81 from the Saite phase, and 46 from the Roman phase. This number also includes twenty post-pubescent juveniles: thirteen from the Saite sample and seven from the Roman. Males outnumbered females in all age groups and in both phases, except in the Senilis group from the Saite phase. Thus, the sex ratio[4] was higher

4 The number of males per 100 females.

	Saite		Roman	
	(n)	%	(n)	%
Infant (<1)	20	12.1	2	3.2
Young Child (1-5)	35	21.2	11	17.5
Older Child (5-12)	18	10.9	3	4.8
Juvenile (12-18)	17	10.3	7	11.1
Young Adult (18-35)	33	20	17	27
Mature Adult (35-50)	24	14.5	12	19
Old Adult (50+)	11	6.7	10	15.9
Adult (18-79)	7	4.2	1	1.6
Total	165		63	

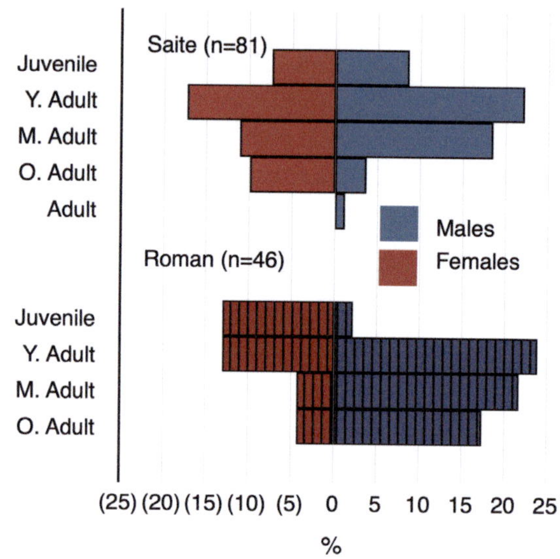

Table 1A (left). Age distribution of the Wall of the Crow sample; 1B (right). Age distribution of adults divided by sex. For clarity, probable male/male (Saite: 9/35; Roman 7/23) and probable female/female (Saite 9/28; Roman 6/10) groups have been combined.

than expected in both temporal phases: 119 for the Saite material, and 188 for the Roman material. Based on census records, Bagnall and Frier have estimated an average sex ratio of 108 for Roman Egypt, though the numbers varied substantially by location (Bagnall & Frier, 1994: 91-96; Kehoe, 2010).

Markers of Skeletal Stress

A number of factors, discussed in further detail below, indicate that neither the Saite nor the Roman populations represented in the Wall of the Crow Cemetery led lives of leisure. Statistically significant physiological stress indicators included a higher rate of degenerative joint disease among Roman males, coupled with lesser sexual dimorphism in stature and higher levels of linear enamel hypoplasia, and taller stature among Saite males. What follows is a discussion of what the differences and similarities in the levels of skeletal stress between the two populations can add to our understanding of the impact of the Roman conquest on the non-elite (Table 2).

Degenerative joint disease (DJD)

Degenerative joint disease (DJD) was common among adults in both phases, though significantly more so in the Roman period: 81.8% of observable adult individuals (n=33) in the Roman sample and 66% of their Saite counterparts (n=61) had osteoarthritic changes to either synovial joints or spine, suggesting cumulative and repetitive physical activity (Larsen, 2015).

In both samples, males were more commonly affected than females. This alone may not necessarily suggest that males were carrying out more physically demanding activities than women, since a recent review of the aetiology of osteoarthritis has shown that prevalence differences between the sexes are often the result of hormone levels, body size and anatomy, rather than activity (Weiss & Jurmain, 2007). However, Schmorl's nodes, which have been shown in modern clinical studies to occur more commonly in individuals who place great stress on their lower spines (Waldron, 1987: 45), occurred only in males in both temporal phases. This likely points to a more physically demanding lifestyle for men than for women, which also fits neatly with the historical narrative of life in Egypt during both the pharaonic and Graeco-Roman periods. There are also some interesting differences in the specific joints affected. Whereas both sexes in the Saite sample and males in the Roman sample were affected by DJD in both upper and lower body joints, the condition was limited to the hip and knee among Roman women, perhaps suggesting a change in gendered labor division from the Saite to the Roman period.

Linear Enamel Hypoplasia (LEH)

Other indicators also imply less than ideal living conditions for the Saite and Roman era Giza population, particularly during childhood. For example, 38% of the Saite individuals (n=71) and 40.7% of the Roman individuals (n=27) with preserved dentitions had at least one hypoplastic line, suggesting they experienced a period of significant stress during childhood. In modern populations from industrialized countries, the prevalence rate of linear enamel hypoplasia (LEH) rarely exceeds ten percent (Cutress & Suckling, 1982). Studies conducted on populations with known low socio-economic status or on populations from developing countries typically report higher prevalence rates, ranging from 30-90% (Goodman

	Saite vs Roman		Males vs Females
	Males	**Females**	**Saite**
LEH	Saite; x2 (1, n=25) = 4.2, p = .041, all incisors pooled	Roman; x2 (1, n=37) = 4.0, p = .046, right lateral incisors	Females (juveniles); right upper canine x2 (1, n=11) = . 4.95, p = .0.026) and the left lower central incisor (x2 (1, n=13) = 4.550, p = .033).
DJD	Roman; x2 (1, n=45) 5.2, p= .023, synovial joints		Males; x2 (1, n=57) = 4.6, p = .032, Schmorl's nodes
PNB		Saite; x2 (1, n=511) = 8.6, p = .003, all long bones	Females; x2 (1, n=511) = . 13.3 p = <.001 (all long bones); x2 (1, n=58) = .049, r tibia, and (1, n=57) = 5.4, p = .02, r fibula
Stature	Saite; (M = 165.3 SD = 5.0), t(48) = 2.3, p = .03.		Not tested

	Adult vs. Subadults		Adult vs. Juveniles
	Saite	**Roman**	**Saite**
LEH			Juvenile females; x2 (1, n=24) = 7.170, p
CO	Subadults; x2 (1, n=97) 8.6, p= .003	Subadults; x2 (1, n=97) 8.6, p= .003	
PNB	Adults; x2 (2, n=127) = 6.534, p = .038.227),		

Table 2. Summary of statistically significant results. Bold italic indicates the group with statistically significant higher frequency of each marker, followed by chi square results (x2 (degress of freedom, N = sample size) = chi-square statistic value, p = p value.), and element examined.

et al., 1987; Goodman & Rose, 1991; Lukacs *et al.*, 2001). Similarly, studies on archaeological materials frequently find LEH in over 50% of the study populations (Goodman & Armelagos, 1988; King *et al.*, 2005; Lovell & Whyte, 1999; Šlaus, 2008). Based on a study of 941 individuals from cemeteries in Egypt and Nubia covering the Predynastic through Christian periods, Hillson (1979) suggests that a 40% prevalence rate should be considered more or less standard in archaeological samples from the region, likely reflecting common nutritional deficiencies and high disease loads. Thus, from the populational prevalence rate alone, it does not appear as if the Giza population was subjected to more stress than other known Egyptian populations. However, the generally common occurrence of LEH in both burial groups at the Wall of the Crow cemetery suggests that living conditions at Giza were not ideal in either the Saite or the Roman period.

The validity of interpreting LEH as a simple marker of populational stress has been called into question (Wood *et al.*, 1992) with the suggestion that high prevalence rates of enamel defects may reflect resilience and biological fitness in a population, rather than elevated morbidity. This proposition stands in marked opposition, however, to the many studies indicating that elevated stress levels during infancy and childhood may have long-term negative consequences for both health and mortality (*e.g.* Armelagos *et al.*, 2009; Barker & Osmond, 1986; Cameron & Demerath, 2002; Goodman, 1996; Humphrey & King, 2000). In the Giza sample, enamel defects were more common among subadults than adults in both the Saite and Roman periods, though the difference was only statistically significant for the Saite period. Further, the mean age-at-death of Saite adults was significantly lower among individuals with LEH compared to individuals without lesions, suggesting that childhood stress did indeed have a negative impact on both mortality and longevity in the Wall of the Crow population, at least in the Saite period.

Cribra Orbitalia

Another stress marker often used as an indicator of elevated levels of childhood stress is *cribra orbitalia*. In the Wall of the Crow population, 26.8% of the Saite individuals with at least one observable orbit (n=97) and 32.5% of their Roman counterparts (n=40) exhibited the lesion. The condition was significantly more common among subadults in both the Saite (n=47) and Roman (n=13) samples, with 40.4% and 53.8% affected, compared to 14% of Saite adults (n=50) and 22.2% of Roman adults (n=27), respectively. In the Saite period, the majority of cases occurred among children between 1 to 5 years of age (n=20), where 55% of individuals exhibited the lesion. In the Roman period sample (n=5), the prevalence among younger children was similarly high, with 60% affected, all exhibiting active lesions. However, the prevalence among Roman juveniles (12-18 years of age, n=4) was even higher, at 75%. Saite period juveniles (n=14) were also commonly affected, with a prevalence rate of 43%. In addition, the majority of severe cases occurred in the Juvenilis age group in both the Saite and Roman periods.

Although the comparison of *cribra orbitalia* rates in the Wall of the Crow sample with those of other published materials (*e.g.* Buzon, 2006; Fairgrieve & Molto, 2000; Nerlich *et al.*, 2000; Raven *et al.*, 2008; Säve-Söderbergh & Troy, 1991; Strouhal & Bareš, 1993; Winkler & Wilfing,

1991) did not show that the prevalence at Giza was unusually high in either the Saite or Roman periods, the results are still valuable for what they may reveal about the so-called "osteological paradox", which holds that bony reactions may be interpreted as representing either resilience or frailty (Wood, 1992). An age-structured approach suggests that the traditional interpretation of skeletal stress, *i.e.*, that skeletal lesions are indicative of poor health, is appropriate at Giza. Not only were orbital lesions significantly more common among subadults, but healed lesions occurred exclusively among adults in both phases, while lesions among younger individuals were predominantly active at the time of death, or to a lesser extent mixed. This suggests that children affected by the stressors that cause *cribra orbitalia* often did not survive, and underscores the supposition that children were especially vulnerable in the Giza population. Similar patterns have been noted in several of the other samples and are not specific to Giza. It has been suggested that the peak in prevalence among younger children may be related to weaning, which we know from both historical and archaeological sources was generally completed around age three (Buzon, 2006; Dupras *et al.*, 2001; Fairgrieve & Molto, 2000; Wheeler, 2009, 2012).

If weanling stress seems a reasonable interpretation of the high percentage of young children with *cribra orbitalia*, the similarly high rate among juveniles is harder to explain. If the lesions had occurred predominantly among females, it would have been tempting to correlate the condition with the pressure of early pregnancy, since textual sources tell us that Egyptian women could (and often did) marry as early as age twelve, and often had their first child soon after (Bagnall & Frier, 1994: 111). However, in the Saite sample the lesions are evenly distributed among males and females, and in the Roman sample, the sole juvenile male and two of the three juvenile females also exhibit the lesion. It is, of course, possible that the young age of male emancipation, which at least in Roman times was 14, exerted comparable pressures on males to those of early marriage on females. Either way, it appears that early childhood and adolescence were both very stressful times for the Wall of the Crow populations.

Periosteal New Bone Formation (PNB)

Although the use of periosteal new bone formation (PNB) as a marker of nonspecific stress has recently been called into question (Weston, 2012), several studies have shown an association between periosteal lesions and elevated risk of mortality (Bullock Kreger, 2010; DeWitte, 2014; DeWitte & Wood, 2008; Grauer, 1993; Lallo *et al.*, 1978; Marklein *et al.*, 2016; Novak & Šlaus, 2010; Usher, 2000). Thus, high frequencies of the condition are still commonly interpreted as a sign of poor community health (Goodman & Martin, 2002; Larsen, 2015: 88-92), often by linking it to increased population density, intensification of agriculture, and unsanitary living conditions (Armelagos *et al.*, 1991; Larsen, 2015: 88). Studies on archaeological populations often report high prevalence of the condition; commonly over 20% of the individuals in such a sample exhibit periosteal lesions, and in some cases the figure is as high as 84% (Boocock *et al.*, 1995; Boylston & Roberts, 1996; DeWitte, 2014; Grauer, 1993; Lallo *et al.*, 1978).

In the Giza Saite sample, 14.7% of the observed individuals (n=156) showed evidence of periosteal new bone formation on any bone, compared to 15.8% of the Roman individuals (n=57). The lesions were more common among adults than subadults: in the Saite population, 23.6% of observable adult individuals (n=72) exhibited periosteal lesions, compared to only 7.1% of subadults (n=84), while in the Roman population 17.9% of observable adult individuals (n=39) and 11.1% of subadults (n=18) were affected.

PNB (on any skeletal element) was more common among females (29.4%, n=32) than males (21.4%, n=37) in the Saite sample, while the reverse was true for the Roman sample, where 17.2% of males (n=28) were affected, but only 13.3% of females (n=14). When periosteal lesions were separated based on location, this pattern remained: both PNB on the long bones and PNB on any other bone excluding the long bones were significantly more common among adults than subadults in the Saite sample. Though the lesions were also more common among adults in the Roman sample, the difference was not statistically significant.

PNB on the long bones was also more common among females (n=32) than males (n=37) in the Saite sample (25% *vs.* 10.8%), while the opposite held true for the Roman sample, where the lesions were entirely absent among females, but occurred in 14.3% of the Roman males (n=28). Roman females did show evidence of general infection, with 7.1% of observable individuals (n=14) exhibiting periosteal lesions on bones other than the long bones, compared to 10.3% of Roman males (n=29). In the Saite population, evidence of general infection was still more common among females (26.5%; n=34) than males (11.9%; n=42).

Given the available information from historical and archaeological sources on living conditions of the non-elite in Saite and Roman period Egypt, higher frequencies of periosteal lesions might be expected in the Giza burials. However, a number of studies involving skeletal samples from Egypt and Nubia, from various time periods, report similarly low occurrences. In the skeletal material from the area around the Ptahshepses mastaba at Abusir, a cemetery close to Giza both temporally and geographically, only 6.3% of individuals (n=159) were affected (9.1% of males, and 2.8% of females), with no evidence of periosteal lesions among subadults (Strouhal & Bareš, 1993: 110). In a sample from the South Tombs Cemetery at Amarna, dating to the New Kingdom, only 8% of adult individuals (n=53) were affected (Rose, 2006). In a sample from Wadi Halfa, Sudan, dating to the X-group period (ca. 350-550 BC), only 12% of individuals exhibited PNB (Armelagos *et al.*, 1981). Finally, in a Roman period sample from the Dakhleh Oasis, none of the adults showed any evidence of infection, although it was commonly encountered among subadults (Cook *et al.*, 1989). Considering that other markers of skeletal stress were common in these materials, the low rate of periosteal lesions is somewhat surprising.

It has been suggested that the unexpectedly scant evidence of infections in Nubian and Egyptian skeletal materials is due to the buffering effect of naturally occurring broad-spectrum antibiotics, stemming primarily from a side-effect of the fermentation process used in beer-production (Armelagos, 2000; Bassett *et al.*, 1980; Mills, 1992). Indeed, thin sections of bone from both Nubia and Egypt showed a pattern of fluorescence indicative of tetracycline labeling, suggesting that these populations were exposed to tetracycline-containing materials during life (Bassett *et al.*, 1980; Cook *et al.*, 1989; Nelson *et al.*, 2010). If tetracycline-laced beer did, in fact, serve as a prophylactic antibiotic, this may explain the apparent low rate of infection in skeletal materials from Egypt and Nubia, considering the prominence of beer in the Egyptian diet (Samuel, 2000).

Stature

Although stature is primarily determined by genotype, extrinsic factors also play a part in human growth (Jantz & Jantz, 1999; Stinson, 2012). Several studies have shown that malnutrition, high disease loads and low socio-economic status correlate with shorter average stature in a population (Cavelaars *et al.*, 2000; Kemkes-Grottenthaler, 2005; Komlos & Baur, 2004; Raxter, 2011:22-25; Schweich & Knüsel, 2003; Steckel, 1995). It has also been suggested that females are better buffered against environmental factors affecting growth than males (Jantz & Jantz, 1999; Stinson, 1985; Vercellotti *et al.*, 2011), meaning that suboptimal living conditions would affect males more than females and thus may result in reduced sexual dimorphism in stature (Greulich, 1951; Tobias, 1975). However, this hypothesis has been incompletely tested and is further complicated because males may have had preferential access to better nutrition in socially stratified societies (Raxter, 2011:88; Stinson, 1985; Zakrzewski, 2003). Nevertheless, variation in stature over time is another aspect that can be potentially informative regarding changes in living conditions in past populations.

Recent studies on body proportions have reported a general decline in stature over time in Egyptian populations (Raxter, 2011; Zakrzewski, 2003), and when the Giza population was compared to other regional samples, this trend still persisted (Table 3). However, when males and females in the Giza material were compared by phase, the difference in stature was statistically significant only between males.

In modern populations, sexual dimorphism in stature (SDS = male height/female height) ranges from approximately 1.04 to 1.10 (Alexander *et al.*, 1979; Holden & Mace, 1999; Pawłowski, 2003; Stini, 1975), meaning that males are between 4-10% taller than females. In the archaeological samples under consideration, SDS scores ranged from 1.05 to 1.08, *i.e.*, males were between 4.6% to 8% taller than females in the same samples. Not surprisingly, sexual dimorphism was greatest in the two samples that also had the tallest average stature among males, namely the Old Kingdom (8%) and Saite (7.6%) samples. Interestingly, the by far smallest difference in height between the sexes was found in the Roman sample from Giza, where males were merely 4.6% taller than females from the same group. Thus, if a reduction in height and sexual dimorphism in stature is indeed related to disadvantageous living conditions and malnutrition, the differences in stature and SDS between the Giza phases may suggest a decline in the standard of living from the Saite to the Roman period.

PERIOD/SITES	MALES			FEMALES			SDS	
	n	Mean	SD	n	Mean	SD	Score	%
Predynastic (Abydos, al-Amrah, Gebelein, Hierakonpolis, Keneh, Mesaeed, Naqada)	255	166.2	5.4	356	155.9	4.9	1.07	6.6
Old Kingdom (Giza, Meidum)	129	166.9	5.5	90	154.6	4.6	1.08	8.0
Middle Kingdom (Gebelein, Lisht, Sheikh Farag)	23	163.3	4.9	21	152.3	4.3	1.07	7.2
New Kingdom (Amarna, Lisht	24	164.0	5.2	38	152.7	5.1	1.07	7.4
Saite (Giza, present study)	29	165.3	5.0	25	153.6	5.6	1.08	7.6
Roman (Giza, present study)	21	161.6	6.3	4	154.5	6.0	1.05	4.6
Roman/Byzantine (al-Hesa, Kharga, Luxor)	37	161.1	4.4	16	150.8	6.4	1.07	6.8

Table 3. Samples used in comparative analysis of stature. With the exception of the samples from the present study (Saite and Roman Giza). Data from Raxter (2011).

Conclusions

A careful study of the Saite and Roman era burials from the Wall of the Crow cemetery at Giza suggests that both groups belonged to the non-elite, and that both were under considerable physiological stress. There were, further, suggestive differences and similarities between the two populations.

The Roman population exhibited higher incidences of skeletal stress markers, indicating a likely increase in physical labor and a decline in living conditions from the Saite to the Roman period, consistent with the historical narrative. Statistically significant indicators included a higher rate of degenerative joint disease among Roman males, coupled with lesser sexual dimorphism in stature and higher levels of linear enamel hypoplasia. Saite males were also significantly taller than Roman males.

However, *cribra orbitalia* was more or less equally common among children in both populations, and linear enamel hypoplasia occurred extensively in the Saite population as well as the Roman. This suggests that children were particularly vulnerable in both eras, and likely reflects relatively stark living conditions at both times. Thus, while the Roman conquest likely added to the vicissitudes of the Giza inhabitants, their circumstances during the Saite period were also far from ideal.

The adverse effects of the Roman takeover, thus, may not have been as far reaching as the textual sources suggest. In particular, there was no statistically significant evidence for a decline in living conditions specifically among women following the Roman conquest. The historical record is clear on the question of diminished legal autonomy for Egyptian women under Roman rule. However, the results of the Wall of the Crow analysis suggest that the decline in the status of women under the Romans may have been less impactful among the non-elite than among the middle class. It is also notable in this context that the higher levels of skeletal stress markers in Saite women when compared with Saite males similarly suggest that legal status had little or no impact on women's physical well-being among the non-elite. Nevertheless, if the non-significant results are taken into account, there may be some suggestion of a curtailment of the social and economic role of women in the Roman period. For example, while degenerative joint disease affected both upper and lower limbs of both women and men in the Saite period, and men in the Roman period, arthritic changes were limited to the knee and hip joints of Roman women. It is possible that this reflects a more physically active role of Saite women, perhaps taking a larger part in agricultural activities, compared to a more restrictive and domestic role during the Roman period.

Overall, the evidence for skeletal stress in both the Saite and Roman samples paints a picture of populations in which both hard work and poor nutrition were commonplace. In terms of the historical narrative of the Roman conquest, the results of the Wall of the Crow study can be interpreted in several ways. It is of course possible that the negative aspects of the Roman conquest have been exaggerated in the historical record. A perhaps more likely interpretation, however, is that the socio-political changes effected by the Romans were more impactful among the middle-class urban population than among the already disadvantaged inhabitants of the Memphite countryside.

Bibliography

Acsádi, G. & J. Nemeskéri. 1970. History of Human Life Span and Mortality. – Budapest, Akadémiai Kiadó.

Alexander, R.D., J.L. Hoogland, R.D. Howard, K.M. Noonan & P.W. Sherman. 1979. Sexual Dimorphism and Breeding Systems in Pinnipeds, Ungulates, Primates and Humans. In: Chagnon, N.A. & W. Irons. Eds. Evolutionary Biology and Human Social Behavior: An Anthropological Perspective. – North Scituate, MA, Duxbury Press: 402-435.

Angel, J.L. 1966. Porotic Hyperostosis, Anemias, Malarias, and Marshes in the Prehistoric Eastern Mediterranean. – Science 153: 760-763.

Armelagos, G.J. 2000. Take Two Beers and Call Me in 1600 Years. – Natural History 109: 50-54.

Armelagos, G.J., A.H. Goodman, K.N. Harper & M.L. Blakey. 2009. Enamel Hypoplasia and Early Mortality: Bioarcheological Support for the Barker Hypothesis. – Evolutionary Anthropology 18: 261-271.

Armelagos, G.J., A.H. Goodman & K.H. Jacobs. 1991. The Origins of Agriculture: Population Growth during a Period of Declining Health. – Population and Environment: 131: 9-22.

Armelagos, G.J., K.H. Jacobs & D.L. Martin. 1981. Death and Demography in Prehistoric Sudanese Nubia. In: Humphreys, S.C. & H. King. Eds. Mortality and Immortality: The Anthropology and Archaeology of Death. – London, Academic Press.

Bagnall, R.S. & B.W. Frier. 1994. The Demography of Roman Egypt. – Cambridge, Cambridge University Press.

Bagnall, R.S. & B.W. Frier. 2006. The Demography of Roman Egypt. Revised Edition. – Cambridge, Cambridge University Press.

Barker, D.J.P. & C. Osmond. 1986. Infant Mortality, Childhood Nutrition and Ischaemic Heart Disease in England and Wales. – Lancet 1: 1077-1081.

Bassett, E.J., M.S. Keith, G.J. Armelagos, D.L. Martin & A.R. Villanueva. 1980. Tetracycline-Labeled Human Bone from Ancient Sudanese Nubia. A.D. 350-550. – Science 209 (4464): 1532-1534.

Bogin, B. 1988. Patterns of Human Growth. – Cambridge, Cambridge University Press.

Boocock, P., K. Manchester & C.A. Roberts. 1995. The Human Remains from Eccles, Kent. – Bradford, Unpublished Skeletal Report.

Boylston, A. & C.A. Roberts. 1996. The Romano-British Cemetery at Kempston, Bedfordshire: Report on the Human Skeletal Remains. – Bradford, Calvin Well Laboratory, Department of Archaeological Sciences.

Brooks, S. & J.M. Suchey. 1990. Skeletal Age Determination Based on the Os Pubis: A Comparison of the Acsádi-Nemeskéri and Suchey-Brooks Methods. – Human Evolution 5: 227-238.

Brothwell, D.R. 1989. The Relationship of Tooth Wear to Aging. In: M.Y. Işcan. Ed. Age Markers in the Human Skeleton. – Springfield, Il, Charles C. Thomas: 303-316.

Buikstra, J. & J.H. Mielke. 1985. Demography, Diet and Health. In: Gilbert, R.I. & J.H. Mielke. Eds. The Analysis of Prehistoric Diets. – New York, Academic Press: 369-422.

Buikstra, J.E. & D.H. Ubelaker. Eds. 1994. Standards for Data Collection from Human Skeletal Remains. – Fayetteville, Arkansas Archeological Survey Research Report (Vol. 44).

Bullock Kreger, M. 2010. Urban Population Dynamics in a Preindustrial New World City: Morbidity, Mortality, and Immigration in Postclassic Cholula. – Unpublished PhD Thesis, Pennsylvania State University.

Buzon, M.R. 2006. Health of the Non-Elites at Tombos: Nutritional and Disease Stress in New Kingdom Egypt. – American Journal of Physical Anthropology 130: 26-37.

Cameron, N. & E.W. Demerath. 2002. Critical Periods in Human Growth and Their Relationship to Diseases of Aging. – Yearbook of Physical Anthropology 45: 159-184.

Capponi, L. 2010. The Roman Period. In: Lloyd, A.B. Ed. A Companion to Ancient Egypt. – Malden, MA, Wiley-Blackwell: 140-159.

Cavelaars, A., A.E. Kunst, J.J. Geurts, R. Crialesi, L. Grötvedt, U. Helmert, E. Lahelma, O. Lundberg, A. Mielck, N.K. Rasmussen, E. Regidor, T. Spuhler & J.P. Mackenbach. 2000. Persistent Variations in Average Height between Countries and between Socio-Economic Groups: An Overview of 16 European Countries. – Annals of Human Biology 27(4): 407-421.

Cohen, J.W. 1988. Statistical Power Analysis for the Social Sciences. – Hillsdale, NJ, Lawrence Erlbaum Associates.

Cohen, M.N. & G.M.M. Crane-Kramer. Eds. 2007. Ancient Health: Skeletal Indicators of Agricultural and Economic Intensification. – Gainesville, University Press of Florida.

Cook, M., J.E. Molto & C. Anderson. 1989. Fluorochrome labelling in Roman Period Skeletons from Dakhleh Oasis, Egypt. – American Journal of Physical Anthropology 802: 137-143.

Cutress, T. & G. Suckling. 1982. The Assessment of Noncarious Defects of Enamel. – International Dental Journal 322: 117-122.

Dasen, V. 1993. Dwarfs in Ancient Egypt and Greece. – Oxford, Clarendon Press.

DeWitte, S.N. 2014. Differential Survival among Individuals with Active and Healed Periosteal New Bone Formation. – International Journal of Paleopathology 7 Supplement C: 38-44.

DeWitte, S.N. & J.W. Wood. 2008. Selectivity of Black Death Mortality with Respect to Preexisting Health. – Proceedings of the National Academy of Sciences of the United States of America 105: 1436-1441.

Dupras, T.L., H.P. Schwarcz & S.I. Fairgrieve. 2001. Infant Feeding and Weaning Practices in Roman Egypt. – American Journal of Physical Anthropology 115: 204-212.

Fairgrieve, S.I. & J.E. Molto. 2000. *Cribra orbitalia* in Two Temporally Disjunct Population Samples from the Dakleh Oasis, Egypt. – American Journal of Physical Anthropology 111: 319-331.

Goodman, A.H. 1991. Health, Adaptation and Maladaptation in Past Societies. In: Bush, H. & M. Zvelebil. Eds. Health in Past Societies: Biocultural Interpretations of Human Skeletal Remains in Archaeological Contexts. – Oxford, British Archaeological Reports: 31-38.

Goodman, A.H. 1996. Early Life Stresses and Adult Health: Insights from Dental Enamel Development. In: Henry, C.J.K. & J. Ulijaszek. Eds. Long Term Consequences of Early Environment: Growth, development and the lifetime developmental perspective. – Cambridge, Cambridge University Press: 163-182.

Goodman, A.H., L.H. Allen, G.P. Hernandez, A. Amador, L.V. Arriola, A. Chávez & G.H. Pelto. 1987. Prevalence and Age at Development of Enamel Hypoplasias in Mexican Children. – American Journal of Physical Anthropology 721: 7-19.

Goodman, A.H. & G.J. Armelagos. 1988. Childhood Stress and Decreased Longevity in a Prehistoric Population. American Anthropologist New Series 90(4): 936-944.

Goodman, A.H. & D.L. Martin. 2002. Reconstructing Health Profiles from Skeletal Remains. In: Steckel, R.H. & J.C. Rose. Eds. The Backbone of History: Health and Nutrition in the Western Hemisphere. – Cambridge/ New York, Cambridge University Press: 11-60.

Goodman, A.H., D.L. Martin, G.J. Armelagos & G. Clarke. 1984. Indications of Stress from Bone and Teeth. In: Cohen, M.N. & G.J. Armelagos. Eds. Paleopathology at the Origins of Agriculture. – Orlando, Academic Press: 13-50.

Goodman, A.H. & J.C. Rose. 1991. Dental Enamel Hypoplasias as Indicators of Nutritional Status. In: Kelley, M.A. & C.S. Larsen. Eds. Advances in Dental Anthropology. – New York, Wiley-Liss: 279-293.

Grauer, A.L. 1993. Patterns of Anemia and Infection from Medieval York, England. – American Journal of Physical Anthropology 91: 203-213.

Graves-Brown, C. 2010. Dancing for Hathor: Women in Ancient Egypt. – London, Continuum Books.

Greulich, W.W. 1951. The Growth and Developmental Status of Guamanian School Children in 1947. – American Journal of Physical Anthropology 91: 55-70.

Hershkovitz, I., L. Bedford, L.M. Jellema & B. Latimer. 1996. Injuries to the Skeleton Due to Prolonged Activity in Hand-to-Hand Combat. – International Journal of Anthropology 6: 167-178.

Hillson, S. 1979. Diet and Dental Disease. – World Archaeology 11: 147-162.

Holden, C. & R. Mace. 1999. Sexual Dimorphism in Stature and Women's Work: A Phylogenetic Cross-Cultural Analysis. – American Journal of Physical Anthropology 110(1): 27-45.

Humphrey, L.T. & T. King. 2000. Childhood Stress: A Lifetime Legacy. – Anthropologie 38: 33-49.

Jantz, L.M. & R.L. Jantz. 1999. Secular Change in Long Bone Length and Proportion in the United States, 1800-1970. – American Journal of Physical Anthropology 110(1): 57-67.

Kaiser, J.E. 2018. Raising the Dead: The Bioarchaeology of the Saite and Roman Period Wall of the Crow Cemetery at Giza. – Unpublished PhD Thesis, University of California Berkeley.

Kehoe, D. 2010. The Economy: Greco-Roman. In: Lloyd, A.B. Ed. A Companion to Ancient Egypt. – Malden, MA, Wiley-Blackwell: 309-325.

Kemkes-Grottenthaler, A. 2005. The Short Die Young: the Interrelationship between Stature and Longevity – Evidence from Skeletal remains. – American Journal of Physical Anthropology 128: 340-347.

King, T., L.T. Humphrey & S. Hillson. 2005. Linear Enamel Hypoplasias as Indicators of Systemic Physiological Stress: Evidence from two Known Age-at-Death and Sex Populations from Postmedieval London. – American Journal of Physical Anthropology 128(3): 547-559.

Kitchen, K.A. 2009. The Third Intermediate Period in Egypt (1100-650 BC). – Oxford, Aris & Phillips.

Komlos, J. & M. Baur. 2004. From the Tallest to One of the Fattest: The Enigmatic Fate of the American Population in the 20th Century. – Economics & Human Biology 21: 57-74.

Lallo, J.W., G.J. Armelagos & J.C. Rose. 1978. Paleoepidemiology of Infectious Disease in the Dickson Mounds Population. – Medical College of Virginia Quarterly 14: 17-23.

Lambert, P.M. 1993. Health in Prehistoric Populations of the Santa Barbara Channel Islands. – American Antiquity 58: 509-521.

Larsen, C.S. 2015. Bioarchaeology: Interpreting Behavior from the Human Skeleton. Second Edition. – Cambridge, Cambridge University Press.

Larsen, C.S., J. Craig, L.E. Sering, M.J. Schoeninger, K.F. Russell, D.L. Hutchinson & M.A. Williamson. 1995. Cross Homestead: Life and Death on the Midwestern Frontier. In: Grauer, A.L. Ed. Bodies of Evidence: Reconstructing History through Skeletal Analysis. – New York, Wiley-Liss: 139-159.

Larsen, C.S., P.L. Walker, R.H. Steckel, P.W. Sciulli, H.D. Klaus, J. Blondiaux, G. Grupe, R. Jankauskas, G. Maatz, A. McGlynn, C. Papathanasiou, C. Roberts, M. Teshler-Nicola, U. Wittwer-Bckofen, A. Agnew, S. Assis, Z. Bereczki, B. Bertrand, T.K. Betsinger, M. Binder, S. Boulter, C. Bourbou, A. Boylsyon, M. Brickley, L. Burli, C. Cooper, A. Coppa, J. Coughlan, A. Drozd, E. During, C. Eliopoulos, J. Eng, F. Engel, S. Fox, M. Furtado, G. Guntis, S. Groves, K. Harkins, P. Holck, M. Holst, G. Hotz, R. Ives, T. Jacob, J. Jennings, H. Justus, K. Kaminska, A. Kjellstrom, C. Knusel, T. Kozlowski, A. Lagia, C. Lopes, S. Manolis, A. Marcsik, C. Marques, C. Moenke, C. Niel, S.A. Novak, F. Novotny, J. Peck, I. Potiekhina, B. Rega, R. Richman, F. Rijpma, J.C. Rose, J. Ruiz, P. Sannen, M. Smith, A. Soficaru, M. Spannagl, R. Storm, G. Stroud, E. Subira, D. Swales, V. Tristaroli, E. Tyler, S. Ulrich-Bochsler, S. Vatteoni, V. Villar, R. Wiggins and L.L. Williams. 2009. History of Degenerative Joint Disease in Europe: Inferences About Lifestyle and Activity. Abstract: Poster Presented at the Annual Meeting of the American Association of Physical Anthropologists, Chicago, April 2. – American Journal of Physical Anthropology 138(S48): 172.

Lewis, N. 1983. Life in Egypt Under Roman Rule. – Oxford, Clarendon Press.

Lovejoy, C.O., R.S. Meindl, R.P. Mensforth & T.J. Barton. 1985. Multifactorial Determination of Skeletal Age at Death: A Method and Blind Tests of Its Accuracy. – American Journal of Physical Anthropology 668: 1.

Lovejoy, C.O., K.F. Russell & M.L. Harrison. 1990. Long Bone Growth Velocity in the Libben Population. – American Journal of Human Biology 2: 533-541.

Lovell, N.C. & I. Whyte 1999. Patterns of Dental Enamel Defects at Ancient Mendes, Egypt. – American Journal of Physical Anthropology 110(1): 69-80.

Lukacs, J.R., G.C. Nelson & S.R. Walimbe. 2001. Enamel Hypoplasia and Childhood Stress in Prehistory: New Data from India and Southwest Asia. – Journal of Archaeological Science 28(11): 1159-1169.

Marklein, K.E., R.E. Leahy & D.E. Crews 2016. In Sickness and in Death: Assessing Frailty in Human Skeletal Remains. – American Journal of Physical Anthropology 161(2): 208-225.

Meindl, R.S. & C.O. Lovejoy 1985. Ectocranial Suture Closure: A Revised Method for the Determination of Skeletal Age at Death Based on the Lateral-Anterior Sutures. – American Journal of Physical Anthropology 668: 57.

Meindl, R.S., C.O. Lovejoy, R.P. Mensforth & R.A. Walker 1985. A Revised Method of Age Determination Using the Os Pubis, with a Review and Tests of Accuracy of Other Current Methods of Pubic Symphyseal Aging. – American Journal of Physical Anthropology 668: 29.

Mensforth, R.P. 1985. Relative Tibia Long Bone Growth in the Libben and Bt-5 Prehistoric Skeletal Populations. – American Journal of Physical Anthropology 68: 247-262.

Mills, J.O. 1992. Beyond Nutrition: Antibiotics Produced through Grain Storage Practices. In: Adams, B. & R. Friedman. Eds. The Followers of Horus: Studies Dedicated to Michael Allen Hoffman. – Oxford, Oxbow Books: 27-36.

Milner, G. R. 1992. Determination of Age and Sex: A Manual Prepared for the Dickson Mound Reburial Team. – Lewiston, Il, Dickson Mounds Museum.

Moorrees, C.F.A., E.A. Fanning & E.E. Hunt. 1963a. Age Variation of Formation Stages for Ten Permanent Teeth. – Journal of Dental Research 42: 1490-1502.

Moorrees, C.F.A., E.A. Fanning & E.E. Hunt 1963b. Formation and Resorption of Three Deciduous Teeth in Children. – American Journal of Physical Anthropology 21: 205.

Nelson, M.L., A. Dinardo, J. Hochberg & G.J. Armelagos. 2010. Brief Communication: Mass Spectroscopic Characterization of Tetracycline in the Skeletal Remains of an Ancient Population from Sudanese Nubia 350-550 CE. – American Journal of Physical Anthropology 143(1): 151-154.

Nerlich, A., A. Zink, H.G. Hagedorn, U. Szeimies & C. Weyss. 2000. Anthropological and Palaeopathological Analysis of the Human Remains from Three Tombs of the Nobles of the Necropolis of Thebes-West, Upper Egypt. – Anthropologischer Anzeiger 58(4): 321-343.

Novak, M. & M. Šlaus. 2010. Health and Disease in a Roman Walled City: An Example of Colonia Iulia Iader. – Journal of Anthropological Sciences 88: 189-206.

O'Connor, D. 1983. New Kingdom and Third Intermediate Period: 1552-664 BC. In: Trigger, B.G., B.J. Kemp, D. O'Connor & A.B. Lloyd. Eds. Ancient Egypt: A Social History. – Cambridge, Cambridge University Press: 183-278.

Ortner, D.J. 2003. Identification of Pathological Conditions in Human Skeletal Remains. Second Edition. – San Diego, Elsevier Academic Press.

Pawłowski, B. 2003. Variable Preferences for Sexual Dimorphism in Height as a Strategy for Increasing the Pool of Potential Partners in Humans. – Proceedings of the Royal Society B: Biological Sciences 270 (1516): 709-712.

Petrie, W.M.F. 1907. Gizeh and Rifeh. – London, School of Archaeology in Egypt.

Phenice, T.W. 1969. A Newly Developed Visual Method of Sexing the Os Pubis. – American Journal of Physical Anthropology 30: 297-301.

Raven, M.J., H.M. Hays, C. Lacher, K. Duistermaat & I. Regulski. 2008. Preliminary Report on the Leiden Excavation at Saqqara, Season 2008: The Tomb of Ptahemwia. – Jaarbericht van het Vooraziatisch-Egyptisch Genootschap Ex Oriente Lux 41: 5-30.

Raxter, M.H. 2011. Egyptian Body Size: A Regional and Worldwide Comparison. – Unpublished PhD Thesis, University of South Florida.

Raxter, M.H., C.B. Ruff, A. Azab, M. Erfan, M. Soliman & A. El-Sawaf. 2008. Stature Estimation in Ancient Egyptians: A New Technique Based on Anatomical Reconstruction of Stature. – American Journal of Physical Anthropology 136(2): 147-155.

Razali, N.M. & Y.B. Wah .2011. Power Comparisons of Shapiro-Wilk, Kolmogorov-Smirnov, Lilliefors and Anderson-Darling Tests. – Journal of Statistical Modeling and Analytics 2(1): 21-33.

Robb, J. 1994. Skeletal Signs of Activity in the Italian Metal Ages: Methodological and Interpretive Notes. – Human Evolution 9: 215-229.

Rose, J.C. 2006. Paleopathology of the Commoners at Tell Amarna, Egypt, Akhenaten's Capital City. – Memórias do Instituto Oswaldo Cruz 101: 73-76.

Samuel, D. 2000. Brewing and Baking. In: Nicholson, P.T. & I. Shaw. Eds. Ancient Egyptian Materials and Technology. – Cambridge, Cambridge University Press: 537-576.

Säve-Söderbergh, T. & L. Troy. 1991. New Kingdom Pharaonic Sites. – Uppsala, Almquist and Wiksell Tryckeri.

Schaefer, M., S. Black & L. Scheuer. 2009. Juvenile Osteology: A Laboratory and Field Manual. – Burlington, MA, Elsevier Academic Press.

Scheidel, W. 2012. Age and Health. In: Riggs, C. Ed. The Oxford Handbook of Roman Egypt. – Oxford, Oxford University Press: 305-316.

Schweich, M. & Knüsel, C. 2003. Bio-Cultural Effects in Medieval Populations. – Economics and Human Biology 1: 367-377.

Scott, E.C. 1979. Dental Wear Scoring Technique. – American Journal of Physical Anthropology 51: 213-218.

Shaw, I. Ed. 2000. The Oxford History of Ancient Egypt. – Oxford, Oxford University Press.

Šlaus, M. 2008. Osteological and Dental Markers of Health in the Transition from the Late Antique to the Early Medieval Period in Croatia. – American Journal of Physical Anthropology 136(4): 455-469.

Smith, B.H. 1984. Patterns of Molar Wear in Hunter-Gatherers and Agriculturalists. – American Journal of Physical Anthropology 63: 39-56.

Steckel, R.H. 1995. Stature and the Standard of Living. – Journal of Economic Literature 334: 1903-1940.

Steckel, R.H., C.S. Larsen, P.W. Sciulli & P.L. Walker. 2006. Data Collection Codebook. – Ohio, Ohio State University.

Steckel, R.H., P.W. Sciulli & J.C. Rose. 2002. A Health Index from Skeletal Remains. In: Steckel, R.H. & J.C. Rose. Eds. The Backbone of History: Health and Nutrition in the Western Hemisphere. – New York, Cambridge University Press: 61-93.

Stini, W.A. 1975. Adaptive Strategies of Human Populations under Nutritional Stress. In: Watts, E.S., F.E. Johnston & G.W. Lasker. Eds. Biosocial Interrelations in Population Adaptation. – The Hague, Mouton: 19-41.

Stinson, S. 1985. Sex Differences in Environmental Sensitivity During Growth and Development. – American Journal of Physical Anthropology 28(S6): 123-147.

Stinson, S. 2012. Growth Variation: Biological and Cultural Factors. In: Stinson, S., B. Bogin & D. O'Rourke. Eds. Human Biology: An Evolutionary and Biocultural Perspective. Second Edition. – Hoboken, NJ, John Wiley & Sons: 587-635.

Strouhal, E. & L. Bareš. 1993. Secondary Cemetery in the Mastaba of Ptahshepses at Abusir. – Prague, Charles University.

Stuart-Macadam, P. 1985. Porotic Hyperostosis: Representative of a Childhood Condition. – American Journal of Physical Anthropology 66: 391-398.

Stuart-Macadam, P. 1987a. Porotic Hyperostosis: New Evidence to Support the Anemia Theory. – American Journal of Physical Anthropology 74: 521-526.

Stuart-Macadam, P. 1987b. A Radiographic Study of Porotic Hyperostosis. – American Journal of Physical Anthropology 74: 511-520.

Stuart-Macadam, P. 1992. Porotic Hyperostosis – a New Perspective. – American Journal of Physical Anthropology 87: 39-47.

Tainter, J.A. 1980. Behavior and Status in a Middle Woodland Mortuary Population from the Illinois Valley. – American Antiquity 45: 308-313.

Tavares, A. & S. Laemmel. 2011. Some Post Old Kingdom Pottery from Giza. In: Aston, D.A., B. Bader, C. Gallorini, P. Nicholson & S. Buckingham. Eds. Under the Potter's Tree: Studies on Ancient Egypt Presented to Janine Bourriau on the Occasion of Her 70th Birthday. – Leuven, Peeters: 949-976.

Temple, D.H. 2007. Human Biological Variation During the Agricultural Transition in Prehistoric Japan. – Unpublished PhD Thesis, Ohio State University.

Tobias, P.V. 1975. Anthropometry among Disadvantaged Peoples: Studies in Southern Africa. In: Watts, E.S., F.E. Johnston & G.W. Lasker. Eds. Biosocial Interrelations in Population Adaptation. – The Hague, Mouton: 287-305.

Todd, T.W. 1920. Age Changes in the Pubic Bone: I. The White Male Pubis. – American Journal of Physical Anthropology 3: 285-334.

Todd, T.W. 1921. Age Changes in the Pubic Bone: II. The Pubis of the Male Negro-White Hybrid; III. The Pubis of the White Female; IV. The Pubis of the Female Negro-White Hybrid. – American Journal of Physical Anthropology 4: 1.

Ubelaker, D.H. 1999. Human Skeletal Remains: Excavation, Analysis, Interpretation. – Chicago, Aldine.

Usher, B.M. 2000. A Multistate Model of Health and Mortality for Paleodemography: Tirup Cemetery. – Unpublished PhD Thesis, Pennsylvania State University.

van Minnen, P. 2000. Agriculture and the 'Taxes-and-Trade' Model in Roman Egypt. – Zeitschrift für Papyrologie und Epigraphik 133: 205-220.

Vandorpe, K. 2002. Apollonia, a Businesswoman in a Multicultural Society. In: Melaerts, H. & L. Mooren. Eds. Pathyris, 2nd-1st Centuries B.C. Le Rôle Et Le Statut De La Femme En Egypte Hellénistique, Romaine Et Byzantine. – Leuven, Peeters: 325-336.

Vandorpe, K. 2012. Identity. In: Riggs, C. Ed. The Oxford Handbook of Roman Egypt. – Oxford, Oxford University Press: 260-276.

Vandorpe, K. & S. Waebens.2010. Women and Gender in Roman Egypt: The Impact of Roman Rule. In: Lembke, K., M. Minas-Nerpel & S. Pfeiffer. Eds. Tradition and Transformation: Egypt under Roman Rule. Proceedings of the International Conference, Hildesheim, Roemer- and Pelizaeus-Museum, 3-6 July 2008. – Leiden, Brill: 415-435.

VanPool, T.L. & R.D. Leonard. 2010. Quantitative Analysis in Archaeology. – Malden, MA, Wiley-Blackwell.

VanPool, T.L. & C.S. VanPool. 2001. Postprocessualism and the Nature of Science: A Response to Comments by Hutson and Arnold and Wilkens. – American Antiquity 662: 367-375.

Vercellotti, G., S.D. Stout, R. Boano & P.W. Sciulli. 2011. Intrapopulation Variation in Stature and Body Proportions: Social Status and Sex Differences in an Italian Medieval Population Trino Vercellese, VC. – American Journal of Physical Anthropology 145(2): 203-214.

Waldron, T. 1987. The Relative Survival of the Human Skeleton: Implications for Palaeopathology. In: Boddington, A., A.N. Garland & R.C. Janaway. Eds. Death, Decay and Reconstruction: Approaches to Archaeology and Forensic Science. – Manchester, Manchester University Press: 55-64.

Waldron, T. 2007. Paleoepidemiology: The Measure of Disease in the Human Past. – Walnut Creek, Left Coast Press.

Waldron, T. 2009. Paleopathology. – Cambridge, Cambridge University Press.

Waldron, T. 2012. Joint Disease. In: Grauer, A.L. Ed. Companion to Paleopathology. – Malden, MA, Wiley-Blackwell: 513-530.

Walker, P. & S.E. Hollimon. 1989. Changes in Osteoarthritis Associated with the Development of a Maritime Economy among Southern California Indians. – International Journal of Anthropology 4: 171-183.

Walker, P.L., R.R. Bathurst, R. Richman, T. Gjerdrum & V.A. Andrushko. 2009. The Causes of Porotic Hyperostosis and Cribra Orbitalia: A Reappraisal of the Iron-DeficiencyAanemia Hypothesis. – American Journal of Physical Anthropology 139(2): 109-125.

Walker, P.L., G. Dean & P. Shapiro. 1991. Estimating Age from Tooth Wear in Archaeological Populations. In: Kelley, M.A. & C.S. Larsen. Eds. Advances in Dental Anthropology. – New York, Wiley-Liss: 169-178.

Wapler, U., E. Crubezy & M. Schultz. 2004. Is *Cribra orbitalia* Synonymous with Anemia? Analysis and Interpretation of Cranial Pathology in Sudan. – American Journal of Physical Anthropology 123: 333-339.

Weiss, E. & R.D. Jurmain. 2007. Osteoarthritis Revisited: A Contemporary Review of Aetiology. – International Journal of Osteoarchaeology 17: 437-450.

Weston, D.A. 2012. Nonspecific Infection in Paleopathology: Interpreting Periosteal Reactions. In: Grauer, A.L. Ed. Companion to Paleopathology. – Malden, MA, Wiley-Blackwell: 492-512.

Wheeler, S.M. 2009. Bioarchaeology of Infancy and Childhood at the Kellis 2 Cemetery, Dakhleh Oasis, Egypt. – Unpublished PhD Thesis, University of Western Ontario.

Wheeler, S.M. 2012. Nutritional and Disease Stress of Juveniles from the Dakhleh Oasis, Egypt. – International Journal of Osteoarchaeology 22: 219-234.

White, T.D., M.T. Black & P.A. Folkens. 2012. Human Osteology. Third Edition. – Burlington, MA, Elsevier Academic Press.

Winkler, E.-M. & H. Wilfing. 1991. Tell El-Dab'a VI: Anthropologische Untersuchungen an den Skelettresten der Kampagnen 1966-69, 1975-80, 1985. – Vienna, Österreichischen Akademie der Wissenschaften.

Wood, J.W., G.R. Milner, H.C. Harpending & K.M. Weiss. 1992. The Osteological Paradox: Problems of Inferring Prehistoric Health from Skeletal Samples. – Current Anthropology 33: 343-370.

Zakrzewski, S.R. 2003. Variation in Ancient Egyptian Stature and Body Proportions. – American Journal of Physical Anthropology 121: 219-229.

Faunal Remains at the Causeway of Sahura

Mohamed Ismail Khaled &
Mohamed Hussein Ahmed

Introduction[1]

In 1994, the Egyptian Mission to Abusir discovered several limestone blocks with superb relief decoration in an area to the north side of the upper part of the causeway of the pyramid complex of King Sahura dates. These new blocks were located directly under the debris from the initial excavation of the pyramid temple by Ludwig Borchardt carried out in 1907 (1910: Figures 2 & 3). Four of these blocks were published and briefly described by Zahi Hawass and Miroslav Verner (1996). This discovery established new evidence that Borchardt never fully excavated the causeway as he did not realise that the causeway blocks might fall from each side, and he never checked the outer sloping level of the causeway. In 2002, the Egyptian mission succeeded in unearthing an additional thirteen huge blocks, from both sides of the causeway of Sahura. Thus far, these are the largest number of polychrome reliefs discovered in any pyramid complex in the Old Kingdom. In 2003-2004, five more huge limestone blocks that originated from the northern wall of the causeway of Sahura were discovered. Each block has four registers with painted relief depicting the funerary domains of Sahura. The discovery of these reliefs provides Egyptology with the most complete set of Lower Egyptian royal funerary domains, together with the depiction of Lower Egyptian nomes from the Old Kingdom. The new scenes comprised the upper part of the north wall of the causeway of Sahura. They show more than 200 funerary domains that the king established in Lower Egypt (el-Awady, 2009: 128; Khaled, 2008: 20-22). It was not clear how many more blocks, with domains, might still be found to the western part of the northern side of the causeway.

In May 2012, the excavation continued in the upper part of the northern side of causeway, where two more blocks, located under Borchardt's excavation debris were discovered. The first block was found broken into two pieces. It originally measured 1.20 m wide and 2 m high, while the second block was buried about 2.5 m deep under a big mound of Borchardt's excavation debris and measured 1.20 m wide and 2.10 m high. This discovery confirmed our assumption that more blocks would be found. Excavations were undertaken in the whole mound of Borchardt's excavation debris to establish the presence of other blocks and to remove and study these (Figure 1).

1 This article was completed within the framework of the DFG Project-ID 389349558 (see https://gepris.dfg.de/gepris/projekt/389349558) *"Archäologie des ägyptischen Staates und seiner Wirtschaft im 3. Jahrtausend v. Chr.: eine neue Untersuchung des Sahurê-Aufwegs in Abusir"*. The authors are indebted to Prof. Salima Ikram for her visit to the site and valuable discussions. Our thanks and appreciation goes also to Prof. Richard Redding for his comments and feedback as well as editing of the present article.

In: Ikram, S., J. Kaiser & S. Porcier. Eds. 2021. *The Ancient Egyptians and the Natural World. Flora, Fauna, and Science.* – Leiden, Sidestone Press: 163-170.

In March 2017 this work was continued; it took more than two months to clean the excavation debris or in other words to re-excavate the excavation debris. On April 6, a wall made of local limestone mortared with mud was uncovered on the northern-eastern side of the mound of debris (Figure 2). We initially thought that Borchardt's workers might have built this wall to support the base of the excavation rail car (Decauville) that was used to transport dirt during excavations. However, to check whether this was an ancient or modern wall, more debris had to be moved. One week later, another wall was unearthed on the western side on a higher level, but their shapes were a little confusing (Figure 3).

During cleaning of the area between these walls many smaller walls were discovered that appear to form the foundation of buildings. After clearance, it became apparent that we had revealed a small settlement. This settlement consists of four huts/houses bisected by a main street measuring 3.70 length × 1.25 width (Figures 4, 5). The

Figure 1. Cleaning the excavation debris left by Borchardt, beginning from the north to the south. Abusir Project. Photograph by M.I. Khaled.

Figure 2. The discovery of the crude limestone wall. Abusir Project. Photograph by M.I. Khaled.

Figure 3. The discovery of second level of the limestone walls. Abusir Project. Photograph by M.I. Khaled.

Figure 4. The northern side of the settlement showing two huts/houses with the main street. Abusir Project. Photograph by M.I. Khaled.

Figure 5. The southern side of the settlement showing the other two huts/houses with the main street. Abusir Project. Photograph by M.I. Khaled.

settlement was built on the foundation of the façade of the outer right (northern) wing of the entrance hall (*pr-wr.w*) of the temple, which is why the huts/houses were at different levels, as they were built on a slope. These huts/houses were constructed of locally available irregular limestone blocks using mud mortar and being chinked with pottery sherds. The walls of the houses were 0.40-0.60 m in height. Based on the study of the pottery, this settlement dates to the first millennium BC. It is noteworthy that Borchardt in his excavation in the pyramid complex of Nyuserra discovered a group of houses that date to the Late Period and may belong to Graeco-Roman times or even later (Borchardt, 1907: 134-142).

The pyramid complex of Sahura was also a cult place for the goddess Sekhmet-Sahura, together with other healing deities. The titles of the priests of this cult as well as the existence of their houses confirm that this local cult continued into the Late Period. It is noteworthy that at the end of the First Millennium BC, a list of three monuments mentions the existence of "priests of Sekhmet Sahura" (Otto, 1956: 6-7; Yoyotte, 1972: 113). This title could explain the existence of such a settlement dated to this period.

Faunal Remains

During the excavation, the team discovered outside the houses, especially to the east, a dumping area or midden used by the residents of the settlement. A sample of animal bone fragments was collected from here. The analysis of these remains was carried out to understand the diet of the small group of people who occupied the uppermost part of the northern side of Sahura's causeway for a short period of time.

Methodology

The bones were collected by hand from each feature as well as being picked out from the sieve (2 mm screen). The bones were initially divided into fish, bird, and mammal. Then the fish fragments were sorted into two piles; one contained potentially identifiable fragments and another that contained unidentifiable fragments. The identifiable fragments were sorted into body part and taxon, and then were counted and weighed. The unidentifiable fish fragments were sorted into three piles: skull vertebra, and post-cranial not vertebra (PCNV). Each of these unidentified quantities of bone was counted and weighted.

Just as with the fish, the fragments with the birds were categorized into two piles: identifiable and unidentifiable. The first pile did contain fragments that were from the first glimpse identifiable, however, it was difficult because a comparative collection was not available to us. Therefore, the identifiable bird fragments were kept separate for later identification. The unidentifiable fragments were sorted into limb, vertebra, rib, sternum/synsacrum, and skull. Each of these unidentifiable quantities of bone was counted and weighted.

The mammal fragments were sorted also into two piles: identifiable and unidentifiable fragments. For the identifiable fragments, they were recorded by taxon and data on body part, fusion/wear, traces of burning were collected. Each identified bone was weighted. The unidentifiable fragments were sorted, large medium and small, into limb, skull, vertebra, teeth, and rib. Each of these unidentified quantities of bone was counted and weighted. Identification of the remains was aided, in some instances, by use of a modern skeletal comparative collection.

In this report we use the number of identified specimens or "NISP", to discuss the abundance of bones. We define NISP as a simple count of the number of bones in each category.

The Faunal from the Late Period Settlement (LPS)

Most of the animal bones from the settlement are probably the remains of human meals. Therefore, they provide direct information about human subsistence behavior. Two important questions are: what kind of animals did the humans consume and did they have preference for certain species (Hussein & Badran, 2018: 22-27).

We examined 1168 fragments of animal bones excavated during the 2019 season. Thirty-one of these fragments were identified as fish, 8 fragments of bird and 1127 fragments as mammal (Figure 6).

Fish

The Nile flood came very close to the site of Abusir during most of its history during pharaonic times, therefore, finding a large number of fish in the sample is not surprising. The 31 fragments in the fish samples weighed a total of 14 grams. Of these, 24 fragments were assigned to a taxon. The other seven fragments could not be identified and were sorted as skull, vertebra, and post-cranial not vertebra fragments. The most common species is a mochokid catfish (*Synodontis* sp.), represented by nine fragments. The second most common species is the Nile catfish (*Clarias gariepinus*), which is represented by seven fragments. The mullet (*Mugil* sp.) is the third most common taxon and is represented by four fragments. Two fragments were identified as Nile Perch (*Lates niloticus*). The other fish taxon identified was a cichlid with two fragments see (Figure 7).

In the descriptions of each taxa presented below the information on habits, food and size are taken from Froese and Pauly (2020).

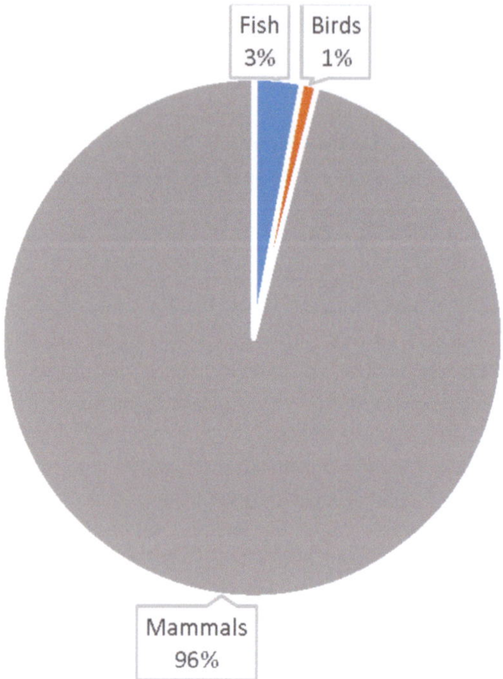

Figure 6. A comparison of the abundance of mammals, birds, and fish from the late period settlement at causeway of Sahura.

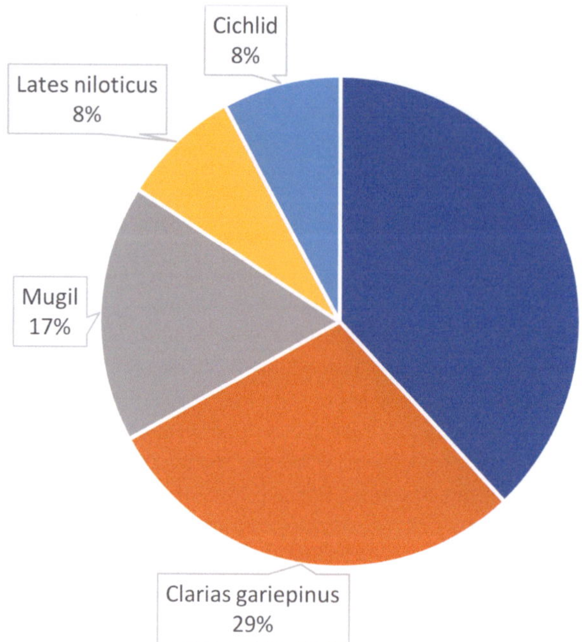

Figure 7. Percent representation of identified fish taxon from the late period settlement at the causeway of Sahura. Based on NISP.

The Taxa

Synodontis sp.

This genus is the most common taxa that was eaten by the inhabitants of the settlement. It is also very common in the Nile in number and species (Brewer & Friedman, 1989: 69-70:67-70). The modern Egyptian names of this catfish are *schall* and *gargoor* (بالعربية: سمك الشال او جرجور). These fish live in fresh water, near to the surface, and are usually caught by net. *Synodontus* sp. feeds on insects, larvae, eggs, and detritus on the surface. Its length ranges between 12 – 17 cm. with a maximum length of 37.0 cm. The maximum published weight is 500g. They live to a maximum age of 12 years. It is a desirable food fish as its flesh is white and soft (Babiker, 1981). In the Giza fish market, a large fish is about 70 LE per kilogram.

Clarias gariepinus

Clarias gariepinus is a species of catfish of the family *Clariidae*, Two species of the genus *Clarias*, C. gariepinus, and *C. anguillaris*, are known to inhabit the Egyptian Nile (Brewer & Friedman, 1989: 60-63). The Nile catfish is called in modern Lower Egypt, *armoot*, and in Upper Egypt, *garmoot* (قرموط او جرموط). It lives in shallow and quiet water of lakes, pools, and canals. It can leave the water and move over the land during the day and night. It is omnivorous, feeding on insects, fish, birds, plants, plankton, and invertebrates. A large *Clarias gariepinus* can weigh up to 60 kg with length of about 170 cm. The maximum reported age is 15 years. It is not a very desirable fish to eat as it is very oily (Babiker, 1981; Brewer & Friedman, 1989: 60-63). In the Giza fish market, one kilogram of *armoot* costs about 25 LE.

Mugil sp.

The *Mugil*, a mullet, is common along the Mediterranean coast of Egypt. Since mugils are euryhaline, they are found in the Nile as far as the south. Mugils in the Nile and Mediterranean are represented by three species (*M. cephalus, M. capito,* and *M. auratus).* Egyptian names include, *bourie, tobar, garan.* (سمك البوري)) Mullets are among the most important commercial fish in Egypt today (Abd el-Mageed, 2015). Fishermen in Damietta, an isthmus where the Nile and Mediterranean meet, can readily recognize the three species found in Egypt. Most of Nile river fishermen, refer to mullets, in general, as *bouri, M. cephalus*, which can be caught near Cairo up to the first cataract (Brewer & Friedman, 1989: 72-73). The largest *Mugil* can weigh up 7.73 kg with length of about 59 cm. The maximum reported age between 4 to 16 years. *M. cephalus* is the most abundant of the three species of mullet found in the Egyptian fresh waters. One kilogram in the fish market costs 70 L.E.

Cichlid

Three species have been identified from the Nile River in Egypt; *Orechromis nilotica, Tilapia galilaea*, and *Tilapia zillii* (Brewer & Friedman, 1989: 70). It is difficult to differentiate between these three taxa, particularly osetologically. The most common species is *Oreochromis nilotica*, whose modern name is *bolti* (سمك البلطي). They are usually found in shallow fresh water, particularly in the delta. They are very easy to be catch by net. The largest *O. nilotica* can weigh up 4.3 kg with length of about 60 cm. The maximum reported age is 9 years. *Bolti* is a very common food fish in modern Egypt and considered very desirable (Babiker, 1981). One kilogram in fish market cost 30 L.E.

Lates niloticus

The Nile perch (*Lates niloticus*), whose Egyptian name is *isher bayad* (قشر بياض), *has* been identified from the Egyptian Nile and Lake Nasser (Brewer & Friedman, 1989: 74). It lives in freshwater channels, lakes and irrigation canals. Adults of the species live in deep water, while juveniles are found in shallow water. The largest adult *L. niloticus* can weigh up 200 kg with a length of about 200 cm. The maximum reported age is 10 years. The juveniles are planktivorous and the adults are predators. *Isher bayad* is the most excellent edible fish in the Nile and an expensive fish (Babiker, 1981). One Kilogram of Nile perch in fish market cost 110 L.E.

Discussion of the Fish Remains

Synodontis sp. and C. gariepinus are the most common taxa in the late period settlement (LPS). The sample may not accurately reflect the diet. *Synodontis* sp. *and C. gariepinus* are probably over-represented in the sample because both species have many hard skull plates that are easily identified and likely to be preserved. This undoubtedly skews the numbers and makes it seem that they are more important in the diet than they actually might have been. The abundance in the sample of elements from desirable fish (*Synodontis schall, Mugil, Lates niloticus and Cichlid)* together makes up 71% of the sample the catfish (Clarias *gariepinus)* makes up 29% of the sample (Figure 7).

Aves (Birds)

Eight fragments of bird bone were collected and all are probably identifiable. They weigh a total of 15.8 grams. Birds comprise 1% of the faunal remains in the Late Period settlement, a very low percentage. One of the fragments was from a crane (*Grus grus*). The seven remaining fragments were not identified because of a lack of comparative material; however, all were from limbs. Future identification of the seven fragments will give us a clearer view of the consumption of birds at the LPS.

The Mammals

The LPS sample included 1127 mammal specimens that weighed 4542 grams. We were able to assign 171 to at least the level of the genus. The sample was composed mostly of pigs (*Sus scrofa*). The counts (NISP) for each taxon are provided in Table 8.

The Taxa

Sus scrofa

Pigs are the most common domestic taxa on the site. Pigs were an important part of the agricultural system of ancient Egypt and were reared in towns and villages to supplement access to cattle, sheep and goats (Redding, 2015; 1991). The 102 fragments identified as pig include eight from meat bearing bones (*e.g.* humerus, radius, ulna, femur tibia, lateral malleolus) and three from non-meat bearing bones (*e.g.* podial, metapodial, and phalanx). The remaining 91 fragments are from the skull. It is the dominance of skull in the sample that is curious. Where are all the limb bones? Were the animals butchered on site and only skulls kept and the meat sent elsewhere, or were skulls only brought onto the site. The very low percentage of limb bone suggests that butchering was not local.

Bos taurus

The ancient Egyptians were very experienced cattle breeders. In their society and economy, cattle played an important role (Arnold, 1995: 51; Ikram, 1995: 316-325). We identified 27 fragments of cattle. Ten of the fragments were classified as meat bearing bones, while the other nine categorized were from non-meat bearing bones (podials, metapodials, and phalanges). In addition, eight skull fragments were recovered. While the sample is small, the nearly equal occurrence of meat and non-meat bearing bones suggests local slaughter and consumption. The age data based on epiphysial fusion for the sample is provided in Table 9. The fusion ages used are from Schmidt (1972). The data are not consistent but do suggest that adult animals, older than four years, were favored.

Taxon	Count (NISP)
Bos taurus	27
Ovis-Capra	26
Sus scrofa	102
Gazella sp.	4
Equus sp.	2

Table 8. Identified mammal presented by taxon for the LPS sample from the Sahura causeway.

Element	Age of fusion (months)	Fused (NISP)	Unfused (NISP)
First phalanx	18	3	2
Second phalanx	18		1
Distal metapodial	24-27	1	
Distal tibia	24-30	2	
Proximal tibia	42-48	4	
Proximal femur	42	1	

Table 9. Fusion data for cattle elements from the LPS.

Ovis aries-/Capra hircus

All of the sheep-goat remains must be from domestic animals as the wild ancestors of sheep and goat did not occur in Egypt. A total of 26 fragments from the sheep or goat were identified of which 15 fragments were from meat bearing bones while five fragments were from non-meat bearing bones. The remaining six fragments were from the skull. These data suggest that animals may have not been butchered locally. Parts of sheep and goat may have been brought to the area for consumption. The state of fusion could only be determined on two bones, a calcaneum and a first phalanx, and both were unfused. This is a very small sample but suggests a preference for young animals, probably younger than 12 months.

Gazella sp.

Six species of gazelles are known to have occurred in ancient Egypt (Osborn & Osbornova, 1998: 186, 193). Since the most reliable way to differentiate these species is by horn core shape (Osborn & Helmy, 1980: 486-487, 492), and since no horn cores were recovered, we could not identify the species. All 4 of the fragments discovered at the settlement were meat bearing. This included two femur fragments, one tibia fragment, and one fragment of a sternum. We suggest that gazelle meat was carried into the area.

Equus sp.

Only two fragments were identified as equid. The elements recovered were both from metapodia. Based on the size of the fragments they were both from small equids, most probably a donkey (*Equus asinus*). The ancient Egyptians had domesticated donkeys by the Early Dynastic period (Osborn & Osbornova, 1998: 175-177). Equid metapodials were frequently used to make bone tools as the compact bone is very thick; however, these showed no sign of work.

Discussion of the Mammal Remains

Pigs are the dominant taxon in the LPS sample. Using the NISP, the ratio of pigs to sheep-goat, is 4:1; surprisingly, the ratio of pigs to cattle is also 4:1. Nevertheless, each *Bos taurus* provides 183.4 kg of edible product, and a sheep/goat 15 kg (Redding, 2020). So, the ratio of cattle to sheep-goat, in terms of contribution to the human diet, is actually 48.9:1. A pig provides 22 kg of edible product (Redding, 2016: 183) so the ratio of cattle to pig contribution to the human diet is 33.3:1. The ratio of pig to sheep goat contribution to the human diet is 5.9:1. In terms of contribution to the diet, cattle were far more important than pig and sheep-goat, and pig was much more important than sheep-goat. At Giza in a structure occupied by priests during the 5th Dynasty (2435-2306 BC), called the Silo Building Complex, the counts (NISP) for pig, cattle and sheep-goat were, 91, 47 and 31, respectively (Redding, 2019). These are very similar to the abundances of the LSP.

If we look at the body part distributions for, cattle, sheep-goat and pigs, we find that the skull fragments among cattle represented 30% of all cattle bones, while in sheep-goat, they represent 23% of all sheep-goat bones, and 89% of all pig bones. There is not a big difference among the percentage of skull fragments in cattle and sheep-goat but the high percentage of skull fragments of the pigs (*Sus scrofa)* is unusual. At Giza in a structure occupied by priests during the 5th Dynasty, the Silo Building Complex, pig bones are more common than cattle, sheep, and goat fragments, and head fragments from pigs form 40% of the sample of pig.

If we look at the relative abundance of non-meat bearing and meat bearing limb fragments, we find that the largest percentage of non-meat bearing bones is in the cattle, assemblage 33%, in sheep-goat 19%, and in pig 3%. There is a big difference in the percentage of non-meat bearing fragments among the taxa. If the whole animals were brought to the area for slaughter then 66% should be non-meat bearing fragments. In cattle and sheep-goat the percentage of non-meat, bearing fragments is much less than expected, only 33% on cattle and 19% in sheep-goat nonetheless, the percentage of non-meat bearing fragments of the pigs, 3% is unusual. This under-representation of non-meat bearing bones in the three taxa, especially in the samples of pigs, has two possible explanations. The first cause could be a sample size, which is small, the second may be that the animals were killed somewhere else and parts of them were brought to the site.

Summary

In the Late Period settlement of the uppermost part of the northern causeway of Sahura at Abusir, mammals were the dominant source of meat. Among the mammals pigs are represented by the most bone fragments, but cattle are the dominant contributor of edible product to the diet. Unusual is the large number of skull fragments representing pigs.

Desirable fish (*Synodontis schall, Mugil, Lates niloticus* and *Cichlid)*, which together make up 71% of the sample, dominate the fish fauna. The catfish (*Clarias gariepinus)* makes up 29% of the sample but this percentage is probably high due to the abundance, robusticity, and identifiability of Nile catfish dermal plates.

Given the dominance of cattle, the similarity of the relative abundance of the domestic taxa between the LPS and the Silo Building Complex at Giza, and the dominance of desirable fish, the LPS was occupied by high status residents, most likely priests who often derived their food from offerings (Ikram, 1995). This confirms the textual evidence for the site at the time. The meat consumed was brought from another site, maybe near Abusir, perhaps in the form of offerings.

Our study of the diet of the people who settled this area of the causeway adds to our knowledge of human subsistence behavior and the economic structure of the Late Period.

Bibliography

Abd el-Mageed, R.N. 2015. A Report on the Faunal Remains from the AA Bakery. In: Sadarangani, F. & A. Witsell. Eds. Settlement and Cemetery at Giza, Papers from the 2010 AERA-ARCE Field School. – Boston, Ancient Egypt Research Associates (AERA): 229-236.

Arnold, D. 1995. An Egyptian Bestiary. – The Metropolitan Museum of Art Bulletin (New Series) 52(4): 7-64.

El-Awady, T. 2009. Abusir XVI: Sahure – the Pyramid Causeway: History and Decoration Program in the Old Kingdom. – Prague, Charles University in Prague.

Babiker, M.M. 1981. Dietary Nile Fishes: A Reclassification According to Nutritional Merit. – Sudan Notes and Record 62: 161-170.

Brewer, D.J. & R.F. Friedman. 1989. Fish and Fishing in Ancient Egypt. – Warminster, Aris & Phillips (The Natural History of Egypt 2).

Borchardt, L. 1907. Das Grabdenkmal Des Königs Ne-User-Re'. – Leipzig, J.C. Hinrichs.

Borchardt, L. 1910. Das Grabdenkmal Des Königs S'ahu-Re. – Leipzig, J.C. Hinrichs.

Hawass, Z. & M. Verner. 1996. Newly Discovered Blocks from the Causeway of Sahure. – Mitteilungen des Deutschen Archäologischen Instituts Kairo 52: 177-186.

Hussein, M.A. & M.R. Badran. 2018. Learning Animal Bones: AERA-ARCE Field School Training. – AERAgram 19 (1): 22- 27.

Ikram, S. 1995. Choice Cuts: Meat Production in Ancient Egypt. – Leuven, Peeters (Orientalia Lovaniensia Analecta 69).

Khaled, M.I. 2008. The Old Kingdom Royal Funerary Domains: New Evidence from the Causeway of the Pyramid Complex of Sahure. – Unpublished PhD Thesis, Charles University in Prague.

Osborn, D.J. & I. Helmy. 1980. The Contemporary Land Mammals of Egypt (Including the Sinai). – Chicago, Field Museum of Natural History (Fieldiana, Zoology, New Series 5).

Osborn, D.J. & J. Osbornova. 1998. The Mammals of Ancient Egypt. – Warminster, Aris & Phillips (Natural History of Egypt IV).

Otto, E. 1956. Eine Memphitische Priesterfamilie Des 2. Jh. v. Chr. – Zeitschrift Für Ägyptische Sprache und Altertumskunde 81: 109-129.

Redding, R.W. 1991. The Role of the Pig in the Subsistence System of Ancient Egypt: A Parable on the Potential of Faunal Data. In: Crabtree, P.J. & K. Ryan. Eds. Animal Use and Culture Change. – Philadelphia. MASCA, the University Museum of Archaeology and Anthropology, University of Pennsylvania: 20-30.

Redding, R.W. 2015. The Pig and the Chicken in the Middle East: Modeling Human Subsistence Behavior in the Archaeological Record Using Historical and Animal Husbandry Data. – Journal of Archaeological Research 23(4): 325-368.

Redding, R.W. 2016. The Faunal Remains from Kom el-Hisn. In: Wenke, R.J., R.W. Redding, and P. Buck. Eds. An Old Kingdom Village in The Nile Delta: Excavations at Kom el-Hisn 1984-1988. – Atlanta, Lockwood Press: 139-203.

Redding R.W. 2019. Report on the Fauna from Gallery III.3 and KKTE+ (SBC). – Unpublished Report on File in Ancient Egypt Research Associates Server, Giza, Egypt.

Redding, R.W. 2020. Cattle in Egypt. – Unpublished draft.

Schmidt, E. 1972. Atlas of Animal Bones for Prehistorians, Archaeologists and Quaternary Geologists. – Amsterdam, Elsevier.

Yoyotte, J. 1972. La Localisation de Ouenkhem. – Bulletin de l'Institut Français d'Archéologie Orientale 71: 1-10.

Interactions between Teeth and Their Environment: A Study of the Effect on Adult Dental Age Estimation

Casey L. Kirkpatrick

Introduction

Teeth are the only elements of the human skeleton that interact directly with the environment. These interactions often leave evidence on the dentition that can be investigated through a variety of specialized methods to learn more about lives of ancient people and the world in which they lived. Dental wear is one of the most commonly documented conditions resulting from the interaction of teeth with the environment. This is understandable given the ease with which dental wear can be scored, and the information it can provide about diet, behaviour, and age-at-death.

This paper investigates the interactions between ancient Egyptian teeth and their environment, and how these interactions can affect age estimation based on dental wear. It begins with a discussion of the aetiology and mechanisms of dental wear in ancient Egypt. This section reveals an often overlooked mechanism of dental wear that may have contributed to the severity of dental wear in ancient Egyptians: dental erosion. This mechanism of dental wear is currently being investigated as it pertains to ancient Egyptians; however, details of these studies are beyond the scope of this paper. Regardless of the specific aetiology, the fact that ancient Egyptian dental wear is characteristically severe compared to many other ancient populations suggests that bioarchaeologists should not rely on universally-applied adult dental age estimation standards (*e.g.* Brothwell, 1963). Given that the Miles' (1962) method for population-specific dental age estimation is time-intensive and requires a significant subadult cohort for calibration, it has not been widely applied in Egypt. Consequently, a large-scale study aimed at creating user-friendly region-specific adult dental age estimation standards is underway. This project is being developed with dental wear data derived from a population sample from the Kellis 2 cemetery (al-Dakhla Oasis, Egypt), with future plans to expand this sample within and beyond the Kellis 2 cemetery (Kirkpatrick, 2019). This paper presents some of the initial results from this study, including: the investigation and testing of a new method for quantifying dental wear through photogrammetry, and tests of intra-oral dental wear symmetry for consideration in the development of new dental age estimation standards.

On the Aetiology and Mechanisms of Dental Wear in Ancient Egypt

Dental wear is a complex process in which dental tissues are lost as a result of the interactions of teeth with each other and their environment. Four mechanisms of wear have been identified: attrition, abrasion, erosion, and abfraction. These mechanisms are interdependent in "*an extremely complex process that involves mechanical, thermal, and*

In: Ikram, S., J. Kaiser & S. Porcier. Eds. 2021. *The Ancient Egyptians and the Natural World. Flora, Fauna, and Science.* – Leiden, Sidestone Press: 171-182.

chemical reactions" (Zhou *et al.*, 2013: 43) and they are defined as follows:

- Dental attrition (or "two-body abrasion" in tribology; d'Incau *et al.*, 2012): The gradual wear of a tooth's surface as a result of tooth-on-tooth contact, whether it is a result of mastication, bruxism, or even slight friction between adjacent teeth (Hillson, 1996; Wolpoff, 1971). In the absence of other mechanisms of wear, attrition facets appear flat with no dentine cupping (Kaidonis, 2008). The manner in which dental attrition progresses and contributes to the loss of dental tissue is not yet completely understood. However, evidence is growing to support the hypothesis that friction and contact load under nominally elastic contact conditions produce microfractures that contribute to the separation of enamel microfragments (Arsecularatne & Hoffman, 2010). The ground enamel particles then form a microscopic layer on the occlusal surfaces, which then acts as a third-body in three-body abrasion (Kaidonis *et al.*, 1998; Zheng & Zhou, 2007). These hypotheses are not mutually exclusive, nor do they represent all of the possible means for attrition-related tissue loss. Consequently, they may just explain different elements of the dental attrition process:
- Dental abrasion (or "three-body abrasion" in tribology; Zhou *et al.*, 2013): The wear produced by foreign objects and particles forced across the dental surface. This mechanism of wear can result in the progressive creation of occlusal facets, fine polishing of dental crown surfaces, the dulling of perikymata (external manifestations of Lines of Retzius resulting in microscopic grooves), the creation of microscopic scratches and pits, and microscopic surface fragmentation (Arsecularatne & Hoffman, 2012; Kaidonis, 2008).
- Dental abfraction (Grippo, 1991): A type of non-carious cervical lesion caused by high occlusal loading, which causes flexure and failure of enamel and dentine, and ultimately tissue loss, usually in the labial or buccal cervical region of the tooth. Abfractions typically present as sharp-edged wedge-shaped lesions at, or near, the cementoenamel junction. Maxillary and mandibular first and second premolars are most frequently affected by abfraction, and the size, depth, and number of abfraction lesions are positively correlated with age (Bernhardt *et al.*, 2006; Levitch *et al.*, 1994).
- Dental/acid erosion (or "corrosion" in tribology; Grippo *et al.*, 2004; Zhou *et al.*, 2013): A chemical mechanism of dental wear in which contact with acidic substances, excluding acids derived from cariogenic bacteria, result in the dissolution of the mineral aspect of the tooth. Acidity in the oral environment may be caused by extrinsic (*e.g.* acidic diet) or intrinsic factors (*e.g.* hyperemesis gravidarum, bulimia nervosa, gastroesophageal reflux). Dental erosion often results in rounded and polished enamel cusps, edges, and/or borders, It also typically gives rise to more tissue loss from dentine than from enamel, which introduces a shiny scooped out/cupped dentine element to the occlusal surface (see Figure 1). Dental erosion has also been shown to soften the dental enamel, making the tooth more vulnerable to abrasion, and thus contributing to more severe dental wear in some cases (Eisenberger *et al.*, 2003; Ganss *et al.*, 2002; Lussi *et al.*, 2004; Wu *et al.*, 2015; Zhang *et al.*, 2015; Zheng *et al.*, 2011).

Studies of ancient Egyptian dental wear have traditionally focused on the first two mechanisms of wear; attrition and abrasion. Consequently, the characteristic severity of ancient Egyptian dental wear has been invariably attributed to these two mechanisms. Ruffer (1920) described dentine cupping and severe dental wear in ancient Egyptian dentition, and attributed it to dental abrasion resulting from the mastication of copious amounts of fibrous foods. Although he acknowledged that foreign particles, such as sand and grinding-stone particles, perhaps had a role in dental wear as well, he argued that "*to explain the attrition it is not necessary to assume, as has been done, that the Egyptians ate earth or that the food was contaminated with sand*" (Ruffer, 1920: 356). In contrast, Leek (1972) asserted that the severity of ancient Egyptian dental wear may be explained solely by dental abrasion resulting from the inclusion of sand and grit in foods (*e.g.* dirt from harvest or storage, fragments from grinding stones or harvesting tools, grit added to

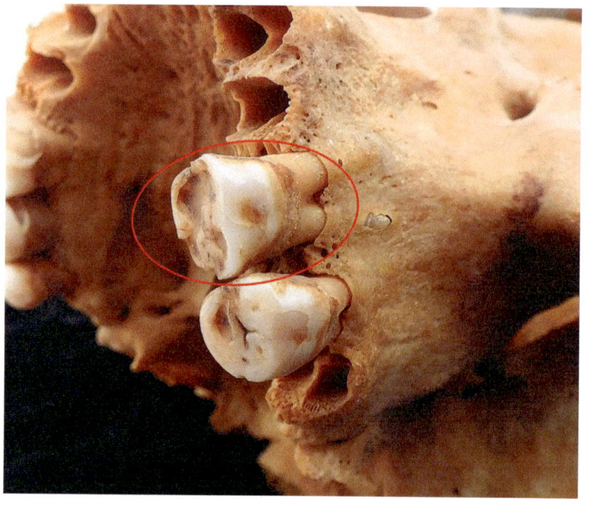

Figure 1. An example of dentine cupping surrounded by rounded and polished enamel edges. Note the enamel border that has broken away post-mortem without affecting the more deeply worn adjacent dentine (lower left in the photo). Photograph by C.L. Kirkpatrick.

facilitate the grinding of grain, and wind-blown or surface contaminants such as sand and dust). He made no mention of dentine cupping. Although Puech and colleagues (1983) mentioned enamel erosion, they gave little weight to the evidence of erosion and, never mentioned the possible role of acid in increasing dental wear rates. Instead, they attributed the observed dentine cupping, rounded enamel borders, and blunt striations solely to high occlusal pressure and dental abrasion due to sand, dust, phytoliths, vegetable masticatories, foods preserved with salt, and/or fibrous foods.

Despite the prior foci on dental attrition and abrasion in ancient Egyptians, advancements in the field of dental biotribology reinforce the need to reconsider the role of dental erosion in ancient Egyptian dental wear. For example, dental erosion is now known to form the smooth, shiny enamel surfaces with rounded cusps, smoothed edges, and the scooped out/cupped appearance of exposed dentine that is commonly observed in ancient Egyptian dentition (Lussi *et al.*, 2011; see Figure 1). Although these modifications are not pathognomonic for dental erosion (since shallow dentine cupping can be caused by dental abrasion), the possible role of dental erosion must be investigated before any conclusions can be reached. Consequently, further studies to ascertain the causative wear mechanism(s) for dentine cupping in ancient Egypt are underway. As previously noted, acidic foods and beverages have been shown to soften dental enamel, making it more vulnerable to abrasion (Ganss *et al.*, 2002; Lussi *et al.*, 2004; Wu *et al.*, 2015; Zhang *et al.*, 2015; Zheng *et al.*, 2011). Consequently, in cases where dental attrition or abrasion occur simultaneously with, or immediately following, dental contact with certain acidic substances, dental wear can progress significantly faster than it does with any one of these dental wear mechanisms alone (Eisenberger *et al.*, 2003; Kaidonis *et al.*, 1998; Wu *et al.*, 2015; Zheng *et al.*, 2011). This resulting acceleration could help explain the relative severity of dental wear typically observed in ancient Egyptians. Furthermore, this hypothesis may be supported by the fact that the critical pH level of dental enamel (pH = 5.5) is surpassed by several ancient Egyptian dietary staples, including selected fruits, fermented foods/beverages, and possibly even bread (Kirkpatrick, 2019). Although details of these findings are beyond the scope of this paper, the erosive potential of the Egyptian diet is detailed in a forthcoming paper.

Regardless of the specific aetiology of dental wear in ancient Egyptians, this discussion demonstrates the complexity of the dental wear process and dietary influence on rates of dental wear. This, in turn, contraindicates the use of universally-applied age estimation standards based on dental wear (*e.g.* Brothwell, 1963). The alternative, population-specific Miles' (1962) method for dental age estimation, is recognized for its relative accuracy (Kieser *et al.*, 1983; Lovejoy *et al.*, 1985); however, it has not been widely applied in Egypt due to its lengthy process and need for a significant subadult cohort for calibration. Consequently, a large-scale study aimed at creating user-friendly region-specific adult dental age estimation standards is now underway (*cf.* Kirkpatrick, 2019). To this end, the following sections present intra- and inter-observer test results for a relatively new method for quantifying dental wear. This method is then used to investigate whether there are any intra-oral patterns of dental wear that are significant on a population scale, for consideration in the development of the new dental age estimation standards.

Materials and Methods

Materials: Digital Dental Photos from Kellis 2 Cemetery

al-Dakhla is one of five oases or huge depressions, located in Egypt's Western Desert (Tocheri *et al.*, 2005). Studying the past human-environmental interactions in al-Dakhla has been the research theme of the Dakhleh Oasis Project (D.O.P.) since its inception in the late 1970s (Molto, 2001). Analysis of human skeletal remains is the domain of the bioarchaeology team who, since the early 1990s, have focused on cemeteries associated with the ancient centre of Kellis which is located in the central part of al-Dakhla Oasis. Kellis, was a small but important urban centre (Bagnall, 1993), which was occupied from Ptolemaic to late Roman times. It was abandoned circa 450 AD (Bowen, 2003; Hope, 2001). At its zenith in the late 3rd century AD, Kellis may have housed over 2000 people (Molto, 2001). In the 1st century AD, the people of Kellis shifted their burial program from family crypts carved into the hills northwest of Kellis (*aka* Kellis 1) to a large cemetery slightly northeast of Kellis proper, called Kellis 2 (Molto, 2001). The Kellis 2 (K2) burials were single, extended interments, oriented east-west, with heads to the west. The exact orientation of the burials however varies slightly according to the seasonal solar alignment (Williams, 2008). AMS radiocarbon dates (N =42) show that K2 was in use from 50-450 AD (Stewart *et al.*, 2003), and was organized in family groups around superstructures (Keron, 2015; Molto, 2001; Molto *et al.*, 2019).

Since the beginnings of excavations in the early 1990s, over 700 usually well-preserved skeletons have been excavated and analyzed of the 3000-4000 burials estimated to be interred in K2 (Molto, 2001). The excellent preservation of the skeletons is due to the hyperaridity of the climate and low soil acidity (Bleuze *et al.*, 2014). Although K2 burials generally had limited grave offerings, many were looted, often with detrimental effects to the head and neck region (Wheeler, 2009; Williams, 2008). Approximately 35% of the excavated individuals are adults, with the remaining 65% identified as juveniles (Wheeler, 2009). Given the extraordinary preservation at this site and the observed paleodemographic pattern, it has been suggested that the mortality profile at K2 is similar to that expected in a natural mortality distribution in preindustrial populations (Tocheri *et al.*, 2005). The current climate in the al-Dakhla Oasis has little rainfall (0.3 mm/year) and moderately low humidity, and is believed to be similar to the climate endured by the Roman Period inhabitants of the Kellis townsite, though there was likely more precipitation at that time (Bleuze *et al.*, 2014; Dupras & Schwarcz, 2001; Giddy, 1987; Stewart *et al.*, 2003; Sutton, 1947). Isotopic and documentary studies indicate that the population interred at K2 consumed an omnivorous diet that included animal proteins, C3 and C4 crops, garden vegetables, fruit, nuts, cow/goat dairy, honey, and various herbs/seasonings (Dupras, 1999). Dietary differences were observed between males and females, with males consuming more millet or ^{13}C enriched meat (specifically goat or cow) and females consuming relatively more C3 grains (such as wheat and barley) (Dupras, 1999). Apart from dietary changes related to infant weaning, which occurred between the ages of approximately six months and three years (Dupras & Tocheri, 2007), isotopic studies did not indicate any change in diet between age cohorts in the K2 population (Dupras, 1999). Therefore, it may be assumed that diet remained relatively stable throughout the lives of Kellis inhabitants.

The large number of well-preserved burials from various age groups excavated from the Kellis 2 cemetery makes this site's population ideal for the examination of adult dental wear standards. Although the original research design involved the analysis of the al-Dakhla dentitions in the regional storage magazine, security concerns in the Western Desert prevented the realization of these plans. Consequently, pre-existing images of adult occlusal dentition in Kellis 2 individuals were used for dental wear analysis. Many of the images analyzed in this project were provided by Dr. Scott Haddow. Additional photographs taken by Dr. Molto and Dr. Sheldrick were digitized from slide format for this project.

Data used in this paper were derived from the aforementioned photographic assemblage collected for the purpose of developing a new age estimation standard based on dental wear. This assemblage was limited to individuals whose occlusal dentition had been previously photographed and whose infra-cranial remains allowed for multifactorial age and sex estimation. Accordingly, individuals included in the photographic assemblage were limited to those with skeletal age estimates above 17 years in order to ensure the possibility for skeletal sex estimation. Individuals with visible evidence of leprosy were also excluded from the assemblage as this disease is known to affect the dentition. Lastly, the assemblage was limited to photographs taken approximately perpendicular to the occlusal surface of the examined teeth with sufficient clarity of enamel-dentine borders.

Skeletal sex and age estimates for the selected population sample were determined through the independent and blind analysis of separate skeletal indicators by members of the al-Dakhla Oasis Project's Kellis 2 bioarchaeological team. For adult skeletons, factors considered in the estimation of age-at-death included: iliac crest and medial clavicular epiphyseal fusion (Webb & Suchey, 1985), S1-S2 vertebral epiphyseal fusion (McKern & Stewart, 1957), symphysis pubis morphology (Brooks & Suchey, 1990; Suchey & Katz, 1986), auricular surface (Lovejoy *et al.*, 1985; Meindl & Lovejoy, 1989), and rib morphology (Iscan & Loth, 1986), dental attrition (Brothwell, 1963), cranial suture fusion (Buikstra & Ubelaker, 1994), antemortem tooth loss, bone mineralization and degenerative joint disease. Factors contributing to the estimation of sex included: os coxae ventral arc, subpubic concavity, and ischiopubic ramus ridge morphology (Phenice, 1969), as well as greater sciatic notch, preauricular sulcus morphology (Buikstra & Ubelaker, 1994), and nuchal crest, mastoid process, supraorbital margin, glabella, and mental eminence morphology (Acsádi & Nemeskéri, 1970; Buikstra & Ubelaker, 1994).

Dental Wear Data Collection Methods and Methodology

Although the documentation of dental wear has traditionally been performed through visual observation (*e.g.* Broca, 1879; Brothwell, 1963; Gustafson, 1950; Hrdlicka, 1939; Kim *et al.*, 2000; Miles, 1962; Molnar, 1971; Murphy, 1959a, b; Scott, 1979; Smith, 1984; Yun *et al.*, 2007; Zurht, 1955), there are more technologically advanced macroscopic methods that have shown promise. These include the measurement of cusp height (*e.g.* Molnar *et al.*, 1983; Tomenchuk & Mayhall, 1979), crown height (*e.g.* Mays *et al.*, 1995), and a combination of crown height and angle of wear (*e.g.* Walker *et al.*, 1991). Unfortunately, these methods are somewhat subjective as they either assume a standard original cusp height or require the researcher to estimate the original crown or cusp height in order to contextualize their results.

Another method that has received some attention is the measurement of occlusal tooth surfaces, wear facets or dentine exposures using a planimeter (*e.g.* LeBlanc & Black, 1974; Walker *et al.*, 1978). Richards and Brown (1981) reported the percentage of the occlusal surface with exposed dentine through the use of a digitizer that was programmed for use as a planimeter. Digital photogrammetry was introduced to the three-dimensional measurement of dental wear by Teaford (1983), who measured the area of exposed dentine based on the length, width and depth of the planes of dentine exposures in macaques and langurs. A variety of two- and three-dimensional methods for the measurement of occlusal wear have since been investigated (*e.g.* Deter, 2009; Kaifu *et al.*, 2003; Kambe *et al.*, 1991; Krejci *et al.*, 1994; Kullmer *et al.*, 2009; Lambrechts *et al.*, 1989; Mayhall & Kageyama, 1997; Richards, 1984, 1990; Richards & Miller, 1991; Smith, 1984; Teaford & Oyen, 1989; van der Bijl *et al.*, 1989).

One method, developed by Phillips-Conroy *et al.* (2000) for the two-dimensional photogrammetric quantification of dental wear in baboons, has shown some promise. It was also successfully applied to humans by Deter (2006, 2009). With a nod to Richards and Brown's (1981) calculations of relative area of exposed dentine, the method by Phillips-Conroy *et al.* (2000) uses the NIH-provided photogrammetry program, IMAGE v.1.5.7., to measure the areas of a tooth's occlusal surface and its exposed dentine, to determine the percentage of the occlusal surface with exposed dentine. Deter's (2006, 2009) method similarly uses image analysis software to determine the relative area of dentine exposure in relation to occlusal surface through precise counts of pixels within these respective areas through the use of SigmaScan Pro 5. These methods were based on dentine exposures because they are more visible in photographs and amenable to quantification than wear facets (Hillson, 1996). Due to the precision and simplicity of Deter's (2006, 2009) pixel-counting method for quantifying selected areas on digital photographs, it was chosen for this study of dental wear; however, the free online software, FIJI (is just imageJ), was used instead of SigmaScan Pro 5 as suggested by Rebecca J. Whiting (personal communication, 2016).

When a digital photo is opened in the Fiji (Fiji Is Just ImageJ) program, there is a freehand selection tool that was used to outline the area of the occlusal surface. This area was then measured in pixels using the "Measure" tool under the "Analyze" tab (Figure 2). Although it is possible to calibrate digital measurements with real scales included in the original images, this was not necessary in this case as the collected data were used to calculate percentages, and so pixel-counts were sufficient. The use of pixel-counts also reduced any challenges that might be associated with scales missing from photographs, illegible from glare, or positioned in a manner that might make calibration difficult.

Following measurement of the area of the occlusal surface, areas of exposed dentine were similarly outlined and measured. Pressing the SHIFT key when making selections allowed for the selection and measurement of multiple areas at one time (Figure 2). The resulting measurements for the area of the occlusal surface and the summed areas of exposed dentine were then transferred to an Excel sheet to calculate the percentage of the occlusal surface with exposed dentine.

For the purpose of this study, this method was applied to first molars included in the photographic assemblage that were not visibly affected by malocclusion, occlusal or gross dental caries, post-mortem crown breakage, ante-mortem or post-mortem loss, or abnormal wear. For the purpose of this study, the lower left first molars (LLM1) provided the "primary M1" data to which isomere and antimere data were compared. For simplicity, lower left first molars were also analyzed in the intra- and inter-observer error tests. Individuals were excluded from this study if reliable data could not be collected from their LLM1 due to caries, breakage, malocclusion, antemortem

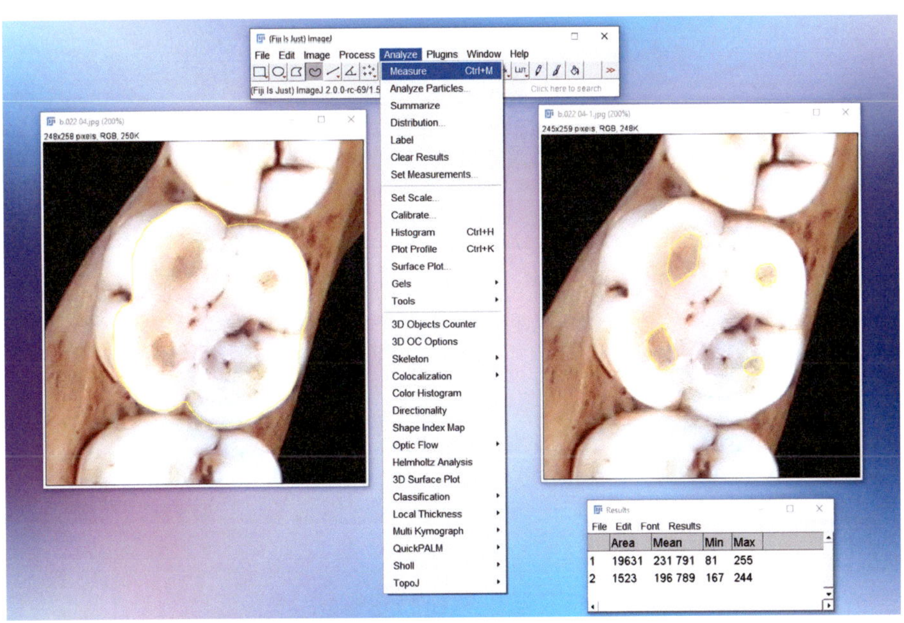

Figure 2. Screen shot illustrating the method for quantification of the occlusal and exposed dentine surfaces through Fiji (Fiji Is Just ImageJ). Quantified areas are isolated with yellow thresholds.

tooth loss, or post-mortem tooth loss. Likewise, if reliable data could not be collected from an individual's lower right first molar or upper left first molar, they were excluded from the antimere or isomere studies, respectively. For the sake of simplicity, the details of test-specific methods for sample selection and data collection are summarized along with their respective results.

Data Analysis

Following the initial data collection, intra- and inter-observer data were collected and compared through intra class correlation (ICC) tests conducted in RStudio. Paired t-tests were then used to examine the differences between first molar antimeres and isomeres (*i.e.* opposing left and right first molars, and occluding upper and lower first molars, respectively). In the case of the isomeres, whose t-test results showed statistically significant directional differences at a 90% confidence interval [as per the statistical reporting convention recommended by Sterne and Smith (2001) and Dahiru (2008)], a simple plot of differences was graphed to visualize the differences between isomeres (Figure 3).

Results

Intra-observer Error Study of Dental Wear Quantification Method

Google's random number generator was used to select 30 individuals with quantified LLM1 wear. For these individuals (n = 30), the percentages of exposed dentine were re-measured and re-calculated by the author and compared with the original data. Intra-class correlation (ICC) was then calculated in RStudio to assess the strength of the correlation between the original and secondary (intra-observer) data. This test indicated that there is a very strong correlation (99%) between the original and intra-observer data at a statistically significant level of $p<0.0001$.

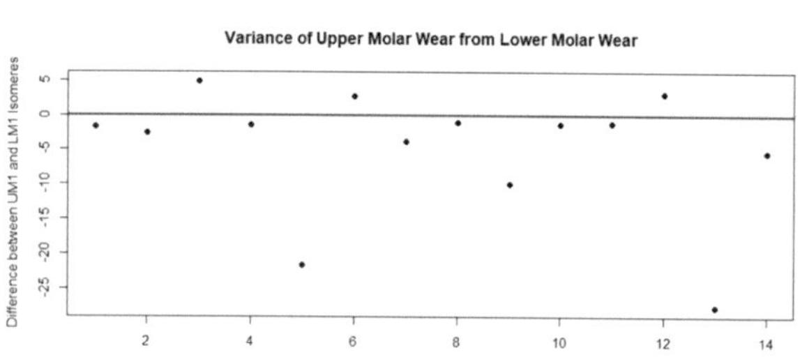

Figure 3. Simple Plot of Differences between First Molar Isomeres (n=14).

Inter-observer Error Study of Dental Wear Quantification Method

Similar to the above intra-observer test, Google's random number generator was used to select 30 individuals with quantified LLM1 wear. A volunteer was then trained to identify the edges of occlusal surface and dentine exposures and to apply the photogrammetric quantification method. Following this training, the volunteer independently quantified and calculated the percentage of the occlusal surface with exposed dentine from dental photographs of the randomly-selected individuals (n = 30). These data were recorded and compared to the original data. Intra-class correlation (ICC) was then calculated in RStudio to assess the correlation between the original and inter-observer data. These test results indicate that there is a very strong correlation (98%) between the original and inter-observer data at a statistically significant level of $p<0.0001$.

Wear in Upper vs. *Lower Left First Molars*

Paired t-tests were conducted on male, female, and combined isomere data (n = 14) in RStudio to determine if there is a significant difference in dental wear between opposing maxillary and mandibular first molars.

t-test for Male Isomeres
Mean of the differences: 6.214286, $t(6) = 1.5916$, p-value = 0.1626

t-test for Female Isomeres
Mean of the differences: 3.344286, $t(6) = 1.0248$, p-value = 0.345

t-test for Combined Male and Female Isomeres
Mean of the differences: 4.779286, $t(13) =1.9297$, p-value = 0.07576

These paired t-tests showed a statistically significant difference between isomeres in the combined male and female data at a 90% confidence interval [as per the statistical reporting convention recommended by Sterne and Smith (2001) and Dahiru (2008)]. This difference was weighted toward heavier wear in the lower molars, with only three cases in which the upper molar had a higher percentage of exposed dentine (see Figure 3). The remaining t-test results were not statistically significant, likely because of their small sample sizes.

Wear in Lower Left vs. *Lower Right First Molars*

Paired t-tests were conducted on male, female, and combined antimere data (n = 31) in RStudio to determine if there was a significant difference in dental wear between the left and right sides of the dentition.

Male Antimeres Dental Wear
Mean of the differences: 4.85, $t(11) = 1.1186$, p-value = 0.2872

Female Antimere Dental Wear
Mean of the differences: 2.25, $t(18) = 1.0178$, p-value = 0.3223

Combined Male and Female Antimere Dental Wear
Mean of the differences: 3.256452, $t(30) = 1.5319$, p-value = 0.136

Results of the above paired t-tests indicate that there is no evidence of a statistically significant directional difference between dental wear in lower left and right first molars in males, females, or combined samples at a 95% confidence interval. Following this test, the absolute differences between lower first molar antimeres were also subjected to a paired t-test.

Absolute Differences between Combined Male and Female Antimere Dental Wear
Mean of the differences: 6.155806, $t(30) = 3.2402$, p-value = 0.002919

The t-test of the absolute differences between antimeres indicates that there is a statistically significant difference between left and right dental wear, independent of direction, at a 95% confidence interval.

Discussion and Conclusions

Beginning with a discussion of the aetiology and mechanisms of dental wear in ancient Egyptians, this paper showed the complexity of the dental wear process, and that dietary factors can significantly influence rates and patterns of dental wear. In particular, the interaction of acid erosion and dental abrasion was briefly considered as a possible explanation for the relatively severe dental wear observed in ancient Egyptians. Although determining the specific aetiology of ancient Egyptian dental wear is beyond the scope of this paper, this discussion brings attention to the relative severity of dental wear in ancient Egyptians compared to many other populations. This oft-noted characteristic in ancient Egyptians contraindicates the use of universally-applied age estimation standards based on dental wear (*e.g.* Brothwell, 1963). Unfortunately, the alternative (and more accurate) population-specific Miles' (1962) method of dental age estimation has not been widely applied in Egypt due to its lengthy process and need for a significant subadult cohort for calibration. Consequently, the author is conducting a large-scale study aimed at creating user-friendly region-specific adult dental age estimation standards (*cf.* Kirkpatrick, 2019).

To this end, a photogrammetric method for quantifying the percentage of a tooth's occlusal surface occupied by exposed dentine was identified and tested. This method, introduced by Phillips-Conroy *et al.* (2000) and modified by Deter (2006; 2009), uses image analysis software to count the number of pixels within an outlined area (*i.e.* the exposed dentine patches and the occlusal surface) which are then used to calculate the percentage of the occlusal surface occupied by exposed dentine. This is the first time that this method of dental wear quantification is being investigated for use in human age estimation.

For the purpose of this study, pixel-counts for the aforementioned areas were collected using the free online software Fiji (Fiji Is Just ImageJ) for lower left first molars (and their respective isomeres and antimeres whenever possible) from a photographic sample of human occlusal dentition from the Kellis 2 cemetery. The resulting data were copied to an Excel sheet, which calculated the percentage of the occlusal surface with exposed dentine. All subsequent calculations, plots, graphs and models were performed in RStudio.

Intra- and inter-observer tests were conducted through the use of intra-class correlation on randomly selected samples of first molar data to determine the reliability of this method. Results showed a 99% correlation between intra-observer data and a 98% correlation between inter-observer data. These data represent an improvement on the Brothwell (1963) scoring method, for which the intra- and inter-observer error rates are both 80% (Alayan *et al.*, 2018). This is not surprising, given the subjectivity inherent in categorical- and atlas-type dental wear scoring methods.

Although this new method for quantifying dental wear has little inter- and intra-observer error, it does have some limitations. For example, in certain lighting conditions it can be difficult to identify the boundaries of the occlusal surface. Bias can also be introduced when photographs are not taken from directly above the occlusal surface. Furthermore, in its present state, this method is still rather time-consuming and work-intensive. However, many of these challenges can be mitigated through standardized photographic protocols and the automation of the quantification process. Since this study demonstrated the reliability of this photogrammetric method for quantifying dental wear, it was deemed appropriate for use in the development of a new dental age estimation standard, described in more detail at the end of this paper.

After confirming the reliability of the new method for quantifying dental wear, the dental wear data collected from first molar antimeres and isomeres (*i.e.* opposing left and right first molars, and occluding upper and lower first molars, respectively) were compared. This was done to determine if there were any discernible patterns of intra-oral dental wear asymmetry within the population that might affect the applicability or accuracy of dental age estimation standards.

T-tests of isomere data revealed a statistically significant difference between wear in upper and lower first molars at a 90% confidence interval [as per the statistical reporting convention recommended by Sterne and Smith (2001) and Dahiru (2008)]. A simple plot of differences showed that lower first molars tended to have more severe wear than upper first molars. This difference may be explained by greater bucco-lingual dimensions and enamel thickness in the upper molars compared to lower molars (Grine, 2005; Smith *et al.*, 2006). Hypothetically, differences in dental morphology may also contribute to asymmetrical rates of dentine exposure in occluding teeth. This finding is significant as it confirms that the relative amounts of exposed dentine are not equal in occluding upper and lower molars. These data support the observations of Murphy (1959a) and Pal (1971) and indicate that, in order to maximize accuracy, age estimation methods based on relative dentine exposure in ancient Egyptians should include separate standards for upper and lower molars. Although several of the existing dental wear scoring methods recognize differences in the patterns of dentine exposure between upper and lower anterior teeth, none of the adult dental age estimation methods account for these differences in the rate of dentine exposure between upper and lower molars.

T-tests of antimere data revealed a statistically significant difference between dental wear in lower left and right molars, independent of direction [at a 90% confidence interval, as per the statistical reporting convention recommended by Sterne and Smith (2001) and Dahiru (2008)]. These findings stand in stark contrast to the longstanding belief that there is little difference in wear between left and right sides of the dentition (Campbell, 1925; Hillson, 1996). However, significant asymmetry in dental wear among antimeres can result from a number of factors: differences in occlusal relationship, differences in patterns of mastication/bruxism, differences in orofacial musculature, arbitrary preference for a chewing side, oral habits affected by handedness, a tendency to sided chewing as a result of dental pain or discomfort, or a tendency to one-sided chewing to compensate for lost chewing surfaces through dental disease or antemortem tooth loss (Kirkpatrick, 2019). Regardless of the specific causes for the observed dental wear asymmetry between molar antimeres, the results of this study contraindicate the common practice of preferentially scoring the left side of the dentition, with the right side acting as an alternative only when the left cannot be scored (Hillson, 1996). This revelation is especially pertinent to the development of dental age estimation standards as it implies that age estimates based on exposed dentine may vary based on the side of the dentition examined. Consequently, this intra-oral dental wear asymmetry must be considered in the development of new standards for dental age estimation based on dental wear. Two possibilities for mitigating these non-directional discrepancies are to average out the relative area of exposed dentine in molar antimeres, or the resulting age estimates, where possible.

In conclusion, all the results from this study were instrumental in the development of a new sex- and region-specific age estimation standard based on quantified dental wear in ancient Egyptians (which will be published in the near future). In the future, the reference data for these standards will be expanded through further study of the Kellis 2 population. The standards will also be tested on geographically and chronologically diverse populations in the regions of Egypt and the Middle East. These studies would enable the systematic expansion of the standards' reference data, and therefore applicability, with information from populations with similar rates of dental wear. In populations shown to have especially deviant rates of dental wear, these tests may prompt the development of population-specific correction factors or new standards for estimating age in that particular population. In addition to contributing to dental age estimation methodology, the study of dental wear across time and space would allow for the identification of significant changes in food production and processing methods, or even differences in food preferences and availability, as observed in dental wear throughout the region.

In the meantime, a trainable weka segmentation program (artificial intelligence software) is currently being trained and tested to establish if it can accurately automate the delineation of areas requiring measurement. If this artificial intelligence program is found to accurately identify the edges of occlusal surfaces and dentine exposures, it would enable the instantaneous analysis of large collections of photographs, which would significantly decrease the time and effort involved in this method of dental age estimation. A software pipeline may also be developed to automate the whole process from photo upload to the completion of wear and age calculations. With these modifications, this process would be more user-friendly and less time-consuming than the Miles (1962) method for population-specific dental age estimation, and it would not require a significant subadult cohort for calibration. Furthermore, it would be less subjective than the current dental wear standards and the sex- and region-specific standards for age estimation will produce more accurate and precise age estimates than the Brothwell (1963) method. Overall, these improved methods for dental age estimation could result in much better osteobiographic, paleo-epidemiologic, and paleo-demographic reconstructions, as dental wear has the greatest potential, among macroscopic aging methods, to rectify the problem of age underestimation in older cohorts.

Acknowledgements

I would like to express my sincere gratitude to the Egyptian Ministry of Tourism and Antiquities afor making this research possible; Dr. Eldon Molto, for his support and supervision of my research; Dr. Scott Haddow, for providing many of the dental photographs used in this study; Dr. Peter Sheldrick and the rest of the Dakhleh Oasis Project, for their excellent work and welcoming nature; the Ontario Graduate Scholarships that funded much of this research; members of the conference committees and volunteers who made the BAE 2019 conference possible; the BAE 2019 Travel Grant for enabling my presentation of this research in Egypt; and the reviewers of this paper for their kind feedback.

Bibliography

Acsádi, G. & J. Nemeskéri. 1970. History of Human Life Span and Mortality. – Budapest, Akadémiai Kiadó.

Alayan, I., M.S. Aldossary & A. Santini. 2018. Validation of the Efficacy of Age Assessment by the Brothwell Tooth Wear Chart, Using Skulls of Known Age at Death. – Journal of Forensic Dental Sciences 10: 18-21.

Arsecularatne, J. & M. Hoffman. 2010. On the Wear Mechanism of Human Dental Enamel. – Journal of the Mechanical Behavior of Biomedical Materials 3: 347-356.

Arsecularatne, J. & M. Hoffman. 2012. Ceramic-Like Wear Behaviour of Human Dental Enamel. – Journal of the Mechanical Behavior of Biomedical Materials 8: 47-57.

Bagnall, R.S. 1993. Egypt in Late Antiquity. – Princeton, Princeton University Press.

Bernhardt, O., D. Gesch, C. Schwahn, F. Mack, G. Meyer, U. John & T. Kocher. 2006. Epidemiological Evaluation of the Multifactorial Aetiology of Abfractions. – Journal of Oral Rehabilitation 33: 17-25.

Bleuze, M.M., S.M. Wheeler, T.L. Dupras, L.J. Williams & J.E. Molto. 2014. An Exploration of Adult Body Shape and Limb Proportions at Kellis 2, Dakhleh Oasis, Egypt. – American Journal of Physical Anthropology 153: 496-505.

Bowen, G.E. 2003. Some Observations on Christian Burial Practices at Kellis. In: Bowen, G.E. & C.A. Hope. Eds. The Oasis Papers 3: Proceedings of the Third International Conference of the Dakhleh Oasis Project. – Oxford, Oxbow: 167-182.

Broca, P. 1879. Instructions relatives à l'étude anthropologique du système dentaire. – Bulletins et Mémoires de la Société d'Anthropologie de Paris 2: 128-163.

Brooks, S. & J.M. Suchey. 1990. Skeletal Age Determination Based on the Os Pubis: A Comparison of the Acsádi-Nemeskéri and Suchey-Brooks Methods. – Human Evolution 5: 227-238.

Brothwell, D.R. 1963. Digging up Bones: The Excavation, Treatment and Study of Human Skeletal Remains. – London, British Museum.

Buikstra, J.E. & D.H. Ubelaker. 1994. Standards for Data Collection from Human Skeletal Remains. – Fayetteville, Arkansas Archeological Survey (Arkansas Archeological Survey Research Series).

Campbell, T.D. 1925. Dentition and Palate of the Australian Aboriginal. – Adelaide, Hassell Press.

Dahiru, T. 2008. P-Value, a True Test of Statistical Significance? A Cautionary Note. – Annals of Ibadan Postgraduate Medicine 6: 21-26.

Deter, C. 2006. Dental Wear Patterns of Hunter-Gatherer and Agriculturists: The Impact of Behavioural Changes Accompanying the Transition. – University of London, University College London.

Deter, C.A. 2009. Gradients of Occlusal Wear in Hunter-Gatherers and Agriculturalists. – American Journal of Physical Anthropology 138: 247-254.

d'Incau, E., C. Couture & B. Maureille. 2012. Human Tooth Wear in the Past and the Present: Tribological Mechanisms, Scoring Systems, Dental and Skeletal Compensations. – Archives of Oral Biology 57: 214-229.

Dupras, T.L. 1999. Dining in the Dakhleh Oasis, Egypt: Determination of Diet Using Documents and Stable Isotope Analysis. – McMaster University.

Dupras, T.L. & H.P. Schwarcz. 2001. Strangers in a Strange Land: Stable Isotope Evidence for Human Migration in the Dakhleh Oasis, Egypt. – Journal of Archaeological Science 28: 1199-1208.

Dupras, T.L. & M.W. Tocheri. 2007. Reconstructing Infant Weaning Histories at Roman Period Kellis, Egypt Using Stable Isotope Analysis of Dentition. – American Journal of Physical Anthropology 134: 63-74.

Eisenburger, M., R. Shellis & M. Addy. 2003. Comparative Study of Wear of Enamel Induced by Alternating and Simultaneous Combinations of Abrasion and Erosion In Vitro. – Caries Research 37: 450-455.

Ganss, C., J. Klimek & N. Borkowski. 2002. Characteristics of Tooth Wear in Relation to Different Nutritional Patterns Including Contemporary and Medieval Subjects. – European Journal of Oral Sciences 110: 54-60.

Giddy, L.L. 1987. Egyptian Oases: Bahariya, Dakhla, Farafra & Kharga during Pharaonic Times. – Oxford, Aris & Phillips.

Grine, F. 2005. Enamel Thickness of Deciduous and Permanent Molars in Modern *Homo sapiens*. – American Journal of Physical Anthropology 126: 14-31.

Grippo, J.O. 1991. Abfractions: A New Classification of Hard Tissue Lesions of Teeth. – Journal of Esthetic and Restorative Dentistry 3: 14-19.

Grippo, J.O., M. Simring & S. Schreiner. 2004. Attrition, Abrasion, Corrosion and Abfraction Revisited: A New Perspective on Tooth Surface Lesions. – The Journal of the American Dental Association 135: 1109-1118.
Gustafson, G. 1950. Age Determinations on Teeth. – The Journal of the American Dental Association 41: 45-54.
Hillson, S. 1996. Dental Anthropology. – Cambridge, Cambridge University Press.
Hope, C.A. 2001. Observations on the Dating of the Occupation at Ismant El-Kharab. In: Marlow, C. & A. Mills. Eds. The Oasis Papers 1: The Proceedings of the First Conference of the Dakhleh Oasis Project. – Oxford, Oxbow: 43-59.
Hrdlička, A. 1939. Practical Anthropometry. – Philadelphia, Wistar Institute of Anatomy and Biology.
Iscan, M.Y. & S.R. Loth. 1986. Estimation of Age and Determination of Sex from the Sternal Rib. – Forensic Osteology: 68-89.
Kaidonis, J.A. 2008. Tooth Wear: The View of the Anthropologist. – Clinical Oral Investigations 12: 21-26.
Kaidonis, J.A., L.C. Richards, G.C. Townsend & G.D. Tansley. 1998. Wear of Human Enamel: A Quantitative In Vitro Assessment. – Journal of Dental Research 77: 1983-1990.
Kaifu, Y., K. Kasai, G.C. Townsend & L.C. Richards. 2003. Tooth Wear and the "Design" of the Human Dentition: A Perspective from Evolutionary Medicine. – American Journal of Physical Anthropology 122: 47-61.
Kambe, T., K. Yonemitsu, K. Kibayashi & S. Tsunenari. 1991. Application of a Computer Assisted Image Analyzer to the Assessment of Area and Number of Sites of Dental Attrition and its Use for Age Estimation. – Forensic Science International 50: 97-109.
Keron, J.R. 2015. The Use of Point Pattern Analysis in Archaeology: Some Methods and Applications. – University of Western Ontario.
Kieser, J., C. Preston & W. Evans. 1983. Skeletal Age at Death: An Evaluation of the Miles Method of Ageing. – Journal of Archaeological Science 10: 9-12.
Kim, Y.-K., H.-S. Kho & K.H. Lee. 2000. Age Estimation by Occlusal Tooth Wear. – Journal of Forensic Science 45: 303-309.
Kirkpatrick, C. 2019. Getting Better All the Time: Re-Evaluating Macroscopic Dental Age Estimation Standards in Egypt. – University of Western Ontario.
Koshoji, N.H., S.K. Bussadori, C.C. Bortoletto, R.A. Prates, M.T. Oliveira & A.M. Deana. 2015. Laser Speckle Imaging: A Novel Method for Detecting Dental Erosion. – PloS One 10: e0118429.
Krejci, I., T.H. Reich, W. Bucher & F. Lutz. 1994. Eine neue Methode zur dreidimensionalen Verschleißmessung. - Schweiz Monatsschr Zahnmed 104: 160-169.
Kullmer, O., S. Benazzi, L. Fiorenza, D. Schulz, S. Bacso & O. Winzen. 2009. Occlusal Fingerprint Analysis: Quantification of Tooth Wear Pattern. – American Journal of Physical Anthropology 139: 600-605.
Lambrechts, P., M. Braem, M. Vuylsteke-Wauters & G. Vanherle. 1989. Quantitative In Vivo Wear of Human Enamel. – Journal of Dental Research 68: 1752-1754.
LeBlanc, S.A. & B. Black. 1974. A Long Term Trend in Tooth Size in the Eastern Mediterranean. – American Journal of Physical Anthropology 41: 417-422.
Leek, F.F. 1972. Teeth and Bread in Ancient Egypt. – The Journal of Egyptian Archaeology 58: 126-132.
Levitch, L., J. Bader, D. Shugars & H. Heymann. 1994. Non-Carious Cervical Lesions. – Journal of Dentistry 22: 195-207.
Lovejoy, C.O., R.S. Meindl, R.P. Mensforth & T.J. Barton. 1985. Multifactorial Determination of Skeletal Age at Death: A Method and Blind Tests of Its Accuracy. – American Journal of Physical Anthropology 68: 1-14.
Lussi, A., T. Jaeggi & D. Zero. 2004. The Role of Diet in the Aetiology of Dental Erosion. – Caries Research 38: 34-44.
Lussi, A., N. Schlüter, E. Rakhmatullina & C. Ganss. 2011. Dental Erosion-An Overview with Emphasis on Chemical and Histopathological Aspects. – Caries Research 45: 2-12.
Mayhall, J.T. & I. Kageyama. 1997. A New, Three-Dimensional Method for Determining Tooth Wear. – American Journal of Physical Anthropology 103: 463-469.
Mays, S., C. de la Rua & T. Molleson. 1995. Molar Crown Height as a Means of Evaluating Existing Dental Wear Scales for Estimating Age at Death in Human Skeletal Remains. – Journal of Archaeological Science 22: 659-670.
McKern, T.W. & T.D. Stewart. 1957. Skeletal Age Changes in Young American Males Analysed from the Standpoint of Age Identification. – Natick, Quartermaster Research and Engineering Command.
Meindl, R.S. & C.O. Lovejoy. 1989. Age Changes in the Pelvis: Implications for Paleodemography. In: Iscan, M.Y. Ed. Age Markers in the Human Skeleton. – Springfield, Charles C. Thomas: 137-168.
Miles, A. 1962. Assessment of the Ages of a Population of Anglo-Saxons from their Dentitions. – Proceedings of the Royal Society of Medicine 55: 881-886
Molnar, S. 1971. Human Tooth Wear, Tooth Function and Cultural Variability. – American Journal of Physical Anthropology 34: 175-189.
Molnar, S., J. McKee, I. Molnar & T. Przybeck. 1983. Tooth Wear Rates among Contemporary Australian Aborigines. – Journal of Dental Research 62: 562-565.

Molto, J. 2001. The Skeletal Biology and Paleoepidemiology of Population Samples from Tirghi and Kellis. Dakhleh, Egypt. In: Marlow, C. & A. Mills. Eds. The Oasis Papers: Proceedings of the First International Symposium of the Dakhleh Oasis Project. – Oxford, Oxbow: 81-100.

Molto, J.E., C.L. Kirkpatrick & J. Keron. 2019. The Paleoepidemiology of Sacral Spina Bifida Occulta in Population Samples from the Dakhleh Oasis, Egypt. – International Journal of Paleopathology 26: 93-103.

Murphy, T. 1959a. Gradients of Dentine Exposure in Human Molar Tooth Attrition. – American Journal of Physical Anthropology 17: 179-186.

Murphy, T. 1959b. The Changing Pattern of Dentine Exposure in Human Tooth Attrition. – American Journal of Physical Anthropology 17: 167-178.

Pal, A. 1971. Gradients of Dentine Exposure in Human Molars. – Journal of the Indian Anthropological Society 6: 67-73.

Phenice, T.W. 1969. A Newly Developed Visual Method of Sexing the Os Pubis. – American Journal of Physical Anthropology 30: 297-301.

Phillips-Conroy, J.E., T. Bergman & C.J. Jolly. 2000. Quantitative Assessment of Occlusal Wear and Age Estimation in Ethiopian and Tanzanian Baboons. In: Whitehead, P.F. & C.J. Jolly. Eds. Old World Monkeys. – Cambridge, Cambridge University Press: 321-340.

Puech, P.-F., C. Serratrice & F.F. Leek. 1983. Tooth Wear as Observed in Ancient Egyptian Skulls. – Journal of Human Evolution 12: 617-629.

Richards, L. 1990. Tooth Wear and Temporomandibular Joint Change in Australian Aboriginal Populations. – American Journal of Physical Anthropology 82: 377-384.

Richards, L. & S. Miller. 1991. Relationships between Age and Dental Attrition in Australian Aboriginals. – American Journal of Physical Anthropology 84: 159-164.

Richards, L.C. 1984. Principal Axis Analysis of Dental Attrition Data from Two Australian Aboriginal Populations. – American Journal of Physical Anthropology 65: 5-13.

Richards, L.C. & T. Brown. 1981. Dental Attrition and Age Relationships in Australian Aboriginals. – Archaeology in Oceania 16: 94-98.

Ruffer, S.A. 1920. Study of Abnormalities and Pathology of Ancient Egyptian Teeth. – American Journal of Physical Anthropology 3: 335-382.

Scott, E.C. 1979. Dental Wear Scoring Technique. – American Journal of Physical Anthropology 51: 213-217.

Smith, B.H. 1984. Patterns of Molar Wear in Hunter-Gatherers and Agriculturalists. – American Journal of Physical Anthropology 63: 39-56.

Smith, T.M., A.J. Olejniczak, D.J. Reid, R. Ferrell & J.-J. Hublin. 2006. Modern Human Molar Enamel Thickness and Enamel-Dentine Junction Shape. – Archives of Oral Biology 51: 974-995.

Sterne, J.A. & G.D. Smith. 2001. Sifting the Evidence – What's Wrong with Significance Tests? – Physical Therapy 81: 1464-1469.

Stewart, J.D., J.E. Molto & P.J. Reimer. 2003. The Chronology of Kellis 2: The Interpretative Significance of Radiocarbon Dating of Human Remains. In: Bowen, G.E. & C.A. Hope. Eds. The Oasis Papers 3: Proceedings of the Third International Conference of the Dakhleh Oasis Project. – Oxford, Oxbow: 373-378.

Suchey, J.M. & D. Katz. 1986. Skeletal Age Standards Derived from an Extensive Multiracial Sample of Modern Americans. – American Journal of Physical Anthropology 69: 269-269.

Sutton, L. 1947. Rainfall in Egypt: Statistics, Storms, Run-Off. – Cairo, Government Press.

Teaford, M.F. 1983. Differences in Molar Wear Gradient between Adult Macaques and Langurs. – International Journal of Primatology 4: 427-444.

Teaford, M.F. & O.J. Oyen. 1989. In Vivo and In Vitro Turnover in Dental Microwear. – American Journal of Physical Anthropology 80: 447-460.

Tocheri, M.W., T.L. Dupras, P. Sheldrick & J.E. Molto. 2005. Roman Period Fetal Skeletons from the East Cemetery (Kellis 2) of Kellis, Egypt. – International Journal of Osteoarchaeology 15: 326-341.

Tomenchuk, J. & J.T. Mayhall. 1979. A Correlation of Tooth Wear and Age among Modern Igloolik Eskimos. – American Journal of Physical Anthropology 51: 67-77.

Van der Bijl, P., J. De Waal, I. Botha & W. Dreyer. 1989. Assessment of Occlusal Tooth Wear in Vervet Monkeys by Reflex Microscopy. – Archives of Oral Biology 34: 723-729.

Walker, A., H.N. Hoeck & L. Perez. 1978. Microwear of Mammalian Teeth as an Indicator of Diet. – Science 201: 908-910.

Walker, P.L., G. Dean & P. Shapiro. 1991. Estimating Age from Tooth Wear. In: Kelley, M.A. & C.S. Larsen. Eds. Advances in Dental Anthropology. – New York, Wiley: 77-119.

Webb, P.A.O. & J.M. Suchey. 1985. Epiphyseal Union of the Anterior Iliac Crest and Medial Clavicle in a Modern Multiracial Sample of American Males and Females. – American Journal of Physical Anthropology 68: 457-466.

Wheeler, S.M. 2009. Bioarchaeology of Infancy and Childhood at the Kellis 2 Cemetery, Dakhleh Oasis, Egypt. – London, University of Western Ontario.

Williams, L.J. 2008. Investigating Seasonality of Death at Kellis 2 Cemetery Using Solar Alignment and Isotopic

Analysis of Mummified Tissues. – London, University of Western Ontario.
Wolpoff, M.H. 1971. Interstitial Wear. – American Journal of Physical Anthropology 34: 205-227.
Wu, Y.-Q., J.A. Arsecularatne & M. Hoffman. 2015. Effect of Acidity upon Attrition-Corrosion of Human Dental Enamel. – Journal of the Mechanical Behavior of Biomedical Materials 44: 23-34.
Yun J.I., J.Y. Lee, J.W. Chung, H.S. Kho & Y.K. Kim. 2007. Age Estimation of Korean Adults by Occlusal Tooth Wear. – Journal of Forensic Sciences 52: 678-683.
Zhang, Y., J.A. Arsecularatne & M. Hoffman. 2015. The Effects of Three Different Food Acids on the Attrition-Corrosion Wear of Human Dental Enamel. – Journal of Physics D: Applied Physics 48: 285401.
Zheng, J., H. Huang, M. Shi, L. Zheng, L. Qian & Z. Zhou. 2011. In Vitro Study on the Wear Behaviour of Human Tooth Enamel in Citric Acid Solution. – Wear 271: 2313-2321.
Zheng, J. & Z. Zhou. 2007. Study of In Vitro Wear of Human Tooth Enamel. – Tribology Letters 26: 181-189.
Zhou, Z.-R., H.-Y. Yu, J. Zheng, L.-M. Qian & Y. Yan. 2013. Dental Biotribology. – New York, Springer.
Zuhrt, R. 1955. Stomatologische Untersuchungen an spatmittelterlichen Funden von Reckkahn (12-14jh.). I. Die Zahnkaries und ihre folgen. – Deutsche Zahn-, Munds-, und Kieferheilkunds 25: 1-15.

Discovery of an Unexpected Textile Fiber in a Fish Mummy from the Musée des Confluences (Lyon) Collection

Fleur Letellier-Willemin

Introduction

The identification of a new textile fibre on an animal mummy enables us to stress the importance of textiles within mummy studies, and more generally their position as archaeological objects. The present study aims to grasp all the issues related to this textile's presence, in the specific religious context of animal mummies.

The fish mummy considered here is one amongst the hundred or so fish mummies in the collection of the Musée des Confluences, formerly the Guimet collection. These have been studied by Alain Charron (2019). All the fish mummies originate from Esna and belong to the same species, *i.e.* the Nile perch or *Lates niloticus* (Figure 1). This freshwater fish was found in great numbers in Upper Egypt.

Gaston Maspero sent the mummies to the museum. He excavated specifically in Esna to respond to a request made by Lortet and Gaillard in 1902, who required well-preserved fish mummies for their collection. Another necropolis of fish mummies was found in Gurob, at the south-eastern end of the Fayum, next to the sanctuary devoted to the goddess Neith (Gozo, 1999: 361). W.M. Flinders Petrie excavated the site between 1888 and 1890, and this necropolis is considered to date to the 19th Dynasty. Unfortunately, it is not very useful for our scholarly research, because these mummies have not been precisely described or studied.

The Mummies

The fish mummies from Esna were found in the necropolis to the west of the city, where they were buried in pits over a large flat area. The necropolis exclusively shelters this variety of animal mummy. Other fish mummies were uncovered in the human necropolis located closer to the foothills of the Libyan mountain range, which was partly excavated in 1907 and 1908 by Henri de Morgan (1912), who met Gaston Maspero in Esna and Lortet in Thebes. He describes more precisely two tombs of this mainly Ptolemaic and Roman necropolis. Tomb 35 displayed *"a very important group of sepultures ... There was an accumulation of human, animal and fish bones"*, and he notably found several scarabs inside, one of them bearing the name of Thutmose III and another that of queen Hatshepsut. Tomb 38 also contained fish mummies. According to Henri de Morgan, the necropolis was in use since the 12th Dynasty, but flourished mainly during the Ptolemaic and Roman era.

As a place of worship, Esna was particularly significant in the Ptolemaic era, as the Ptolemies refurbished the previous temples at great expense (Moreno-Garcia & Agut, 2016: 702, 704, 714-721). Ptolemy VI Philometor started the building of the edifice, and

In: Ikram, S., J. Kaiser & S. Porcier. Eds. 2021. *The Ancient Egyptians and the Natural World. Flora, Fauna, and Science.* – Leiden, Sidestone Press: 183-190.

Figure 1. A *Lates Niloticus* N. Inv. 90001350, without textile, from the collection of the musée des Confluences. Courtesy of the CCEC, Lyon. Photograph by F. Letellier-Willemin.

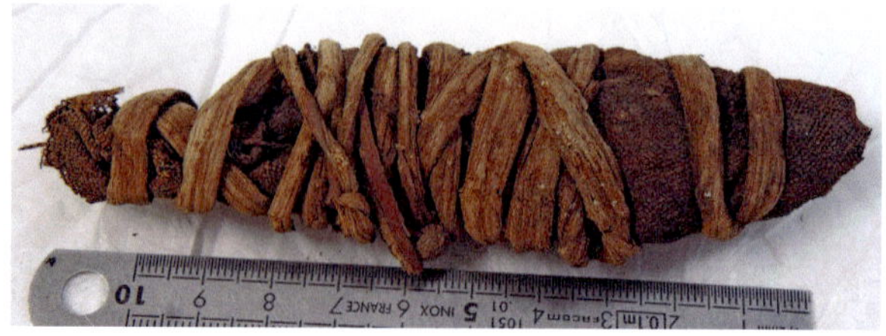

Figure 2. The fish mummy N. Inv. 90002234 with cotton textile. Photograph by F. Letellier-Willemin.

Ptolemy VIII Euergetes II made considerable investments in it (Moine-Bianchi, 2018: 176). Esna was known as Latopolis and became the capital of the 3rd Upper Egyptian Nome, and in the Roman Period, the *Lates* was depicted on coins of the Latopolite Nome. Work on the temple was carried out until the Roman period, without it ever really being finished. Esna comprised several temples, but only one survives: the great temple devoted to Khnum, of which only the hypostyle hall remains. This escaped destruction because it was used as a cotton storage facility (Champollion, 2001). Its walls are inscribed with mythological and liturgical texts, where the theme of creation in relation to Khnum and Neith predominates (Dunand & Zivie-Coche, 1991: 225). The chief god was Khnum, mainly associated with Neith, a very ancient goddess of Sais, in the western Delta; their child-god Heka (magic) was also worshipped in the mammisi. These cults featured many celebrations, listed in a festival calendar in the temple, which also included an extensive hymn to Neith and her arrival in Sais (Sauneron, 1959: 25-28).

As mentioned above, the fish mummies of Esna are only of the Nile perch (Lortet, 1903: 185-186). Found in great numbers in the river, it is a dark-coloured fish that favours deep living water. Rather aggressive, its Egyptian name, *ꜥḥꜣw*, meant "fighter"; the Greeks assimilated Neith with Athena, both of whom are goddesses of warfare (Vernus & Yoyotte, 2005: 264-265). The perch was often associated with the light-coloured tilapia, which dwells in more quiet waters and is a mouth-brooder, associated with resurrection (Brewer & Friedman, 1990). Through mummification, the bodies of these fish were rendered divine/sacred (Charron, 2014: 277). The fish were mummified in a curious way, according to Lortet: after macerating in brine, they were covered with a coat of salty mud made of clay and sand (Lortet, 1903: 186-187). They were then consecrated to Neith and then interred.

The fish mummy no. 90002234 considered here (Figure 2) is 16 cm long and 5 cm wide, and comprises a single, well-preserved fish. It is wrapped in a single piece of textile forming a curious thick fold. The textile is held in place by transversely and longitudinally wrapped pieces of papyrus pith. The textile is unfolded at the head, which helps to identify the fish as a *Lates*. The scales are preserved, as well as part of the mud coat. The wrappings comprise a single piece of thick, dark-brown cloth, which has clearly been torn from a larger piece/garment. A (now) dark brown substance was applied over the surface of the textile, and the threads are not deeply soaked. The textile folds form an unusual and very conspicuous hump, which was probably held with some difficulty by the thongs. Only a few of these are still in place, but their impressions can still be seen around the tail. The fabric shows a simple plain weave pattern, *i.e.* one warp thread for one weft thread (Figure 3). The yarns display a 'S'-twist and are irregular, their angle of twist ranging from 60 to 90 degrees, and their diameter from 0.5 to 0.8 mm. The fibres are thus poorly managed during spinning, and the weaving reveals these irregularities. There are also weaving errors, where two warp threads are woven together, as the craftsman had trouble in weaving the threads one by one with his shuttle. These threads which caused some problems during spinning and weaving are made of cotton, as confirmed by an analysis carried out by Christophe Moulherat, who also observed that the fibres used are mainly mature, therefore making for a high-quality thread (Figure 4).[1] This identification of cotton in a fish mummy from the Nile Valley was worthy of ^{14}C dating, which was carried out by Pascale Richardin and Fiona Lobo. The resulting calibrated dates range from 150 BC to AD 55.

Other fish mummies in the collection have already been dated (Richardin *et al.*, 2017): No. 90002254, calibrated dates: 114 BC-AD 52 and No. 90001178, calibrated dates: 206 BC-AD 50.

It is interesting to note that the dating of 63 animal mummies from the Lortet and Gaillard collection,

1 C. Moulherat (personal communication 13 November 2017, cc. Didier Berthet, F. Letellier-Willemin, T. Bajon-Bouzid).

Figure 3. The cotton textile of the mummy, a plain weave pattern, with irregular yarns. Photograph by F. Letellier-Willemin.

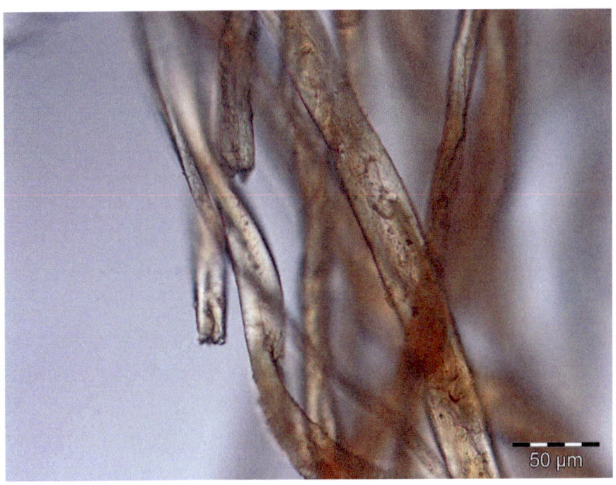

Figure 4. Cotton yarns enlarged 1600 x 1200, Image Tv2. jpg. Photograph by C. Moulherat.

carried out by Pascale Richardin, showed a peak from the 3rd century to the 1st century BC (Richardin *et al.*, 2017: 595-607) (Figure 5).

In most cases, the fish mummies are wrapped in a single layer of textile. Less frequently, there are two or three layers. It is important to stress that all the other textiles associated with the animal mummies in the Lyon collection, regardless of species, are made of linen. Generally, these bandages do not consist of narrow strips of cloth, but rather, fairly wide single pieces. All the textiles on the other fish mummies are made of linen, thus mummy no. 90002234 is unique in being wrapped in cotton fabric. As a whole, the linen yarns display a good quality, with an S-twist and a regular diameter (Figure 6).

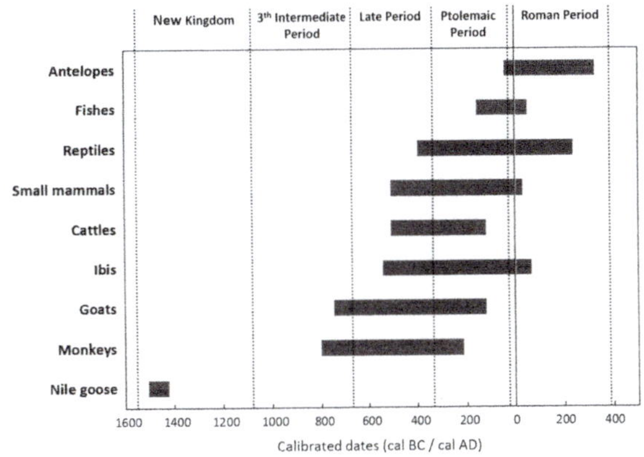

Figure 5. Datations of 63 animal mummies (mostly with their textiles) from the Lortet and Gaillard collection. From Richardin *et al.* (2017 : fig. 3).

Figure 6. A linen textile wrapping a fish mummy of the collection, a medium good quality. N. Inv. 90002222. Photograph by F. Letellier-Willemin.

The features distinguishing warp and weft respect the very ancient Egyptian tradition, with more numerous warps (dominating warp), finer and of higher twist than the wefts. The plain weave fabrics show a generally consistent quality. There are a few basket-weave patterns (two weft threads and two warp threads), a Gros de Tours (one warp thread for two weft threads) (Figure 7), as well as square settings (where the number of warp and weft threads per cm is the same). These pieces of textile display the quality of objects of daily use: these mummy wrappings were all reused, as was generally the case for human mummy wrappings (Ikram & Dodson, 1998), and none of them was new. Their daily use is also obvious due to the presence of a few seams (Figure 8), hems, selvedges (simple), traditional weft decoration (several weft threads are inserted simultaneously, giving the piece a ribbed aspect, also traditional and ancient), and fringes.

Figure 7. A linen textile wrapping a fish mummy with a basket weaving, and a well-known traditional weft decoration (several weft threads inserted simultaneously). N. Inv. 90001283. Photograph by F. Letellier-Willemin.

Figure 8. A linen textile wrapping a fish mummy with stitches, N. Inv. 90002221. Photograph by F. Letellier-Willemin.

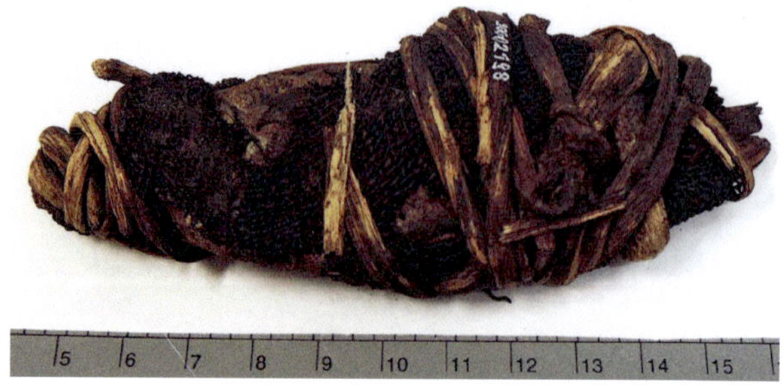

Figure 9. A fish mummy, with dark products on the textile, which is not usual. N. Inv. 90002198. Photograph by F. Letellier-Willemin.

Figure 10. A bunch of juvenile fishes wrapped in a linen textile. N. Inv. 90001176. Photograph by F. Letellier-Willemin.

The textile wrappings are held in place by papyrus thongs of various thicknesses. It is surprising to see that the small fishes are frequently tied with many of these thick thongs, almost hiding the textile underneath. Is this due to the state of preservation of the fish, a deliberate aesthetic effect, or perhaps the signature of a specific craftsman, in a specific period? They were most often tied with a weaver's knot. Some of these also have linen threads used as ties in addition to the linen, a practice already observed on many other animal mummies. As mentioned above, a substance or substances are smeared over the surface of the textiles. These do not penetrate in the thickness of the threads, but rather, dye them brown and evoke the resinous-oily substances used in human and animal mummification. Work should be carried out on identifying this material in the future. Some of the textiles were apparently wet (maybe impregnated by a natron solution?) when they were wrapped around the fish perhaps in order to ensure a better wrapping of the fish. Less frequently, some of the textiles used on the other fish mummies display a clearly different aspect, with a black tint[2] and as if stiffened by product substance, which shows a different composition (Figure 9) to the one used on the fish mummy in question. One should also note the presence, in one case only,[3] of a jerit (palm rib) used as a stake to keep the general shape of the fish, an object traditionally and frequently used in the preparation of animal and human mummies. Finally, the mud in which the fishes were dipped before the wrapping is sometimes clearly visible. Although most of these fish mummies, which were X-rayed by Roger Lichtenberg, contain one fish only, there are some examples of containing single mummy bundle consisting of an adult and a juvenile fish, two fishes or even a bunch of juvenile fishes wrapped in a linen textile held by thongs (Figure 10).[4] One must also stress that the collection is not representative of the necropolises as a whole, since

2 N.Inv. 90002237, N.Inv. 90007289, N.Inv. 90006084, N.Inv. 90002187.

3 N.Inv. 90003343.

4 N.Inv. 90007446, N.Inv. 90002174, N.Inv. 90002184 (bunch of juvenile fishes), N.Inv. 90002226 (two fishes), N.Inv. 90001178 (a fish and a juvenile).

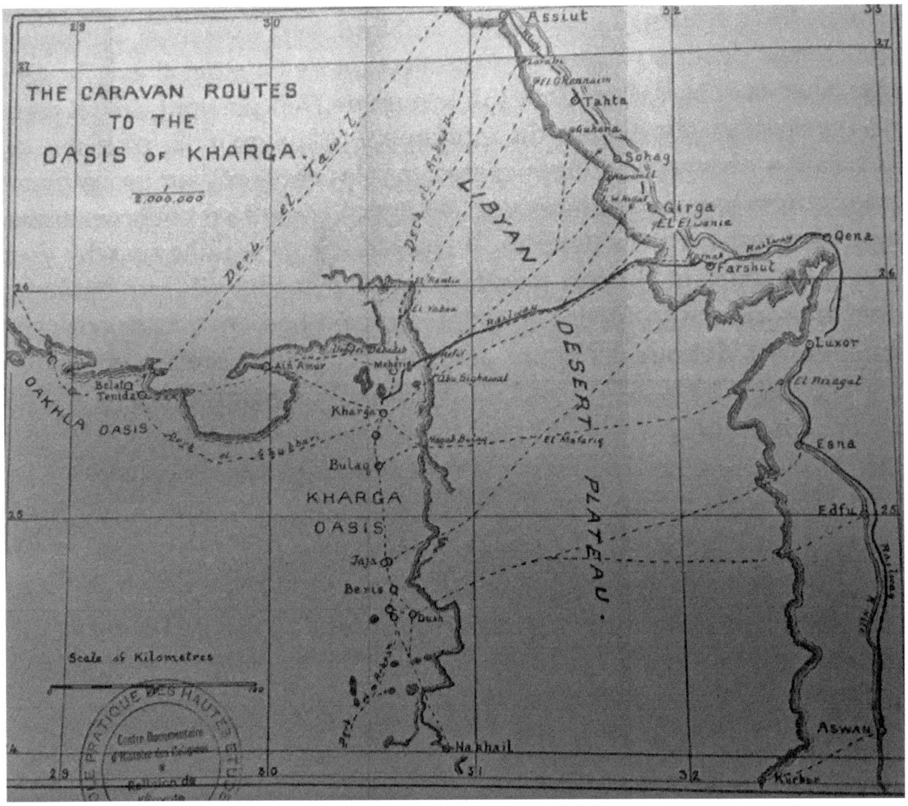

Figure 11. The roads between the Great Oasis, Dakhla and Kharga, and the valley. Courtesy of the Centre de documentation d'histoire des religions, Ecole Pratique des Hautes Etudes. Map by Beadnell.

Maspero selected the mummies because of their good state of preservation.

All of the mummies in the collection display a high-quality mummification method, as shown by the treatment of the body, the textile wrappings and the careful winding of the papyrus thongs. The two sides are not identical: one of them was supposed to be seen, and the other one not, and thus it received all the textile folds and the knotted lashes. This difference between the two sides of a given mummy is a common feature seen on other animal mummies in other locations. The mummy was showcased on one side so as to be chosen and bought by a worshipper, in the context of temple activities. As an aside, we also find the aspect of the fish mummies from Esna in Deir al-Medina, on a wall painting in the tomb of Khabekhnet (TT2, 19th Dynasty, being prepared for publication by Anne-Claire Salmas), where Anubis is embalming the deceased depicted as a large fish. In this case, we are not dealing with a *Lates* but with the so-called *Lepidote* or *Barbus bynni*, a fish chosen "*to symbolize the Osirian dogma of rebirth to eternal life*" (Bruyère, 1929: 46-47).

The Textile

As noted above, all the other textiles pertaining to the animal mummies in the collection, whatever the species, are made of linen. A study of funerary textiles used for humans and animals from El Deir (Figure 11) a contemporaneous date to the fish mummy that is the subject of the article, provides general information about the textiles that are used in mummification during this period. Notable about the El Deir material is that a few of the wrappings/padding of the human mummies are made of cotton (Gradel *et al.*, 2012: 119-141) (Figure 12), although none of the dogs were prepared using this cloth (Letellier-Willemin, 2017: 187-205). This use of cotton fabric in mummies seems to be a feature of Kharga Oasis (Jones, 2018), although it is possible that extensive textile studies in other necropoleis will change this picture. About ten ^{14}C dates, unpublished thus far, have been carried out so far on the funeral material from Kharga, and the earliest date ranges from the late 1st century BC to the early 1st century AD.

The use of cotton instead of linen is unusual, but possibly within the parameters of Egyptian funerary practice. The cotton, like linen, is naturally white, which is a religious criterion for textiles used in cult practices (Sauneron, 1988: 31, 90, 100). However, it is possible that linen being used in funerary contexts was a significant marker of religious/cultural identity until the advent of Christianity. Thus, if a priest used cotton, an unusual textile, this innovation must have been deliberate, although the reasons for this are unknown.

The main question is: why was cotton used to wrap this fish mummy (and the few other human mummies from

Figure 12. A cotton textile from El Deir, Oasis de Kharga, Data basis : ED.F022-ind.3817. Photograph by F. Letellier-Willemin.

Kharga)? White/natural cotton is similar to linen, and as the textiles used in mummification were reused, it is possible that amongst the textile offerings presented to the temple, cotton, a novelty, figured (Dunand, 1991: 207-213; Préaux, 1939: 481, 484). The new fabric might be valued for its novelty, though it is also possible that it arrived in the temple as it was woven here, or indeed, gathered by the embalmers from cast-offs of the well-to-do.

Cotton was known in Egypt at this time, as has been established by the examples from Kharga (Letellier-Willemin, 2019). Furthermore, a few Egyptian sources mention cotton, which are rare but significant. According to Herodotus, the pharaoh Amasis (Ahmose II) offered linen textiles decked with cotton embroidery to some of the Greek temples in Asia Minor (Barguet, 1964: III: 47). Earlier evidence for cotton is still disputed. A small cotton ball was found between the bandages of the anonymous male mummy known as PUM II, a mummy dating by the linen to 170 BC ± 70 year (Cockburn & Cockburn, 1998: 87-88). This find was considered so unusual at the time of its discovery in 1980, that it was deemed as a modern contamination introduced during its transportation (Cockburn & Cockburn, 1998: 87-88). The question remains open, as we do not have a dating for this cotton ball.[5]

That cotton was available in Egypt is known from Kharga and Qasr Ibrim (Clapham & Rowly-Conwy, 2007). It is possible that this cotton originated from Nubia – according to Elsa Yvanez, in Meroitic society cotton was apparently a highly valued plant (personal communication), which had close trade affiliations in both Kharga and Qasr Ibrim (Clapham & Rowly-Conwy, 2007; Rossi & Ikram, 2018). Additionally, it is notable that Esna had access to the port of Berenike and goods coming from there. Berenike was a key importer of Indian goods, possibly even cotton. Indeed, cotton textiles have been found at Berenike and are roughly contemporaneous with the fish mummy (Sidebotham, 2011: 39-53; Wild & Wild, 2002: 9-16). An Indian origin for the cotton used in Esna remains a hypothesis, but the Berenike yarns show a Z-twist, unlike Egyptian cotton that usually shows an S-twist, as is the case for the Esna cotton, suggesting that it is of local production.

Conclusion

To conclude, it seems as if the Esna fish mummy is, thus far, the only animal mummy to be wrapped in cotton. During the Ptolemaic/early Roman era only a handful of mummies were wrapped in cotton, and these originate from Kharga Oasis. This result is possibly due to the paucity of extensive textile studies carried out on mummified remains; more extensive investigation might alter the picture. Why the Esna fish mummy is wrapped in cotton is unclear. It might be that cotton was either being offered to the temple, or even woven there, or the cloth was expressly brought by someone who was conscious of its uniqueness, and thought it a fitting gift to the gods, or someone who had access to raw cotton or even cotton cloth (though this would not be 'S'-spun). Its novelty and rarity might have rendered it a special offering, and thus set this fish mummy in a more special and/or elite category than its peers.

Acknowledgements

I would like to warmly thank Stéphanie Porcier, Didier Berthet, Alain Charron, Pascale Richardin and Salima Ikram for their listening and collegial spirit, and Pierre Meyrat for his translation. I am also grateful to Elsa Yvanez for her remarkable overview of Nubian textiles. Finally, my respectful admiration goes to John Peter Wild and Felicity Wild for their seminal researches and contributions regarding cotton.

Bibliography

Barguet, A. 1964. Herodote. Histories III, 47. L'enquête Livre I. – Paris, Gallimard.

Bleiberg, E. 2013. Animal Mummies: The Souls of the Gods. In: Bleiberg, E., Y. Barbash & L. Bruno. Eds. Soulful Creatures. – Brooklyn, Brooklyn Museum: 91-105.

Brewer D.J. & R.F. Friedman. 1990. Fish and Fishing in Ancient Egypt. – Cairo, AUC Press.

Bruyère, B. 1929. L'enseigne de Khabekhnet. – Bulletin de l'Institut Français d'Archéologie Oriental 28: 46-47.

Charron, A. 2019. L'étude des momies animales du musée des Confluences à Lyon. L'exemple des momies de poissons. In: Porcier, S., S. Ikram & J. Kaiser. Eds. Creatures of Earth, Water, and Sky. Essays on Animals

5 Meryl Johnson is convinced that the piece of cotton was originally there, found between bandages and glued by the fluids, and that it could be an amulet. I warmly thank her for sharing her view on the question.

in Ancient Egypt and Nubia. – Leiden, Sidestone Press. 137-144.

Charron, A. 2014. Des momies par millions. In : Guichard, H. Ed. Des animaux et des pharaons (Le règne animal dans l'Egypte ancienne). – Paris, Somogy, Louvre Lens: 277.

Clapham, A.J. & P.A. Rowley-Conwy. 2009. The Archaeobotany of Cotton (*Gossypium* sp. L.) in Egypt and Nubia with Special Reference to Qasr Ibrim, Egyptian Nubia. In: Fairbairn, E. & E. Weiss. Eds. From Foragers to Farmers: Papers in Honour of Gordon C. Hillman. – Oxford, Oxbow Books: 244-253.

Cockburn, A. & E. Cockburn, E. 1998 [1980]. Mummies, Diseases & Ancient Cultures. – Cambridge, Cambridge University Press: 87-88.

De Morgan, H. 1912. Report on Excavations made in Upper Egypt During the Winter 1907-1908. – Annales du Service des Antiquités de l'Egypte 12: 25-50.

Dunand, F. & C. Zivie-Coche. 1991. Dieux et hommes en Égypte 3000 av. J.-C., 359 ap. J.-C. – Paris, Colin.

Gozo, D. 1999. Gurob. In: Bard, K.A. Ed. Encyclopedia of the Archaeology of Ancient Egypt. – London, Routledge: 361.

Gradel, C., F. Letellier-Willemin & G. Tallet. 2012. Une laine bien plus belle et douce que celle des moutons, à El Deir (Oasis de Kharga, Egypte) : le coton au cœur de l'économie oasienne à l'époque romaine. In: Guédon, S. Ed. Entre Afrique et Egypte : relations et échanges entre les espaces au sud de la Méditerranée à l'époque romaine. – Bordeaux, Ausonius Éditions (Collection Scripta Antiqua 49): 119-141.

Ikram, S. & A. Dodson. 1998. The Mummy in Ancient Egypt: Equipping the Dead for Eternity. – London, Thames & Hudson.

Jones, J. 2018. The Textiles. In: Rossi, C. & S. Ikram. Eds. The North Kharga Oasis Survey: Explorations in Egypt's Western Desert. – Leuven/London, Peeters/ British Museum (British Museum Publications on Egypt and Sudan 5): 541-549.

Letellier-Willemin, F. 2017. Les textiles. In: Dunand, F., R. Lichtenberg, C. Callou, F. Letellier-Willemin. Eds. El-Deir Nécropoles IV, Les chiens momifiés d'El-Deir. – Paris, Éditeur Cybèle – Librairie Decitre : 187-205.

Letellier-Willemin, F. 2019. Le coton d'El-Deir. Premières observations sur l'existence d'une nouvelle fibre textile dans l'oasis de Kharga (désert occidental égyptien, Ve siècle AEC- Ve siècle EC). Revue d'ethnoécologie n° 15/2019. In: Bouchaud, Ch. & E. Yvanez. Eds. Cotton in the World: Domestication, Cultivation, Uses and Exchange. – Paris, Laboratoire Eco-anthropologie et Ethnobiologie (https:/journals.openedition.org/ethnoecologie/4063 : DOI :https://doi.org/10.4000/ethnoecologie.4283).

Lortet, L.C E. & C. Gaillard. 1903. Poissons. In: Georg, H. Ed.. La Faune momifiée de l'ancienne Égypte. – Lyon, Archives du Museum d'Histoire naturelle de Lyon (Tome huitième): 185-186.

Moine-Bianchi, D. 2018. Analyses stylistiques de quelques scènes du temple d'Esna. – Egypte Nilotique et Méditerranéenne 11: 176.

Moreno-Garcia, J-C. & D. Agut. 2016. L'Égypte des Pharaons, De Narmer à Dioclétien 3150 av. J.- C. – 284 ap. J.- C. – Paris, Belin.

Préaux, C. 1939. L'économie royale des Lagides. – Bruxelles, Fondation Égyptologique Reine Élisabeth.

Riggs. C. 2014. Unwrapping Ancient Egypt. – London, Bloomsbury.

Rossi, C. & S. Ikram. Eds. 2018. The North Kharga Oasis Survey: Explorations in Egypt's Western Desert. – Leuven/London, Peeters/British Museum (British Museum Publications on Egypt and Sudan 5).

Richardin, P., G. Louam, D. Berthet, S. Ikram & S. Porcier. 2017. Cats, Crocodiles, Cattle, and More: Initial Steps toward Establishing a Chronolgy of Ancient Egyptian Animal Mummies. – Radiocarbon 59 (2): 595-607.

Sauneron, S. 1954, La justice à la porte des temples. – Bulletin Français d'Archéologie Orientale 54: 117-127.

Sauneron, S. 1959. Esna I. – Le Caire, Institut Français d'Archéologie Orientale.

Sauneron, S. 1988 [1957]. Les prêtres de l'ancienne Egypte. – Paris, Perséa Éditeur.

Sidebotham, S.E. 2011. Berenike and the Ancient Maritime Spice Route. – Berkeley/Los Angeles, University of California Press.

Thiers, C. 1997. La stèle de Pithom et les douanes de Philadelphe. – Göttinger Miszellen 157: 95-101.

Vernus, P. & J. Yoyotte. 2005. Bestiaire des pharaons. – Paris, Agnès Viénot Éditeur: Éditeur Perrin.

Wild, J.P. 1997. Cotton in Roman Egypt: Some Problems of Origin. – Al Rafidan XVIII, Special volume in commemoration of the 70th Birthday of Professor Hidea Fuji, Tokyo: 287-298.

Wild, J.P. & F.C. Wild. 2002. The Webbing from Berenike: A Classification. – Archaeological Textiles Newsletter 31: 9-16.

Women's Health Issues Reflected in Case Studies from Theban Tomb 16

Suzanne Onstine, Jesus Herrerín, Miguel Sanchez & Rosa Dinarès

Theban Tomb 16 (TT16) is located in the New Kingdom necropolis associated with the ancient Egyptian capital of Thebes, now modern Luxor. It was made for Panehsy and Tarenu, a priestly couple of the 19th Dynasty (ca. 1250 BC). It consists of two decorated rooms and an undecorated winding corridor leading to their rough-cut burial chamber. It is in this corridor that the secondary burials were found looted and in disarray during archaeological clearance. Thousands of bones and partially mummified body parts were analyzed by the physical anthropology team between 2012 and 2019.[1] It is estimated that between 100 and 200 individuals are present.[2] The volume of funerary objects and broken cartonnage dating to the Third Intermediate Period suggests an intense period of reuse followed by sporadic reuse up to the Graeco-Roman Period. Unfortunately, the bodies were unwrapped when the tomb was looted, so it is impossible to assign a specific date to any single piece of human remains. This gives a date range to all the human remains of about 1000 BC to the 2nd century AD, with the bulk of the material dating to the Third Intermediate Period (1000-750 BC) based on the high number of funerary objects from that era. During the Third Intermediate Period at Thebes the practice of reusing existing New Kingdom tombs for numerous secondary burials was very common.[3] The secondary burials and frequent access to the tomb did not, however, lead to a usurpation or redecoration of the painted 19th Dynasty reliefs.

Although the jumbled state of the burials is a terrible reminder of the fragility of the archaeological record, we can still ask some pertinent questions of the material. How do burial practices change through time? What were the general and specific health conditions of Theban people in the first millennium BC? Does their health correlate with social class (as interpreted through artifacts generally)? Can we see gendered activities and conditions?

Specifically, we address this last point and present some observations from TT16 that have implications for the study of gender in ancient Egypt. In addition to the specific issues related to reproduction and sexual genitalia that are the most common facets of such studies, we should be considering all diagnoses through the lens of gender even when those diagnoses are not related to genitalia or sexual reproduction. This includes a recognition that male and female bodies can present similar pathologies differently, and

1 The physical anthropology team includes Jesus Herrerín, Miguel Sanchez, Rosa Dinarès, Nataša Šarkić, Casey Kirkpatrick, and Amr Shahat.

2 The team is in the process of making a more specific estimation.

3 See for example TT65 Bács *et al.* (2009).

In: Ikram, S., J. Kaiser & S. Porcier. Eds. 2021. *The Ancient Egyptians and the Natural World. Flora, Fauna, and Science.* – Leiden, Sidestone Press: 191-198.

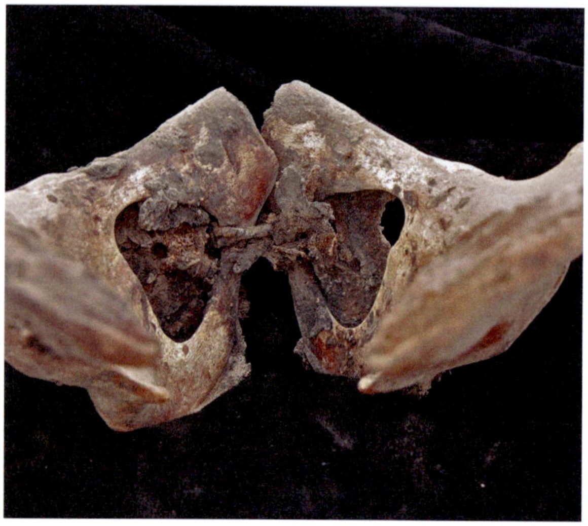

Figure 1. Pelvis tied with rope. Photograph by V. Reckard.

that male and female bodies may follow different patterns of experience for the same condition. These differences can be biologically and culturally based. The following cases represent a sample of the human remains found that speak to these issues.

Both cases of reproductive health presented here relate to possible death of the mother during childbirth or shortly thereafter. The first case consists of a pelvis (Figure 1). Although the rest of the body is not present, it can be determined that the pelvis belonged to a female aged 20-25. Sex was estimated according to the pelvis morphology (great sciatic notch, pubis topography, open subpubic angle) by the methods of Bruzek (1991) and Ferembach *et al.* (1979). These clearly represent a female pelvis. The age at death was estimated using the morphology of the auricular surface (Buckberry & Chamberlain, 2002; Lovejoy *et al.*, 1985) and the morphology of the pubic symphysis (Brooks & Suchey, 1990). No significant arthritis or trauma is present, but the two halves of the pelvis are tied together with rope that goes through the obturator foramen and is tied from a posterior position, meaning it was reached from inside the body cavity. This was likely executed during the mummification process although normal mummification procedures did not generally interfere with this area of the body (Ikram & Dodson, 1998; Smith & Dawson, 1991). A larger abdominal incision would have been necessary to reach this position. The question then is why? A possible hypothesis requiring intervention in this area is the removal of a fetus after the death of the mother. Although we do not know the actual cause of the woman's death, the young age is suggestive of the all-too-common occurrence of maternal death related to bearing children. Our theory is that the embalmers separated the two coxae, breaking the ligaments that join them at the height of the pubic symphysis, in order to remove the fetus. Tying it back together with a rope restored its anatomical position so that in the afterlife the body's functionality would also be restored. It is unfortunate that this find was not discovered *in situ*. Nevertheless, it brings up the question of whether or not this is a practice that could be found in other young women's mummies. This practice follows a general ideology we have called "prostheses for the afterlife" in which the body is restored to include fixtures that may address infirmity or loss of limbs in the mummification process (Herrerín *et al.*, 2014). The purpose of these procedures would be to provide optimal conditions for the afterlife, intervening during the process of mummification by repairing or restoring parts of the body affected by pathologies or adding artifacts, like palm sticks, inside the body of the individual (Herrerín *et al.*, 2014). Since motherhood was an important aspect of women's identities this repair might have been deemed an important part of her mummification.

The second case is a larger fragment consisting of an abdominal/pelvic cavity and part of the spinal column. The most interesting aspect of this case is the preserved female genitalia, including the labia and clitoris which are visible and well preserved (Figure 2). Significantly, the vagina appears to be in an unusual distended state. The diameter of the vaginal opening measures 32 x 43 mm. This wide opening suggests a time of death within 24 hours of a birth event. The vagina should shrink back to a smaller size within this time frame. Based on the spinal measurements, and due to a lack of arthritis in the spinal column it is estimated that this woman was between 20 and 25 years old. The pelvic cavity was packed with a sandy concrete-like hard substance. The reason for the introduction of this packing is unclear and has been found in mummies of both sexes in TT16. In addition to an aesthetic reason (to give a natural volume to the anatomical area once the thoracic and abdominal organs are removed during mummification) in this case there may be a reason related to the delivery of a child, and the packing replaced the fetus during the mummification process.

The next topic under consideration is arthritis, which was widespread in the tomb's population. Many individuals had typical manifestations of arthritis consistent with work, injuries, and age – in the spine, knees, elbows, wrists, and ankles. There are, however, some examples of arthritis that can possibly be tied to cultural practices. The causes of arthritis are not well known, although certain factors (the term is less precise than cause) are known. Metabolic (De Castro, 1992), endocrinological, hereditary (Stecher, 1959), climatic (Robles, 1997) and local factors, such as dislocations, infections or traumatic lesions (Jurmain, 1999), are all recognized as exacerbating factors for arthritis, but it is age (Ortner, 2003) and mechanical overload that are the most widely accepted factors

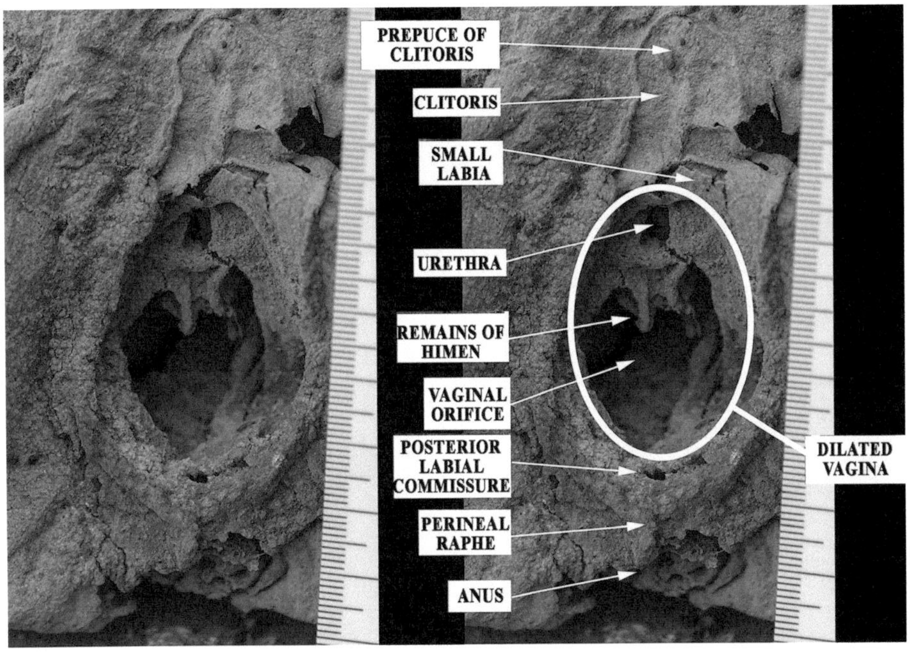

Figure 2. Genitalia of a young woman. Photograph by V. Reckard.

(Bennike, 1985; De Castro, 1992; Herrerín, 2011; Klaus *et al.*, 2009; Law, 1950; Lince, 1978; Waldron, 2007). Despite all of the above, the relationship between the effort required by certain physical activities and osteoarthritis is sometimes difficult to establish (Anderson & Duthie, 1963; Lovell, 1994; Merbs, 1983).

These four examples show pathologies that are consistent with carrying weight on one's head (Figure 3). The lesions observed in the atlanto-occipital joint show an osteoarthritic pattern that consists of the formation of osteophytic lips around the contour of the articular surface or restricted to the posterior half of its perimeter. The presence of pores and holes shows the destruction of the superficial and subarticular bone tissue. The atlanto-occipital joint is certainly stable and is not usually among the most affected one by osteoarthritis, but can suffer chronic mechanical irritation (Boleaga-Durán *et al.*, 2006). These arthritic signs have been related to postures that involve pressure in this area, such as carrying weights on the head or movements of the head backwards when lifting the chin (Baxarias & Herrerín, 2008; De Miguel *et al.*, 2019; Rihuete, 2003; Robles, 1997; Yoldi *et al.*, 2009). The high percentage of female skulls with these lesions found in TT16 opens the possibility that the lesions reflect physical activity that predominantly involves women. Carrying items on the head is a well-known cultural practice in modern Egypt, especially among women, and from ancient Egypt we have numerous artistic representations of women engaging in the practice, such as the female offering bearer statuettes from Middle Kingdom tombs.[4] Although this activity cannot be interpreted as having strict gendered categories, there seems to be a predominance in art for women to be engaged in carrying things this way.

Metabolic conditions that affect women disproportionally are also present in TT16. Bone loss or thinning in the parietal region is indicative of a loss of estrogen and can be seen in one case. Thinning of the parietal bone is a disease described already in the eighteenth century. Sandifort (1783) is credited with the first detailed description and drawing, while the first radiologic findings were described in 1926 by Greig (Herrerín *et al.*, 2017).

Although this condition has been known for a long time there is still no consensus in terms of its etiology. Some have considered it an anatomical variant (Phillips, 2007; Pratap, 1969) or a non-progressive congenital dysplasia of the díploe (Shepherd, 1893; Tsutsumi *et al.*, 2008; Wilson, 1944), others consider it an acquired and progressive disease, some associate it with senility, or osteoporosis (Cederlund *et al.*, 1982), others blame growth defects or even constant pressure on this area of the skull as its cause, and finally, there are researchers that relate it to acute inflammatory atrophy associated with trauma (Satoshi *et al.*, 2008), primary and metastatic tumors, the Gorham-Stout disease (Satoshi *et al.*, 2008), diabetes

4 For example the lovely "estate figure" in the Metropolitan Museum of Art (20.3.7) [https://www.metmuseum.org/art/collection/search/544210 accessed June 8, 2020] or the women of the Bersha Procession from the Museum of Fine Arts (21.326) [https://collections.mfa.org/objects/143592 accessed June 8, 2020].

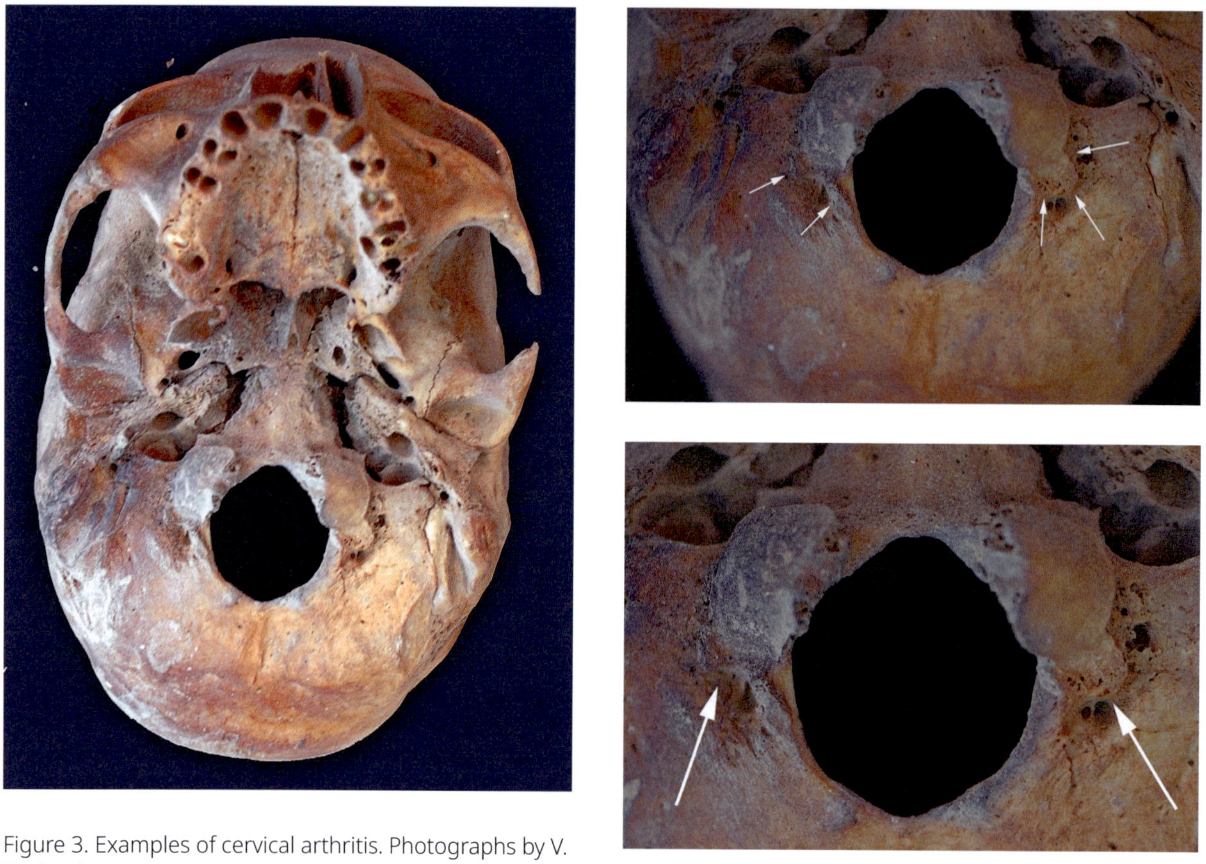

Figure 3. Examples of cervical arthritis. Photographs by V. Reckard.

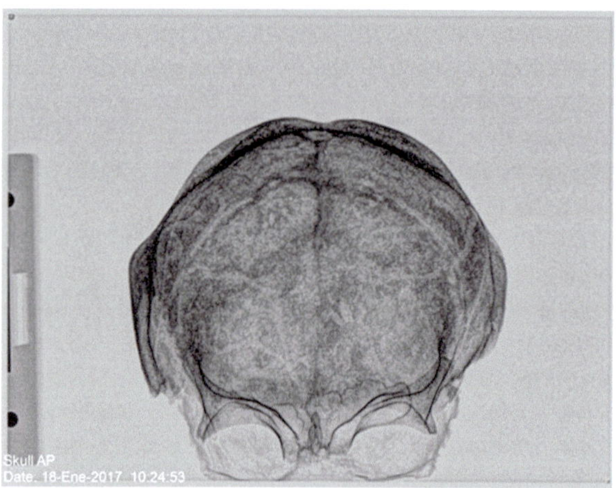

Figure 4. X-ray image of parietal thinning. Imaging by R. Dinarès, J. Herrerín, M. Sanchez.

mellitus, and, in actual clinical practice, a prolonged steroid therapy (Takata *et al.*, 2008).

Parietal thinning is a rare pathology, with a prevalence rate estimated at 0.25-0.8% according to Bruyn and Bots (1978) or 0.4-1.3% according to Breitinger (1982). It is more common in women than in men (Lim & Sohn, 2001). The male/female sex ratio is 1:1.9 (Cederlund *et al.*, 1982), or even 1:2.5 according to some researchers (Bruyn & Bots, 1978). It occurs mostly in older people, over 60 years old (Humphry, 1858). Even though many authors have suggested a genetic element in the etiology (Shepherd, 1893; Wilson, 1944), the lesion seems to be predominantly sporadic (Bruyn & Bots, 1978). It shows no special association with ancestry or geographical area (Yilmaz *et al.*, 2015).

Histopathological studies have shown a lack of osteoclasts, suggesting that Parietal Thinning can be related to osteoporosis due to a decrease in bone formation rather than increased bone destruction (Cederlund *et al.*, 1982). In actual clinical practice, hormone treatments that increase bone mineral metabolism in older patients with Parietal Thinning have given a good result (Takata *et al.*, 2008). These results suggest that acceleration of bone resorption is one of the causes of Parietal Thinning in osteoporotic patients.

The most probable hypothesis is that parietal thinning is a consequence of osteoporosis acceleration of bone resorption, which is in most of the cases connected with older age and post-menopausal hormonal changes in women (but see Hauser & DeStefano, 1989 and Phillips, 2007 for opposing views). Other unknown factors, such as a lack of calcium, physical labor that includes carrying weight on the head, or continuous pressure on that area of the parietal bone, may increase the severity of the disease and make it more visible.

Hyperostosis frontalis interna is present in two cases in the tomb. This condition consists of a thickening of the internal table of the frontal bone and in both cases thickening from extra new bone growth is clearly visible.

In current clinical practice, the condition usually affects elderly women due to endocrine imbalance (Campillo, 2001; Cuesta *et al.*, 2010; Ortner, 2003; Subirana *et al.*, 2012). The predilection for its occurrence in the frontal zone can be considered related to the blood supply (Cuesta *et al.*, 2010; Hershkovitz *et al.*, 1999). The etiology of *Hyperostosis frontalis interna* is unknown but is considered associated with menopausal hormonal changes (Hershkovitz *et al.*, 1999) and in elderly men with hypogonadism (Yamakawa *et al.*, 2006). It is also found in eunuchs (Belcastro *et al.*, 2011). It was first described in the 18th century by Morgagni after assessing internal frontal thickening in the autopsy of a woman with obesity and hirsutism (Moore, 1955), and was initially linked to Morgagni-Stewart-Morel syndrome, the symptoms of which included migraine headaches, vertigo, hirsutism, menstrual disorders, galactorrhea, obesity, depression, irritability, fatigue, transient hemiplegia, hearing disorders, nerve paralysis cranial, muscle weakness, vestibular dysfunction and epilepsy (Nallegowda *et al*, 2005; Subirana *et al.*, 2012). In recent years, however, several studies have shown that it is an independent condition associated primarily with post-menopausal women, which does not generally cause significant clinical disease on its own (She & Szakacs, 2004).

It is not clear that in current populations the frequency of HFI is higher than that of historical populations (Anton, 1997; Cuesta *et al.*, 2010; Mulhem *et al.*, 2006), although there are authors who thus think about it (Barber *et al.*, 1997; Bebel-Nowak & Golijewskaja, 2015; Rühli *et al.*, 2004). This can be, at least partially, explained by underestimation related to diagnostic issues and because the endocranium is not often inspected in complete crania. In cases of skeletons with complete crania no CT or RTG are used in standard examination and therefore HFI is mostly observed in fragmented remains (Bebel-Nowak & Golijewskaja, 2015).

In conclusion, we continue to refine our analysis of the remains of the TT16 population and will continue to present the findings in a variety of forums, both Egyptological and anthropological in order to maximize cross disciplinary discussions about women's health issues and their biological and cultural dimensions.

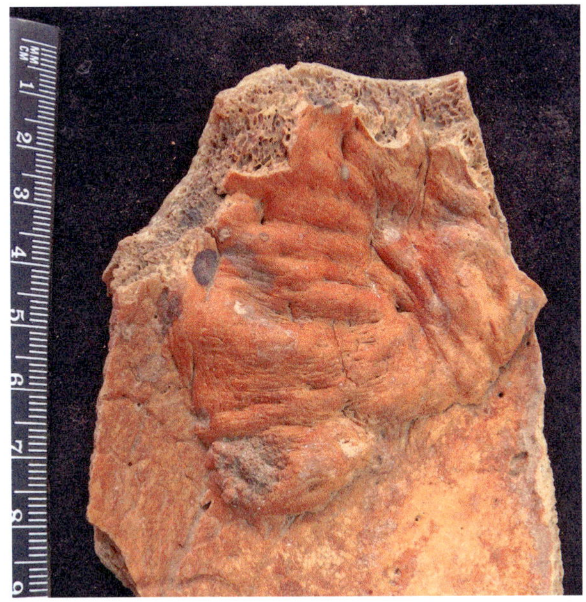

Figure 5. Example of *Hyperostosis frontalis interna* with cross section. Photographs by J. Herrerín and V. Reckard.

Bibliography

Bács, T.A., Z.I. Fábián, & G. Schreiber. 2009. The Third Intermediate Period on el-Khokha and Sheikh abdel-Gurnah. In: Bács, T. A., Z.I. Fábián, G. Schreiber & L. Török. Eds. Hungarian Excavations in the Theban Necropolis. A Celebration of 102 Years of Fieldwork in Egypt. – Budapest, Eötvö Loránd University: 95-97.

Belcastro, M.G., A. Todero, G. Fornaciari & V. Mariotti. 2011. Hyperostosis Frontalis Interna (HFI) and Castration: The Case of the Famous singer Farinelli (1705-1782). – Journal of Anatomy 219(5): 632-637.

Breitinger, E. 1982. Depressio biparietalis circumscripta. Zwei Fälle aus dem awarenzeitlichen Graberfeld v on Zwolfaxing bei Wien. – Archaologia Austriaca 66: 141-153.

Brooks, S. & J.M. Suchey 1990. Skeletal Age Determination Based on the Os Pubis: A Comparison of the Acsádi-Nemeskéri and Suchey-Brooks Methods. – Human Evolution 5: 227-238.

Bruyn, G.W. & G.T.A.M. Bots. 1978. Biparietal Osteodystrophy. – Clinical Neurology and Neurosurgery 80(3): 125-148.

Bruzek, J. 1991. Fiabilité des procédés de détermination du sexe à partir de l'os coxal, Implications à l'étude du dimorphisme sexuel de l'Homme fossile. – Thèse de Doctorat du Muséum National d'Histoire Naturelle, Institut de Paléontologie Humaine, Paris.

Buckberry, J.L. & A.T. Chamberlain. 2002. Age Estimation from the Auricular Surface of the Ilium: A Revised Method. – American Journal of Physical Anthropology 119(3): 231-239.

Cederlund, C.G., L. Andren & H. Olivecrona. 1982. Progressive Bilateral Thinning of the Parietal Bones. – Skeletal Radiology 8: 29-33.

Ferembach, D., I. Schwidetzky & M. Stloukal. 1979. Recommandations pour déterminer l'âge et le sexe sur le squelette. – Bulletins et Mémoires de la Société d'anthropologie de Paris 6(1): 7-45.

Hauser, G. & G.F. DeStefano. 1989. Epigenetic Variants of the Human Skull. – Stuttgart, Schweizerbartsche Verlagsbuchhandlung.

Herrerín, J., M. Sanchez, S. Onstine, V. Reckard, E. Warkentin & T. Redman. 2014. Prosthesis for the Afterlife in TT16, Luxor, Egypt. – Journal of the American Research Center in Egypt 50: 127-145.

Hershkovitz, I., C. Greenwald, B.M. Rothschild, B. Latimer, O. Dutour, L.M. Jellema & S. Wish-Baratz. 1999. Hyperostosis Frontalis Interna: An Anthropological Perspective. – American Journal of Physical Anthropology 109(3): 303-325.

Humphry, G. 1858. The Human Skeleton. – Cambridge, Macmillan: 242-243.

Ikram, S. & A. Dodson. 1998. The Mummy in Ancient Egypt: Equipping the Dead for Eternity. – London, Thames & Hudson.

Jurmain, R. 1999. Stories from the Skeleton: Behavioral Reconstruction in Human Osteology. – Amsterdam, Gordon and Breach.

Lim, S.T. & M.H. Sohn. 2001. Bilateral Symmetric Photon Defects in the Parietal Bones on Tc-99m MDP Bone Scintigraphy – Bilateral Parietal Thinning. – Clinical Nuclear Medicine 26: 570-571.

Lovejoy, C.O., Meindl, R. S., Pryzbeck, T. R., & Mensforth, R. P. 1985. Chronological Metamorphosis of the auricular Surface of the Ilium: A New Method for the Determination of adult skeletal Age at Death. – American Journal of Physical Anthropology 68: 15-28.

Lovell, N. C. 1994. Spinal arthritis and physical stress at Bronze Age Harappa. – American Journal of Physical Anthropology 93: 149-164.

Merbs, C. F. 1983. Patterns of activity-induced Pathology in a Canadian Inuit population. National Museum of Man Mercury Series. – Ottawa, National Museums of Canada (Archaeological Survey of Canada, Paper 119).

Moore, S. 1955. Hyperostosis Cranii. – Springfield Illinois, Charles C Thomas.

Nallegowda, M., U. Singh, M. Khanna, S.L. Yadav, A.R. Choudhary, & A. Thakar. 2005. Morgagni Stewart Morel Syndrome-Additional Features. – Neurology India 53(1):117-119.

Ortner, D. J. 2003 [1985]. Identification of Pathological Conditions in Human Skeletal Remains. – San Diego, Elsevier Academic Press.

Phillips, S. R. 2007. Cranial Anomaly, Pathology, or Normal Variant? Thin Parietal Bones in Ancient Egyptian Remains. – Unpublished PhD Thesis, University of Pennsylvania, Philadelphia.

Rühli, F.J., T. Böni & M. Henneberg. 2004. *Hyperostosis Frontalis Interna*: Archaeological Evidence of Possible Microevolution of Human Sex Steroids? – HOMO – Journal of Comparative Human Biology 55: 91-99.

She, R. & J. Szakacs. 2004. *Hyperostosis Frontalis Interna*: Case Report and Review of Literature. – Annals of Clinical & Laboratory Science 34(2): 206-208.

Shepherd, F.J. 1893. Symmetrical Depressions on the Exterior Surface of the Parietal Bones (With Notes of Three Cases). - Journal of Anatomy and Physiology 27(4): 501-504.

Smith, G.E. & W.R. Dawson. 1991. Egyptian Mummies. – London, Kegan Paul International.

Waldron, T. 2007. Paleoepidemiology: The Measure of Disease in the Human Past. – Walnut Creek, Left Coast Press.

Wilson, A.K. 1947. Thinness of Parietal Bones. – American Journal of Roentgenology and Radiology Therapy 58: 724-725.

Yamakawa, K, K. Mizutani, M. Takahashi, M. Matsui & T. Mezaki. 2006. *Hyperostosis Frontalis Interna* Associated with Hypogonadism in An Elderly Man. – Age and Ageing 35(2): 202-203.

Yilmaz, M.B., E. Egemen, B. Özbakır & A. Tekiner. 2013. Epidural Hematoma after Minor Trauma on Patient with Biparietal Osteodystrophy. – Journal of Korean Neurosurgical Society 53(1): 57-58.

Appendix

Case 1. Pelvis Attached

The etiology of this separation of the symphysis could have been premortem (in moments close to delivery) or postmortem, during embalming maneuvers. Traumatic diastasis of the pubic symphysis is a rare complication, including 9% maternal mortality and 35% fetal mortality (Cueto-Callejón *et al.*, 2010). It occurs due to a separation of the pubic symphysis > 10-13 cm during delivery due to endocrine changes (Agustín-Oliva *et al.*, 2017; Yoo *et al.*, 2014), which can cause the rupture of the ligaments that bind the pubis (Cueto-Callejón *et al.*, 2010; Reyes-Méndez *et al.*, 2014). Its etiology is not well established, but there are associated factors that favor pathological diastasis, such as the age of the mother older than 35 years (Pires *et al.*, 2015), multiparous women (Dunbar & Ries, 2002; Taylor & Sonson, 1986), multiple births (Urraca-Gesto *et al.*, 2015), the weight of the fetus (Urraca-Gesto *et al.*, 2015), the significant increase in the weight of the mother during pregnancy (Jain *et al.*, 2011), cephalopelvic disproportion (Yoo *et al.*, 2014) or intense uterine contractions (Reyes-Méndez *et al.*, 2014).

The McRoberts maneuver, used in a shoulder dystocia, has been described as a risk factor of diastasis of the pubic symphysis during delivery due to hyperflexion of the legs and increased traction forces on the pubic symphysis (Fidan *et al.*, 2013; Joosoph & Kwek, 2007; Urraca-Gesto *et al.*, 2015).

Maternal bone changes can also favor the appearance of a diastasis of the pubic symphysis, such as: osteomalacia, chondromalacia, rickets, pubic arthritis or bone tuberculosis (Dunbar & Ries, 2002; Hierholzer *et al.*, 2007; Moiety & Azzam, 2014).

In this case, we have not found any bone abnormalities in the pelvis, after an intense macroscopic and radiological analysis. The estimated age of this woman (20-25 years) also separates her of the highest risk group of suffering traumatic diastasis of the symphysis. The rest of the factors are impossible to determine, considering that only the pelvis was preserved.

In current clinic, patients feel pain after delivery, increasing with movement, weight lifting. Instability and inability to sit and ambulate are not uncommon. Most of the cases have a favorable evolution with easy conservative treatment (Reyes-Méndez *et al.*, 2014). Only a very small percentage needs surgery (Reyes-Méndez *et al.*, 2014).

For all these reasons, although it is impossible to completely rule it out, we believe that the etiology of this separation of the symphysis in this young woman should be sought in postmortem manipulation of the corpse during embalming and not in a natural process of rupture of ligaments during delivery.

Whatever the etiology (premortem, due to symphysis diastasis or postmortem, due to mummification), there is no doubt that the pelvic knot is postmortem and that embalmers attempted to place the pubic symphysis in its correct anatomical position.

This maneuver requires an exhaustive knowledge of the anatomy of the pelvis, since during the usual emptying of the viscera in mummification, the pubic symphysis is not usually manipulated by embalmers. Therefore, it seems clear that they had to specifically search for the anatomical site and carried out the maneuver from inside the abdominal cavity.

Case 2. Prolapse of the Vagina

The prolapse of the vagina, uterus or rectum, in addition to being caused by a weakening of the pelvic ligaments and especially the musculature often after delivery, can be a consequence of fermentation processes with gas formation that result in intra-abdominal activity pressure increase, forcing viscera out of the available orifices, over all vagina and rectum (Aufderheide, 2003). However, in this case, we can see that the abdominal cavity is completely empty of viscera and replaced by a mixture of sand, resin and sawdust. In this case, fermentation of the viscera can be ruled out as the cause of the prolapse.

Evisceration was one of the first steps of the mummification process, immediately after the death of the individual. For this reason, we believe that it would be a protracted delivery as main diagnosis. It should be remembered that the action of gas-producing bacteria on the corpse (methane, CO_2, hydrogen sulfide, and ammonia) generally begins after 72 hours until 7 days after death (Gamarra, 2015).

Case 3. Occipito-Atlantoid Osteoathritis

The causes of arthritis are not well known, although certain factors (the term is less precise than cause) are known. Metabolic (De Castro, 1992), endocrinological, hereditary (Stecher, 1959), climatic (Robles, 1997) and local factors, as dislocations, infections or traumatic lesions (Jurmain, 1999) are recognized like exacerbate arthritis factors, but is the age (Ortner, 2003) and the mechanical overload the most widely accepted factors (Bennike, 1985; De Castro, 1992; Herrerín, 2011; Klaus *et al.*, 2009; Law, 1950; Lince, 1978; Waldron, 2007). Despite all of the above, the relationship between the effort required by certain physical activities and osteoarthritis is sometimes difficult to establish (Anderson & Duthie, 1963; Lovell, 1994; Merbs, 1983).

The lesions observed in the occipito-atlantoid joint show an osteoarthritic pattern that consists of the formation of osteophytic lips around the contour of the articular surface or restricted to the posterior half of its perimeter. The presence of pores and holes shows the destruction of the superficial and subarticular bone tissue.

The atlantoid occipital joint is certainly stable and is not usually among the most affected one by osteoarthritis, but they can suffer chronic mechanical irritation (Boleaga-Durán *et al.*, 2006). These arthritic signs have been related to postures that involve pressure in this area, such as carrying weights on the head or movements of the head backwards when lifting the chin (Baxarias & Herrerín, 2008; De Miguel *et al.*, 2019; Rihuete, 2003; Robles, 1997; Yoldi *et al.*, 2009).

The high percentage of female skulls with these injuries found in TT16, opens the possibility that it is a reflection of physical activity that preferably involves women.

Case 4. Parietal Thinning

Parietal thinning is defined as external thinning of the parietal bone of the skull (Takata *et al.*, 2008). It is manifested by a partial or complete absence of the díploë of the calvarium at the site and by a corresponding thinness of the skull in the involved portion. It can be unilateral or, more often, bilateral, and its most constant position is just above the temporal ridge and about equidistant from the coronal and lambdoid sutures (Hauser & DeStefano, 1989). The inner table is usually intact.

Although a patient feels no pain in the area, the presence of thinness is thought to facilitate skull trauma with brain injury, and even death following fracture through the thinned areas (Cederlund *et al.*, 1982; Tsutsumi *et al.*, 2008). Progression of this pathology may result in perforations (Lim Seok Tae & Sohn Myung-Hee, 2001). In cases of slowly progressive bony thinning, in actual clinical practice, a cranioplasty may be required to prevent exposure of the brain to atmospheric pressure (Tsutsumi *et al.*, 2008).

Some authors considered parietal thinning an anatomical variant (Phillips, 2007; Pratap, 1969), while others thought it is related to some kind of trauma (Satoshi *et al.*, 2008), metastatic tumours or Gorham-Stout disease (Satoshi *et al.*, 2008). But follow-up work of Cederlund *et al.*, (1982), which was done for 14 years on x-rays of skulls, shown that slow progress of the disease was noted on 10 of 25 patients, which were proof that it is not congenital, but a progressive anomaly. The same research confirmed that this pathology is often connected to old age, as previously pointed out by Humphry (1874), as the average age for women was 72 years, and 63 for men (Herrerin *et al.*, 2017).

Case 5. Hyperostosis Frontalis Interna

There are only 13 published cases of HFI from Egypt and Nubia including the remains of both females and males of different ages and social statuses (Armelagos & Chrisman, 1988; Baker, 2013; Bebel-Nowak & Golijewskaja, 2015; Dequeker *et al.*, 1997; Jakob, 2007; Mant, 2014; Nielsen, 1970; Rösing, 1990; Shahin *et al.* 2014; Watrous *et al.* 1993). There are 7 affected females (6 cases over 40 years), 4 males (between 25 and 54 years of age), one adult of unknown sex, and one of unknown age-at-death and sex (Bebel-Nowak & Golijewskaja, 2015).

The occurrence of HFI in archaeological remains is comparatively low in contrast to modern populations with a ratio of up to 70% of women after 40 years of age affected in modern times and 1-4% in archaeological crania (Barber *et al.*, 1997; Bebel-Nowak & Golijewskaja, 2015). This is can be, at least partially, explained by underestimation related to diagnostic issues and because the endocranium is not often inspected in complete crania. In cases of skeletons with complete crania no CT or RTG are used in standard examination and therefore HFI is mostly observed in fragmented remains (Bebel-Nowak & Golijewskaja, 2015).

Analyse des gazelles momifiées de Kom Mereh/Komir (Haute Egypte) conservées au Musée des Confluences (Lyon, France)

Stéphanie Porcier & Louis Chaix[1]

Introduction

Les momies de gazelles découvertes à ce jour sont peu nombreuses relativement aux autres animaux momifiés tels les bovins, les canidés, les félidés ou encore les ibis. On en dénombre seulement quelques centaines provenant de quelques nécropoles et temples: Kom Mereh/Komir, Kom Ombo, Dendera, Tuna al-Gebel, Louqsor, al-Achmunein (Hermopolis) et Saqqara (Kessler, 1989: 18-26)[2]. Peu d'études leur ont été consacrées (Lortet & Gaillard, 1903: 78-86 ; Strandberg, 2009), ceci s'expliquant très probablement par la rareté des vestiges archéologiques mis au jour. En cela, les pratiques techniques et rituelles en lien avec ces animaux demeurent méconnues et de nombreuses questions subsistent.

Le musée des Confluences à Lyon conserve une quarantaine de gazelles momifiées, en grande partie inédites, provenant de Kom Ombo, Tuna al-Gebel et Kom Mereh/Komir. Toutes font l'objet d'une étude pluri- et interdisciplinaire dans le cadre du programme de recherche MAHES (Momies Animales et Humaines EgyptienneS)[3]. Pour cette contribution nous nous sommes attachés à étudier un seul groupe, celui des gazelles de Kom Mereh en Haute Égypte qui constitue l'ensemble le plus important (Figure 1). D'autres gazelles momifiées provenant de Kom Mereh sont conservées au Musée du Caire : CGC 29511, 29513, 29514, 29515, 29528, 29661, 29662, 29664, 29665 et 29671 (Gaillard & Daressy, 1905) mais, n'ayant pu faire l'objet d'une campagne de radiographie et d'une analyse archéozoologique, elles n'ont pas été prises en compte dans la présente étude.[4] Les autres

1 Cette étude s'inscrit dans le cadre du programme de recherche MAHES (Momies Animale et Humaines EgyptienneS) financé par l'Agence Nationale de la Recherche au titre du programme "Investissement d'Avenir" program ANR-11-LABX-0032-01 LabEx ARCHIMEDE. Elle a notamment pour objectif de documenter les actes et les contextes rituels de fabrication et d'usage des momies animales, de même que les systèmes de représentations symboliques qui ont sous-tendu ces pratiques de momification.

2 Des gazelles associées à des inhumations humaines ont également été découvertes, voir par exemple celles mises au jour à Hérakonpolis et datant de la période pré-dynastique (Adams, 2004 ; Friedman, 2004 : 132ff) ou encore deux autres spécimens plus tardifs provenant de Deir al-Bahari (la gazelle d'Isetemkheb D, 21e dynastie : Dodson & Hilton, 2004: 206 ; Ikram, 2003: 79-80 ; Maspero, 1889: pl. XXI B ; la gazelle d'Ankh-Shepenwepet, 23e dynastie : Winlock, 1924: 30, Fig. 35).

3 L'ensemble des gazelles momifiées a été radiographiée et a fait l'objet d'une analyse archéozoologique. 2 ont fait l'objet d'une analyse au scanner médical (MHNL 90001291, 90001623), 2 d'une analyse à la lumière synchrotron (MHNL 9002282, 90002285), 1 d'une analyse des baumes de momifcation (MHNL 90001211), 7 de datation (MHNL 51000043, 90001211, 90001291, 90001623, 90002281, 90002282, 90002285 ; Porcier *et al.*, 2019 ; Richardin *et al.*, 2017).

4 CG 29661 a été publié par Ikram & Iskander 2002 avec la radiographie.

In: Ikram, S., J. Kaiser & S. Porcier. Eds. 2021. *The Ancient Egyptians and the Natural World. Flora, Fauna, and Science.* – Leiden, Sidestone Press: 199-212.

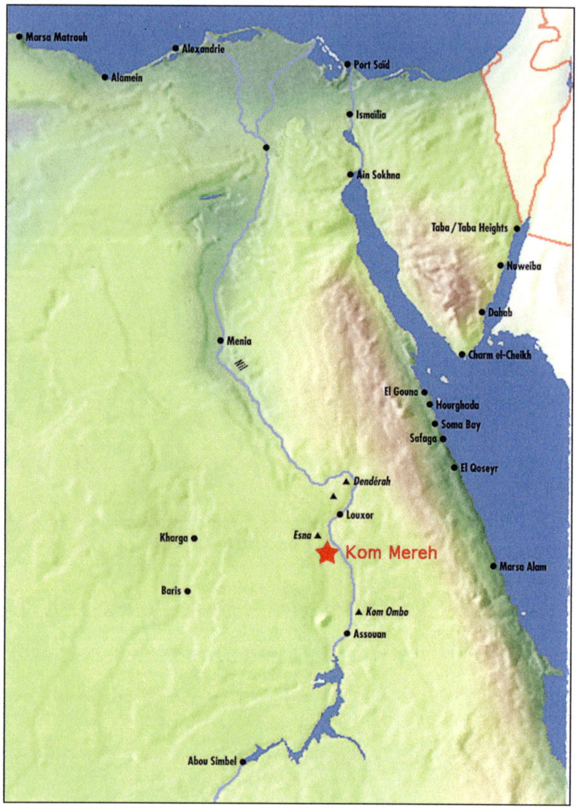

Figure 1. Localisation de Kom Mereh /Komir.

spécimens de la collection de Lyon, issus de Kom Ombo et Tuna al-Gebel, de même que celles dont la provenance est inconnue, feront l'objet d'une autre publication au sein d'un ouvrage en préparation consacré à l'ensemble des gazelles de la collection du musée des Confluences.

Après une première étude réalisée par Fleur Letellier-Willemin (2019) portant sur les textiles des spécimens de la collection, nous nous intéressons ici non plus au contenant mais au contenu des momies. Sont donc présentés dans cet article les résultats de l'analyse archéozoologique des gazelles.

La campagne de fouille menée en 1882 par Gaston Maspero près du village de Kom Mereh/Komir, à 13 km au Sud-Est d'Esna, a permis de mettre au jour des gazelles momifiées datant de l'époque romaine[5] associées à la déesse locale Anoukis-Nephthys (Lortet & Gaillard, 1903: 78-81 ; Valbelle, 1981). 29 spécimens ont été expédiés en 1901 au Muséum d'Histoire Naturel de Lyon (aujourd'hui musée des Confluences) pour y être étudiés par Louis Lortet assisté de Claude Gaillard (spécimens arrivés au Muséum le 06 avril 1901 d'après le Journal d'Entrée du musée). Dans une lettre adressée à Louis Lortet datée du 9 août 1901, Gaston Maspero livra quelques renseignements sur le contexte de découverte et d'inhumation des gazelles envoyées à Lyon :

> « *Kom-Méreh. – Lorsque le petit temple de Kom-Méreh me fut signalé pour la première fois en 1882, une des chambres en était remplie de momies de gazelles, entassées là à l'époque romaine, vers la fin du Ier siècle après Jésus-Christ au plus tôt. De nombreuses momies étaient enterrées à même le sable ou dans des puits peu profonds dans la plaine qui borde la montagne libyque à l'ouest du village. C'est de là que viennent les momies que je vous ai envoyées. Elles sont peut-être plus anciennes que celles qu'on trouvait dans le temple, probablement du Ier siècle avant Jésus-Christ.* » (Lettre provenant du fonds ancien du musée des Confluences, sans numéro d'inventaire).

Ceci est confirmé par Louis Lortet et Claude Gaillard dans leur publication sur la faune momifiée, les gazelles de la collection lyonnaise proviendraient donc, non pas de l'une des chambres du petit temple, mais de la plaine qui borde la montagne libyque à 3 km au Sud-Ouest du village de Kom Mereh. Elles ont été découvertes dans des fosses quadrangulaires de 3 à 4 m de profondeur et d'une surface de 10 à 20 m^2, entassées les unes sur les autres, sans aucune séparation et en désordre apparent (Lortet & Gaillard, 1903: 81). Aucun autre document n'ayant été retrouvé, ce sont donc les seules informations dont nous disposons à l'heure actuelle.

Des mouflons à manchettes momifiés étaient également inhumés parmi elles (Lortet & Gaillard, 1903: 78, 103-106) : l'un est conservé à Lyon (MHNL 51000069), les autres, disparus, ne figurent pas au Catalogue Général des Antiquités Égyptiennes du Caire (Gaillard & Daressy, 1905 ; Ikram & Iskander, 2002).

Matériel et méthode

Le corpus étudié est conservé au musée des Confluences à Lyon et comprend 23 *Gazella dorcas* et *Gazella sp.* 7 d'entre elles ont conservé leurs enveloppes textiles alors que 16 individus sont représentés par des éléments du squelette (essentiellement des crânes + 4 squelettes complets ; Table 1).[6] 5 momies de très jeunes gazelles et des restes de pattes isolées n'ont pas permis d'attribution spécifique certaine. Une confusion est possible avec la

5 Les gazelles attribuées à la période gréco-romaine par Lortet et Gaillard (1903: 78-81 ; voir aussi Gaillard & Daressy, 1905: 13 ; Strandberg, 2009: 178-179) datent en fait uniquement de la période romaine et sont comprises entre le début du 1er s. et le début du 4e s. de n.è. (datations ^{14}C réalisées par Pascale Richardin du C2RMF, voir Porcier *et al.*, 2019 ; Richardin *et al.*, 2017).

6 Les ossements de la momie MHNL 90001208 étant remontés et maintenus par des fils en acier, ces derniers n'ont pas pu faire l'objet d'une analyse ostéométrique pour la présente étude.

	No	Espèce	Attribution spécifique Lortet & Gaillard (1903)	Attribution spécifique base de données du musée	Sexe	Conservation Anatomie	Age	Stade
1	51000062	*Gazella dorcas*		*Gazella sp.*	mâle	crâne	1.5 - 3 ans	9
2	51000067	*Gazella dorcas*		*Gazella sp.*	mâle	crâne	> 8 ans	11
3	51000063	*Gazella dorcas*		*Gazella sp.*	mâle	crâne et mandibule	> 8 ans	11
4	51000064	*Gazella dorcas*	*Gazella dorcas*	*Gazella sp.*	mâle	crâne et mandibule	> 8 ans	11
5	51000068	*Gazella dorcas*		*Gazella sp.*	mâle	crâne et mandibule	3 - 5 ans	10
6	51000041	*Gazella dorcas*			femelle	squelette complet	1.5 - 3 ans	9
7	51000044	*Gazella dorcas*	*Gazella isabella*	*Gazella gazella*	femelle	squelette complet	1.5 - 3 ans	9
8	51000045	*Gazella dorcas*		*Gazella dorcas*	femelle	squelette complet	3 - 5 ans	10
9	51000169	*Gazella dorcas*		*Gazella sp.*	femelle	crâne, mandibule, hyoide, atlas, axis	3 - 5 ans	10
10	51000059	*Gazella dorcas*	*Gazella dorcas*	*Gazella sp.*	femelle	crâne et mandibule	> 8 ans	11
11	51000060	*Gazella dorcas*		*Gazella sp.*	femelle	crâne et mandibule	> 8 ans	11
12	51000061a	*Gazella dorcas*		*Gazella sp.*	femelle	crâne et mandibule	> 8 ans	11
13	51000061b	*Gazella dorcas*		*Gazella sp.*	femelle	crâne et mandibule	> 8 ans	11
14	51000413	*Gazella dorcas*	*Gazella sp.*	*Gazella sp.*	femelle	crâne, mandibule, atlas, axis, métacarpien, tibia, métatarsien	> 8 ans	11
15	90002282	*Gazella sp*		*Gazella sp.*	indet	momie enveloppée	0-3 mois	
16	90002283	*Gazella sp*		*Gazella sp.*	indet	momie enveloppée	0-3 mois	
17	90002284	*Gazella sp*		*Gazella sp.*	indet	momie enveloppée	> 7 mois	
18	90002285	*Gazella sp*		*Gazella sp.*	indet	momie enveloppée	0-3 mois	
19	90001208	*Gazella dorcas*		*Gazella sp.*	mâle	squelette complet	18-20 mois	
20	90001209	*Gazella dorcas*		*Gazella sp.*	femelle	momie enveloppée	9 mois	
21	90001211	*Gazella dorcas*	*Gazella isabelle puis dorcas*	*Gazella sp.*	femelle	momie enveloppée	18 mois	
22	90001291	*Gazella sp*		*Gazella sp.*	indet	momie enveloppée	7-9 mois	
23	90010003	*Gazella sp*		*Gazella sp.*	indet	pattes avec chaire sans textile		

Table 1. Corpus des gazelles momifiées conservées au musée des Confluences à Lyon.

gazelle leptocère (*Gazella leptoceros*) également présente en Égypte (Kindon *et al.*, 2013 ; El Alqamy & Baha el Din, 2006). L'attribution à la gazelle dorcas a été établie à partir de critères ostéométriques et morphologiques (Gentry, 1964; Osborn & Helmy, 1980). Les ossements ont été mesurés suivant les normes préconisées par Angela von den Driesch (1976), et reprises par Desse *et al.* (1986).

L'âge des gazelles a été estimé sur la base de l'état de la dentition et de l'usure des dents, mais il est possible que l'alimentation affecte le schéma d'usure des dents. Il n'existe pas, pour la gazelle dorcas, de table disponible pour ces observations. Nous avons donc utilisé celle établie pour une espèce voisine présente essentiellement au Proche-Orient, *Gazella gazella* (Davis, 1980 ; 1983; Munro *et al.*, 2009). En outre, le développement du squelette postcrânien (dimensions, fusion des épiphyses) permet aussi une estimation de l'âge (Davis, 1980; Hakker, 1986).

En l'absence de coefficients utilisables pour les gazelles, la taille au garrot a été estimée à partir des coefficients de Schramm (1967) établis pour la chèvre. Nous avons choisi la longueur du tibia qui s'avère la mieux corrélée avec la taille au garrot.

En parallèle, nous avons utilisé des données comparatives provenant de gazelles dorcas conservées au Muséum de Paris, étudiées et mesurées par nos soins et auxquelles nous avons ajouté quelques données issues de la littérature, essentiellement tirées du travail de J. Peters (1989). Ce corpus complémentaire se compose de 36 mâles et 21 femelles (Table 2, voir Annexe). Les individus de la collection parisienne proviennent essentiellement d'Algérie, du Maroc, de Tunisie, du Niger et du Tchad. Les gazelles des autres collections proviennent du Soudan et de Somalie.

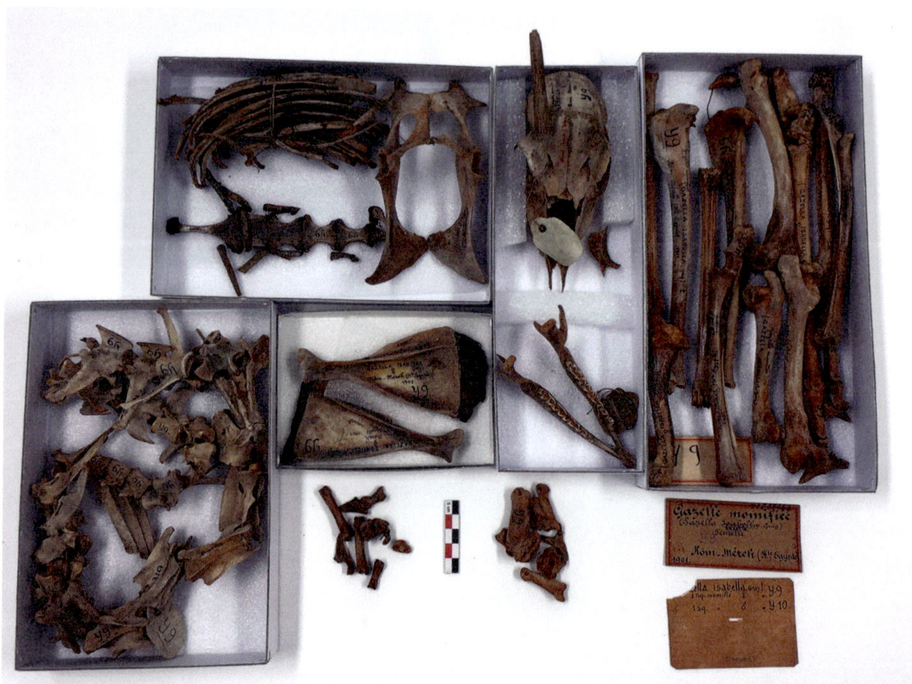

Figure 2. Gazelle momifiée conservée à l'état de squelette, musée des Confluences à Lyon MHNL 51000044. Photographie des auteurs.

Les momies dont l'enveloppe textile et/ou charnelle est encore conservée ont fait l'objet d'une campagne de radiographie réalisée en juin 2013 à l'ancien Muséum d'histoire Naturel de Lyon, 2 rue Morelet. Les clichés obtenus au moyen d'un appareil radiographique mobile de type Mobilett Mira Max gracieusement mise à notre disposition par la société Siemens ont été réalisés par Roger Lichtenberg, Stéphanie Porcier et Didier Berthet. L'état différentiel de conservation (momies complètes, corps ayant conservés les parties molles et éléments squelettiques ; Figure 2) s'explique en partie par le fait que ces momies ont été étudiées au début des années 1900, époque durant laquelle la radiographie étaient rarement utilisée. Pour mener à bien leurs recherches qui visaient en partie à appuyer la théorie de Darwin controversée à cette période en France, L. Lortet et Cl. Gaillard ont alors désenveloppé et décharné une partie des momies (Lortet & Gaillard, 1903: VI, 78-86).

Résultats et discussion

Dans un premier temps, il semble utile de rappeler ici quelques caractéristiques générales des gazelles dorcades. Les deux sexes portent des cornes, annelées et lyriformes, plus petites chez les femelles. Ce sont des gazelles graciles, bâties pour la course et possédant un long cou ainsi que des membres allongés, en particulier les métapodes. Leur taille au garrot varie entre 58 et 67 cm alors que leur poids oscille entre 12 et 25 kg. Leur alimentation est exclusivement herbivore, leur préférence allant vers les feuillages et les herbes riches en protéines. Elles trouvent également de l'eau dans les végétaux ingérés, ce qui leur permet de ne pas boire durant de longues périodes.

Les gazelles dorcades vivent en groupes familiaux de 2 à 8 individus ou en bandes de plusieurs dizaines d'individus. C'est une espèce nomade qui exploite les zones éphémères de végétation dans des environnements très arides (Lafontaine *et al.*, 2005 ; Yom-Tov *et al.*, 1995).

Gazella dorcas est une espèce saharo-arabique, dont la distribution s'étend aujourd'hui du Maroc au Soudan, ainsi qu'au nord de l'Éthiopie, en Somalie et au Tchad. Plusieurs sous-espèces ont été décrites dans la littérature mais plusieurs d'entre elles n'ont pas fait l'objet d'analyses génétiques (Lange, 1972; Abaigar, 2005).

Les découvertes fossiles témoignent d'une répartition plus orientale de l'espèce (Uerpmann, 1987; Yom-Tov *et al.*, 1995) mais aussi d'une extension méridionale le long de la vallée du Nil jusqu'au sud de Khartoum (Chaix, 2003) (Figure 3).

Attribution spécifique

Les mesures crâniennes ainsi que la forme des os nasaux, de la fosse sub-orbitaire et des étuis cornés nous ont permis d'attribuer l'ensemble les individus identifiés à *Gazella* dorcas (Linnaeus 1758). Le corpus étudié se compose de 17 gazelles dorcas et de 6 *Gazella sp.* de sexe indéterminé (Table 1).

Dans leur premier volume sur la faune momifiée regroupant des spécimens provenant de plusieurs sites, Lortet & Gaillard (1903: 82) identifiaient 2 espèces : *G. dorcas* et *G. isabella*. Dans la collection lyonnaise, deux crânes, ceux d'une femelle (MHNL 51000044 ; provenance

Kom Mereh) et d'un mâle (MHNL 51000042 ; provenance Haute-Égypte) ont été attribués par les auteurs à *Gazella isabella*. Or, les études récentes montrent que cette gazelle est une sous-espèce de *Gazella dorcas* et non une espèce à part entière (Abaigar 2005). En outre, les travaux récents indiquent que les limites géographiques des trois sous-espèces actuelles (*G.* d. *dorcas*, *G. d. isabella et G. d. beccarii*) sont impossibles à définir (East, 1998 ; Bedeuls *et al.*, 2006). En revanche, les données génétiques semblent montrer qu'il n'existe qu'une sous-espèce au sud du Sahara (*G. d. pelzeni*, Lange, 1972). On peut néanmoins remarquer que les gazelles de la collection lyonnaise proviennent, pour la plupart, du site de Kom Mereh, dans la zone de répartition orientale de *G.* d. *dorcas*.

L'espèce de plusieurs autres gazelles de la collection (MHNL 51000041, 51000044, 51000062, 51000059, 51000060, 51000061, 51000063, 51000064, 51000067, 51000068, 51000069, 51000413, 90001208, 90001209, 90001211) n'avait pas, jusqu'à présent, été déterminée, ces dernières étant enregistrées dans la base de données du musée en *Gazella sp.*, leur étude récente effectuée par nos soins a permis de les attribuer à des *Gazella dorcas*.

L'attribution spécifique des jeunes individus encore enveloppés de leur textile est difficile à préciser à la seule lecture des radiographies. Seuls deux individus adultes ont été identifiés comme étant des *G. dorcas*.

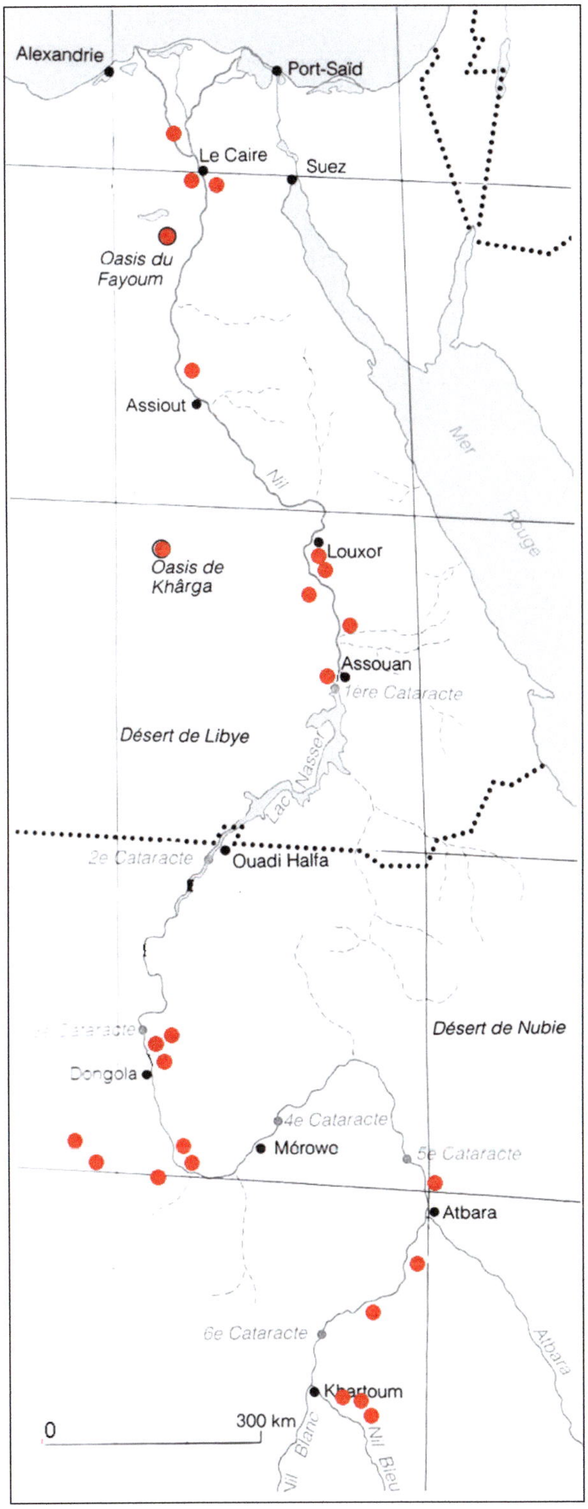

Figure 3. Carte de répartition des découvertes de *G. dorcas* en contexte archéologique en Égypte et au Soudan.

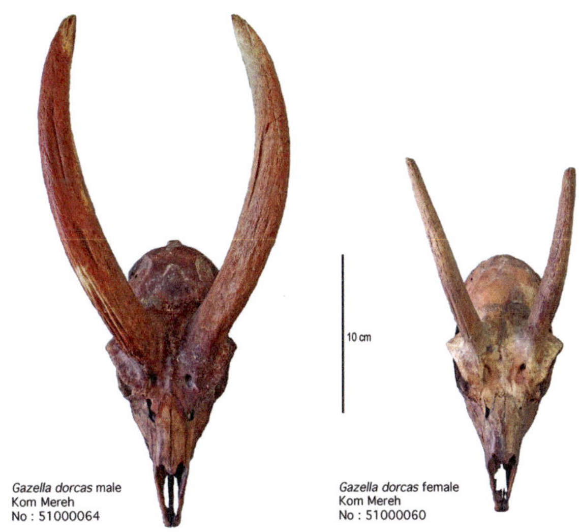

Figure 4. Crânes de *G. dorcas* mâle et femelle de Kom Mereh illustrant les différences morphologiques et métriques. Photographie des auteurs.

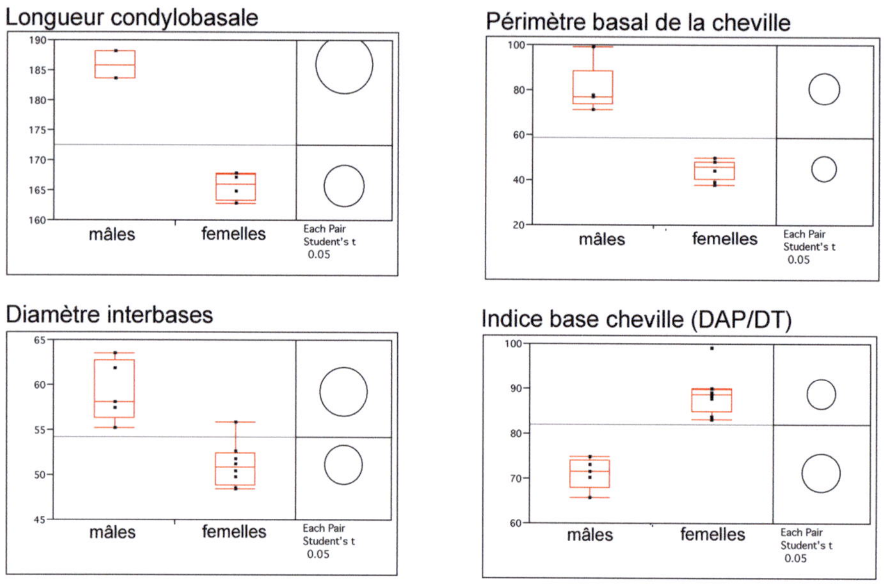

Figure 5. Graphique illustrant les principales mesures discriminantes entre mâles et femelles (à gauche); Graphique montrant la différence entre chevilles osseuses femelles et mâles de Kom Mereh et des individus actuels (ci-dessous).

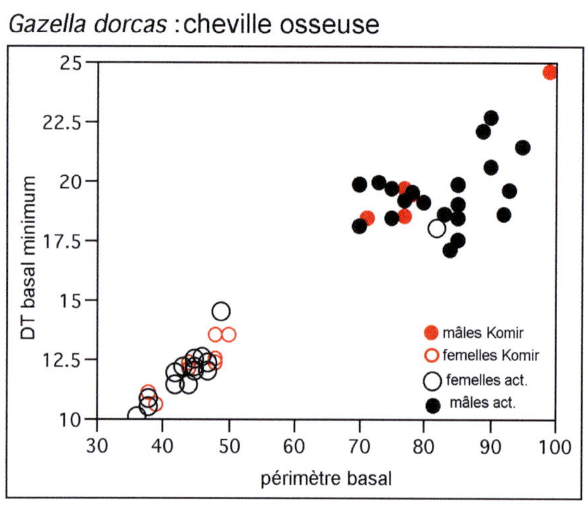

Dimorphisme sexuel

Le dimorphisme sexuel est assez marqué chez la gazelle dorcas. Les mâles sont plus grands, avec une taille au garrot entre 58 et 67 cm alors que celle des femelles varie entre 53 à 62 cm. Il en va de même du poids, qui oscille entre 12 et 16 kg pour les femelles alors que les mâles peuvent peser jusqu'à 25 kg. Les cornes, présentes chez les deux sexes, montrent un dimorphisme marqué (Figure 4) (Haltenorth & Diller 1980; Groves 1981; Yom-Tov *et al.* 1995).

Pour les gazelles de Kom Mereh, nous avons mis en évidence 4 mesures ou indices crâniens qui témoignent d'une différence significative entre mâles et femelles. Parmi elles, les valeurs liées à la taille des chevilles osseuses sont les plus discriminantes (périmètre basal, DAP basal et DT basal) (Figure 5). Les os longs sont trop peu nombreux à Kom Mereh (1 mâle et 3 femelles) pour permettre une discrimination sexuelle fiable.

Le dimorphisme sexuel de deux femelles jeune-adultes enveloppées de textile (MHNL 90001209, 90001211) a été établi à partir de la forme des cornes sur la base de radiographies. Comme l'attribution spécifique, le sexe des jeunes individus encore enveloppés de leur textile n'a pu être déterminé à la seule lecture des radiographies. Seuls le sexe des deux individus adultes ont été identifiés comme étant des femelles.

La collection lyonnaise compte donc 6 mâles, 11 femelles et 6 individus dont le sexe est indéterminé (Table 1).

Âge

L'éruption et l'usure dentaire ainsi que la fusion des épiphyses nous ont permis d'estimer l'âge des gazelles de Kom Mereh (Figure 6). Les 16 spécimens conservés à l'état d'ossements correspondent à des individus adultes alors que la radiographie des 7 spécimens ayant conservé leurs textiles montre qu'il s'agit, pour 5 d'entre eux, de très jeunes et jeunes individus dont l'âge est compris entre 0 et 9 mois (Table 1) (Figure 7).

Il nous a semblé intéressant de comparer la distribution des âges des gazelles de Kom Mereh à un ensemble cohérent de gazelles chassées par les habitants du site néolithique de Ain Mallaha en Palestine. Le site a livré 382 restes de gazelles chassées dont l'espèce n'a pu être déterminée (Ducos, 1968). Nous avons regroupé les âges des gazelles de Kom Mereh dans les catégories utilisées par P. Ducos. Les données comparatives sont visibles sur la figure (Figure 8). En dépit du nombre beaucoup plus réduit, la distribution des âges des gazelles de Kom Mereh diffère clairement de la courbe de chasse néolithique. La répartition observée est relativement homogène alors que la courbe de chasse montre une concentration d'individus

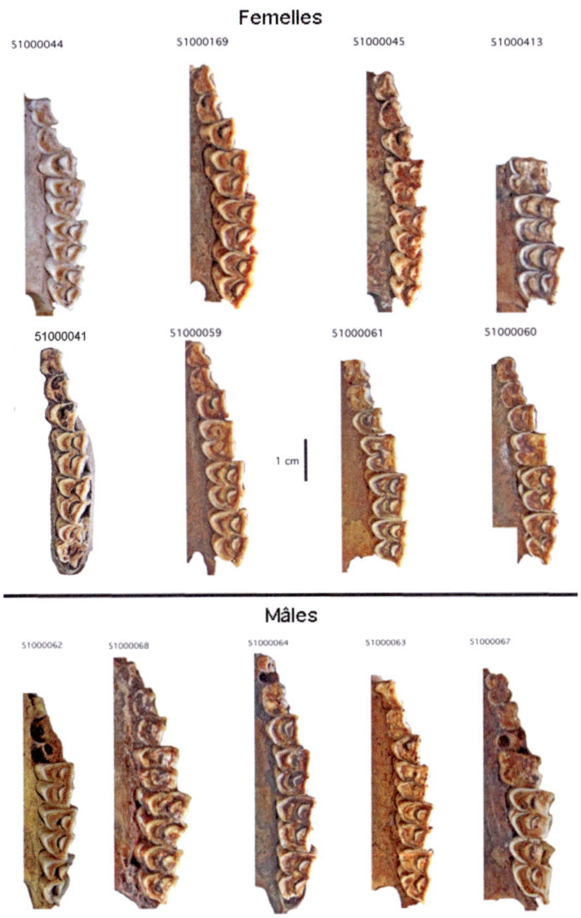

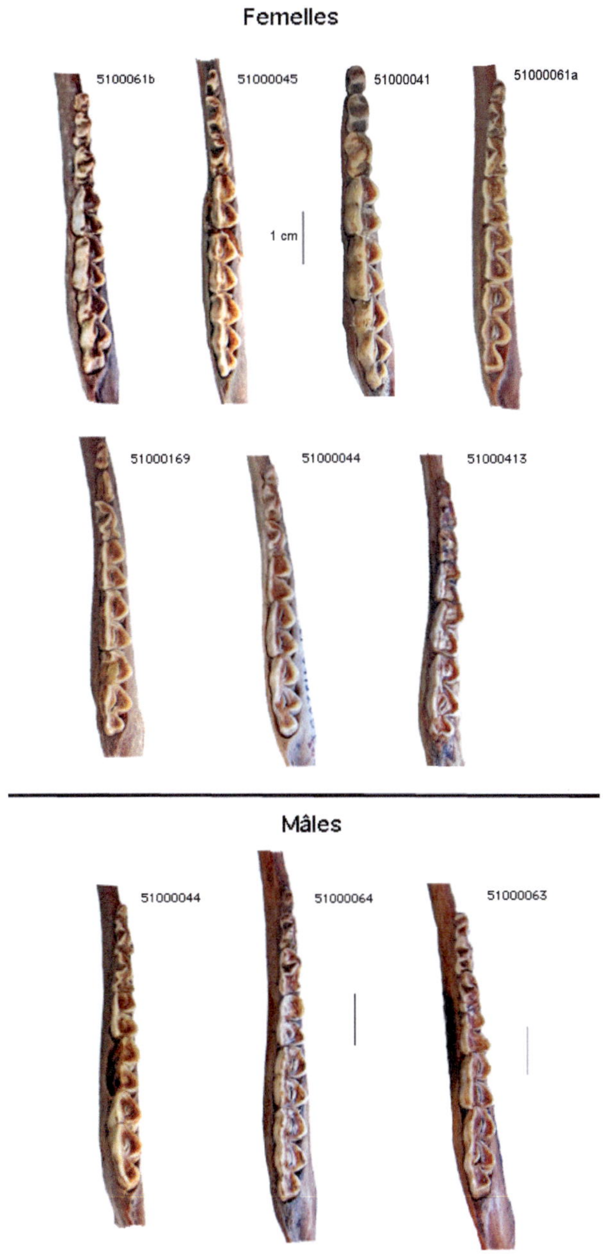

Figure 6. Stades d'usure des maxillaires (à gauche) et des mandibules des *G. dorcas* de Kom Mereh (à droite). Photographie des auteurs.

entre 1 et 3 ans (Baharav, 1974 ; Davis, 1983). La courbe observée à Kom Mereh pourrait plutôt correspondre à un groupe retenu dans un lieu clos et où les animaux pouvaient notamment se reproduire.

Taille

Les os du squelette postcrânien, nécessaires au calcul de la hauteur au garrot, sont mal représentés dans la collection de Lyon (1 mâle [MHNL 90001208] et 3 femelles [MHNL 51000041, 51000044, 51000045], contrairement aux crânes (5 mâles [MHNL 51000062, 51000063, 51000064, 51000067, 51000068] et 6 femelles [MHNL 51000059, 51000060, 51000061a, 51000061b, 51000169, 51000413]

La comparaison entre les gazelles de Kom Mereh et le corpus de référence montre que les gazelles de la collection lyonnaise sont plus grandes que les gazelles actuelles, sans que les différences soient statistiquement significatives, excepté pour quelques mesures comme la longueur condylobasale du crâne et les longueurs maximum du radius et des métapodes (Figure 9).

Il faut rappeler ici que les femelles sont plus nombreuses à Lyon alors que les femelles du corpus de référence sont moins nombreuses (36.8 %), ce qui accentue encore cette différence de taille.

Il nous a semblé intéressant d'estimer la taille au garrot des gazelles lyonnaises (Figure 10). On note que les gazelles de Kom Mereh sont dans la marge supérieure des tailles des gazelles actuelles. Pour les gazelles dorcades de la collection les données sont comprises entre 60,2 et 63,2 cm alors que les données sur les gazelles actuelles donnent

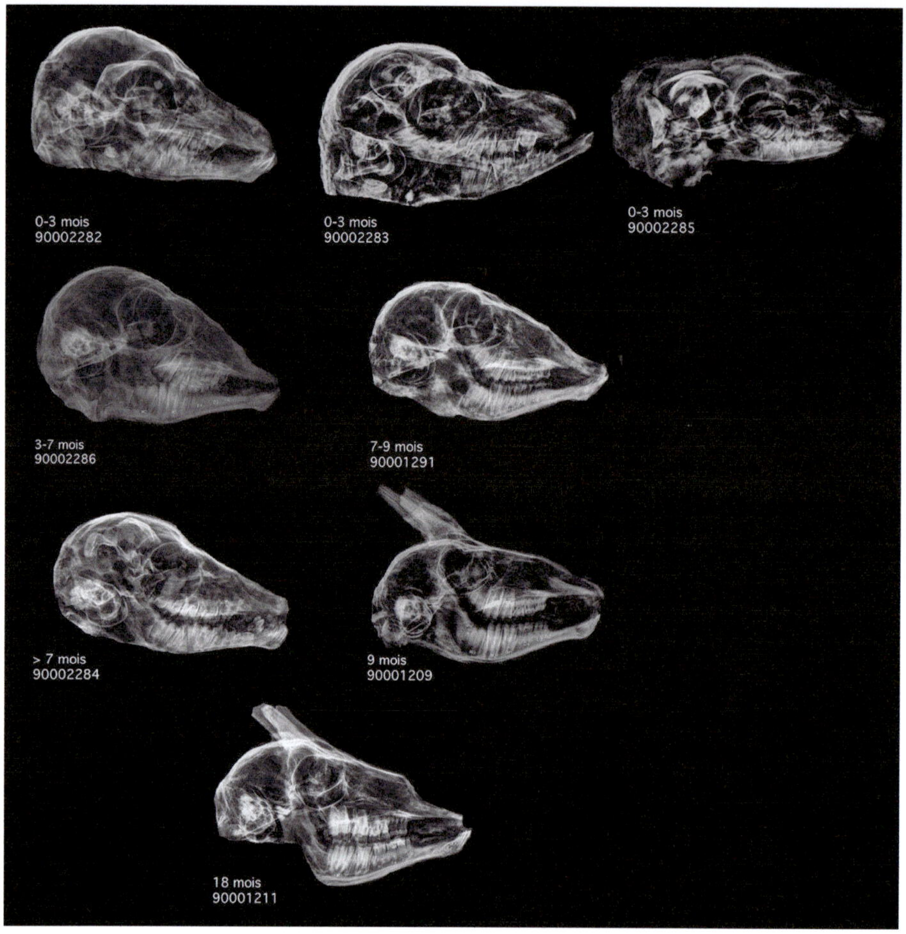

Figure 7. Radiographies des crânes de jeunes *G. dorcas* et *G. sp.* provenant de Kom Mereh (R. Lichtenberg/Programme MAHES). Les crânes ont été disposés par classes d'âges du plus jeune au plus âgé. Assemblage radiographique des auteurs.

Age (années)	Kom-Mereh NI	Mallaha NI
0-1	6	10
1-3	5	25
3-5	3	20
5-7	1	13
> 7	8	7

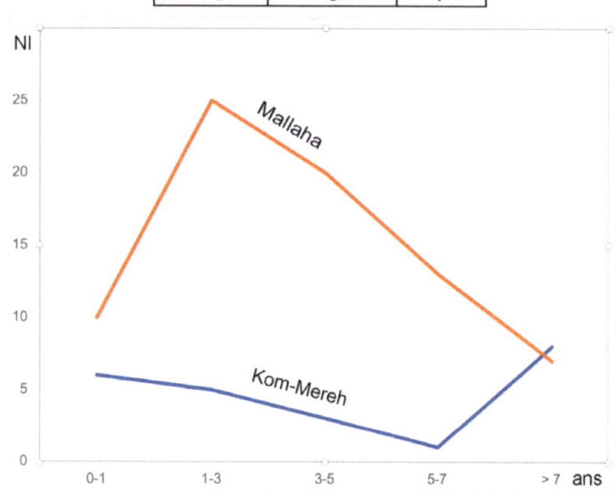

Figure 8. Comparaisons entre les âges des *G. dorcas* de Kom Mereh avec ceux des *Gazella* sp. sauvages du site natoufien d'Aïn Mallaha. D'après Ducos (1968 : 73).

des valeurs entre 55 et 65 cm au garrot (Haltenorth & Diller, 1980 ; Hoath, 2003).

Pathologies et cause de la mort

Aucune trace de mise à mort n'a été observée sur l'ensemble des gazelles de Kom Mereh.

Quelques traumatismes ont été observés à partir des radiographies. Les gazelles MHNL 90001209, 90001291, 90002282, 90002283, 90002284 et 90002285 présentent une luxation au niveau des vertèbres cervicales, qui sont probablement à imputer à la manipulation des cadavres lors de la momification alors que ces derniers étaient dans un état de décomposition avancé.

Une femelle adulte (MHNL 51000044) présente une fracture de la zone distale du métacarpien gauche. Cette fracture s'est ressoudée en entraînant une infection, source d'importants ostéophytes.

Techniques de momification et assemblages

Plusieurs momies ont été débarrassées de leur enveloppe textile au début du 20e siècle afin d'être étudiées. À l'examen des ossements des gazelles analysées par Louis Lortet

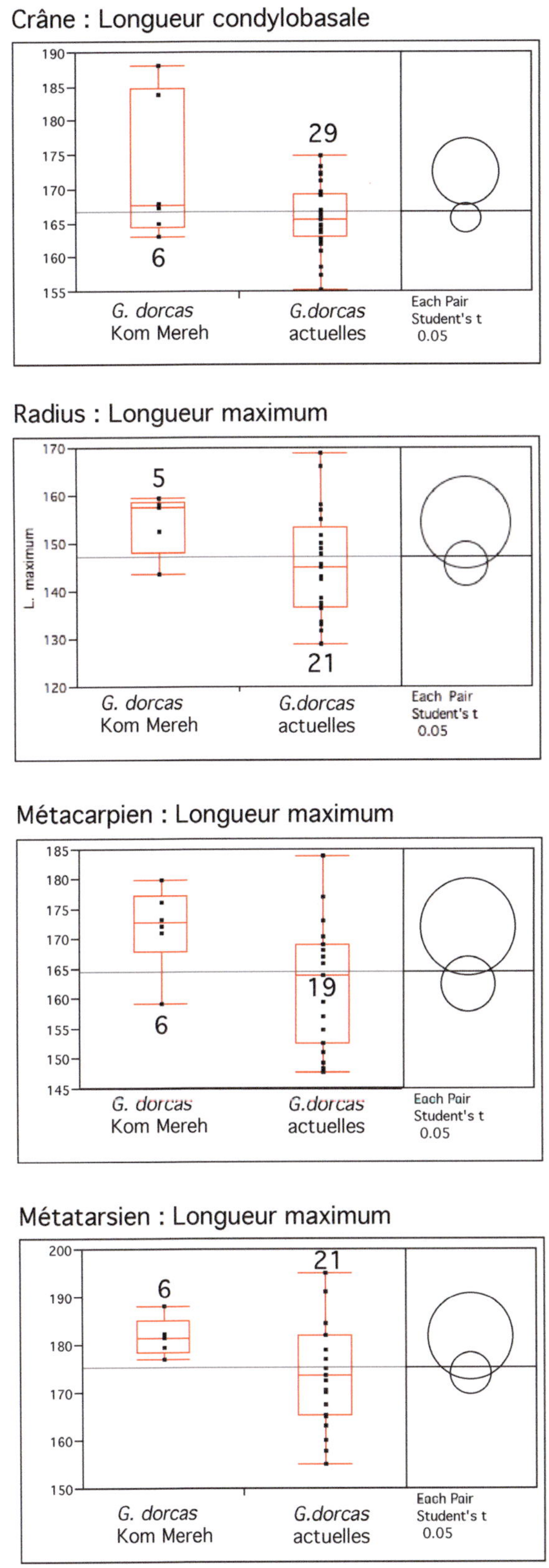

Figure 9. Graphique illustrant la taille importante des *G. dorcas* de Kom Mereh par rapport à leurs congénères actuels.

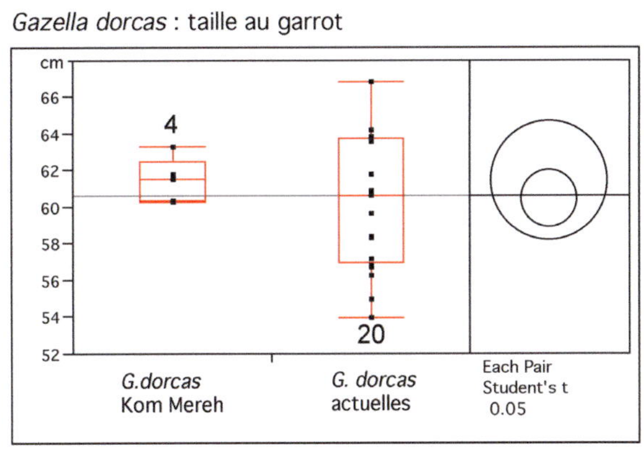

Figure 10. Comparaison de la taille au garrot des *G. dorcas* de Kom Mereh avec les gazelles dorcades actuelles.

et Claude Gaillard, nous constatons que ces dernières, à l'inverse, par exemple, des bovins de la même collection, n'ont pas fait l'objet d'un décharnement pré-momification.

Louis Lortet et Claude Gaillard précisent que

> « *les momies renferment en général un seul individu* » mais qu'à « *l'intérieur de chacune on ne trouve pas toujours un animal entier, tantôt elle ne contient que la moitié du corps, tantôt seulement la tête avec les extrémités ou les canons des quatre membres* » : « *l'une de ces dernières renfermait, outre le crâne et les extrémités osseuses des membres, une portion de 20 centimètres de longueur environ de la colonne vertébrale d'un poisson de grande taille, probablement d'un Lates.* » (Lortet & Gaillard, 1903: 78).

Les momies que nous avons étudiés et dont l'enveloppe textile est encore conservée ont un contenu moins original que ceux décrits par Louis Lortet et Claude Gaillard : toutes contiennent un individu complet dont l'état de décomposition est plus ou moins avancé ; excepté la MHNL 90002282 qui ne contient que la portion antérieure d'un individu complet. Notons enfin que les gazelles de Kom Mereh ont fait l'objet d'une momification assez sommaire. Les corps ont, semble-t-il, seulement été desséchés sans traitement particulier (éviscération, excérébration) et cela, alors qu'ils avaient commencé leur processus de décomposition.

Le fait que certaines momies de Kom Mereh aient été façonnées à l'aide de morceaux d'individus (Lortet & Gaillard, 1903: 78 ; MHNL 90002282) ou à partir de plusieurs individus d'espèces différentes (Lortet & Gaillard, 1903: 78) pose la question du statut de ces objets et de ces animaux. L'absence d'un corps complet indique que l'animal momifié n'était a priori pas destiné à un au-delà. De fait, il semble inconcevable que cet ensemble soit à associer aux restes momifiés d'un animal sacré de la déesse Anoukis du type Apis. Les individus dont l'intégralité du corps est préservée pourraient être issus du troupeau sacré, à l'instar de celui des taureaux Apis et de Mnévis (Porcier *et al.*, 2019). Le lieu de découverte, de simples fosses dans la plaine et non au sein du temple lui-même, la momification sommaire dont ils ont fait l'objet, et le fait, qu'aussi bien des femelles que des mâles aient été momifiés, vont dans le sens de cette hypothèse. On ne peut toutefois pas écarter une dernière hypothèse, celle de momies « votives » offertes à la divinité.

Conclusion

A l'instar de leur contenant, les textiles étudiés par Fleur Letellier-Willemin (2019), le contenu des momies de gazelles de Kom Mereh constitue un ensemble cohérent composé de gazelles dorcas, d'individus jeunes et adultes, et femelles pour la plupart.

La distribution des âges des gazelles de Kom Mereh indique qu'il ne s'agit pas d'animaux chassés mais plus vraisemblablement de gazelles capturées puis gardées en captivité dans un espace où elles auraient pu se reproduire. Afin de vérifier cette hypothèse, une analyse des micro-usures dentaires couplée à une analyse des isotopes O et N doivent être réalisées.

Toutes les momies sont composées d'un animal entier ne présentant aucune trace de mise à mort par l'homme. Leur lieu de découverte de même que leur momification sommaire laisse penser que nous n'avons vraisemblablement pas affaire ici à des animaux sacrés de la déesse locale Anoukis du type de celui de l'Apis de Memphis, mais ces animaux seraient selon nous plus vraisemblablement apparentés à son troupeau sacré.

Même si ces dernières constituent des groupes moins importants, nous attendons beaucoup de l'analyse en cours des autres gazelles de la collection lyonnaise provenant de Kom Ombo et Tuna al-Gebel. Contemporaines des gazelles de Kom Mereh, leur étude nous permettra d'appréhender les différentes pratiques en fonction des contextes géographiques.

Remerciements

Nous tenons à remercier ici le Musée des Confluences et plus particulièrement Monsieur Didier Berthet, responsable de la collection d'Égyptologie, pour son accueil chaleureux et son entière disponibilité. Nous remercions également très chaleureusement Monsieur Roger Lichtenberg pour sa précieuse collaboration au sein du Projet MAHES ; de même que la Société Siemens pour le prêt, à titre gracieux, d'une mobilett Mira dernière génération.

Notre gratitude s'adresse enfin à Madame Joséphine Lesur, maître de conférences au Muséum national d'histoire naturelle de Paris, de nous avoir donné accès, une nouvelle fois, à la collection dont elle a la charge.

Bibliographie

Abaigar, T. 2005. Taxonomy of Dorcas gazelle (*Gazella dorcas*) Subspecies. In: Monfort, S. & T. Correll. Eds. Fifth Annual Sahelo-Saharan Interest Group Meeting. – Tunesia, Sousse: 30-32.

Adams, B. 2004. Excavations in the Elite Predyanstic Cemetery at Hierakonpolis Locality HK6: 1999-2000. – Annales du Service des Antiquités de l'Égypte 78: 35-52.

Baharav, D. 1974. Notes on the population structure and biomass of the mountain gazelle, *Gazella gazella gazella*. – Israël Journal of Zoology 23 : 39-44.

Bedeuls, R.C., P. Devillers, R.M. Lafontaine, J. Devillers-Terschuren & M.O. Beudels. Eds 2006. Sahelo-Saharan Antelopes. Status and perspectives. Report on the Conservation Status of the Six Sahelo-Saharan Antelopes, CMS SSA Concerted Action, 2nd Edition. – Bonn, UNEP/CMS Secretariat (CMS Technical Series Publication, no. 11).

Chaix, L. 2003. La faune des sites mésolithiques et néolithique de la zone du Nil Bleu (Soudan Central). In : Fernandez, V. Ed. The Blue Nile Project. Holocene archeology in Central Sudan. – Dossier monográfico de la revista Complutum 14: 373-396.

Davis, S.J.M. 1983. The Age Profiles of Gazelles Predated by Ancient Man in Israel: Possible Evidence for a Shift from Seasonality to Sedentism in the Natufian. – Paléorient 8: 5-15.

Davis, S.J.M. 1980. A Note on the Dental and Skeletal Ontogeny of *Gazella*. – Israel Journal of Zoology 29: 129-134.

Desse, J., L. Chaix & N. Desse-Berset. 1986. « Osteo », base-réseau de données ostéométriques pour l'archéozoologie. – Paris, CNRS (CRA, Notes et Monographies Techniques, no. 20).

Ducos, P. 1968. L'origine des animaux domestiques en Palestine. – Bordeaux, Publications de l'Institut de Préhistoire de l'Université de Bordeaux (Mémoire no. 6).

Dodson, A. & D. Hilton. 2004. The Complete Royal Families of Ancient Egypt. – London, Thames & Hudson.

East, R. 1998. African Antelopes Database 1998. Compiled by Rod East and the IUCN/SSC Antelope Specialist Group. – Occasional Paper, IUCN, no.21.

El Alqamy, H. & S. Baha El Din. 2006. Contemporary Status and Distribution of Gazelle Species (*Gazella dorcas* and *Gazella leptoceros*) in Egypt. – Zoology in the Middle East 39:1: 5-16.

Friedman, R.F. 2004. Elephants at Hierakonpolis. In: Hendrickx, S., R.F. Friedman, K.M. Cialowicz & M. Chlodnicki. Eds. Egypt at its Origins. Studies in Memory of Barbara Adams. – Leuven, Peeters (Orientalia Lovaniensia Periodica 138): 131-168.

Gaillard, C. 1934. Contribution à l'étude de la faune préhistorique de l'Egypte. – Archives du Museum d'histoire naturelle de Lyon 14: 1-126.

Gaillard, C. & G. Daressy. 1905. La faune momifiée de l'antique Égypte. Catalogue général des antiquités égyptiennes du Musée du Caire, no. 29501-29733 et 229751-29834. – Le Caire, Institut français d'archéologie orientale.

Gentry, A.W. 1964. Skull Characters of African Gazelles. – Annals and Magazine of Natural History 7, 13: 353-382.

Groves, C.P. 1981. Notes on the Gazelles. 3. The Dorcas Gazelle of North Africa. – Anali del Museo Civico di Storia Naturale di Genova 83: 455-471.

Hakker D. 1986. The Determination of Sex and Species in Gazelles with the Use of Stepwise Discriminant Analysis. In : Van Wijngaarden-Bakker L.H. Eds. Database Management and Zooarchaeology. Pact, 14: 173-184.

Haltenorth, T. & H. Diller. 1980. Mammals of Africa including Madagascar. – London, Collins.

Hoath, R. 2003. A Field Guide to the Mammals of Egypt. Cairo, The American University in Cairo Press.

Ikram, S. 2003. Death and Burial in Ancient Egypt. – London, Longman.

Ikram, S. & N. Iskander. 2002. Non-human Mummies. Catalogue général of Egyptian Antiquities in the Cairo Museum. – Cairo, The Supreme Council of Antiquities.

Kessler, D. 1989. Die heiligen Tiere und der König. – Wiesbaden, Harrasssowitz.

Kingdon, J., D. Happold, T. Butynski, M. Hoffmann, M. Happold & J. Kalina. 2013. Mammals of Africa. – London, Bloomsbury.

Lafontaine, R.M., R.C. Beudels-Jamar, P. Devillers & T. Wacher. 2005. Gazella dorcas. In: Beudels-Jamar, R.C., P. Devillers, R.M. Lafontaine, J. Devillers- Terschuren & M.-O. Beudels. Eds. Les Antilopes Sahélo-Sahariennes. Statut et Perspectives. Rapport sur l'état de conservation des six Antilopes Sahélo-Sahariennes. Action Concertée CMS ASS. 2d edition (CMS Technical Series Publication N° 11, 2005). – Bonn, UNEP/CMS Secretariat, Institut Royal des Sciences Naturelles de Belgique: 93-126.

Lange, J. 1972. Studien an Gazellenschädeln. Ein Beitrag zur Systematik der kleineren Gazellen, *Gazella* (De Blainville, 1816). – Säugetierkundliche Mitteilungen 20: 193-249.

Letellier-Willemin, F. 2019. Appréhender les momies autrement. L'étude des textiles des momies de gazelles du musée des Confluences à Lyon. In: Porcier, S., S. Ikram & S. Pasquali Eds. Creatures of Earth, Water, and Sky. Essays on Animals in Ancient Egypt and Nubia. – Leiden, Sidestone Press: 221-230.

Lortet, L.C. & C. Gaillard. 1903. La faune momifiée de l'Ancienne Egypte. – Archives du Museum d'Histoire Naturelle de Lyon 8, 2: I-VIII et 1-206.

Lortet, L.C. & C. Gaillard. 1907. La faune momifiée de l'Ancienne Egypte. Bovidés. – Archives du Museum d'histoire naturelle de Lyon 9: 54-82.

Lortet, L.C. & C. Gaillard. 1909. La faune momifiée de l'Ancienne Egypte et recherches anthropologiques. – Archives du Museum d'histoire naturelle de Lyon 10: 1-336.

Maspero, G. 1889. Les momies royales de Deir-el-Bahari. – Mémoires publiés par les membres de la mission archéologique française au Caire 1, 4: 511-790.

Munro, N.D., G. Bar-Oz & A.-J. Stutz. 2009. Aging Mountain Gazelle (*Gazella gazella*) : Refining Methods of Tooth Eruption and Wear and Bone Fusion. – Journal of Archaeological Science 36: 752-763.

Osborn, D.L. & I. Helmy. 1980. The Contemporary Land Mammals of Egypt (Including Sinai). – Fieldiana Zoology 5: 1-579.

Peters, J. 1989. Osteomorphological Features of the Appendicular Skeleton of Gazelles, Genus *Gazella* Blainville 1816, Bohor Reedbuck, *Redunca redunca* (Pallas, 1767) and Bushbuck, *Tragelaphus scriptus* (Pallas, 1766). – Anatomia, histologia, embryologia 18: 97-113.

Porcier, S., P. Richardin, G. Louarn, S. Ikram & D. Berthet. 2019. Datations par le carbone 14 de 63 momies animales du musée des Confluences à Lyon (France). In: Porcier, S., S. Ikram & S. Pasquali. Eds. Creatures of Earth, Water, and Sky. Essays on Animals in Ancient Egypt and Nubia. – Leiden, Sidestone Press: 283-292.

Richardin, P., S. Porcier, S. Ikram, G. Louarn & D. Berthet. 2017. Cats, Crocodiles, Cattle, and More: Initial Steps Toward Establishing a Chronology of Ancient Egyptian Animals Mummies. – Radiocarbon 59, 2: 595-607.

Schramm, Z.1967. Long Bones and Height in Withers of Goat. – Roczniki Wyzszej Szkoly Rolniczej w Poznaniu 36 : 89-105.

Strandberg, Å. 2009. The Gazelle in Ancient Egyptian Art: Image and Meaning. – Uppsala, Uppsala University (Studies in Egyptology 6).

Uerpmann, H. P. 1987. The Ancient Distribution of Ungulates Mammals in the Middle East. Fauna and Archaeological Sites in Southwest Asia and Northeast Africa. – Wiesbaden Dr. Ludwig Reichart (Beihefte sum Tübinger Atlas des Vordeeren Orients Reihe A (Naturwissenschaften Nr. 27R, Reihe A, no. 27).

Valbelle, D. 1981. Satis et Anoukis. - Mainz am Rhein, Philipp von Zabern (Sonderschriften des Deutschen Archäologischen Instituts Kairo 8).

Vandier, J. 1964. Manuel d'archéologie égyptienne IV. Bas-reliefs et peintures. Scènes de la vie quotidienne. – Paris, Éditions A. et J. Picard.

Vernus, P. & J. Yoyotte. 2005. Bestiaire des Pharaons. – Paris, Perrin.
von den Driesch, A. 1976. A Guide to the Measurement of Animal Bones from Archaeological Sites. – Cambridge, Peabody Museum Press (Peabody Museum Bulletins 1).
Yom-Tov, Y., H. Mendelssohn & C.P. Groves. 1995. *Gazella dorcas*. – Mammalian Species 491: 1-6.
Winlock, H.E. 1924. The Metropolitan Museum of Art: the Egyptian Expedition 1923-1924. - New York, The Metropolitan Museum of Art (The Metropolitan Museum of Art Bulletin 19, 12): 5-33.

Source	No	Sexe	Anatomie
Muséum national d'histoire naturelle Paris	1932/5237	mâle	crâne
Muséum national d'histoire naturelle Paris	1916/22	mâle	crâne
Muséum national d'histoire naturelle Paris	1995/3153	mâle	crâne
Muséum national d'histoire naturelle Paris	1930/69	mâle	crâne
Muséum national d'histoire naturelle Paris	1930/252	mâle	crâne
Muséum national d'histoire naturelle Paris	1930/247	mâle	crâne
Muséum national d'histoire naturelle Paris	1930/256	mâle	crâne
Muséum national d'histoire naturelle Paris	1893/508	mâle	crâne
Muséum national d'histoire naturelle Paris	1974/113	mâle	crâne
Muséum national d'histoire naturelle Paris	?	mâle	crâne
Muséum national d'histoire naturelle Paris	1908/534	mâle	crâne
Muséum national d'histoire naturelle Paris	1914/695	mâle	crâne
Muséum national d'histoire naturelle Paris	1960/3615	mâle	crâne
Muséum national d'histoire naturelle Paris	sans no	mâle	crâne
Muséum national d'histoire naturelle Paris	2004/304	mâle	crâne
Muséum national d'histoire naturelle Paris	1930/244	mâle	crâne
Muséum national d'histoire naturelle Paris	1936/517	mâle	crâne
Muséum national d'histoire naturelle Paris	1930/482	mâle	crâne
Muséum national d'histoire naturelle Paris	sans no	mâle	crâne
Muséum national d'histoire naturelle Paris	1930/243	mâle	crâne
Muséum national d'histoire naturelle Paris	1930/255	mâle	crâne et mandibule
Muséum national d'histoire naturelle Paris	1960/3689	mâle	crâne et mandibule
Muséum national d'histoire naturelle Paris	?	mâle	crâne et mandibule
Muséum national d'histoire naturelle Paris	1957/478	mâle	crâne et mandibule
Muséum national d'histoire naturelle Paris	1909/491	mâle	crâne et mandibule
Muséum national d'histoire naturelle Paris	1930/482	mâle	crâne et mandibule
Muséum national d'histoire naturelle Paris	1930/245	mâle	crâne et mandibule
Muséum national d'histoire naturelle Paris	1931/740	mâle	crâne, fémur, pied
Muséum national d'histoire naturelle Paris	1961/218	mâle	squelette complet

Source	No	Sexe	Anatomie
British Museum, London	1935A	mâle	squelette postcrânien
British Museum, London	1935B	mâle	squelette postcrânien
British Museum, London	1935C	mâle	squelette postcrânien
British Museum, London	1935D	mâle	squelette postcrânien
British Museum, London	1939C	mâle	squelette postcrânien
Institut des Sciences naturelles, Bruxelles	8596	mâle	squelette postcrânien
Laboratoire de paléontologie, Gand	3700	mâle	squelette postcrânien
Muséum national d'histoire naturelle Paris	1961/997	femelle	crâne
Muséum national d'histoire naturelle Paris	1930/242	femelle	crâne
Muséum national d'histoire naturelle Paris	1911/1757	femelle	crâne
Muséum national d'histoire naturelle Paris	1930/250	femelle	crâne
Muséum national d'histoire naturelle Paris	1925/13	femelle	crâne
Muséum national d'histoire naturelle Paris	1910/252	femelle	crâne
Muséum national d'histoire naturelle Paris	1912/415	femelle	crâne
Muséum national d'histoire naturelle Paris	1931/740	femelle	crâne et squelette
Muséum national d'histoire naturelle Paris	1930/232	femelle	crâne et mandibule
Muséum national d'histoire naturelle Paris	1930/231	femelle	crâne et mandibule
Muséum national d'histoire naturelle Paris	1930/230	femelle	crâne et mandibule
Muséum national d'histoire naturelle Paris	1953/224	femelle	crâne et mandibule
Muséum national d'histoire naturelle Paris	1948/55	femelle	crâne et mandibule
Muséum national d'histoire naturelle Paris	1918/49	femelle	crâne et mandibule
Muséum national d'histoire naturelle Paris	1952/103	femelle	crâne et mandibule
Muséum national d'histoire naturelle Paris	1952/103	femelle	crâne et mandibule
Muséum national d'histoire naturelle Paris	1908/175	femelle	crâne et mandibule
Musée de l'Afrique, Tervuren	31219	femelle	squelette postcrânien
Laboratoire de paléontologie, Gand	3706	femelle	squelette postcrânien
Laboratoire de paléontologie, Gand	3707	femelle	squelette postcrânien
Laboratoire de paléontologie, Gand	3708	femelle	squelette postcrânien

Annexe: Table 2. Corpus comparatif de *G. dorcas* du Muséum National d'Histoire Naturelle de Paris et d'autres sources.

Did Egyptians Eat Donkeys? Reflections from Historical and Archaeological Data

Mathilde Prévost & Joséphine Lesur

Introduction

When one looks at the scenes of parades of domestic animals taken as booty or bred by an institution or a private owner (*e.g.* Figure 1), one realizes that most of the species that are represented were eaten by Egyptians, even when it was not their main function: cattle was bred for its workforce, as well as for its milk and its meat; ovicaprines, for their milk, meat and wool; pigs for their meat (Boessneck, 1988: 67, 72-73, 74-75, 77; Yokell, 2004: 6-7). The only species for which this dietary use is not well established is the donkey.

Indeed, we know that wild asses were hunted from the Predynastic period (Brunner-Traut, 1975: 27; Linseele & Van Neer, 2009: 61, 63, 68; Vandenbeusch, 2010) and probably eaten. On the other hand, domestic donkeys, which probably appeared from the 5th millennium BC at the latest[1], were the main beast of burden in Egypt with oxen (Köpp-Junk, 2015: 107), since horses, which appeared in the Second Intermediate Period, were used to pull chariots in warfare and parades (Boessneck, 1988: 79-81; Köpp-Junk, 2015: 107; Veldmeijer & Ikram, 2018), and camels were apparently not commonly used in Egypt until the 7th or 6th century BC, if then (Agut-Labordère, 2018: 179-180, 181, 183; Köpp-Junk, 2015: 116). One could suppose that donkeys could be eaten when they were too old or too tired to work. However, there is very little data showing that they were eaten by Egyptians (there is no mention of this possible use, for example, in Brunner-Traut, 1975; Ikram, 1995 or Vernus, 2005), so that it is sometimes assumed that there was a taboo prohibiting the consumption of their meat.

Figure 1. Cattle, donkeys, ovicaprines and pigs in the tomb of Paheri (El-Kab, 18th Dynasty). From Tylor & Griffith (1894: pl. III).

1 The first supposed domesticated donkeys are from the site of Omari in the Nile Delta, dated from the 5th millennium (Boessneck & von den Driesch, 1990: 107).

In: Ikram, S., J. Kaiser & S. Porcier. Eds. 2021. *The Ancient Egyptians and the Natural World. Flora, Fauna, and Science.* – Leiden, Sidestone Press: 213-222.

In this paper, we want to examine if Egyptians ate donkeys during the Late Predynastic and pharaonic times, or if there might have been a ban against this practice. Because of the paucity of data, we will use all types of sources that are available, either historical or zooarchaeological.

We will first summarise the reasons why it is believed that Egyptians did not usually eat donkeys and see if it is reasonable to use the concept of taboo. Then we will present the evidence of donkey consumption, after which we will evaluate to what extent donkey eating could have occurred.

Donkey Eating: Absence or Taboo?

As mentioned above, there is very little data suggesting that the Egyptians used donkeys as a meat source: in texts and iconography, there is hardly any mention or representation of domestic donkeys being killed or eaten. For example, butchery scenes in tombs mostly show cattle being slaughtered (*e.g.* Figure 2; Vandier, 1969: 132-186, 234-250, 283-296). It could be argued that the other species - sheep, goats and pigs - are almost never present in this iconographic theme, or are rarely represented at all, but they are numerous in dietary remains. A case in point is that of the pig, which is almost absent in iconography but well attested in archaeology, because it was the main meat eaten by the inferior classes (Bertini & Cruz-Rivera, 2014; Moreno Garcia, 1999: 251-254, 256; Volokhine, 2014: 84-87; 2016: 101) and it was even consumed by the elites (Rossel, 2006). This is the opposite with the donkeys, which are present in the iconography but are poorly represented in faunal assemblages. This is usually the case for the species that are rarely eaten but used for transportation or work (Marshall, 2007: 379).

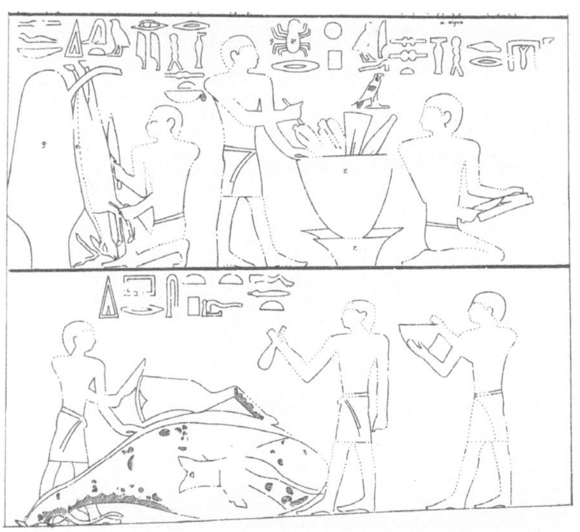

Figure 2. Facsimile of a wall scene from the tomb of Ibi (Deir el-Gebrawi, 6th Dynasty), with a bull being slaughtered and of a goat being cut up. From Davies (1902: pl. IX).

Therefore, it would seem that Egyptians did not usually eat donkeys, as it is the case in some African societies (Blench, 2000: 339, 342-343; Marshall, 2007: 375). But could donkey eating be prohibited, be 'taboo'? This was/is the case in some societies. For example, donkey flesh is considered impure in the Muslim religion, like the pig or the dog, but also in some Christian and traditionalist communities in Africa (Baroin, 1999: 279-281; Blench, 2000: 343). Let us note nonetheless that such a prohibition can also exist for positive reasons: for example, the Mafa people considers that one of its ancestors was a donkey, hence the prohibition of its consumption (Baroin, 1999: 287).

Another reason why this prohibition was assumed for ancient Egypt is the claim by classical sources that the donkey was impure for the Egyptians. A major source for this idea is in Plutarch's *On Isis and Osiris* (2nd century AD). In Chapter 30, the Greek essayist writes that "[the Egyptians] *think that the donkey is a, not pure but demoniac, animal because of its resemblance to him [Typhon/Seth]*" (τὸν ὄνον **οὐ καθαρὸν** ἀλλὰ δαιμονικὸν ἡγοῦνται ζῷον εἶναι διὰ τὴν πρὸς ἐκεῖνον ὁμοιότητα*)*. In Chapter 31, he asserts that this association between the donkey and 'Typhon' was "*as much due to its stupidity and lust as to the colour of its hair*" (διά τὴν ἀμαθίαν καὶ τὴν ὕβριν οὐχ ἧττον ἢ διὰ τῆν χρόαν).

This led some authors to think that it could not be eaten by Egyptians who wanted to remain pure, and to have contact with the divine world, especially if they were working as priests. For example, Smith writes in *The Domestication and Exploitation of Plants and Animals* (Smith, 1969: 307-314), that, in Egypt, pigs and donkeys "*were considered 'Typhonian', but were normally avoided even for sacrifices: eating them was tabued*".

However, this statement implies two pre-suppositions: first, it suggests that the donkey was associated with the god Seth during the entirety of Egyptian history and that this association had a negative connotation. But the association between the donkey and Seth, which is attested from the Middle Kingdom (see *e.g.* dramatic papyrus of the Ramesseum: Griffiths, 1980: 76-77, 163-164; Lorand, 2009: 118-119), and maybe even from the Old Kingdom (see *e.g.* Griffiths, 1980: 75), becomes more common only during the 1st millennium BC, when the murderer of Osiris is reduced to a strictly negative figure. There is no evidence that, in the previous millennia, the donkey was systematically and exclusively associated with Seth and chaotic forces (Te Velde, 1967; Vernus, 2005: 468-470).

On the other hand, Smith's statement (1969: 307-314) implies that taboos, in other terms prohibitions for an entire society for religious reasons or as a sign of belonging (Volokhine, 2016: 91-92), existed in Egypt. Yet the historian and anthropologist Volokhine (2016: 96-104) pointed out that there was no such dietary taboos in ancient Egypt, and that there could only be prohibitions for priests in practice, and only in some contexts, to make sure no impurity affected the contact with the divine:

> "*En aucun cas, un dogme général sur l'alimentation et ses interdits n'est proféré par la classe sacerdotale: tout dépend des circonstances, de la nature des personnes concernées (les prêtres sont en effet les premiers visés), ainsi que de celle des lieux. Car c'est bien le temple qui est le lieu par excellence des interdits*" (Volokhine, 2016: 104).

It is only in the ritual context that animals such as pigs, sheep and fishes could be designated as impure in theological and ritual discourses (Volokhine, 2016: 102).

In light of these preliminary considerations, we have now two questions to answer. First, are there ritual contexts where donkey meat consumption is known or, on the contrary, was prohibited? And secondly, as there was according to Volokhine no dietary ban for all Egyptians, did they ever eat donkeys and in which context(s)? We will start with the question about donkey consumption in daily life.

Daily Context: Did People Eat Donkeys?

It is difficult to find any evidence of donkey eating in daily life. In iconography, as noted above, there are no images showing donkeys being killed or eaten in non-ritual or quotidian contexts.

In archaeology, donkey bones are virtually absent from dietary remains, and, when some are found, their fragmentary state usually prevent from telling if those animals were eaten (*e.g.* Redding, 2016: 175-176, 179-180 table 5.12, 202-203, table 5.17).

In addition, one must deal with the difficulty of distinguishing wild asses from domestic donkeys. The size is often used as a criterion, with the idea that domesticated donkeys are smaller than their wild relatives. However, this size diminution is a very long process, all the more since it is believed that wild males were captured for a long time to copulate with domestic females and to transmit their characteristics to their offspring. Consequently, the differences between wild and domestic donkeys are almost impossible to detect, especially during the Predynastic and the beginning of the pharaonic times (Lesur, 2010: 42; 2018: 66-67; Rossel *et al.*, 2008: 3716, 3719). For example, the morphology of the ten Early Dynastic donkeys found buried in Abydos is a mix of domestic and wild characteristics, bone pathologies being the only proof that they were used as beasts of burden (Rossel *et al.*, 2008).

Nevertheless, some donkey bones are found with characteristic marks (see Table 4). Cut marks, which are the best way to prove that animals were eaten, are rarely detected (*e.g.* for Predynastic sites: Boessneck & von den Driesch, 1990: 107; 1992: 98, 103; Lesur, 2018: 63, 66; for Dynastic sites: Lesur, 2015: 36-37, 42-43), but there are exceptions. For example, at the Predynastic site of Maadi, dated from the first half of the fourth millennium, among more than one hundred donkey bones, "*numerous characteristic cut marks on the bones*" were noted (Boessneck *et al.*, 1989: 122). But the zooarchaeologists could not determine if those bones belonged to domestic or wild donkeys, or both.

They are also sometimes burnt remains, but they do not necessarily indicate cooking. For example, the two found in Tell el-Iswid can be only burnt waste (Lesur, unpublished data).

However, there is at least one site where the consumption of domestic donkeys is certain: it is the site of Ayn Sukhna, on the Red Sea shore, that was used as a base for expeditions to Sinai from the Old Kingdom to the New Kingdom ('Abd el-Raziq *et al.*, 2016; Tallet, 2015). In the faunal collection of the Old Kingdom levels, the donkey is the third most numerous species, after cattle and caprines; it seems that they were all imported on the shore from the Nile Valley. Several dozen domestic donkey remains exhibit cut marks and traces of burning, which strongly suggests the consumption of the species. This idea is reinforced by the skeletal parts distribution, as it shows a majority of long bone fragments that hold the main fleshy parts (Lesur, unpublished data).[2]

Lastly, there is hardly any text suggesting that this practice was ordinary. The Papyrus Chester Beatty 3 (Papyrus British Museum EA 10683), a manual for interpreting dreams written in the 19th Dynasty, contains several dreams dealing with eating an animal. One of them mentions a donkey (rto 2.21): "(if a man sees himself) eating donkey meat: good; it means his (social) elevation" *(jr m33 sw s m rsw.t) ḥr wnm jwf n ꜥ3: nfr; sꜥ3=f pw*) (Bresciani, 2005: 59; Gardiner, 1935: I, 12; II, pl. 5-5A). However, all the other eaten animals are not usual meat sources: there are allusions to the bull, later in the text (rto 7.5) (Bresciani, 2005: 75), but also, twice, to the crocodile (rto 2.22 and 5.17) (Bresciani, 2005: 59; Gardiner, 1935: I, 12; II, pl. 5-5A). And the interpretation of those dreams does not help indicate in which category donkey meat lies, since all of them are positive. Consequently, the Papyrus Chester Beatty 3 cannot be counted as a proof that donkey eating was normal. The same problem can be identified for the first formula of the Lahun gynaecological papyrus (Petrie Museum, UC 32057), where a sick woman must eat a fresh donkey liver (*cf.* infra): as it is a medical recommendation,

2 There is evidence of donkey bones bearing cut marks from the Graeco-Roman era site of Marea at Egypt's north coast (S. Ikram & L. Bertini, personal communication).

it cannot be seen as a proof that donkey livers were usually eaten by Egyptians.

We nevertheless detected one text that may mention an action of donkey eating: it is an administrative document from Deir el-Medina, more precisely a Journal of Accounts, written during the 20th Dynasty on the fragmentary Papyrus Turin 2084+2091; among events such as the bringing of food or business dealt with the vizier, is written the following (II.8-11):

sẖ3w r p3 6 ꜥ3.w n p3 ḫr, jw=tw šd=w m [...]
wnn bn se m-dj Pr-ꜥ3 ꜥ.w.s. p3y=w nb, ḫr bn st f3y [...]
jw=tw šsp=w m nbw, jw wnm=w r-tnw [...]
jw=tw jn=tw jdr jḥ.w

"Memo about the six donkeys of the Tomb, which were taken from [...] Since they were not in the possession of Pharaoh, l.p.h., their master (or owner?), and they were not carrying [...], (then) they were bought (lit. received for gold), and they were eaten every [...] And a herd of cows was brought." (Kitchen, 1983: 604.4-7; transl. after Kitchen, 2012: 440; on the construction *wn/jw*: Neveu, 1996: 188-181).[3]

Despite the poor state of the papyrus, it is clear that the donkeys are the alluded food here. It seems that they were the propriety of Pharaoh, probably at the disposal of the workers of Deir el-Medina, and that the latter bought some unemployed animals in order to be able to eat them.

To this day, this is the only text dealing with donkey eating that we have found. Since the action takes place at the end of the New Kingdom, during the first eight years of the reign of Rameses IX, it is tempting to suggest that eating donkeys was not usual, that the reported case was an exception, and that the workers were driven by scarcity to eat equids, even if we found no sign of such a crisis in the rest of the, yet very incomplete, document.

In conclusion, according to archaeological and historical data, it seems that donkey eating was not an ordinary practice, hence the absence of this action in texts and pictures and of this species in the dietary remains (Lesur, 2010: 41). It is probable that village donkeys, when they died, were carried away from the community site, to keep putrefaction odours and carnivores away from the village. Nonetheless, consumption of donkeys could occur, particularly in contexts where the more usual meat sources were scarce. One can assume that the donkeys in Ayn Sukhna were part of the caravans between the Nile Valley and the Sinai, and that they were eaten, at least during the Old Kingdom, because they were hurt or too tired to complete a trip, and because the members of the expeditions did not want to waste that potential source of meat. Similar practices were evidenced for the Eastern Desert of Egypt in the Graeco-Roman Period (Leguilloux, 2018), and are still witnessed in West Africa (Blench, 2000: 343). It cannot be proved, however, if scarcity drove workers of Deir el-Medina to eat six donkeys at the end of the New Kingdom, although this is quite plausible.

If donkeys could be eaten in some daily contexts, could they be consumed in some ritual contexts as well?

Ritual Context: Could Egyptian Gods, Dead People, and Priests Eat Donkeys?

By ritual contexts, we mean contexts where there is contact with the divine. These can include cases where donkey meat was offered to gods or deceased people, and then sometimes consumed by priests, but also where humans ate a donkey body part for a magical or mythological reason.

It is mostly among pictures and texts that we find allusions to such uses. Indeed, in archaeology, we find the same difficulties of interpretation as for the daily context: on the one hand, donkey bones are not often found in archaeological ritual contexts (temples, tombs, ritual deposits) (*e.g.* Rossel, 2006). And, on the other hand, when there are donkey bones in such contexts, they are not easy to interpret. For example, in the site of Elephantine, less than forty donkey remains were identified in the temple of Satet, dated between the Early Dynastic period and the New Kingdom (Hollmann, 1990: table 3). Among the Early Dynastic remains at the site, Hollmann identified a burnt metatarsus of a wild donkey. However, as noted above, the distinction between wild and domestic donkeys is difficult to determine for that period, so the wild origin of this equid[4] is questionable. Additionally, a burnt bone does not necessarily indicate that is was cooked. Furthermore, one can hardly determine the function of the animals whose remains were exhumed: some of them were found, with bones of other species, in a pit in the temple court, so that they could be interpreted as the remains of offerings (Boessneck & von den Driesch, 1982: 91). But others were found in the residential and storage areas (Boessneck & von den Driesch, 1982: 91), and the reason of their presence there is more equivocal: were there intended as food for people or for the gods? Or were they the remains of beasts of burden that were used for temple work, and were disposed of after their death and were brought in the temple area by carnivores after the site was abandoned?

3 We thank Dr. Chl. Ragazzoli for her suggestions which helped us improve our translation of this text. Helck (2002: 492) translates: *"sondern man nahm sie gegen Geld entgegen und benutzte sie jedes Jahr"*, as such rejecting the idea of eating. But the word, according to the transcription realized by Černy in his Notebooks and used by Kitchen, can hardly, in our opinion, be anything else that *wnm*.

4 Boessneck & von den Driesch (1982: 91) attributed wildness with certainty only to one distal part of the femora found in the residential area of the temple of Satet.

Figure 3. Offering procession for the prince and vizier Nikaoura (Giza, 4th Dynasty). From Lepsius (1972: II, pl. 15b).

Indeed, young as well as adult specimens were identified (Boessneck & von den Driesch, 1982: 91).[5]

Finally, even when it is established that an animal was sacrificed, it is difficult to prove that it was intended as food for the gods, as donkeys do not appear in offering lists. They might be otherwise offered as a way of showing the victory of order (*maat*) over chaos (Sethian forces). Thus, the wild and domestic ass remains in Early Dynastic and Dynastic tombs, like those of Hierakonpolis (Van Neer *et al.*, 2004: 106-107) or Elephantine (Hollmann, 1990: table 5), or in ritual deposits (*e.g.* Ikram, 2004: 42 fig. 2, 45; 2011: 365, 372; Moustafa, 1964: 260-262) could be, or not, seen as some form of food or blood offering for the divine or the dead. For example, Ikram (2004: 46) suggested that an Old Kingdom deposit of animal bones including those of donkeys could symbolize a victory over the forces of chaos. For the deposit found in a 19th Dynasty tomb, she does not exclude the possibility of pets or beasts that were ridden (Ikram, 2011: 365). Consequently, it is very difficult to interpret faunal remains as food for the divine as the contexts are rarely clear.

In iconography, the authors found only one picture that almost certainly alludes to an offering of a donkey as food for the other world. It comes from the tomb of Nikaoura, son and vizier of king Khafra (4th Dynasty): in the classic scene with animals and offerings being brought before the deceased, one can see a fattened bull, a fattened hyena and, according to the facsimile published by Lepsius (1972: II, pl. 15b; see Figure 3), a donkey. The inscription above the latter is lost but an *n* remains, reminiscent of the word *rn*, "young" (Erman & Grapow, 1971: 429.1-5), which usually accompanies animals that are about to be sacrificed.

Thus, this picture supports the idea that donkeys could be included in meat offerings for the deceased, at least during the 4th Dynasty, a time when more exotic and wild animals were part of the offering list, like hyenas (Ikram, 2001: 130-131, 135-138, table 2). Yet one can wonder why it is the only known representation of this ordinary animal in a procession of animals being offered to the dead.[6] On the other hand, one can assume that this donkey, after being offered to the dead prince, was, like any other dietary offering, consumed by funerary priests. But this point, also, can be discussed: to take a similar example, the hyena is represented in some Old Kingdom chapels being force-fed or brought before the deceased, probably to be sacrificed, but its slaughtering is never shown, and there is no evidence of its consumption by people before the New Kingdom (Ikram, 2001; Legge, 2011; Linseele & van Neer, 2009: 68-69, 72).

One text, from many centuries later, could allude to donkeys as offerings to gods but, once more, it is difficult to be affirmative on this point. It is the decree of Nauri established by Seti I: the king forbids that "any supervisor of bulls (*jmy-rꜣ kꜣ.w*), any supervisor of keepers (*jmy-rꜣ mnj.w*), any keeper of the temple of Menmaatre 'Happy is the heart in Abydos' take (jṯꜣ) a bull, a donkey, a pig, a goat from the temple of Menmaatre 'Happy is the heart in Abydos' [to sell it to] another, or to do that it is given (*mꜣ*ʿ) to another god, so that it is not given (*mꜣ*ʿ) to Osiris its master [...]" (l. 58-61) (Griffith, 1927: 202-203, pl. 37-43; Kitchen, 1971: 54.16-55.3; transl. after Kitchen, 1993: 47 (§18)).

The verb *mꜣ*ʿ generally means 'to give' (Erman & Grapow, 1971: 22.8-23.6). More specifically it also can mean 'to offer', 'to sacrifice' (Erman & Grapow, 1971: 22.5-7). It is this last meaning that is usually chosen to translate this text (Griffith, 1927: 203; Kitchen, 1993: 47 (§18); Hafemann, accessed on 14/04/2020). If this choice is correct, it would mean all those species (cattle, but also goats, pigs and donkeys) were to be sacrificed to

5 Most of them were under three and a half, and not only those from the temple and the cemetery (Hollmann, 1990: 72-73).

6 One cannot exclude the possibility of a mistake in the facsimile. Unfortunately, we were not able to check the picture *de visu*.

the god Osiris. Nevertheless, it is not impossible that all those animals were attributed to the temple for diverse purposes: the ones could be sacrificed to the god and the others could be used to provide incomes, with their milk and skins (cattle, caprines), or work (cattle, donkeys).

To summarize, only one image and a handful of texts may indicate that donkeys could be sacrificed to the gods or dead people; unfortunately, they are rarely unambiguous. The only clear occurrence is from the Ptolemaic Period, in a letter from the temple of Elephantine, where the meat of ten donkeys is being prepared for the sacred falcon (Meeks & Favard-Meeks, 1993: 200; Zauzich, 1978, n°13547, pl. 13547).

As for ritual consumption of donkey meat or body parts by humans, the Lahun gynaecological papyrus, from the 12th Dynasty, gives us one clear example of its existence: indeed, in the first formula, a woman with painful eyes must eat a fresh donkey liver, probably for some empiric or symbolic reasons which are not yet identified: "Examination [of a woman whose eyes] are aching so much that she cannot see, on top of aches in her neck: you should say about it: 'it is discharges of the uterus in her eyes'. You shall treat it by fumigating her [...]. You should have her eat a fresh donkey liver *(rd.ḫr=k wnm=s mjs.t n(y).t ʿ3 w3ḏ.t).*" (transl. Collier & Quirke, 2004: 58).

One can finally note that there were also symbolic consumptions of a donkey. On the one hand, there are occurrences where someone eats or drinks something that was named after the animal. For example, in the Papyrus Ebers, from the 18th Dynasty, in a formula dealing with getting rid of *gḫ*-parasites (Ebers 334 (55.16-19)), the sick must drink a beverage made with several ingredients, one of them being a plant named after a donkey body part, the *pḫ.t-ʿ3* (Bardinet, 1995: 302.). Since the word is written with a phallic determinative, the *pḫ.t* could allude to a donkey penis. Probably this item was chosen because this body part was supposed to have some properties and, consequently, also the plant named after it.[7]

On the other hand, humans could eat or drink an object that had the shape of a donkey or of a donkey body part. For the pharaonic period, the only example we found is not about humans, but a cat being fed with a cake in the form of a erect donkey phallus, in the formula 38 of the medical Papyrus British Museum EA 10059: "Spell to be said on the erected phallus <of a donkey> in *dp.t*-bread (*ḏd mdw ḥr ḥnn <ʿ3> r(w)d(w) m dp.t*). It shall be made with the name of [that]

Sites with donkey remains	Period	Context	Cut marks	Burnt marks
Omari	Predynastic	Daily		
Tell el-Iswid	id.	Daily		X
Maadi	id.	Daily	X	
Ibrahim Awad	id.	Daily		
Hierakonpolis	id.	Ritual		
Elephantine	Early Dynastic	Ritual		X
Wadi el-Jarf	Old Kingdom	Daily		
Tell el-Fara'in	id.	Daily		
Dakhla Oasis	id.	Daily		
Abadiya	id.	Daily		
Balat	id.	Daily		
Saqqara	id.	Ritual		
Elephantine	FIP-MK	Ritual		
Kom el-Hisn	Middle Kingdom	Daily		
Mersa Gawasis	id.	Daily		
Ayn Sukhna	id.	Daily	X	X
Elephantine	MK-SIP	Ritual		
Qantir	New Kingdom	Daily		
Elephantine	id.	Daily		
Kom Firin	id.	Daily		
Kom Rebwa	id.	Daily		
Saqqara	id.	Ritual		
Kom Firin	Late Period	Daily		
Tell el-Fara'in	id.	Daily		
Saqqara	id.	Ritual		

Table 4. Table with marks on domestic donkey remains as mentioned in literature (excluding complete inhumed complete skeletons) (Bard *et al.*, 2013: 542; Bertini, 2014: 307 tabl. 1; Bertini & Lindseele, 2011: 282, 283 tabl. 29; Boessneck & van den Driesch, 1982: 9-10, 13, 50 tabl. 15, 91, 137, 140; 1988: 117-122; 1990: 100, 107; 1992: 97-109; 1997: 206-216; Bökönyi 1985: 495 tabl. 1, 497, 498 tabl. 2; Churcher, 1982: 106-107 tabl. 1; Groves, 1986: 37, 53; Hollmann, 1990: 73; Ikram, 2003: 45, 42 fig. 2; 2011: 365; Lesur, 2014: 276-277, 278 fig. 1-3; 2018: 63 tabl. 1, 66-67; unpublished data (Ayn Sukhna, Wadi el-Jarf, Tell el-Iswid); Pantalacci & Lesur, 2012: 296; Redding, 2016: 175-176, 179-180 tabl. 5.12, 202-203 tabl. 5.17; van den Brink, 1992: 43-68; Van Neer, 2002: 532 ; Van Neer *et al.*, 2004: 106-107).

7 For other medical uses of the donkey organs, see Papyrus Ebers 756 (90.4-5) (Bardinet, 1995: 356; Wreszinski, 1913: 185). See also Papyrus British Museum EA 10059: *cf.* infra. On the reputation of the donkey as a hypersexual being see Bohms (2013: 64-66) and Vernus (2005: 464-466). Other plants named after a donkey body part are mentioned in the medical papyri, but they were intended for external application (see *e.g.* Papyrus Ebers 106 (25.15-17), 108 (25.19-20), 770 (92.5-6): Bardinet, 1995: 265, 359; Wreszinski, 1913: 29, 30, 189).

enemy, [the name] of his father, the name of his mother. It shall be put in meat fat, and it shall be given to a cat *(rd(w) n mjy.t)*." (Bardinet, 1995: 489; Leitz, 1999: 68; Westendorf, 1999: 422; Wreszinski, 1912: 197). The mentioned spell invokes the lion-headed goddess Mafdet against the *ꜣʿ*-liquid. As this substance is often evoked as seed (Bardinet, 1995: 121-125), the phallus certainly symbolizes it[8], while the cat represents Mafdet. Thus, the action of giving the bread to a cat clearly aims to parallel the victory of the goddess against the source of danger.

However, cases with humans are known only from the 1st millennium BC through Greek authors: Plutarch, in the 2nd century AD, talks about festivals to Osiris during which Egyptians ate cakes in the form or with an image of a chained up donkey, probably evoking that god's victory over Seth (*De Iside*, §30). He also writes about similar facts for hippopotamus (*De Iside*, §50), and Herodotus, in the 5th century BC, for pigs (Herodotus II.47 [transl. Legrand, 1997]; Closse, 1998: 37).

Conclusion

The authors did not find any explicit evidence for a taboo on consuming donkey meat during the pharaonic period. Generally, donkey remains in the archaeological record are few, and they usually are devoid of butchery or cooking marks (see Table 4). But the absence of a practice does not mean there was a prohibition; it could be merely a cultural avoidance of this practice, like nowadays in the United States of America and in many European countries. Ritual sites where zooarchaeological studies have been carried out are too rare to make any generalisations about this practice. Texts revealed no evidence for banning donkey meat, save from the Classical era. On the other hand, when clearly ritual deposits or remains are found, it is difficult to know if they were intended as food for the gods, or were the result of some other sort of ritual. Only a few textual and images indicate that donkeys might have been offered to gods or deceased, but these are not always clear or dependable (such as the Lepsius drawing).

As for the daily context, faunal remains are also rare. The only relatively clear indication of donkeys that were consumed comes from Deir el-Medina. Therefore, we assume that donkeys, like horses (Meeks, 2005: 53), were usually not eaten by Egyptians, but that they were consumed at least when meat was scarce, such as in Ayn Sukhna. But, clearly, further historical and zooarchaeological data is needed to confidently establish the status of donkeys as food.

8 See note 7 on the sexual connotation of the donkey. There is not necessarily a Sethian connotation here: *contra* Westendorf (1999: 422, Anm. 718).

Bibliography

'Abd el-Raziq, M., G. Castel & P. Tallet. With contributions by V. Le Provost & G. Marouard. 2016. Ayn Soukhna III. Le complexe de galeries-magasins: rapport archéologique. – Cairo, Institut Français d'Archéologie Orientale (Fouilles de l'Institut Français d'Archéologie Orientale 74).

Agut-Labordère, D. 2018. L'introduction du dromadaire dans le désert occidental égyptien au Ier millénaire av. J.-C. In: Tallet, G. & T. Sauzeau. Eds. Mer et désert de l'Antiquité à nos jours : approches croisées. – Rennes, Presses Universitaires de Rennes: 175-195.

Bard, K.A., R. Fattovich & A. Manzo. 2013. The Ancient Harbor at Mersa/Wadi Gawasis and How to Get There: New Evidence of Pharaonic Seafaring Expeditions in the Red Sea. In: Förster F. & H. Riemer. Eds. Desert Road Archaeology in Ancient Egypt and Beyond. – Cologne, Heinrich Barth Institut (Africa Praehistorica 27): 533-556.

Bardinet, T. 1995. Les papyrus médicaux de l'Égypte pharaonique: traduction intégrale et commentaire. – Paris, Fayard.

Baroin, C. 1999. L'âne, ce mal aimé. In: Baroin, C. & J. Boutrais. Eds. L'homme et l'animal dans le bassin du lac Tchad. – Paris, Editions IRD: 277-298.

Bertini, L. 2014. Faunal Remains at Kom Firin. In: Spencer, N. Ed. Kom Firin II: The Urban Fabric and Landscape. – London, The British Museum Press: 306-311.

Bertini, L. & V. Linseele. 2011. Appendix 7: Faunal Report. In: Wilson, P. Ed. Sais I. The Ramesside-Third Intermediate Period at Kom Rebwa. – London, Egypt Exploration Society (Excavation Memoirs 98): 277-285.

Bertini, L. & E. Cruz-Rivera. 2014. The Size of Ancient Egyptian Pigs. A Biometrical Analysis Using Molar Width. – Bioarchaeology of the Near East 8: 83-107.

Blench, R.M. 2000. A History of Donkeys, Wild Asses and Mules in Africa. In: Blench, R.M. & K.C. MacDonald. Eds. The Origins and Development of African Livestock: Archaeology, Genetics, Linguistics and Ethnography. – London, Routledge: 339-354.

Boessneck, J. 1988. Die Tierwelt des Alten Ägypten: Untersucht anhand kulturgeschichtlicher und zoologischer Quellen. – Munich, Beck.

Boessneck, J. & A. von den Driesch. 1982. Studien an subfossilen Tierknochen aus Ägypten. – Munich, Deutscher Kunstverlag.

Boessneck, J. & A. von den Driesch. 1988. Tierknochenfunde vom Tell Ibrahim Awad im Ostlichen Nildelta. In: Brink, van den, E.C.M. Ed. The Archaeology of the Nile Delta, Egypt: Problems and Priorities. – Amsterdam, Netherlands Foundation for Archaeological Research in Egypt: 117-122.

Boessneck, J. & A. von den Driesch. 1990. Tierreste aus der Vorgeschichtlichen Siedlung von El-Omari bei Heluan/ UnterÄgypten. In: Debono, F. & B. Mortensen. Eds. El Omari: A Neolithic Settlement and Other Sites in the Vicinity of Wadi Hof, Helwan. – Mainz am Rhein, Philipp von Zabern (Archäologische Veröffentlichungen 82): 99-107.

Boessneck, J. & A. von den Driesch. 1992. Weitere Tierknochenfune vom Tell Ibrahim Awad im östlichen Nildelta. In: Brink, E.C.M. van den. Ed. The Nile Delta in Transition: 4th.-3rd. Millennium B.C.: Proceedings of the Seminar Held in Cairo, 21.-24. October 1990, at the Netherlands Institute of Archaeology and Arabic Studies. – Tel Aviv, E.C.M. van den Brink: 97-110.

Boessneck, J. & A. von den Driesch. 1997. Tierknochenfunde. In: Way, T. von der. Ed. Tell el-Fara'in; Buto I: Ergebnisse zum frühen Kontext: Kampagnen der Jahre 1983-1989. – Mainz am Rhein, Philipp von Zabern (Archäologische Veröffentlichungen 83): 206-216, pl. XXIII.

Boessneck, J., A. von den Driesch & R. Ziegler. 1989. Die Terreste von Maadi und Wadi Digla. In: Rizkana, I. & J. Seeher. Maadi III: The Non-Lithic Small Finds and the Structural Remains of the Predynastic Settlement. – Mainz am Rhein, Philipp von Zabern (Archäologische Veröffentlichungen 80): 87-128.

Bohms, I. 2013. Säugetiere in der altägyptischen Literatur. – Berlin, Lit.

Bökönyi, S. The Animal Remains of Maadi, Egypt: A Preliminary Report. In: Liverani, M., A. Palmieri & R. Peroni. Eds. 1985. Studi di paleoetnologia in onore di Salvatore M. Puglisi. – Roma, Università di Roma "La Sapienza": 495-499.

Bresciani, E. 2005. La porta dei sogni. Interpreti e sognatori dell'Egitto antico. – Turin, Giulio Einaudi.

Brink, van den, E.C.M. 1992. Preliminary Report on the Excavations at Tell Ibrahim Awad, Seasons 1988-1990. In: Brink, van den. E.C.M. Ed. The Nile Delta in Transition: 4th.-3rd. Millennium B.C.: Proceedings of the Seminar Held in Cairo, 21.-24. October 1990, at the Netherlands Institute of Archaeology and Arabic Studies. - Tel Aviv, E.C.M. van den Brink: 43-68.

Brugsch, H. 1881. A History of Egypt under the Pharaohs: Derived Entirely from the Monuments, to which is added a Discourse on the Exodus of the Israelites. Translated by P. Smith. 2 Vols. – London, J. Murray.

Brunner-Traut, E. 1975. Esel. In: Helck, W. & E. Otto. Eds. Lexikon der Ägyptologie II. – Wiesbaden, Harrassowitz: 27-30.

Churcher, C.S. 1982. Dakhleh Oasis Project – Geology and Paleontology: Interim Report of the 1981 Field Season. – Journal of the Society of the Studies of Egyptian Antiquities 12 (3): 103-114.

Closse, K. 1998. Les ânes dans l'Egypte ancienne. – Anthropozoologica 27: 27-39.

Collier, M., & S. Quirke. Eds. 2004. The UCL Lahun Papyri: Religious, Literary, Legal, Mathematical, and Medical. – Oxford, Archaeopress (BAR International Series 1209).

Davies, N. de G. 1902. The Rock Tombs of Deir El-Gebrâwi, Part 1, Tomb of Aba and Smaller Tombs of the Southern Group. – London, Egypt Exploration Fund (Archaeological Survey of Egypt 11).

Erman, A. & H. Grapow. Eds. 1971. Wörterbuch der Ägyptischen Sprache. Band II (M – H) – Berlin, Akademie-Verlag.

Gardiner, A.H. Ed. 1935. Chester Beatty Gift. 2 vols. – London, British Museum (Hieratic Papyri in the British Museum, III).

Griffith, F.L. 1927. The Abydos Decree of Seti at Nauri – Journal of Egyptian Archaeology 13: 193-206 and pl. 37-43.

Griffiths, J.G. 1980. The Origins of Osiris and His Cult. – Leiden, Brill (Studies in the History of Religions 40).

Groves, C.P. 1986. The Taxonomy, Distribution, and Adaptations of Recent Equids. In: Meadow, R.H. & H.-P. Uerpmann. Eds. Equids in the Ancient World 1. - Wiesbaden, Reichert (Atlas des Vorderen Orients, Reihe A, 19/1): 11-65.

Hafemann, I. Nauri-Felsstele, Abydos-Dekret. In: Thesaurus Linguae Aegptiae. Online: aaew.bbaw.de/tla; accessed on 14/04/2020.

Helck, W. 2002. Die datierten und datierbaren Ostraka, Papyri und Graffiti von Deir el-Medineh. – Wiesbaden, Harrassowitz (Ägyptologische Abhandlungen 63).

Herodotus. Legrand, P.-E. Transl. 1997 [1982]. L'Égypte. Histoires, livre II. – Paris, Les Belles Lettres.

Hollmann, A. 1990. Säugetierknochenfunde aus Elephantine in Oberägypten (8. bis 16. Grabungskampagne, 1978-1987). – Unpublished PhD Thesis, Munich.

Ikram, S. 1995. Choice Cuts: Meat Production in Ancient Egypt. – Leuven, Peeters (Orientalia Lovaniensia Analecta 69).

Ikram, S. 2001. The Iconography of the Hyena in Ancient Egypt. – Mitteilungen der Deutsches Archäeologisches Institut, Kairo 57: 127-140.

Ikram, S. 2004. Typhonic Bones: a Ritual Deposit from Saqqara? In: O'Day, S.J., W. Van Neer & A. Ervynck. Eds. Behaviour Behind Bones: The Zooarchaeology of Ritual, Religion, Status and Identity: Proceedings of the 9th Conference of the International Council of Archeozoology, Durham, August 2002. – Oxford, Oxbow: 41-46.

Ikram, S. 2011. The Animals Remains from Horemheb's Forecourt. In: Raven, M.J., V. Verschoor & R. van Walsem. Eds. The Memphite Tomb of Horemheb, Commander in Chief of Tutankhamun. – Turnhout,

Brepols (Papers on Archaeology of the Leiden Museum of Antiquities 6): 365-374.

Janssen, J.J. 2005. Donkeys at Deir El-Medina. – Leiden, Nederlands Instituut voor het Nabije Oosten (Egyptologische Uitgaven 19).

Kitchen, K.A. 1971. Ramesside Inscriptions, Historical and Biographical. Vol. I. Ramesses I, Sethos I and Contemporaries. Fascicule 2. – Oxford, Blackwell.

Kitchen, K.A. 1983. Ramesside Inscriptions, Historical and Biographical. Vol. VI. Ramesses IV to XI and Contemporaries. – Oxford, Blackwell.

Kitchen, K.A. 1993. Ramesside Inscriptions: Translated and Annotated. Series A: Translations. Vol. I. Ramesses I, Sethos I and Contemporaries. – Oxford, Wiley-Blackwell.

Kitchen, K.A. 2012. Ramesside Inscriptions: Translated and Annotated. Series A: Translations. Vol. VI. Ramesses IV to XI and Contemporaries. – Oxford, Wiley-Blackwell.

Köpp-Junk, H. 2015. Reisen im Alten Ägypten: Reisekultur, Fortbewegungs- und Transportmittel in pharaonischer Zeit. – Wiesbaden, Harrassowitz.

Legge, A.J. 2011. The Hyaena in Dynastic Egypt: Fancy Food or Fantasy Food? – International Journal of Osteoarchaeology 21.5: 613-621.

Leguilloux, M. 2018. L'exploitation des animaux dans les *praesidia* des routes de Myos Hormos et de Bérénice: alimentation, transports et matières premières animales. In: Sidebotham, S., J.-P. Brun, T. Faucher & B. Redon. Eds. Le désert oriental d'Égypte durant la période gréco-romaine: bilans archéologiques. – Paris, Collège de France. Online: http://books.openedition.org/cdf/5175; accessed on 24/03/2021.

Leitz, C. 1999. Magical and Medical Papyri of the New Kingdom. – London, The Trustees of the British Museum (Hieratic Papyri in the British Museum, VII).

Lepsius, R. 1972 [1897]. Denkmäler aus Ægypten und Æthiopien. 7 vols. – Geneva, Éditions de Belles Lettres.

Lesur, J. 2010. La domestication animale en Afrique. – Les Nouvelles de l'Archéologie 120-121: 38-46.

Lesur, J. 2014. La faune de Tell el-Iswid, secteur 1a et 1b. In: Midant-Reynes, B. & N. Buchez. Eds. Tell el-Iswid, 2006-2009. – Cairo, Institut Français d'Archéologie Orientale (Fouilles de l'Institut Français d'Archéologie Orientale 73): 275-280.

Lesur, J. 2015. Exploitation animale à l'Ancien Empire en Égypte: les apports d'Ayn Asil (oasis de Dakhla). – Anthropozoologica 50 (1): 33-47.

Lesur, J. 2018. Fishing and Herding in the Nile Delta during the Predynastic (4th millennium BC). In: Çakirlar, C., J. Chahoud, R. Berthon, S. Pilaar Birch. Eds. Archaeozoology of the Near East XII, Proceedings of the 12th International Symposium of the ICAZ Archaeozoology of Southwest Asia and Adjacent Areas Working Groups, Groningen Institute or Archaeology, June 14-15 2015. – Groningen, University of Groningen: 59-72.

Linseele, V. & W. Van Neer. 2009. Exploitation of Desert and Other Wild Game in Ancient Egypt: The Archaeozoological Evidence from the Nile Valley. In: Riemer, H., F. Förster, M. Herb & N. Pöllath. Eds. Desert Animals in the Eastern Sahara: Status, Economic Significance, and Cultural Reflection in Antiquity: Proceedings of an Interdisciplinary Acacia Workshop Held at the University of Cologne December 14 – 15, 2007. – Cologne, Heinrich Barth Institut: 47-78.

Lorand, D. 2009. Le papyrus dramatique du Ramesseum: étude des structures de la composition. – Leuven, Peeters.

Marshall, F. 2007. African Pastoral Perspectives on Domestication of the Donkey. In: Denham, T., J. Iriarte & L. Vrydaghs. Eds. Rethinking Agriculture: Archaeological and Ethnoarchaeological Perspectives. – Walnut Creek (Calif.), Left Coast Press: 371-407.

Meeks, D. 2005. L'introduction du cheval en Égypte et son insertion dans les croyances religieuses. In: Gardeisen, A. Ed. Les équidés dans le monde méditerranéen antique. Actes du colloque organisé par l'École française d'Athènes, le Centre Camille Jullian, et l'UMR 5140 du CNRS, Athènes, 26-28 Novembre 2003. – Lattes, Éd. de l'Association pour le développement de l'archéologie en Languedoc-Roussillon: 51-59.

Meeks, D. & C. Favard-Meeks. 1993. La vie quotidienne des dieux égyptiens. – Paris, Hachette.

Moreno Garcia, J.C. 1999. "J'ai rempli les pâturages de vaches tachetées..." Bétail, économie royale et idéologie en Égypte, de l'Ancien au Moyen Empire. – Revue d'Égyptologie 50: 241-57.

Moustafa, Y.S. 1964. The Domesticated Animals of the Sekhem-khet Step-Pyramid. – Annales du Service des Antiquités de l'Égypte 58: 255-264.

Neveu, F. 1996. La langue des Ramsès: grammaire du néo-égyptien. – Paris, Khéops.

Pantalacci, L. & J. Lesur. 2012. Élevage et consommation de viande à Balat (oasis de Dakhla). Fin de l'Ancien Empire – Première Période Intermédiaire. – Bibliothèque de l'Institut Français d'Archéologie Orientale 112: 291-315.

Plutarch. Froidefond, C. Transl. 1988 Œuvres Morales. Tome V – 2e partie. Isis et Osiris. – Paris, Les Belles Lettres.

Redding, R.W. 2016. The Vertebrate Fauna from the Excavations at Kom el-Hisn, Giza, and Other Sites. In: Wenke, R.J., R.W. Redding & A. Cagle. Eds. Kom El-Hisn (ca. 2500-1900 BC): an Ancient Settlement in the Nile Delta. – Atlanta, Lockwood Press: 139-205.

Rossel, S. 2006. A Tale of the Bones. Animal Use in the Temple and Town of Wah-Sut. – Expedition 48.2: 41-43.

Rossel, S.F. Marshall, J. Peters, T. Pilgram, M.D. Adams & D. O'Connor. 2008. Domestication of the Donkey: Timing, Processes, and Indicators. – Proceedings of the National Academy of Sciences USA 105: 3715-3720.

Smith, H.S. 1969. Animal Domestication and Animal Cult in Dynastic Egypt. In: Ucko, P.J. & G.W. Dimbledy. Eds. The Domestication and Exploitation of Plants and Animals. Proceedings of a Meeting of the Research Seminar in Archaeology and Related Subjects Held at the Institute of Archaeology, London University, May 18-19, 1968. – London, Gerald Duckworth: 307-314.

Tallet, P. 2015. Les « ports intermittents » de la mer Rouge à l'époque pharaonique: caractéristiques et chronologie. – NeHeT 3: 31-72. Online: https://www.nehet.fr/NEHET3/04-NeHeT%203-02-TALLET-p.31-72.pdf; accessed on 24/03/2021.

Tylor, J.J. & F.L. Griffith. 1894. The Tomb of Paheri at El Kab. – London, The Egypt Exploration Fund/Kegan Paul, Trench, Trübner & Co.

Te Velde, H. 1967. Seth, God of Confusion: A Study of His Role in Egyptian Mythology and Religion. – Leiden, Brill.

Van Neer, W. 2002. Le matériel faunique. In: Midant-Reynes, B. & N. Buchez. Eds. Adaïma I. Économie et habitat. – Le Caire, Institut Français d'Archéologie Orientale (Fouilles de l'Institut Français d'Archéologie Orientale 45): 521-565.

Van Neer, W., V. Linseele & R.F. Friedman. 2004. Burials and Food Offerings at the Elite Cemetery HK6 of Hierakonpolis. In: Hendrickx, S., R.F. Friedman, K. Cialowicz & M. Chlonicki. Eds. Egypt at Its Origins: Studies in Memory of Barbara Adams: Proceedings of the International Conference "Origin of the State, Predynastic and Early Dynastic Egypt", Krakow, 28 August – 1st September 2002. – Leuven, Peeters (Orientalia Lovaniensia Analecta 138): 67-130.

Vandenbeusch, M. 2010. Hunting the Predynastic Donkey. – Nekhen News 22: 14-15.

Vandier, J. 1969. Manuel d'archéologie égyptienne V. Bas-reliefs et peintures. Scènes de la vie quotidienne II. – Paris, A. et J. Picard.

Veldmeijer, A.J. & S. Ikram. Eds. 2018. Chariots in Ancient Egypt: The Tano Chariot: A Case Study. – Leiden, Sidestone Press.

Vernus. P. 2005. Âne. In: Vernus, P. & J. Yoyotte. Eds. Bestiaire des pharaons. – Paris, Perrin: 459-470.

Volokhine, Y. 2014. Le porc en Égypte ancienne: mythes et histoire à l'origine des interdits alimentaires. – Liège, Presses universitaires de Liège (Religions 3).

Volokhine, Y. 2016. À propos des « interdits alimentaires » en Égypte ancienne ». In: Suter, P., Bordessoule, N. & C. Fournier-Kiss. Eds. Regards sur l'interculturalité: un parcours interdisciplinaire – Les Acacias, MétisPresses: 91-108.

Westendorf, Wolfhart. 1999. Handbuch der altägyptischen Medizin. 2 Vols. – Leiden, Brill.

Wreszinski, W. 1912. Der Londoner medizinische Papyrus (Brit. Museum Nr. 10059) und der Papyrus Hearst: in Transkription, Übersetzung und Kommentar (Die Medizin der alten Ägypter 2). – Leipzig, J.C. Hinrichs.

Wreszinski, W. 1913. Der Papyrus Ebers: Umschrift, Übersetzung und Kommentar. – Leipzig, J.C. Hinrichs (Die Medizin der alten Ägypter 3).

Yokell, C. 2004. Modeling Socioeconomic Evolution and Continuity in Ancient Egypt. The Value and Limitations. – Oxford, Archaeopress (BAR International Series 1315).

Zauzich, K.-T. 1978. Papyri von der Insel Elephantine. – Berlin, Akademie-Verlag.

What I Have Learned: Assumptions Bad, Intersections Good

Richard W. Redding

Introduction

I have been doing archaeobiology, specifically archaeozoology and ecological anthropology, for 52 years, and I have been working in Egypt for 39 years. What are the great lessons that I have learned about doing archaeobiology? My greatest epiphany is how important a scholar's training and paradigm is in influencing the questions asked and the research undertaken. I often tell my students that they should read an article like it is a sales brochure – what is the writer trying to sell you? I advise students that they should know from reading the introduction and methods section what questions a researcher see as interesting and the paradigms guiding the researcher. An understanding of the interests and paradigms of a researcher help students evaluate the strengths, or weaknesses, of the arguments being made. In the interest of full disclosure, I will layout my intellectual history, research interests, and paradigms before I continue so that the reader can evaluate my statements about what I have personally learned in 52 years of research in archaeozoology.

I am trained as an anthropological archaeologist; I am not an Egyptologist. I am interested in cultural processes and how culture evolved. I use the material culture of the past to understand and explain the origins and evolution of human culture. I have developed a theoretical framework, a paradigm, in which human subsistence behavior/decisions are based on risk reduction and that a goal of reducing risk has led to major changes in human culture (Redding, 1986; 1993). Since human subsistence decisions are so important, I have focused on archaeozoological and archaeobotanical data. I have attempted to answer two big questions with targeted research: 1) the origins of food production and, 2) the evolution of cultural complexity. My work in Turkey, on the sites of Hallan Çemi and Demirkoy; Iran, at Chogha Bonut; and Egypt, on sites in the Fayyum; focused on the origins and evolution of food production. My work in Iran, Tepe Farukhabad and Tepe Sharafabad; and Egypt, Kom al-Hisn and Giza; focused on the development of cultural complexity. I have tried to tease out the relationship between subsistence decisions and socio-economic infrastructure.

Beyond my epiphany, two personal lessons stand out from my experiences. The first is that assumptions are bad. It is not that assumptions are innately bad: they are necessary for science to advance. The problem is when assumptions become normalized. The second lesson is that intersections of all kinds are good. The more we intersect with other fields of research and other researchers, the more productive, innovative and interesting our research is. I would also argue it becomes more consequential and significant.

In: Ikram, S., J. Kaiser & S. Porcier. Eds. 2021. *The Ancient Egyptians and the Natural World. Flora, Fauna, and Science.* – Leiden, Sidestone Press: 223-232.

Assumptions Bad

> "*Assumptions are made and most assumptions are wrong.*" Attributed to Albert Einstein

> "*Your assumptions are your windows on the World. Scrub them off every once in a while, or the light won't come in.*" Alan Alda (Commencement Address, 1980)

We all make assumptions about the world and assumptions are a pillar of the science. Assumptions can be very useful as they simplify the real world and help to focus research. The question is not if assumptions are being made – they are. The question is what are the assumptions, the need to be made explicit, and what is the consequence of being wrong?

Archaeologists make some big assumptions, like 1) The past is knowable; 2) Material culture, the human made or manipulated remains in the archaeological record, can be used as evidence about the past or, 3) The patterning of material remains in an archaeological site is the result of the patterned behavior of members of extinct societies.

But we make small ones every day that condition our research. I would like to examine three assumptions that have been made about the archaeological record. The origin of wet cooking in the Paleolithic, the use of bone density to explain bias in representation of mammal bones in archaeological samples, and, a very personal assumption I have made, distal, non-meat bearing limb elements were not consumed but were discarded.

The Origin of Wet Cooking

The normative view is that wet-cooking (stewing or boiling) did not appear until the Upper Paleolithic when we find fire-cracked rocks on archaeological sites (Speth, 2015). Fire-cracked rock is evidence of heated rocks being placed in a flammable container (*e.g.* animal hide, baskets, tree bark, and wooden vessels) to heat water and cook meat and vegetables in a stew. In Egypt the appearance of fire-cracked rock is later than in other areas of the world. The earliest evidence of fire cracked rock in Egypt is from Fayyum and Western Sahara sites that date to Saharan Neolithic (8800-5100 BC) (Barich *et al.*, 2012; Holloway *et al.*, 2017; Koopman *et al.*, 2016; McDonald, 2009; McHugh *et al.*, 1989; Nelson, 2001; Wenke *et al.*, 1988). So, do we conclude that stewing and boiling appeared very late in Egypt? Could it have occurred earlier in Egypt and even earlier in the rest of the world?

Why is wet-cooking important? It allows humans access to fats and nutrients in animals that they cannot access with roasting. Hunters and gatherers can access marrow by simply breaking bones, but wet-cooking the bones makes the trabecular marrow and fats and the cartilage and connective tissue associated with the ends of the limb bone accessible.

Bone, and the marrow and fats they contain, is an excellent source, "*of minerals, amino acids, vitamins, healthy fats, collagen, gelatin, and nourishing compounds called Glycosaminoglycans…*" (https://www.precisionnutrition.com/encyclopedia/food/bones). Beef marrow is a high value food resource. Each 100gm contains as much as 84.4gm of fats and 6.7gm of protein. Given the fat content, it is not surprising that 100gm of beef marrow can yield up to 786 kcal. Minerals found in bone include calcium, phosphorus, magnesium and potassium. The minerals obtained in boiling are critical in humans to bone growth and re-growth, nerve impulse transmission, regulating the heartbeat and muscle contractions, cell maintenance, carbohydrate and fat catabolism, fluid balance in the body (https://medlineplus.gov/). Magnesium alone is needed for over 300 biochemical reactions in the body (MedlinePlus).

Cartilage and connective tissue are a rich source of proteins and the collagen when boiled releases amino acids including glycine, proline, and glutamine (https://www.precisionnutrition.com/encyclopedia/food/bones). Glycosaminoglycans (GAGs) are found in cartilage and contain chondroitin sulfate, hyaluronic acid and glucosamine (https://www.precisionnutrition.com/encyclopedia/food/bones). These compounds are used in treatment of osteoarthritis and help in, "*growth and repair of joint tissue*" (https://www.precisionnutrition.com/encyclopedia/food/bones).

Wet cooking of plants would also have impacted human diet and nutrition. It would have increased the availability and digestibility of starches from plants by breaking down cell walls (Speth, 2015: 57).

Could wet cooking have appeared much earlier than the evidence of fire-cracked rock suggests? Speth (2015) has reviewed the ethnographic literature, which has largely been ignored by archaeologists, regarding boiling in flammable containers directly over a heat source. The epiphany for Speth was an episode of Survivorman on the discovery channel, 2008, Season 2, Episode 4, Part 3, Day 3, African Plains. Survivorman, Les Stroud, is a survival expert that films himself living in the wilderness for several days with a minimum amount of equipment.

> "*What caught my attention on that particular occasion was that Stroud had decided to boil polluted water to make it potable, but the only container he had was an empty plastic water bottle. To my utter astonishment, he filled it with water, suspended it over an open fire with the flames licking at the bottom of the bottle, and proceeded to bring the contents to a rolling boil without destroying the container, noting in passing that so long as the portion of the bottle that came in contact with the flames was filled with liquid the bottle might*

blacken and deform somewhat but it would not burn or melt. Stroud needed neither heated stones nor a pit to accomplish this." (Speth, 2015: 58)

I have found a You Tube a video of boiling water in a paper cup that is well worth viewing. (https://www.youtube.com/watch?v=I9gKzea3Cno). The paper cup filled with water is placed over a Bunsen burner. The water boils and the paper cup is only slightly blackened. Water has a very high specific heat due to hydrogen bonding and the water absorbs heat rapidly without the paper cup reaching its kindling temperature. I have boiled water in a plastic water bottle over a flame at the Ancient Egypt Research Associates (AERA) Villa at Giza.

It is certainly possible that wet cooking began much earlier than we have previously thought. Animal skins, wooden vessels, tree bark, and baskets could have been used for wet cooking. The early development of wet cooking probably antedates FCR in Egypt. This is important because of the effects on human diet and nutrition and the range of items that humans could have consumed.

Bone Density as an Explanation of Bias in Archaeozoological Samples

In collections of animal bones from archaeological sites some bones, particularly limb elements, are consistently under-represented in the samples. For example, the proximal humerus, distal femur, and proximal tibia do not occur in the expected proportions. Specifically, we do not find a proximal humerus for each distal humerus; we do not find a proximal tibia for each distal tibia. This pattern has been a focus of taphonomic research in archaeozoology since the work of Brain (1967). Limb bones vary in the thickness of compact bone and in the amount of trabecular (spongey) bone. Even individual limb bones vary along their length and the distal end can be quite different than proximal end. Some bones appear "softer" than others (Gifford-Gonzalez, 2018: 455). Brain (1981) calculated specific gravity of different elements of bone by using water displacement and used these data to explain differential survivorship of bone elements. Since the work of Brian, other researchers have calculated bone densities using x-ray densitometry (Lyman, 1984). These studies have been replaced by using Computed Tomography scanning to estimate bone mineral densities (Lam & Pearson, 2004; 2005; Lam *et al.*, 1998).

All of the studies using bone density or bone mineral density assume that these measures are a proxy for bone survivorship. Indeed, bone density has proven useful and has shown correlations with archaeological and ethnographic faunal remains as well as with non-human mammal accumulations of bone (Marean & Spencer, 1991; Lyman, 1984; 1994; Marean *et al.*, 1992). However, I have concerns about using bone density as a proxy for survivability of animal bone. My concern is based on three facts. First, large areas of cancellous bone function to absorb force and disperse it. A low density in a limb bone end, based on large amounts of cancellous bone in an element, may mask a high ability to resist force. Second, von Meyer (1867) recognized that trabeculae in limb bone ends are not random but highly organized to disperse force. Von Meyer wrote extensively about what he called the architecture of trabecular bone. Finally, researchers on osteoporosis in humans have become dissatisfied with bone density as a measure of susceptibility to fracture and are developing a measure called "bone quality" (Fonesca *et al.*, 2014; Licata, 2009; Saito & Marumo, 2010; Seeman, 2008).

A direct measure of an area, end or shaft, of a limb bone's ability to resist fragmentation may best be found in the mechanical stimulus that fragments it. It is force, acting via human, animal, or environmental/geological actions, that fragments bone and leads to variation from expected representation of skeletal parts in the faunal record that we observe and need to explain. Can we measure the force necessary to fracture an element? How do we describe it? I have begun a research program on the force necessary to break different skeletal elements of the domestic sheep. In cooperation with Andrea Poli of the Department of Mechanical Engineering at the University of Michigan, we have been using the drop tower (Figure 1, 2) in the "Breaker Space". The drop tower is a 2 kg weight enclosed in a plastic tube that can be raised to a predetermined height and dropped onto an object, in our case a limb bone. The limb bone sits on a piezoelectric sensor that sends a record of the force, in Newtons/millisecond, being applied to the area of the limb bone being struck. Each drop was considered a test. The drop tower has been used to fracture limb bone shafts and ends from five sheep. All the sheep were females of the Suffolk breed. Each test produced a curve that tracks the force applied to the bone or element over time (Figure 3). The sudden drop in force corresponds to a fracture. We frequently saw multiple peaks reflecting a complex breakage pattern. The tests were recorded using a high-speed camera so we could associate each break with a peak We used two measures in the analysis. The first is the maximum force, in newtons (N), that is needed to fracture the area being tested. The second is the area under the curve, force over time, referred to as the Impulse (I).

I have compared the data on the maximum force necessary to break an element and the element densities published for sheep by Lyman (1994) and by Ioannidou (2003) in Table 4. The data on impulse from our tests are compared to the published densities by Lyman and Ioannidou in Table 5.

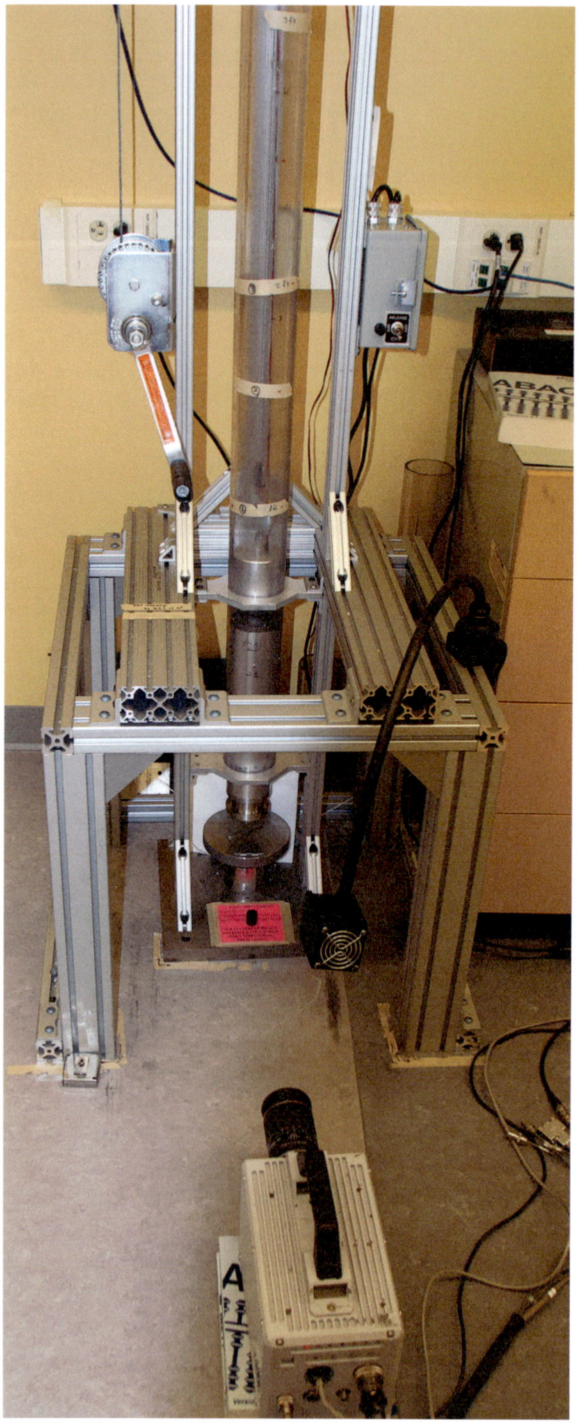

Figure 1. Drop tower in the Breaker lab at the University of Michigan. Photograph by R. Redding.

So, what is the better measure, force or impulse, of an elements ability to resist all the mechanical causes of attrition that affect faunal remains. We used Pearson's r to test the relationship between maximum force and impulse. We found a highly significant correlation (r= 0.659, df=19). We are concerned that the maximum force is a single value that gives an incomplete measure and ignores pressure, or force over time. Given that impulse smooths out some of the extreme variation seen in maximum force, we suggest that it is the better measure as we proceed with future testing. It is a measure of force over time.

We tested the relationship between the published bone densities and the impulse measures using Pearson's r. We found no significant correlation between Ioannidou's values for bone density and impulse (r=-0.261, df=16). We also found no significant correlation between Lyman's density values and impulse (r=0.009, df=18). These results suggest that the assumption that

Figure 2. Set up for test 124 on the drop tower. The sheep first phalanx sits on a the piezo-electric sensor. The weight is lined up to hit the bone mid-shaft. Photograph by R. Redding.

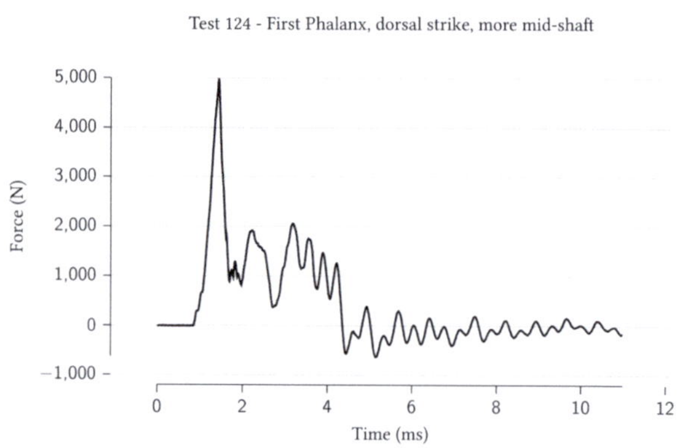

Figure 3. A curve of force (N) over time for test 124 of a mid-shaft strike on a sheep first phalanx. Prepared by R. Redding.

Element	Density Ioannidou (2003)	Density Lyman (1984)	Mean Maximum Force (N)
Radius, shaft	0.9	0.52	4249
Femur, proximal	0.8	0.28	2038
Tibia, Shaft	0.75	0.49	3227
Metacarpal, shaft	0.71	0.67	2186.5
Metatarsal, shaft	0.62	0.68	2292
Humerus, shaft	0.52	0.42	2862
Astragalus	0.46	0.63	9688
Metacarpal, distal	0.37	0.50	1906
Metatarsal, distal	0.36	0.39	5710.5
Tibia, distal	0.34	0.28	5113
Radius, proximal	0.32	0.35	4073.5
Metacarpal, proximal	0.32	0.40	2681
Metatarsal, proximal	0.32	0.43	2336
Humerus, distal	0.31	0.34	3652.5
Ulna, proximal	0.31	0.26	3982
Femur, shaft	0.30	0.36	2043.5
Calcaneum	0.29	0.56	2910
Radius, distal	0.27	0.21	2406
Tibia, proximal	0.17	0.16	2678
Humerus, proximal	0.l5	0.13	892
Femur, distal	0.14	0.22	3460
First phalanx		0.42	2684
Second phalanx		0.39	2087

Table 4. Elements of sheep arranged by density based on the work of Ioannidou (2003). These densities are compared to the densities published by Lyman (1984) and the mean of the maximum force necessary to break the elements from a sample of two sheep.

Element	Density Ioannidou (2003)	Density Lyman (1984)	Mean Impulse (e+03)
Radius, shaft	0.9	0.52	4.1086
Femur, proximal	0.8	0.28	2.3206
Tibia, Shaft	0.75	0.49	2.6359
Metacarpal, shaft	0.71	0.67	1.6323
Metatarsal, shaft	0.62	0.68	1.6786
Humerus, shaft	0.52	0.42	1.5746
Astragalus	0.46	0.63	7.5195
Metacarpal, distal	0.37	0.50	9.0907
Metatarsal, distal	0.36	0.39	8.4594
Tibia, distal	0.34	0.28	2.3848
Radius, proximal	0.32	0.35	6.3699
Metacarpal, proximal	0.32	0.40	1.2204
Metatarsal, proximal	0.32	0.43	1.4413
Humerus, distal	0.31	0.34	6.2451
Ulna, proximal	0.31	0.26	1.8009
Femur, shaft	0.30	0.36	2.6234
Radius, distal	0.27	0.21	6.3121
Tibia, proximal	0.17	0.16	3.4950
Humerus, proximal	0.l5	0.13	3.2959
Femur, distal	0.14	0.22	6.0631
First phalanx		0.42	4.2883
Second phalanx		0.39	3.1542

Table 5. Elements of sheep arranged by density based on the work of Ioannidou (2003). These densities are compared to the densities published by Lyman (1984) and the impulse necessary to break the elements from a sample of three sheep.

bone density is a proxy for the ability of a bone to survive force is not a good assumption.

Shorbet Kawara: Death of an Assumption

During my 52 years in studying faunal remains from archaeological sites, like most researchers, I have developed a system of recording and study that reflects many of my assumptions. One of my assumptions has been that the distal limb elements of mammals, what I called the "non-meat bearing bones" (*i.e.*, the podials, metapodials and phalanges) are not of interest to humans and were discarded in slaughter. I have used the relative occurrence of "non-meat bearing bones" to "meat-bearing bones" (*i.e.*, humerus, radius, ulna, femur, patella, tibia and lateral malleolus) to see if slaughter was local or whether packages of meat were being brought to the area of the site where the fauna was recovered.

In early March 2018, the AERA laboratory at Giza began to receive materials from our re-excavation of the Kromer Dump. The Kromer Dump, first excavated by Kromer in the 1970's (Kromer, 1978), is west of the Heit al-Ghurab site over the Gebel al-Qibli (Figure 6). AERA's excavations in 2018 (Witsell, 2018) explored the extent of Kromer's Dump and sampled previously unexcavated strata. Animal bone from the new excavations was coming into the laboratory in such large quantities that the site crew put the bone in black rubber buckets, guftas, produced from tires that we use on site to move dirt (Figure 7). While my two students, Mohamed Hussein and Mohamed Raouf, and I were laying out and sorting one of the guftas, I commented that the material was unlike anything I had seen from Giza as it was filled with shafts of distal limb bones, many of them

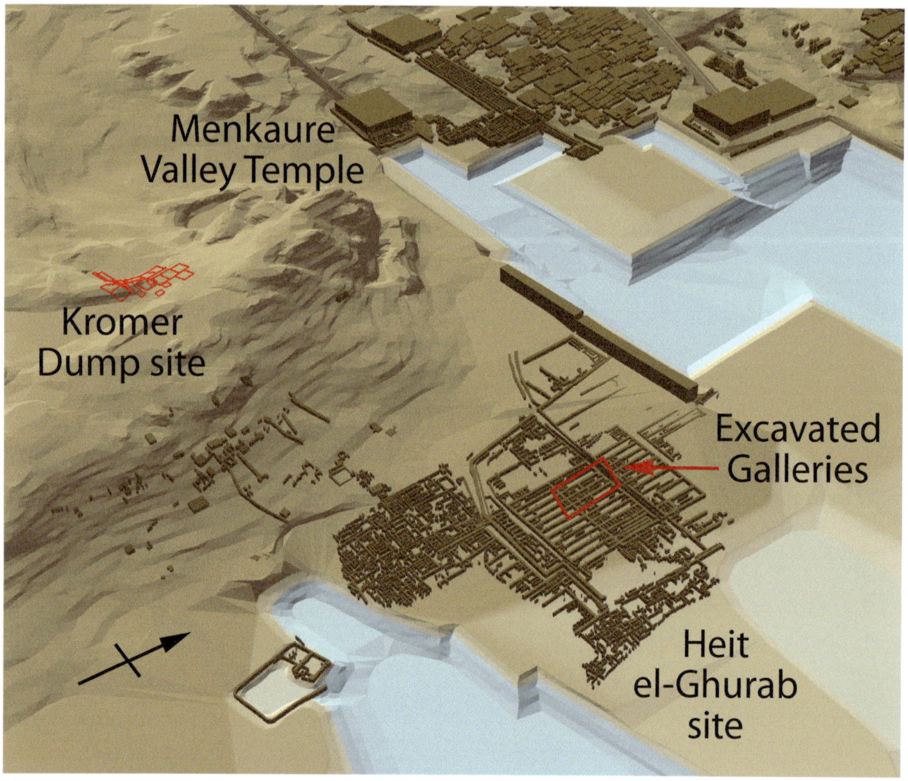

Figure 6. A 3-D map of the spatial relation between the Heit al-Ghurab and the Kromer Dump site. The excavated barracks (Galleries III.3 and III.4) are labeled in the Heit al-Ghurab site. Map prepared by Rebekah Miracle.

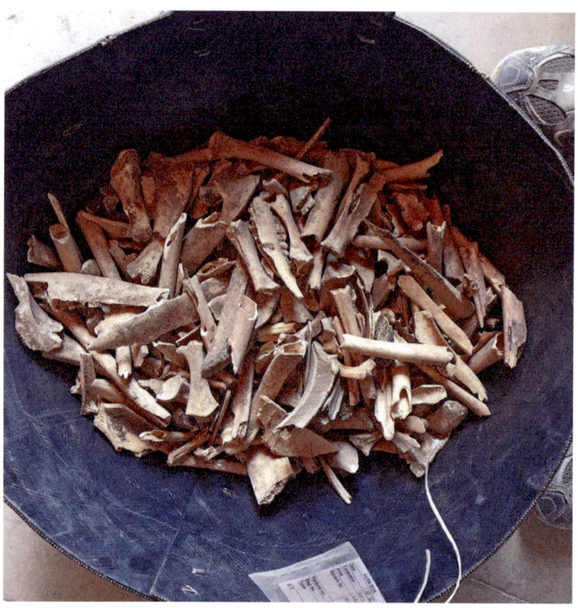

Figure 7. A ***gufta*** filled with bone fragments from AERA excavations at the Kromer Dump in 2018. Note large number of sheep-goat limb bone shafts missing epiphyses. Photograph by R. Redding.

my "non-meat bearing bones". I was puzzled. I asked my two students what they thought and they looked at each other and said, in unison, "*Shorbet Kawara*!"

Shorbet Kawara is a stew or soup made from the feet of cattle, sheep or goats and it is very popular in Egypt at present and considered a hearty meal. The dish is high in calories, 519 kcal/100g, high in protein, 20g/100g, and high in fat, 65.5g/100g (Pellet & Shadarevian, 2013: 20). *Shorbet Kawara*, because it includes boiled collagen, is high in amino acids, including glycine, proline, and Glycosaminoglycans (GAGs). The bone marrow contained in the distal metapodials and the phalanges releases healthy fats Omega 3, an anti-inflammatory, and linoleic fatty acids (Redding, 2020). This makes *Shorbet Kawara* an ideal food for individuals engaged in heavy labor.

Limb elements that I had always assumed were waste from slaughter and not part of the subsistence of the residents whose garbage we were recovering, may actually be an important component of the diet. Perhaps a perfect diet for the laborers who moved the stone to build the pyramids. So, instead of an indication of butchery or disposal, the non-meat bearing bones may indicate status and occupation. Could an assumption that non-meat bearing bones are indicative of consumption, by wet-cooking, of a high calorie, nutrient rich food replace the assumption that non-meat bearing bones are indicators of disposal in my research and analysis paradigm?

First, it is unlikely that workers in the Pharaonic Egypt were eating *Shorbet Kawara*. The residents of the Old Kingdom site of the Heit al-Ghurab at Giza were most likely consuming stews, with meat, fish vegetables all thrown in a large pot (Redding, 2020). Illustrations of cooking in the Old Kingdom almost exclusively show cuts of meat simmering a large pot, with ducks, which are shown being grilled, the only exception (Ikram, 1995; Redding, 2020). Stews are the most efficient way to feed large numbers of people at once, stretches ingredients, and effectively accessing the marrow and fat in animal bones.

I have begun a review of the analyses and publication of the fauna from the Heit al-Ghurab using this new assumption. As an example, I had been puzzled in my initial analysis of the faunal remains from two barracks buildings (Galleries III.3 and III.4; Figure 6) at the Heit al-Ghurab by the occurrence of 217 cattle elements. The reason for my confusion is that cattle are a higher status food item than sheep and goat (Redding, 2010). The barracks, which we assume were occupied by workers and not higher ranking individuals, should not have had much, if any, cattle. When I re-examined the raw data, I found that 90.1% of the cattle fragments were from Kawara elements; metapodials and phalanges. The residents of the barracks were eating sheep and goats, but they supplemented these meat sources with a stew of cattle feet.

Intersections Good

> *"You must go where these very changes are occurring – at the intersection of industries, cultures fields and disciplines"* Frans Johansson (2017)
>
> *"... the action most worth watching is not at the center of things but where the edges meet... shorelines, weather fronts, international borders..."* Anne Fadiman (1998)

The three stories I have presented above, which question assumptions, all have a common thread. The assumption is challenged with new research at the intersections of fields, or as the result of intersections with colleagues who have new or different ideas/knowledge. The exposure to new ideas and data from fields and people outside traditional archaeology will yield insight as research has shown that advances take place at the intersection of fields of study (National Academy of Sciences, 2010).

My research paradigm has at its core the exploration of the intersections between fields of study. I completed an inter-departmental doctoral program in anthropological archaeology and biological sciences. I used theory from biological sciences, specifically evolutionary ecology, in my dissertation to model human decision making in structuring flocks of sheep and goats. I also very quickly learned of the vast literature on unimproved breeds of sheep and goats and how it could be used to create models of human decision making that are applicable to archaeological data. My interest in the data from animal husbandry literature continues to the present. I have just completed a series of herd growth models for Egyptian cattle using data for unimproved breeds; specifically the Baladi breed. An analysis of these models makes it most likely that female cattle reproduced biennially in Pharaonic Egypt (Redding, 2020). Different females would be on different schedules so calves would be born every year, but a two-year cycle of reproduction would be necessary to assess the productivity of cattle herds. Does this have anything to do with the biennial cattle counts?

In my first example on assumptions, John Speth's challenge to the normative view of the antiquity of wet cooking and its dependence upon fire-cracked rock, had its root in a television show on surviving in the wild. He then searched the ethnographic literature and found several examples of wet cooking in flammable containers.

My challenge to the use of bone density started as a graduate student when I saw a longitudinal section of a femur. The trabecular bone in the distal and proximal ends exhibited a pattern that suggested stress lines. I later encountered the work of von Meyer. The idea took several decades to mature and then I did not know how to proceed. How could I directly test the resistance of bone to fracture. The real break-through was the discovery of the equipment in the Breaker Space at the University of Michigan School of Engineering that could be used to analyze the elements of a bone's potential to fracture.

The work on bone density is also an example of the importance of intersections with colleagues. My fortuitous meeting with Andrea Poli in the Breaker Space led to a research project and a new world of techniques and equipment. My discussions with Poli have also widened my understanding of problems in bone taphonomy and, in the process, I introduced Poli to archaeology. While work was underway at the Breaker Space, I met and had discussions with doctors at the Mayo Clinic who were working on human osteoporosis. They introduced me to new research that questioned the use of bone density as a proxy for human osteoporosis.

An important lesson on intersections with colleagues is my introduction to *Shorbet Kawara* by Mohamed Hussein and Mohamed Rauf. They possessed cultural information that I had not encountered, while I was limited by my Midwestern American preconceptions of what is "edible." The introduction to the culinary and nutritional value of distal limb elements was a revelation to me. In a recent paper with Mohamed Hussein I have put this lesson to action in a reconsideration of some of the faunal data from the Heit al-Ghurab (Hussein & Redding, 2020). Mohamed Hussein has also sent me an enlightening film of an itinerant butcher near Abydos preparing a sheep for a market, including the preparation of the distal limb for *Shorbet Kawara*. The exchange of information and data is a two-way street. My intersections with Mohamed Hussein have proved most enlightening for me.

Conclusion

> *"To raise new questions, new possibilities, to regard old problems from a new angle, requires creative imagination and marks real advance in science."* Albert Einstein (1938: 92)

So, what is my advice? First, question assumptions. We are all, sometimes, blinded by our assumptions. We need to re-examine them constantly and check the data for patterns that call our assumptions into question.

Second, explore for new techniques and resources to attack problems. Explore the literature and data from other fields. Is there information or data that could help you? What do other fields say about the problem or even aspects of it? How did they attack a similar problem? Are there models or theories you can use?

Third, explore the ethnographic record of Egypt, the Middle East and North Africa. These are a written record of problem solving in the present and past and may provide new approaches and data.

Fourth, intersect with other researchers in archaeology and other fields. Importantly, intersect with people that have different cultural experiences. A wonderful world of information, research, and ideas is available. Universities are a source of labs filled with equipment and researchers. Most of whom I have encountered, are keen to try out their techniques on problems in archaeology.

Fifth, we should all be engaging in experimental archaeology.

Bibliography

Ahmed, M.H. & R.W. Redding. 2020. Faunal Remains from Excavations in 2019 at the Menkaure Valley Temple (MVT-W). – Journal of the American Research Center in Egypt 56: 17-34.

Barich, B.E., G. Lucarini, M. Gallinario & M. Hamadan. 2012. Sheikh/Bir El Obeiyid: Evidence of Sedentism in the Northern Farafra Depression (Western Desert, Egypt). – Prehistory of Northeastern Africa: Studies in African Archaeology 10: 259-383.

Brain, C.K. 1967. Hottentot Food Remains and Their Bearing on the Interpretations of Fossil Bone Assemblages. – Scientific Papers of the Namib Desert Research Station 32: 1-11.

Brain, C.K. 1981. The Hunters or the Hunted? An Introduction to South African Cave Taphonomy. – Chicago, University of Chicago Press.

Einstein, A. & L. Infeld. 1938. The Evolution of Physics: The Growth of Ideas from Early Concepts to Relativity and Quanta. – Cambridge, Cambridge University Press.

Fadiman, A. 1998. The Spirit Catches You and You Fall Down: A Hmong Child, Her American Doctors, and the Collision of Two Cultures. – New York, Macmillan.

Fonseca, H., D. Moreira-Gonçalves, H-J. A. Coriolano & J. A. Duarte. 2014. Bone Quality: The Determinants of Bone Strength and Fragility. – Sports Medicine 44:37-53.

Gifford-Gonzalez, D. 2018. An Introduction to Zooarchaeology. – Cham, Springer

Holloway, S. J., R. Phillips, A. Koopman, V. Linseele & W. Wendrich. 2017. The L Basin Archaeological Record. In: S.J. Holloway & W. Wendrich. Eds. The Desert Fayum Reinvestigated: The Early to Mid-Holocene Landscape Archaeology of the Fayum North Shore, Egypt. – Los Angeles, UCLA Cotsen Institute of Archaeology Press: 51-98.

Ikram, S. 1995. Choice Cuts: Meat Production in Ancient Egypt. – Leuven, Peeters (Orientalia Lovaniensia Analecta 69).

Johansson, F. 2017. The Medici Effect: What Elephants and Epidemics Can Teach Us About Innovation. – Boston, Harvard Business School Publishing.

Koopman, A., S. Kluiving, S. Holdaway & W. Wendrich. 2016. The Effects of Holocene Landscape Changes on the Formation of the Archaeological Record in the Fayum Basin, Egypt. – Geoarchaeology 31 (1): 17-33.

Kromer, K. 1978. Siedlungsfunde aus dem Frühen Alten Reich in Giseh. – Wien, Österreichischen Akademie der Wissenschaften Philosophisch-Historische Klasse Denkschriften, 136).

Lam, Y.M., X. Chen, C.W. Marean & C.J. Frey. 1998. Bone Density and Long Bone Representation in Archaeological Faunas: Comparing Results from CT and Photon Densitometry. – Journal of Archaeological Science 25 (6): 559-570.

Lam, Y.M. & O.M. Pearson. 2004. The Fallibility of Bone Density Values and their Use in Archaeological Analyses. – Journal of Taphonomy 2 (2): 99-115.

Lam, Y. M. & O.M. Pearson. 2005. Bone Density Studies and the Interpretation of the Faunal Record. – Evolutionary Anthropology: Issues, News, and Reviews 14 (3): 99-108.

Licata, A. 2009. Bone Density vs Bone Quality: What's a Clinician to Do? – Cleveland Clinic Journal of Medicine 76 (6): 331-336.

Lyman, R.L. 1984. Bone Density and Differential Survivorship in Fossil Classes. – Journal of Anthropological Archaeology 3 (4): 259-299.

Lyman, R.L. 1994. Vertebrate Taphonomy. – Cambridge, Cambridge University Press.

Marean, C.W. & L.M. Spencer. 1991. Impact of Carnivore Ravaging on Zooarchaeological Measures of Element Abundance. – American Antiquity 56 (4): 645-658.
Marean, C.W., L.M. Spencer, R.J. Blumenschine & S.D. Capaldo. 1992. Captive Hyaena Bone Choice and Destruction, the Schlepp Effect and Olduvai Archaeofaunas. – Journal of Archaeological Science 19 (1): 101-121.
McDonald, M.M.A. 2009. Increased Sedentism in the Central Oases of the Egyptian Western Desert in the Early to Mid-Holocene: Evidence from the Peripheries. – African Archaeology Review 26: 3-43.
McHugh, W.P., G.G. Schaber, C.S. Breed & J.F. McCauley. 1989. Neolithic Adaptation and the Holocene Functioning of Tertiary Palaeodrainages in Southern Egypt and Northern Sudan. – Antiquity 63: 320-336.
von Meyer, G.H. 1867. Die Architectur der Spongiosa. – Reichert und Du Bois-Reymond's Archiv 8: 615-628.
National Academy of Sciences. 2010. Research at the Intersection of the Physical and Life Sciences. – Washington, The National Academies Press.
Nelson, K. 2001. Site E-75-8: A Slice Through Time. In: Wendorf, F. & R Schild. Eds. Holocene Settlement of the Egyptian Sahar: The Archaeology of the Nabta. – Boston, Springer: 386-411.
Pellett, P.L. & S. Shadarevian. 2013. Food Composition: Tables for Use in the Middle East, 3rd Edition. – Beirut, American University of Beirut.
Redding, R.W. 1986. A General Explanation of Subsistence Change: From Hunting and Gathering to Food Production. – Journal of Anthropological Archaeology 7: 56-97.
Redding, R.W. 1993. Subsistence Security as a Selective Pressure Favoring Increasing Cultural Complexity. – Journal of Sumerian Agriculture 7: 77-98.
Redding, R.W. 2010. Status and Diet at the Workers' Town, Giza, Egypt. In: Campana, D., P. Crabtree, S.D. deFrance, J. Lev-Tov & A. Choyke. Eds. Anthropological Approaches to Zooarchaeology: Complexity, Colonialism, and Animal Transformations. – Oxford, Oxbow: 65-75.
Redding, R. 2020. Cattle in Egypt. – Unpublished Manuscript.
Saito, M. & K. Marumo. 2010. Collagen Cross-Links as a Determinant of Bone Quality: A Possible Explanation for Bone Fragility in Aging, Osteoporosis, and Diabetes Mellitus. – Osteoporosis International 21: 195-214.
Seeman, E. 2008. Bone Quality: The Material and Structural Basis of Bone Strength. – Journal of Bone Mineral Metabolism 26: 1-8.
Speth, J.D. 2015 When Did Humans Learn to Boil? – PaleoAnthropology 2015: 54-67. Online URL: http://paleoanthro.org/media/journal/content/PA20150054.pdf
Wenke, R J., J.E. Long & P.E. Buck. 1988. Epipaleolithic and Neolithic Subsistence and Settlement in the Fayyum Oasis of Egypt. – Journal of Field Archaeology 15 (1): 29-51.
Witsell, A. 2018. Kromer 2018: Basket by Basket. – AERAgram 19 (1): 1-9.

Biomolecular Stable Isotope and Carbon-14 Dates of Ancient Egyptian Food Offerings: A Case Study from a Provincial Cemetery of Deir al-Ballas

Amr Khalaf Shahat & Victoria Jensen

Introduction

In this paper we[1] concentrate on five tomb contexts of Cemetery 1-200 at Deir al-Ballas (Figures 1, 2) and discuss the results of carbon dates and stable carbon isotope analysis on the findings of dom (*Hyphaene thebaica* Mart.), domesticated watermelon (*Citrulus lanatus* (Thunb.) Matsum. & Nakai), Phoenician juniper (*Juniperus phoenicea* L.), grape (*Vitis vinifera* L.), and pomegranate (*Punica granatum* L.) to differentiate between local and imported species and answer questions on social history of cultural interactions. Our report on these five contexts includes the architecture, human remains,[2] and material culture present in each tomb, however, it should be noted that the available evidence varies from tomb to tomb. Information was gleaned from the unpublished notes, photographs, and maps from the 1900-1901 Hearst Expedition, as well as examination of the artifacts now housed at the Phoebe A. Hearst Museum of Anthropology (PAHMA) at the University of California, Berkeley. The ^{14}C dates are compared to the material culture found in each tomb and the chronological implications are discussed.

The population buried in the cemetery is designated as "non-elite" as distinguished from royal or high-level courtiers and priests whose decorated tombs at Thebes and Saqqara have been a prominent source of Egyptological information on funerary food offerings.[3] The population that inhabited the area after the abandonment of the palace show a lack of access to texts with only a handful of individuals' names recorded and almost no titles that are a key component of elite self-presentation used to display their

1 The authors thank the editors of this volume and the anonymous reviewers for their helpful comments and feedback. We also extend our deep thanks to Peter Lacovara, the director of the Deir al-Ballas expedition; Leslie Freund, Collections Manager at the Phoebe A. Hearst Museum of Anthropology for facilitating our access to the botanical remains held at the museum; and Christine Hastorf, Director of the McCown Archaeobotany Laboratory and Archaeological Research Facility, UC Berkeley, and Todd Dawson, Director, and Stefania Mambelli, Lab Scientist, of the Dawson Isotope Lab, UC Berkeley for their mentorship regarding the research experiments.

2 Human remains are reported based on Reisner's excavation notes, rather than analysis of the human remains whose location post-excavation is not known.

3 For the New Kingdom, the term "elite" in Egyptology is often mainly defined with a bias towards what burials look like in the Theban region; there is less discussion of what a provincial elite or non-elite burial may have looked like in other regions (K. Cooney, personal communication).

In: Ikram, S., J. Kaiser & S. Porcier. Eds. 2021. *The Ancient Egyptians and the Natural World. Flora, Fauna, and Science.* – Leiden, Sidestone Press: 233-252.

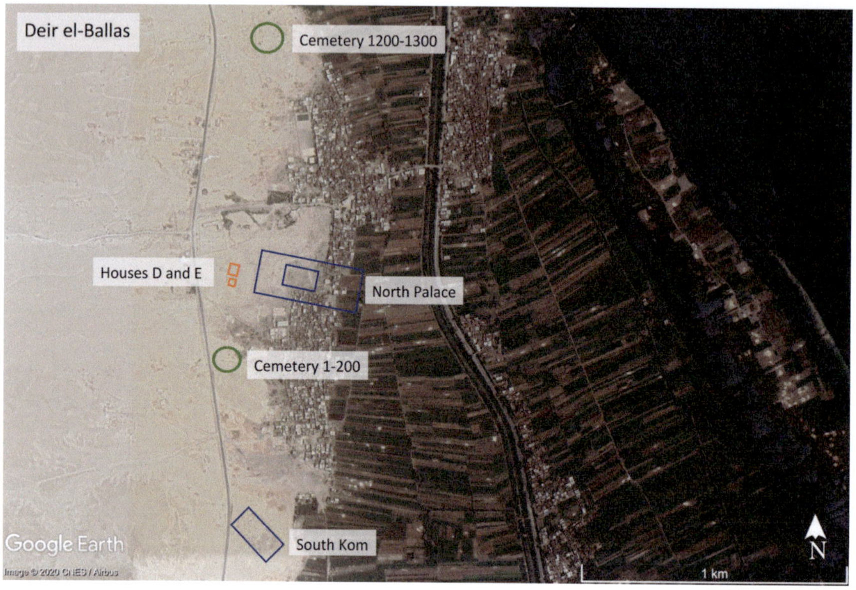

Figure 1. Deir al-Ballas site overview. Google Earth image annotated by Victoria Jensen.

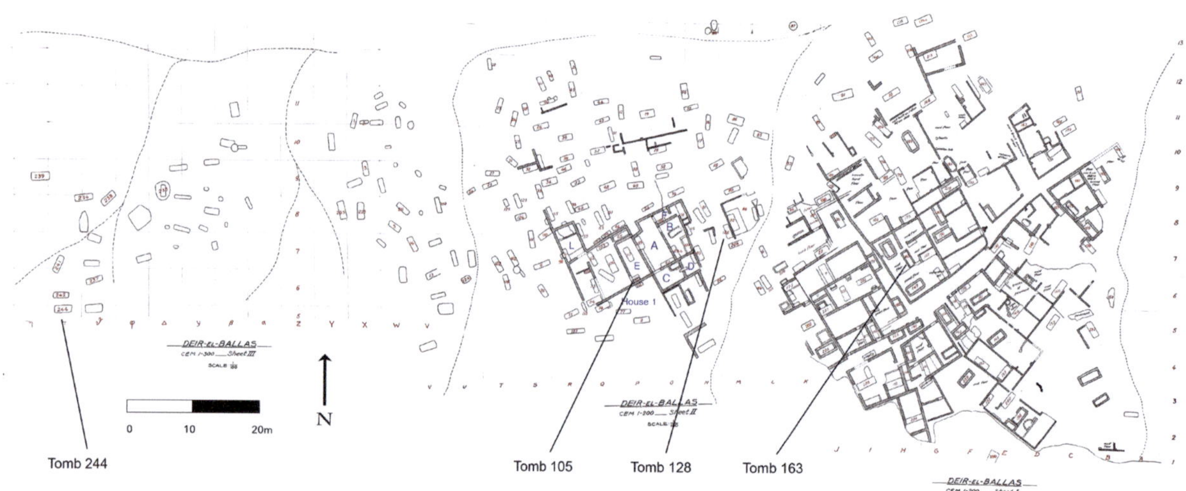

Figure 2. Map of Cemetery 1-200. Hearst Expedition maps (Sheets I, II, and III). Courtesy of the Museum of Fine Arts, Boston. Edited and annotated by Victoria Jensen. Tombs 255 and 257 are not identified on this map because the Hearst expedition did not record their locations.

proximity to the king (Frood, 2010: 476-477).[4] However, the community exhibits a range of socioeconomic hierarchy as reflected in the graves that show differential access to material culture, indicating that some individuals might have been better-off than others (Jensen, 2019: 431-443). For instance, both Tomb 128 and 163 discussed in this paper exhibited a higher than average diversity and quantity of material culture including stone cosmetic jars, bronze razors, and royal name scarabs, as well as rarely attested fruits, including imported juniper. Archaeobotany may contribute to the discernment of social status by showing who had access to imported fruits versus not.

The botanical samples from Cemetery 1-200 had not been published since they were stored in the University of California at Berkeley's museum over 100 years ago. A visual analysis using a stereomicroscope was conducted on all botanical remains known from the site by Shahat in 20x magnification at the UC Berkeley McCown Archaebotany

4 Cemetery 1-200 features simple tomb architecture that primarily consisted of undecorated shafts or shafts with loculi (Jensen, 2019). In Cemetery 1-200, of approximately 200 tombs there were only two stelae discovered: one dating to the early 18th Dynasty in which the deceased is given no title, and one dating to the late 18th -early 19th Dynasty that mentions a *nb.t pr* offering to a *wꜥb*-priest, which is the lowest level in the priestly hierarchy.

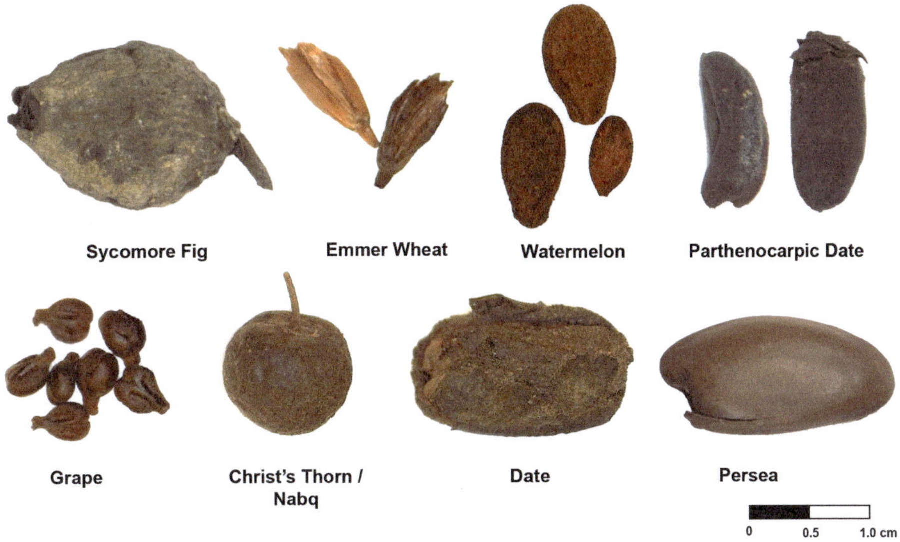

Figure 3. Example of desiccated botanical materials found in cemetery 1-200 from Deir el Ballas. This includes Sycomore fig, emmer wheat, watermelon, parthenocarpic (unpollinated) dates, grapes, christ's thorn, whole date, and persea seed. Photo Credit: Amr Khalaf Shahat (UCLA), Venicia Slotten and Leslie Freund (UC Berkeley).

Laboratory. 245 specimens were identified are based on morphology, including 13 different species belonging to 11 different families (Shahat & Jensen, in press). This analysis showed regular use of native wild fruits such as dom, date (*Phoenix dactylifera* L.*)*, sycomore fig (*Ficus sycomorus* L.), desert date (*Balanites aegyptiaca* L.), persea (*Mimusops laurifolia* Forssk.), and *nabq* (*Ziziphus spina-christi* (L.) Willd) along with domesticated wheat (*Triticum turgidum ssp. dicoccon*, Schrank ex Schübl*)* and barley (*Hordeum vulgare* L.). Moreover, funerary food offerings also included foreign species such as domesticated watermelon, which was originally a West African plant, but could have been grown locally by this time having come to Egypt through interaction with Nubia, modern Sudan (Chomicki & Renner, 2015), and pomegranate and juniper berries from the eastern Mediterranean (Asouti, 2018: 1-38) (Table 1, Figure 3). Although these plant species were found in a funerary context, their presence at the site indicates that they were available to the population living at Deir al-Ballas in the early 18th Dynasty. Moreover, except for pomegranate, below we discuss that all of these species have been also encountered in the settlement excavations at Deir al-Ballas and/or the Workmen's Village at Amarna, and thus almost certainly were not reserved for exclusive use as grave goods but were consumed by the living.

Aims and Objectives

Since the site of the cemetery was originally a settlement associated with the palace at Deir al-Ballas before its reuse as a necropolis, it was necessary to conduct ^{14}C AMS dating on the archaeobotanical materials to ensure that they are not intrusive or a result of post-depositional processes and to determine the date of tombs lacking material culture. Carbon dating was also helpful to verify the ancient dates of the plants which have been rarely reported to have been found prior to this cemetery's investigation. In particular, the earliest encounter of domesticated watermelon in Egypt continues to be much debated in Egyptian archaeobotany. The earliest securely dated samples come from this site at Deir el Ballas, predating the finding at the Amarna workmen's village (Stevens & Clapham, 2013), so one specific objective was to discover the date ranges of the samples buried as food offerings at Deir al-Ballas to contribute to *C. lanatus* seeds discussion.

One alert for us regarding post-depositional concerns in these reused contexts comes from archaeobotanical materials excavated by the same mission at another site – the predynastic cemetery of Nag ed-Deir. In one case, the carbon dating of the botanical remains was at odds with the material culture. Samples of the imported species *Pistacia lentiscus* (n=24 raw seeds) are registered at PAHMA as coming from Tomb 7626 at Nag al-Deir, which contained four bodies. This context contains material culture dated to early Predynastic such as a Cross-line Ware vessel, a rhomboidal slate palette, and an ivory comb (Lythgoe, 1902-1904, Naga ed-Der Tomb Cards, PAHMA archives). Based on these objects, the entire assemblage including the botanical remains is attributed to Nagada Ic in the museum records, but surprisingly, the pistacia seeds were carbon-dated to between 2494-2460 calBC with 75.3% certainty (Shahat, 2019) suggesting a reuse of the burial pit for a later internment approximately 1000 years after the

Tomb #	*Family*	*Genus*	*species*	Common Name	Count	Plant part	PAHMA museum number*
Pit east of tomb 59	*Moraceae*	*Ficus*	*sycomorus*	Sycomore fig	1	Whole fruit dried	6-8128
Pit east of tomb 59	*Arecaceae*	*Phoenix*	*dactylifera*	Date	2	Date pit one with date skin preserved	6-8128
80	*Arecaceae*	*Hyphaene*	*thebaica* Mart.	Dom Fruit	3	Fruit endocarp	6-6435
89	*Arecaceae*	*Phoenix*	*dactylifera*	Dates	21	Parthenocarpic whole dates	6-6469
105	*Arecaceae*	*Phoenix*	*dactylifera*	Dates	1	Whole date with skin preserved	6-6507
105	*Arecaceae*	*Hyphaene*	*thebaica* Mart.	Dom Fruit	6	Fruit endocarp fragments; one endocarp punched with hole	6-6507
105	*Arecaceae*	*Hyphaene*	*thebaica* Mart.	Dom Fruit	33	5 whole fruit, 2 large & 26 smaller fragments	6-6509
128	*Rhamnaceae*	*Ziziphus*	*spina-christi* (L.) Willd	Christ's thorn	7	Whole fruit with pedicel	6-6626
128	*Vitaceae*	*Vitis*	*vinifera*	Grape	21	Grape pips	6-6626
128	*Cupressaceae*	*Juniperus*	*phoenicea*	Juniperus berry	36	Whole fruits	6-6626
128	*Lythraceae*	*Punica*	*granatum*	Pomegranate	52	Small fragments of rind	6-6626
128	*Arecaceae*	*Phoenix*	*dactylifera*	Dates	6	Seed pits	6-6626
163	*Arecaceae*	*Hyphaene*	*thebaica* Mart.	Dom Fruit	2	Large endocarp	6-6895
163	*Arecaceae*	*Phoenix*	*dactylifera*	Dates	1	Date skin mostly preserved without the seed pit	6-6896
163	*Vitaceae*	*Vitis*	*vinifera*	Grapes	34	Grape pips, some with skin preserved	6-6895
190	*Arecaceae*	*Hyphaene*	*thebaica* Mart.	Dom Fruit	1	Whole fruit	6-7043
190	*Balanitaceae*	*Balanites*	*aegyptiaca* (L.) Delile	Desert date	2	Fruit pits	6-7049
192	*Poaceae*	*Hordeum*	*vulgare ssp. Distichon*	Barley	1	Spikelet	6-7070
205	*Balanitaceae*	*Balanites*	*aegyptiaca* (L.) Delile	Desert date	1	1 fruit exocarp and mesocarp including 4 fragments	6-8895
244/255	*Vitaceae*	*Vitis*	*vinifera*	Grapes	2	Grape pips	6-7482
244/255	*Cyperaceae*	*Cyperus*	*esculentus*	Tiger nut	1	1 whole rhizome	6-7482
244/255	*Cucurbitaceae*	*Citrullus*	*lanatus*	Watermelon	11	5 whole seeds and 6 fragments	6-7482
244/255	*Arecaceae*	*Hyphaene*	*thebaica* Mart.	Dom Fruit	1	Mesocarp	6-7482
257	*Lythraceae*	*Punica*	*granatum*	Pomegranate	1	Fragmented fruit endocarp and red seeds	6-7517
257	*Rhamnaceae*	*Ziziphus*	*spina-christi* (L.) Willd	Christ's thorn	4	Whole fruits with part of the fruit pedicel preserved	6-7517
257	*Vitaceae*	*Vitis*	*vinifera*	Grape	7	Grape pips	6-7517
257	*Arecaceae*	*Phoenix*	*dactylifera*	Dates	3	Two dates with fruit skin mostly preserved and one date pit	6-7517
257	*Poaceae*	*Triticum*	*turgidum ssp. dicoccon*	Emmer wheat	8	2 whole spikelets (one of them is the apical spikelet) and 6 chaff	6-7517
257	*Poaceae*	*Hordeum*	*vulgare*	Barley	1	Spikelet	6-7519
257	*Poaceae*	*Triticum*	*turgidum ssp. dicoccon*	Emmer wheat	1	Spikelet	6-7519
257	*Sapotaceae*	*Mimusops*	*laurifolia*	Persea seeds	30	14 whole seeds 16 fragments	6-7519
uncertain prov.	*Arecaceae*	*Phoenix*	*dactylifera*	Date	25	Date pits	6-9176
uncertain prov.	*Moraceae*	*Ficus*	*sycomorus*	Sycomore fig	1	Whole fruit	6-8297
uncertain prov.	*Balanitaceae*	*Balanites*	*aegyptiaca*(L.) Delile	Desert date	7	Endocarp	6-8297
uncertain prov.	*Poaceae*	*Triticum*	*turgidum ssp. dicoccon*	Emmer wheat	2	Mass of cereal husk and fine chaff, only two spikelet forks	6-8455
uncertain prov.	*Arecaceae*	*Hyphaene*	*thebaica* Mart.	Dom Fruit	3	Dom fruit, 1 complete and 2 mostly intact	6-8492

Tomb #	*Family*	*Genus*	*species*	Common Name	Count	Plant part	PAHMA museum number*
uncertain prov.	*Arecaceae*	*Hyphaene*	*thebaica* Mart.	Dom Fruit	1	Dom fruit, mesocarp preserved	6-8711
uncertain prov.	*Arecaceae*	*Hyphaene*	*thebaica* Mart.	Dom Fruit	1	Whole dom fruit	6-8712
uncertain prov.	*Arecaceae*	*Hyphaene*	*thebaica* Mart.	Dom Fruit	1	Macro-fossilized dom fruit (completely preserved)	6-8276
uncertain prov.	*Arecaceae*	*Hyphaene*	*thebaica* Mart.	Dom Fruit	2	1 fruit significantly large 7.5X5.0cm & 1 exocarp	6-6510
uncertain prov.	*Arecaceae*	*Hyphaene*	*thebaica* Mart.	Dom Fruit	1	One whole fruit and tiny fragments	6-8259

Table 1. Archaeobotanical remains from Deir al-Ballas at the Phoebe A. Hearst Museum of Anthropology.

Predynastic internments (Savage, 1998).[5] So, this is an alert for researchers to conduct ^{14}C dating on archaeobotanical materials whenever dealing with reused, insecure or mixed contexts for more accurate contextualization and interpretation.

The species we investigated were chosen as part of a long-term project to create a baseline dataset of ^{13}C values to serve the community of researchers in Egyptian archaeobotany and bioarchaeology. The question for us is to understand the ancient food ecology and how much imported fruit arrived into these communities, identified by different paleoecological conditions compared to those that were grown along the Nile. The species measured here are the commonly encountered wild fruit found in the cemetery in Deir al-Ballas as well as cultivated species, such as grape, as well as the rarer findings of what were originally (ca. 3000 BC) non-indigenous plants such as pomegranate, watermelon, and juniper. Measuring the carbon stable isotope traditionally helps determine the photosynthesis pathway of the plant, whether it is C_3 such as cereals, fruits, and nuts in Egypt or C_4 (*e.g.* sorghum, millet) which were more common in sub-Saharan Africa than in Egypt during the pharaonic period (Thompson *et al.*, 2008). The value of the stable isotopes to archaeobotany is thus helpful to bioarchaeologists to differentiate between the Egyptian plant-dietary input, which was predominantly C_3 versus the Nubian diet that was higher in C_4 (Thompson *et al.*, 2008).

In this study, a further step is taken to provide additional archaeobotanical data. The ^{13}C isotope values provide information not only regarding whether each species is a C_3 or C_4 plant but also about the ecophysiological conditions that affected the plant and the paleoenvironmental history of the sample. For example, in this study, the ^{13}C of the juniper berries had a carbon value of -19.5‰ showing that the plant grew in ecological conditions of limited water, which differs from local Egyptian plants that grew along the Nile and were also found in the same cemetery. Thus, the use of ^{13}C isotope analysis captures important information on the paleoecology of the plant, particularly on the efficiency of water use efficiency by plants and the climate conditions under which the plant grew and helps identification regarding whether it is local or imported. Therefore, this project contributes a dataset that serves bioarchaeologists and zooarchaeologists by adding to a baseline of stable isotope data on specific plant species that were part of ancient Egyptian flora and were part of their food culture and funerary food offerings. The dataset demonstrates the impact of climate change and damming of the river on its isotopic composition. To communicate the stable isotope results of these plant species to bioarchaeologists, the key argument here is that bioarchaeological literature to date has focused on C_3 domesticates such as wheat and barley as dietary staples in ancient Egypt, but stable isotope tests on many archaeobotanical remains provide more fine-grained information of the detailed "menu" that contributed to such isotopic values we gain from human bodies. Local wild plants found among the funerary food offerings as well as the settlement of Deir al-Ballas, such as dom and sycomore fig, are infrequently discussed by bioarchaeologists dealing with past diet and isotopic data even though these were a wide range of C_3 plants contributing to the isotopic input of the Egyptian diet.[6] As such, these plants we have analyzed contributed to the ^{13}C isotope values observed in the analysis of ancient Egyptian

5 In Lythgoe's field notes for this tomb (Naga ed-Der Tomb Cards, PAHMA archives, 1902-1904), he recorded three Predynastic burials (A-C) and one later, intrusive burial (D) that was placed at the edge of the pit. He did not attempt to date Burial D, only noting that it was placed 20 cm above the other three burials and had partially disturbed the closest skeleton, Burial C. Burial D also was later disturbed. Lythgoe states that the seeds were found near the elbow of Burial B, but their carbon date indicates that they must have entered the tomb approximately 1000 years later, most likely during the placement or the disturbance of Burial D.

6 The ^{13}C isotope value for dom coming from the cemetery is presented below in this paper, while the value of wild sycomore fig from a modern type collection is -26.76‰ (Shahat lab report, 2019).

human bones and teeth and should be included in future interpretations of their dietary interpretation.[7]

The Site and Its Excavation History

Located 40 km north of Luxor on the west bank of the Nile, Deir al-Ballas features a royal palace dating to the reigns of Seqenenre Taa (ca. 1553 BC) and Nebpehtyra Ahmose (ca. 1550-1525 BC) ("North Palace") (Figure 1). Immediately west of the palace's enclosure wall are several building complexes (Houses A-F), one of which (House D) appears to have been a bakery while another (House E) shows evidence of textile production (Lacovara, 1990). At the southern end of the site, a high hill was transformed into a huge two-level platform (South Kom/South 'Palace') with a monumental staircase connecting the two levels.[8] While its function remains enigmatic, this structure may have served as an observation platform to safeguard the surrounding desert and the Nile. Approximately 70 large houses were scattered across the settlement, administrative facilities have been identified, and a compact Workers' Village stood on a hillside overlooking the North Palace from the south (Lacovara, 1990: 1-5; 1997: 7-15).

Three cemeteries have been identified at the site: Cemetery 1-200 where over 200 tombs were dug among the walls of the former Workers' Village, Cemetery 500 consisting of 14 tombs directly west of the palace, and Cemetery 1200-1300 consisting of approximately 60 tombs at the northern end of the site.[9] There are only a few tombs that may possibly complement the 17th Dynasty settlement. After Ahmose's court moved to Thebes along with most of the attendant population, there may have been a brief period of complete abandonment followed by "squatter occupation" (Lacovara, 1997: 15) or perhaps a small contingent of individuals remained at the site and moved into the abandoned homes. In either case, the area of the workers' village was reused as a cemetery, beginning early in the 18th Dynasty (Jensen, 2019: 431-443; Lacovara, 1981; 1997: 83-84), with a resurgence in the reigns of Hatshepsut/Thutmose III (Jensen, 2019: 404-413). A few burials can be dated by their material culture to the late 18th-early 19th Dynasty, after which no datable activity occurred until the Coptic era when the palace was reused as a monastery (Jensen, 2019: 25, 409-410). Such reuse of domestic spaces as a necropolis is rarely published but not unknown (Jensen, 2019: 70-73). The walls may have served as convenient, ready-made dividers delineating cultic space for funerary rites, and the fact that the settlement faced the abandoned palace may have been a factor in selecting this area for use as a necropolis (Jensen, 2019: 6-10).

The excavation history of Deir al-Ballas began in the late 19th century. Georges Daressy found a stone lintel in the village near the North Palace with the cartouche of Seqenenre Taa (1894: 44). Two years later, James E. Quibell recorded that he *"turned over a considerable part of the ruins"* but noted that the site *"had been thoroughly plundered, worked by a dealer at Qeneh as well as others."* (Petrie & Quibell, 1896: 1). Quibell published no list of finds or further description of his work. In 1900-1901, a more systematic excavation was undertaken with funding from Phoebe Apperson Hearst on behalf of the University of California, led by George Andrew Reisner assisted by Albert M. Lythgoe and Frederick W. Green, but their work was never published, we only have field notes housed in the Hearst Museum.

The Hearst Expedition only recorded two of the three cemeteries at the site well enough to describe many of the tombs' contents. Reisner oversaw the excavations in Cemetery 1-200 while Lythgoe supervised the work at Cemetery 1200-1300. In both cemeteries, the most common grave good was pottery, with a few burials preserving stone kohl jars, bronze toilet implements, scarabs, amulets, and beads made of faience, stone, and glass, all modest burial goods (Jensen, 2019: 64-296, 308-356).

At Deir al-Ballas, Reisner began his procedure of taking photographs of tombs as they were being excavated as well as the assemblages of objects from particular tombs, but he was not consistent in doing this (Jensen, 2019: 48-50; Lacovara, 1997: 7). However, Reisner was usually careful to record his observations of botanical remains from the tombs in his field notebook. Lythgoe's recording of Cemetery 1200-1300 was less thorough than Reisner's. Our study of the archaeobotanical remains is confined to the samples retrieved from Cemetery 1-200 that were sent to the University of California, Berkeley by Reisner. Given the limited standards of recording in 1900 used by the Hearst Expedition, we cannot reconstruct the specific stratigraphy of Cemetery 1-200 to differentiate between the workers' village settlement and the later burial shafts that were found throughout these houses. Information regarding human remains was very cursorily recorded in Reisner's records and the current location of the remains is unknown (Jensen, 2019: 78-80), so detailed analysis

7 One major caveat here is our inability to study the human skeletal remains from Deir al-Ballas to correlate the stable isotope results between food remains and human bones. Reisner did not record where he stored the human remains and they were not sent to Berkeley.

8 The monument was termed "South Kom" by Reisner (1904: 1) and the actual excavator, Lythgoe (unpublished field notes), acknowledging the uncertain function of this monumental two-level platform. The South Kom was later termed a palace by Reisner (1908: V-VI) and Stevenson Smith (1998: 159-160), a term that was followed by Lacovara but enclosed in quotation marks, as the latter notes that this is certainly a misnomer for the monument (1981: 121).

9 As of this writing, Cemetery 1-200 has been almost entirely bulldozed due to the expansion of the modern village.

of sex and age and their relationship to the grave goods including the botanical offerings is not possible.

Peter Lacovara worked at the site in the 1980s and published a preliminary report (Lacovara, 1990) as well as an analysis of the architecture in comparison with other royal cities (Lacovara, 1997). He cleared and mapped several areas of the site including part of the North Palace, several houses, and a possible chapel. As part of this project, Wilma Wetterstrom conducted a flotation of organic samples from one rubbish deposit in the settlement and discovered domesticated wheat, barley, and flaxseeds (*Linum usitatissimum* L.) (Wetterstrom, 1990: 25). The cemetery areas were not revisited by Lacovara's team. Since 2017, Lacovara has resumed work at the site to restore and protect certain areas, particularly the two large mudbrick monuments (North Palace and South Kom), and several settlement structures located to the west of the palace. Additional archaeobotanical remains were retrieved from one of these buildings, House E, in the 2019-2020 season by the author; the species recovered included emmer wheat, six-row barley, flax, sycomore fig, dates, dom, grape, bean (*Vicia faba* L.) and acacia pod (*Acacia nilotica* (L.) Del.) (Shahat, 2020).

Other researchers have investigated sectors of the Deir al-Ballas cemetery material held at PAHMA over the years (Aston, 1994; Bourriau, 1982: 78-81; 1990; Lilyquist & Brill, 1993: 24; Merrillees, 1968: 199, Plate XXXIII; Minor, in press). Victoria Jensen's dissertation (2019) is the first holistic publication and analysis of a large quantity of the cemetery material, synthesizing the Hearst Expedition's archival notes, maps, and photographs as well as information gleaned from examination of the artifacts held at PAHMA. A thorough publication of the Hearst Expedition work in the cemeteries is currently being prepared by a team led by Lacovara that includes the present authors.

Archaeobotanical Remains

Reisner's field notes for Cemetery 1-200 record the presence of food offerings in 66 of the nearly 200 tombs recorded in the cemetery (33%) (Jensen, 2019: 287; Shahat & Jensen, in press). The most frequently-mentioned offerings were "grain husks", "vegetable husks" or "vegetable matter" that were found in at least 87 small beaker-shaped jars (Holthoer's "wine decanter" type (1977: 171-172, Plate 41 185/0:1 and 185/427:1) and occasionally in a larger storage jar. Unfortunately, we were not able to analyze these remains as they were discarded. It is uncertain at what point in time the husks were emptied from the containers. This must have happened between their discovery, when the botanical remains were recorded, and the present, given the empty state of the vessels in the museum. Because the museum has retained and catalogued other botanical remains and disintegrated wood samples sent from the site, we hypothesize that the loss occurred in 1901 as the ceramics were being prepared for shipment to California from Egypt.

Another common food offering were dom fruits (recorded in 32 tombs). Twelve tombs contained dates, while just a few tombs had a wider variety of foods such as grapes (number of tombs=4), pomegranate (n=3), sycomore fig (n=3), nabq (n=2), desert date (n=2), juniper (n=1), persea nuts (n=1), watermelon (n=1), tiger nut (*Cyperus esculentus* L.) rhizome (n=1), and a honeycomb (n=1) (Jensen 2019: 287-295; Shahat & Jensen, in press).

Of the botanical remains that were recorded in the field notes, some, but not all have been matched with samples held at PAHMA. Conversely, there are some botanical remains registered at the museum that have lost their exact provenience so they cannot be linked with a specific tomb or another context. Table 1 provides the museum registration numbers for the botanical samples from Deir al-Ballas and the precise provenience (if known) that we have investigated.

Materials and Methods

All archaeobotanical materials were found during Reisner's excavations in 1900 and sent to the University of California, Berkeley in the early 1900s. The botanical samples are preserved at PAHMA in a desiccated condition.

AMS dating was conducted on the five botanical samples noted in the table, showing the following the pretreatment protocol steps of chemical cleaning, combustion, and graphitization. The samples were treated by an Acid-Base-Acid chemical cleaning process to remove the calcareous contamination, humic and fulvic acids (Olsson, 1986: 273-312). Humic and fulvic acids can remain in botanical samples coming from residing in a dry desert, in which case they would not require chemical cleaning with a base, however, if the humic and fulvic acids come from the soil the resulting dates will be biased towards a more recent age. Preparation sequence:

1. Chemical cleaning: Since all samples were preserved by desiccation and not charred, the protocol commonly applied on charred botanicals was modified to a less rigorous pre-treatment process to allow for the fragility of the desiccated botanical materials from a desert environment. The samples were thus treated with less severe base solutions and heat as noted in Table 2: 1N HCl was applied for 30 minutes at an elevated temperature of 70-80 °C.
2. The next step was conducting a single rinse with 0.25N NaOH at room temperature for 10-30 minutes. 1N HCl was applied again for 30 minutes at room temperature or elevated temperature according to the integrity of the samples. Neutralization of the samples was conducted by rinsing several times using ultrapure milliQ water. When the samples were treated with

an acid solution, they turned yellow, which indicates contamination with fulvic acid, and when treated with base solution they turned yellow or brown. In either case, they were completely cleaned by the end of the application of the final acid solution.

3. ombustion: For the combustion process, 2 mg were taken from each dried sample mixed with cupric oxide as an oxygen source and put in a quartz tube. The tube was sealed under vacuum with a gas torch. Then it was exposed to combustion for 3 hours at 900 °C temperature. The samples were then cryogenically cleaned and placed in vials with an iron-powder catalyst in them.
4. Graphitization: After the combustion process, the samples were graphitized under gaseous state using a hydrogen-reduction method by which the samples were heated for 3 hours at 525 °C. The graphites were then placed in an aluminum sample pellet to be analyzed in an accelerator mass spectrometer (AMS) at UC Irvine (Table 3)
5. ^{13}C stable isotope analysis (Table 3): conducted on the remaining fragments from the cleaned and graphitized gaseous samples of the botanical remains. An aliquot of the gaseous sample was collected from the vacuum line and placed in a Fisons NA-1500NC elemental analyzer connected with a Delta-Plus stable isotope mass spectrometer (IRMS). We confirm that these are the ^{13}C values reported here in Table 3. This is not to be confused with the conventional ^{13}C measures taken by the AMS spectrometer to correct for fractionation factors during the ^{14}C dating process. The latter are usually less precise and give drifted values as much as 3‰ (T. Dawson, personal communication, 2018) and are therefore not reported here. Carbon isotope results are reported in part per thousand, permil values) $\delta^{13}C$‰ relevant to the international standard VPDB (Vienna Pee Dee Belemnite).[10] The isotopic composition of carbon in the plant specimens is expressed following the standard equation (Faure & Mensing, 2005: 753).

$$\delta^{13}C = \left(\frac{(^{13}C/^{12}C)\text{ sample} - (^{13}C/^{12}C)\text{ standard}}{(^{13}C/^{12}C)\text{ standard}} \right) X1000$$

Also, it should be noted that both the grape and juniper samples were run in a second lab (Dawson Lab) without chemical pretreatment to test the precision of our results, and these gave similar carbon isotopic values as those run in the first lab.

Results

Results of the isotopic analysis and carbon dating from the five tombs are discussed below, relating the archaeobotanical samples to their archaeological context.

Tomb 105

This was a small, shallow shaft 75 cm deep that was dug into the corner of a room of a house, oriented E-W (Figure 2). A several-courses high mudbrick wall was built at the bottom of the shaft along the east and north sides, with only a few centimeters of space that were cleared

Progression of pretreatment steps ===>		HCl(1)	NaOH(2)	HCl(3)	milliQ(4)
PAHMA #	**Context**	**Sample**			
6-6509	Tomb 105	dom fruit fragment			
6-6626	Tomb 128	juniper berry seed fragment			
6-6895	Tomb 163	grape seed fragment			
6-7482	Tomb 244 or 255	watermelon seed fragment			
6-7517	Tomb 257	pomegranate rind fragment			

(1)5 mL 1N HCl for >30 min. @ 70-80 °C
(2)5 mL 0.2N NaOH for 5-10 min. @ room temperature
(3)5 mL 1N HCl for 10 min. @ 70-80 °
(4)2 or 3 rinses of 5 mL ultrapure milliQ water for 10 min. @ 70-80 °C

Table 2. Chemical treatment summary.

Sample ID	UCI AMS #	Yield % C	$\delta^{13}C$ ‰	Modern fraction	$D^{14}C$‰	^{14}C age (^{14}C yr BP)	Calibrated date (calBC)
Hyphaene thebaica 6-6509	219088	47	-25.0 ± 0.1	0.6830 ± 0.0012	-317.0 ± 1.2	3060 ± 15	1395 – 1334 (51.9%) 1326 – 1265 (43.5%)
Juniperus phoenicea 6-6626	219089	50	-19.5±0.1	06726 ± 0.0012	-327.4 ± 1.2	3185 ± 15	1498 – 1428 (95.4%)
Vitis vinifera 6-6895	219090	54	-24.93	0.6715 ± 0.0011	-328.5 ± 1.1	3200 ± 15	1501 – 1435 (95.4%)
Citrullus lanatus 6-7482	219091	56	-23.5 ± 0.1	0.6740 ± 0.0011	-326.0 ± 1.1	3170 ± 15	1496 – 1471 (22.1%) 1465 – 1413 (73.3%)
Punica granatum 6-7517	219092	53	-22.3 ± 0.1	0.6695 ± 0.0011	-330.5 ± 1.1	3225 ± 15	1526 – 1447 (95.4%)

Table 3. AMS dating and $\delta^{13}C$ results of desiccated archaeobotanical materials from Deir al-Ballas.

10 Pee Dee Belemnite standard named after Pee Dee river in South Carolina, USA (T. Dawson, personal communication, March 2019).

behind these walls. The width at the mouth of the shaft was 80 cm but the bottom of the shaft was just 37 cm wide. The depth of the shaft was 103 cm, and the northern mudbrick wall reduced the usable space to 95 cm. Human remains were not recorded, and the only objects found included fragments of dom fruit and bits of wood and plaster; a fragment of wood was found clinging to the bricks on the north side. Given the size of the shaft, we propose that the wood may have been from a child's coffin. No pottery or other material culture was recorded, but the ^{14}C dates obtained from these dom fragments have a range of 1395-1265 calBC, the period from Amenhotep III into the early reign of Ramesses II (Figure 4). Only a few other tombs in the cemetery can be dated to this same late 18th to early 19th Dynasties, based on their material culture, so this ^{14}C data is useful in identifying another tomb of that period that otherwise would have remained a mystery in terms of its dating.

The dom fruit's ^{13}C stable isotope value was -25.0 ± 0.1‰, identifying it as a C_3 plant growing in good water conditions. Since dom palms grow in Egypt along the Nile valley in Upper Egypt, this sample may serve as a baseline for non-water stressed C_3 plants locally grown in the region at this time.

Tomb 128

This tomb was oriented north-south along the outside of a wall (Figure 2). No dimensions were recorded but the field note sketch indicates that it had one loculus, or long niche carved out along the western edge of the bottom of the shaft. Two seemingly intact burials were found in the tomb (Figure 5, 6). The burial along the eastern side at the base of the shaft was supine and had its head to the south; a biconical necked jar with pendant line decoration, alabaster kohl jar, bronze needle, and bronze blade as well as a beaker containing grain husks was found near the head while two large dishes and part of an offering stand were near the feet. The burial in the loculus had its head to the north, with the head turned to the side. Eight beakers and a jar were clustered near the head of this burial, while a stone kohl jar, biconical ceramic jar, white-painted Nile silt jar containing fruit, two beakers, a bowl, and a dom fruit were placed near the feet. A scarab (6-8767) inscribed "*Ḏḥw.ty-ms*" (Thutmose) was found in the tomb but its precise findspot was not recorded. However, it provides possible *terminus post quem* evidence that this tomb dated as early as the 18th Dynasty. While the scarab could have been an heirloom of indeterminate age, some of the beakers found in the tomb had black-painted rims which is a decoration that is not attested after the reign of Thutmose III (Bourriau, 1982: 78), indicating that the

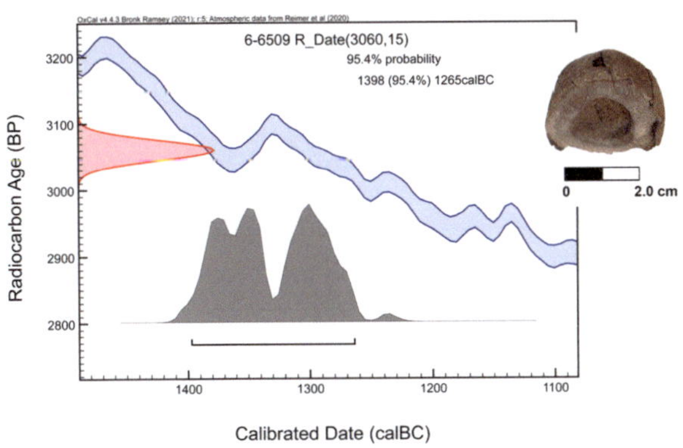

Figure 4. Sample #6-6509 *Hyphaene thebaica* dom fruit, Tomb 105, Cemetery 1-200, Deir al-Ballas. Figure presents calibrated radiocarbon results using OxCal 4.3.2. Double lines denote the IntCal13 terrestrial calibration curve with 1-σ envelope (Reimer *et al.*, 2013). Inset denotes the results of Bayesian statistical analysis. Graphic by Brian Damiata, Cotsen Institute of Archaeology, UCLA.

Figure 5. Tomb 128, south end, showing head end of burial in shaft, legs of burial in loculus (Hearst Expedition photograph B-1042). Courtesy of the Phoebe A. Hearst Museum of Anthropology and the Regents of the University of California.

scarab must be inscribed for Thutmose I, II, or III but not Thutmose IV.

The white-painted Nile silt jar (6-6621; Figure 7) contained a variety of fruit species including *nabq*, juniper berries, dates, pomegranate rind fragments, and grapes. Notably, there is a combination of wild local plants (*nabq* and dates), domesticated pomegranate and grapes, as well as imported juniper berries. AMS ^{14}C dating was run on a fragment of juniper (6-6626) yielding 50% of carbon with a calibrated date range between 1498-1428 BC with 95.4% certainty (Figure 8). Aston's study of archaeological, astronomical, textual, and radiocarbon results and their implications for New Kingdom regnal lengths concludes that the most likely accession date for Thutmose III should be 1493 BC (Aston, 2012/2013: 309-310; Bronk Ramsey *et al.*, 2010); this king is known to have died in his Year 54 (Aston, 2012/2013: 293), which would be 1440 BC according to this chronology. Thus, the radiocarbon dates for the tomb's juniper berries align well with the reign of Thutmose III, as does the ceramic evidence-based with the presence of black-painted rims that disappears from use after this reign. The scarab could have been an heirloom inscribed for one of his Thutmoside predecessors (Thutmose I or II), or it may have been inscribed for Thutmose III making it contemporary with the tomb's organic contents.

The ^{13}C stable isotope analysis on the juniper berries gave a significantly higher value of $\delta^{13}C= -19.5 \pm 0.1‰$ compared with the other samples from the site. The implications of this higher value for understanding the ecological environment in which the plant grew are explored in the Discussion section below.

Figure 6. Tomb 128, north end, head end of burial in loculus (Hearst Expedition photograph B-1069). Note that some pottery that was recorded in the notes as being found in this end of the tomb had already been removed at the time the photograph was taken. Courtesy of the Phoebe A. Hearst Museum of Anthropology and the Regents of the University of California.

Figure 7. Tomb 128 tomb assemblage (Hearst Expedition photograph C-1739). Note the globular jar with flaring mouth that held the fruit (top row, sixth vessel) and the intact dom fruit (bottom row, to the right of the offering stand). Missing from this photograph are the tall storage jar (visible in Figure 4), stone kohl jars, bronze razor, scarab, and other small finds. Courtesy of the Phoebe A. Hearst Museum of Anthropology and the Regents of the University of California.

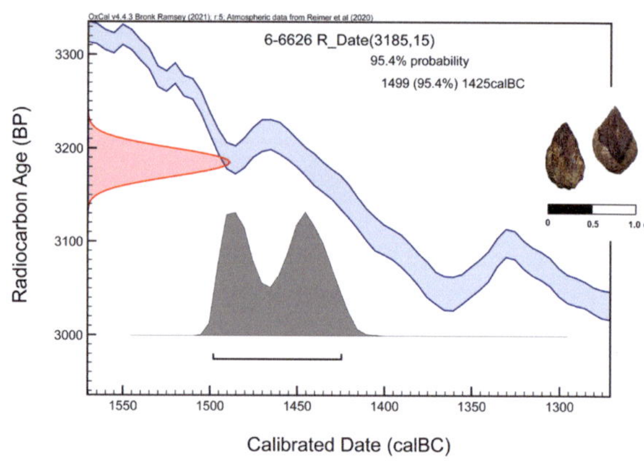

Figure 8. Sample #6-6626 *Juniperus Phoenicea* juniper berry, Tomb 128, Cemetery 1-200, Deir al-Ballas. Figure presents calibrated radiocarbon results using OxCal 4.3.2. Double lines denote the IntCal13 terrestrial calibration curve with 1-σ envelope (Reimer *et al.*, 2013). Inset denotes the results of Bayesian statistical analysis. Graphic by Brian Damiata, Cotsen Institute of Archaeology, UCLA.

Tomb 163

This was a shaft tomb that was oriented North-South and was located in the center of a room in the worker's village (Figure 2). The shaft was 210 cm deep, 70 cm wide and 230 cm long. Reisner recorded finding the lower legs of a human skeleton on a "rotten board", presumably the floor of a wooden coffin, the rest of which had decayed completely. There were twenty-two pottery vessels found in this tomb. Reisner noted the presence of roots inside a beer jar (6-8101, Holthoer's BB-3 type) (Holthoer, 1977: 86-88, Plate 18; Reisner, 1900: 173). However, at some point, the contents of this jar were emptied and the vessel as it is currently preserved at PAHMA does not contain any visible residue. Two dom fruit endocarps, one whole date and 34 grape pips (6-6895), along with a scarab inscribed with a spiral design (6-8798) was found within one of the beakers. At least 13 of these grape pips have part of the fruit skin preserved. A sample weighing 30 mgC was taken from grape pip 6-6895 for carbon dating. The sample yielded 54% of organic carbon. The AMS-^{14}C dates calibrated results fall in the range between 1501-1435 BC with 95.4% certainty (Figure 9). The sample size was too small to obtain δ^{13}C results. However, the measure of a powdered yet chemically untreated duplicate sample of grape pip was measured in the Dawson lab, yielding 28.11% C, which is almost half of the yield from the treated sample. The δ^{13}C value = -24.93‰ which falls in the range of regular C_3 plants in non-water stress conditions or irrigated. The nitrogen isotopic composition in this grape pip sample is d ^{15}N =+14.84‰ with C: N % ratio is 9.699%, a high value that reflects either the use of natural fertilizers from animal dung or richer soil fertility in Upper Egypt where this grape was grown (Szpak & Chiou, 2019; Araus *et al.*, 2014: 4), as opposed to modern depleted soil fertility reflected in modern type collections of similar species grown in the same region, due primarily to the impact of damming along the river, and the subsequent use of fossil-fuel based fertilizers which are depleted (close or below to 0‰) in their ^{15}N isotopic composition (Sharp, 2017: 9-7).[11]

The objects in Tomb 163 revealed an assemblage that was relatively more numerous and diverse than most tombs in the cemetery, despite the simple architectural style of the tomb. Small finds from the tomb whose exact findspot is uncertain include bronze tweezers, a necklace made of thin faience discs, penannular earrings, an ear stud, finger rings, a bone needle, as well as amulets made of glass and faience that include the deities Taweret, Sekhmet and Bes (Jensen, 2019: 115-119, Fig. 31). As shown in an expedition photograph (Figure 10), clustered around the lower legs at the foot end of the shaft were three stone kohl pots, two small biconical ceramic jars with painted decoration, offering dishes, and numerous small beaker jars, including the one mentioned above that contained the fruit and scarab. Near this beaker, beside the right leg a bronze razor and two scarabs were found, one of which was inscribed with the name of Hatshepsut's high official Senenmut (6-8795), giving his title as Steward of Amun (Jensen, 2019: 102, 104, 468-469, Appendix 3).[12] This helps to narrow down the carbon dating results for the grape seeds (1501-1435 BC), as Senenmut is thought to have been given this title around the time of Hatshepsut's coronation in Year 7 of Thutmose III's reign (Dorman, 1988: 119-120; Eichler, 2000: 11ff., 17 ff., 217; Shirley, 2014: 191), *i.e.* 1486 BC.[13] Senenmut's last known date in the office is Year 16 (1477 BC) although he may have lived a few more years longer (Dorman, 1988: 176-177; Hayes, 1960: 42-43; Shirley, 2014: 192). This date range for Senenmut's tenure as Steward of Amun helps to pinpoint the production date of the scarab to no earlier than ca. 1486 BC and thus indicates the earliest possible date of the botanical sample, but of course, the tomb owner might have been buried with this object after Senenmut's career had ended, therefore the AMS dating of the grapes is important to pinpoint the *terminus ante quem* date of burial as 1435 BC, some 40 years after Senenmut disappears from the historical record.

Tomb 244/255

A collection of diverse botanical remains in the Hearst Museum were registered under number 6-7482 and were attributed to Tomb 244 as their provenience. These included cereal chaff, a dom fruit mesocarp fragment, one tiger nut, and the intriguing findings of 7 seeds of domesticated watermelon. However, a point of confusion is that Reisner's field notebook did not include any notes on botanical findings from Tomb 244. Normally, he carefully numbered and described the material from each tomb (including plant remains) but in this case, he only sketched the tomb elevation and wrote "19 pots." However, he did record finding "melon and date seeds" in another tomb, Tomb 255. Since this was the only mention of melon seeds in all the excavation notes by Reisner, it is not certain whether the watermelon seeds held at PAHMA belong to

11 For example, nitric acid emission from farming using modern industrialized fertilizer and fossil fuel emission causes lower values of ^{15}N (Durka *et al.*, 1994: 765). This is extremely important in the case of Egypt especially after the increased use of industrial fertilizer after losing the naturally fertile soil deposited by the annual floods due to the construction of the high dam in Aswan, Egypt in 1971.

12 The complete inscription reads *ỉmy-rꜥ pr Ỉmn šnnmwt ms n ḥꜣt-nfr* (Overseer of the House of Amun, Senenmut, born of Hatnofer) (Jensen, 2019: 104). Photos of the scarab can be found on the Hearst Museum website at https://portal.hearstmuseum.berkeley.edu/catalog/3525bed0-5cd7-4d4c-ac90-8eeb00584777.

13 Following Aston's dating of the coronation of Thutmose III in 1493 BC, this would place Hatshepsut's coronation seven years later in 1486 BC and Year 16 of the joint reign would equate to 1477 BC (Aston, 2012/2013: 309-310).

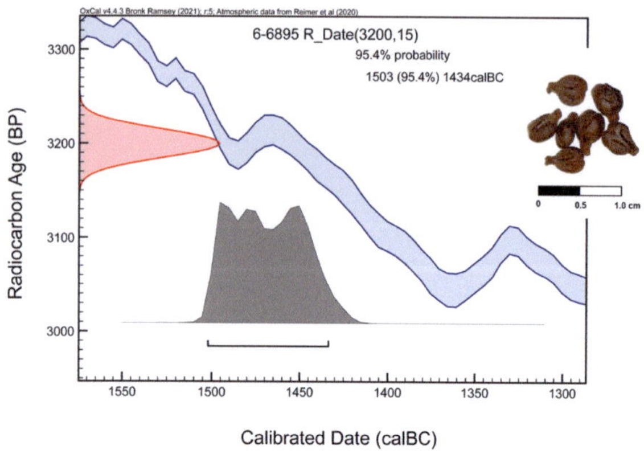

Figure 9. Sample #6-6895 *Vitis vinifera* grape pip fragment, Tomb 163, Cemetery 1-200, Deir al-Ballas. Figure presents the calibrated radiocarbon results using OxCal 4.3.2. Double lines denote the IntCal13 terrestrial calibration curve with 1-σ envelope (Reimer *et al.*, 2013). Inset denotes the results of Bayesian statistical analysis. Graphic by Brian Damiata, Cotsen Institute of Archaeology, UCLA.

Figure 10. Tomb 163, south end (Hearst Expedition photograph B-1061). The dom endocarp is visible in the mouth of the beaker on the right side of the photograph, under the large open dish. Courtesy of the Phoebe A. Hearst Museum of Anthropology and the Regents of the University of California.

Tomb 244 or 255. Nonetheless, both contexts can be dated to the early 18th Dynasty based on their similar pottery assemblages.

Tomb 244 was located in the open area west of the workers' village (Figure 2). It consisted of a vertical shaft with a loculus carved out at the bottom. No dimensions were recorded, but the sketch indicates that the loculus was separated from the shaft by a mudbrick partition wall that was apparently intact. No human remains were listed in the field notes, but Reisner took five photographs of the tomb, in some of which disarticulated human bones can be seen.[14] Although the contents of Tomb 244 were not described in Reisner's notes in detail, ten ceramic vessels at PAHMA have been associated with this tomb, thanks to the original field numbers preserved on the pottery. These include a biconical necked jar (6-7480) that is decorated with concentric red- and black-painted lines; such bichrome ware came into vogue in the reign of Thutmose III (Hope, 1987: 109).

Tomb 255 was described as being a pit with its entrance bricked up, which was covered with a 1 m deep layer of rock chips. The expedition did not record its location on the map. Human remains were not mentioned, but Reisner recorded the presence of a coffin that was painted red, black, yellow, and white. No information about the pattern or designs of the coffin were noted. Like Tomb 244, Tomb 255 contained pottery of the early-mid 18th Dynasty such as beakers with black-painted rims, one of which (6-7499) contained grain husks that are now lost. Tomb 255 also held a tall ovoid storage jar (6-7498) that was red-slipped and decorated with concentric black bands. This type of decorated jar finds parallels in Aston's New Kingdom Phase I, from the reigns of Ahmose through Thutmose II (Aston, 2003: 141, Fig. 1b) but also continues into the reign of Hatshepsut/Thutmose III, particularly in the Theban region (*e.g.*, Galán, 2014: 254).

The ^{14}C dates on the Deir al-Ballas watermelon seed revealed a date between 1465-1413 calBC with 73.3% certainty making it the earliest reliably identified watermelon found in Egypt, possibly introduced through interaction with Nubia, modern Sudan (Figure 11). According to Aston's suggested high chronology of 18th Dynasty reigns, this 52-year time period begins about 30 years into the reign of Thutmose III (1493-1440 BC) and extends into the reign of his successor, Amenhotep II (1440-1409 BC) (Aston, 2012/2013: 309). This date range aligns well with the Tomb 244 pottery, given the bichrome ware that appears in the reign of Thutmose III and continues into Aston's Phase 2B (Amenhotep II-Thutmose IV), but it also could match Tomb 255 if the actual date of the watermelon falls in the earlier half of the range suggested by the carbon dating, *i.e.*, the reign of Thutmose III.

14 The photographs are B-1054, B-1055, B-1056, B-1057, and B-1058.

The carbon isotope ratio δ^{13}C of the domesticated watermelon sample is -23.5 ± 0.1‰ which is that of a C_3 plant.

Tomb 257

This tomb revealed the greatest diversity of food offerings of this population, but unfortunately it has the least archaeological information that can be associated with its botanical remains. Under the number 257 in Reisner's field notebook, there is no description or sketch of a tomb and it was not identified on the cemetery map. The only items he recorded under this number were persea, *nabq*, emmer wheat, date pits, pomegranate, etc. botanical remains and one Nile silt beaker that was burnished vertically and has a black-painted rim (6-7520). There were 14 whole seeds and 15 fragments of persea found in this tomb, while the beaker contained two emmer wheat spikelets, four *nabq* fruits, seven grape pips, three date pits, as well as a large number of pomegranate rind and mesocarp fragments, one of which still held one red seed of pomegranate with the color preserved. A fragment of pomegranate rind (6-7517) was sampled for carbon dating, revealing dates between 1526-1447 cal BC with 95.4% certainty (Figure 12). The relative dating suggested by the ceramic vessel with its black-painted rim would place this context between the late 17th Dynasty through the reign of Thutmose III (Bourriau, 1982: 78). The absolute carbon date range obtained from the pomegranate sample helps narrow this window to the early 18th Dynasty, excluding the late 17th Dynasty and the reign of Ahmose.

The carbon isotope ratio δ^{13}C was -22.3 ± 0.1‰ indicating that the pomegranate is a C_3 plant. There was no sign of drought but the ratio indicates slightly water-stressed conditions compared with local species, perhaps implying that the pomegranate was imported. However, it is not straightforward to say whether it was grown locally in Egypt or imported without conducting further isotope analysis of oxygen and strontium.

Discussion

The composition of food offerings in the cemetery shows the significance of wild Upper Egyptian fruits such as dom, persea, *nabq*, desert dates, palm dates, and sycomore figs in addition to domesticated cereal grains (barley, wheat) as well as local and imported fruits (juniper berries). While it can be argued that finding these plant species in a burial context is not evidence that they were consumed or were a standard part of ancient Egyptian diet, the recent discovery of several of these fruit species (dom, date, sycmore fig, and grape) in the Deir al-Ballas settlement itself, refutes the idea that such plants were reserved exclusively for funerary use (Shahat, 2020).[15] However, specific foods may have had different meanings in the funerary context. For instance, dom fruit, which was the most common grave offering in Cemetery 1-200 after cereal grains, can rather be interpreted through the hymn to Thoth, in which the dom is mentioned to contain water as a blessing from Thoth to refresh the deceased (Gardiner, 1937: 85-86).

Meanwhile, the concept of offerings for the afterlife was not limited to traditional plants that were locally available. Imported or newly introduced species were also added to the food offerings across Egyptian history, and we see this introduction of novel species echoed in as part of cultural interactions between Egypt and other regions. Supportive evidence of cultural interaction to corroborate with the

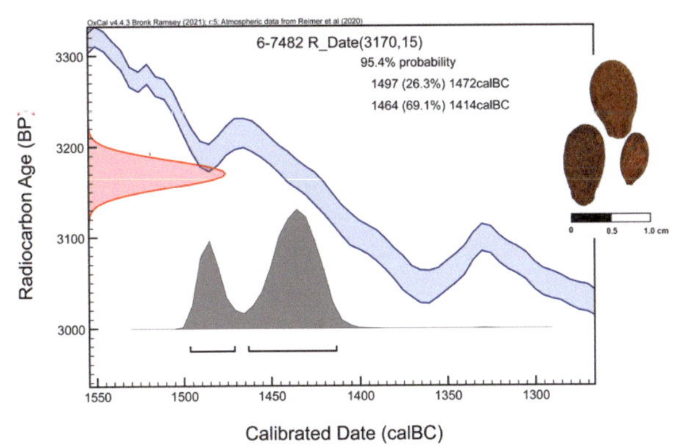

Figure 11. Sample #6-7482 *Citrullus lanatus* domesticated watermelon seed fragment, Tomb 244 or 255, Cemetery 1-200, Deir al-Ballas. Figure presents the calibrated radiocarbon results using OxCal 4.3.2. Double lines denote the IntCal13 terrestrial calibration curve with 1-σ envelope (Reimer *et al*., 2013). Inset denotes the results of Bayesian statistical analysis. Graphic by Brian Damiata, Cotsen Institute of Archaeology, UCLA.

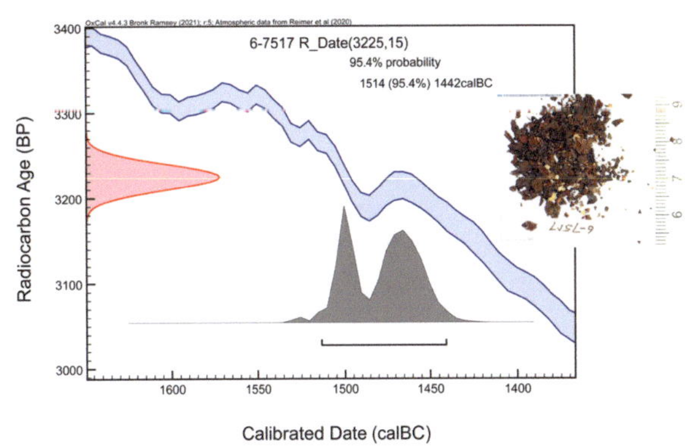

Figure 12. Sample #6-7517 *Punica granatum* pomegranate rind fragment, Tomb 257, Cemetery 1-200, Deir al-Ballas. Figure presents calibrated radiocarbon results using OxCal 4.3.2. Double lines denote the IntCal13 terrestrial calibration curve with 1-σ envelope (Reimer *et al*., 2013). Inset denotes the results of Bayesian statistical analysis. Graphic by Brian Damiata, Cotsen Institute of Archaeology, UCLA.

botanical remains comes from findings of Kerma pottery and Syrian vessels from both grave and domestic contexts at Deir al-Ballas. Textual evidence, albeit scarce, is found in ostraca found at the site that provide names of Canaanite and Nubian origins (Lacovara, personal communication 2021). It is beyond the scope of this paper to delve into an analysis of the socio-economic hierarchy at Deir al-Ballas and a thorough comparison of this provincial site with Thebes. However, we assert that while imported fruits may have been largely restricted to people of relatively high status, they were not entirely unreachable (albeit in small quantities) to people of lower social status as they became indigenized.[16] Indeed, all of the species observed in the graves of Cemetery 1-200, except pomegranate, have been encountered and published from the workmen's village in Amarna (Stevens & Clapham, 2014). The non-native species: watermelon, pomegranate, and juniper berries will now be discussed in turn.

Watermelon

The earliest finding of domesticated watermelon in Egypt is an area of debate in Egyptian archaeobotany. In 2004, Wasylikowa and van der Veen noted that the earliest reliable published identification of domesticated watermelon in Egypt came from the tomb of Tutankhamun (Cairo Museum # sp. 2792) (2004: 215), although other, early samples might belong to this species such as examples of *C. vulgaris* found in the gut of a mummy from the predynastic site of Nag' ed-Deir that were published by Netolitsky in 1943.[17]

More recently, the domestic watermelon species along with *C. colocynth* (sour melon) was published among the botanical discoveries from the vicinity of the workmen's village at Amarna (Stevens & Clapham, 2014: 157), suggesting that this species was not reserved for elite usage. To contribute new evidence to the debate on the appearance of watermelon in Egypt, we conducted AMS carbon dating on one of the Deir al-Ballas watermelon seed fragments, revealing a date between 1465-1413 calBC with 73.3% certainty. Identification of the seeds in our study as being the domesticated watermelon was confirmed by Claire Newton, Christine Hastorf, and Marijke van der Veen. Thus, while we cannot assert that the Deir al-Ballas sample is the earliest domesticated watermelon known in Egypt because of the disputed Predynastic sample mentioned above, it is between 60-100 years older than the confirmed *C. lanatus* specimens identified from the Amarna settlement.

The carbon isotope ratio $\delta^{13}C$ of the domesticated watermelon sample is -23.5 ± 0.1‰ which is that of a C_3 plant. At present, we cannot determine whether the Deir al-Ballas samples were grown locally or were imported but this study provides baseline stable isotope measures that we hope will be useful in answering this question as more samples from other regions are tested and published.

Meanwhile, results of oxygen isotope experiment shows huge impact of climate change and damming of the river on altering the oxygen isotope of the Nile water and plants irrigated with the Nile water. The average δ 18O of ancient plant remains= +29.53‰ showing better water availability and relative humidity conditions than in modern botanical samples which have average δ 18O of +40.29‰ indicating increased aridity and evaporative enrichment of 18O as a consequence of climate change and dams along the river (Shahat lab report 2019).

Pomegranate

Pomegranate was another non-native species that became integrated into Egyptian food culture as well as a funerary offering. The earliest report of pomegranate in Egypt was a specimen from Dra Abu al-Naga (Western Thebes) in what may have been a 12th Dynasty context (Schweinfurth 1884: 314),[18] but which is known securely from Tell ed-Dab'a in the Second Intermediate Period (Thanheiser, 2004: 378). The fruit became indigenized and was absorbed into Egyptian cultural foodways in the 18th Dynasty (Murray, 2000: 625; see Pearsall & Hastorf, 2011: 181-183 on indigenization of food). The fruit's acceptance into Egyptian culture seems to have been complete by the Thutmosid period, as witnessed by the botanical depictions of pomegranate trees and fruit in Thutmose III's Akh-Menu at Karnak Temple and elite tombs of the 18th Dynasty (Murray, 2000: 625). Several models of pomegranate-shaped faience were excavated by Loret in 1898 in KV 35, the tomb of

16 Although the example of the juniper berries in Tomb 128 co-occurred with a scarab inscribed with the name Thutmose, indicating that the individual may have been of a higher social status compared to other individuals in the cemetery by having access to imported fruit and material culture often associated with elites, the quantity (n=21) and quality of the fruits (small cones) are much less than those found in elite Theban tombs (*e.g.* Kha and Merit). Additionally, the Deir al-Ballas graves lack any meat and fowl among the food offering assemblages, items that are oftentimes found in elite Theban tombs and even the tombs of the craftsmen of Deir al-Medina.

17 Netolitsky was given the samples by Elliott Smith, who had received them from George Reisner. Netolitsky had some of the samples in his laboratory at Czernowitz which was destroyed in World War I, while others were sent to a colleague at Wageningen who died before publishing any results (Netolitsky, 1943: 5-7).

18 Schweinfurth refers to seeing pomegranate displayed in a case in the Boulaq Museum (the forerunner of the Egyptian Museum at Tahrir Square). These offerings were said to be from a 12th Dynasty vault from Dra Abu al-Naga. However, there is no way to verify this date. If the ceramics were the basis for this date, it should be remembered that Nineteenth century Egyptologists did not have as firm a grasp on the relative dating of pottery as is currently available and any dating assertions should be reviewed before they are accepted (C. Redmount, personal communication, May 2019).

Amenhotep II in the Valley of the Kings, currently in the Egyptian Museum in Cairo (JE 32452) (Loret, 1898). The discovery of pomegranates in both Tomb 128 and Tomb 257 at Deir al-Ballas indicates that this non-native fruit had become available (at least to a limited extent) to provincial members of Egyptian society by the early 18th Dynasty. In addition to these two tombs, according to Reisner's field notes, Tomb 102 contained a small, intact pomegranate but it has not been identified at PAHMA. However, one of the differences between elite and non-elite grave goods is that non-elite graves contained fewer of these food items than elite graves.

The carbon isotope ratio $\delta^{13}C$ was -22.3 ± 0.1‰ indicating that the pomegranate is a C_3 plant. The carbon isotope ratio indicates slightly water-stressed conditions."pomegranate actually shows also more arid conditions than the one grown locally. This is also confirmed by the 18O isotope value which is 30.85‰, indicating more arid condition and lower relative humidity where the pomegranate come from. However, it is not straightforward to say whether it was grown locally in Egypt or imported without conducting further isotope analysis of oxygen and strontium.

Juniper Berries

The juniper berries from Tomb 128 were confirmed to be ancient with ^{14}C dates of 1498-1428 BC. The sample had a significantly higher $\delta^{13}C$ value of -19.5 ± 0.1‰ which is very enriched for a C_3 plant and indicates that the berries grew in a water-limited condition in which the plant leaves closed the stomata longer than normal to avoid water loss (Escudero *et al.*, 2008: 705-713; Hartman *et al.*, 2010: 837-852).[19] In plentiful water conditions, C_3 plant leaves open their stomata more and discriminate against ^{13}C thus resulting in lower values for the ^{13}C isotope. The results of the ^{13}C isotope of the Deir al-Ballas juniper berries thus show clear evidence of water stress and ecological differences compared with plants that grew along the Nile that we have found in the same cemetery (*e.g.*, the dom fruit from Tomb 105 had a $\delta^{13}C$ value of -25.4 ± 0.1‰ (Shahat, 2019). Thus, the carbon isotope information not only speaks to the type of plant photosynthesis as a C_3 or C_4 crop but also presents information about a plant's water-use efficiency and hence provides valuable ecological information on the ancient environment in which the plant grew. It is noteworthy that the $\delta\ ^{15}N$ of this same ancient juniper berry is extremely high, with a value of +28.58‰. While the enriched carbon isotopic value indicated increased stomatal closure and hence drought conditions, $\delta\ ^{15}N$ provided additional evidence for the drought conditions as the $\delta\ ^{15}N$ values in plants increase with decreasing mean annual precipitation (Niespolo *et. al.*, 2020: 4). Therefore, we conclude that the juniper berries from these tombs grew in a very water-stressed environment.

Juniperus grows along with coastal areas throughout the Aegean including Crete, the Peloponnese, and the Cyclades, and is attested in mainland Greece since the Paleolithic (Asouti, 2018: 1-38; Asouti *et al.*, 2015: 1569; M. Ntinou, personal communication, November 2018). The species also flourishes in mountainous and water-limited areas in the Northern Levant; the closest location to the Nile Valley. It currently grows in north Sinai where it is an endangered species (El-Bana *et. al.*, 2010: 171). The carbon isotope of these archaeobotanical remains is therefore of significant implication for Egyptologists to understand food offerings and identify whether these plants are local to the Nile environment or were imported from other regions within or outside of Egypt. Ethnographically, it is known as *'Ar'ar sury*, Syrian juniper, and remains an import in Egyptian markets today (Shahat, ethnographic observation, 2019). We conclude that these were probably traded in from the eastern Mediterranean.

A combination of the carbon dating as well as the carbon stable isotope measurements help compare the juniper specimens with similarly dated archaeobotanical samples from other regions in the ancient Near East. Samples closer to our date discussed by Riehl and colleagues show similar high ^{13}C values for rain-fed C_3 crops from the Near East, corresponding to the well-documented drought conditions of the "4.2 k Event" that hit the Levant between (3000-1200 BC). This drought event had led to the fluctuation of climate conditions during the Bronze Age (Riehl *et al.*, 2008: 1011-1022). The sample of juniper berries that absolute dated to 1498-1428 cal BC corresponds to the later years of this drought event. This interpretation is based on the assumption that the ancient juniper samples retrieved at Deir al-Ballas may have originated in the northern Levant. Stable isotope analysis of strontium will be conducted in the future for further verification. It is clear that carbon isotope analysis of archaeobotanical remains contributes significant implications for Egyptologists not only to understand food offerings but also opens new areas of research about ancient foodways and past food ecology to determine whether these plants were local to the Nile environment or were imported from other regions within or outside of Egypt. In this case, we conclude that the juniper berries were imported and were not grown locally.

Implications for Bioarchaeology

The data-rich and informative study on human mummies by Touzeau *et al.* (2014), whose results agreed with results published by Iacumin *et al.* (1996) from the Nile valley and

19 The stomatal closure in drought conditions is driven by the plant hormone ABA or abscisic acid. ABA is known to have a very important "control" on stomatal (conductance) and thus play a role under water deficits that are either slowly imposed or caused by chronic water limitations (Lambers *et al.*, 2019: 60).

Dupras and Tocheri's study in the Dakhleh Oasis (2007), all show that the ancient Egyptian diet has as its major component, C_3 plants with a minor (<10%) contribution of C_4-derived foods, from sub-Sahara (Touzeau *et al.*, 2013: 122). Touzeau and colleagues interpreted the consistency of the C_3 carbon isotope levels by stating that *"ancient Egyptians had a relatively basic diet with a restricted number of food items"* (2013: 119). However, they also acknowledge that *"the C_3 plant group is by far the most diverse and comprises the majority of vegetables, cereals, and fruits, while C_4-plants are rare, and limited to millet and sorghum in Africa"* (2013: 115). Touzeau and colleagues briefly consider other food sources beyond cereals that contribute to the C_3 isotope values, noting that *"only indirect inferences can be made by considering the salaries paid in kind to pyramid workers and craftsmen from the King's Valley"* about vegetables and legumes (2013: 115). Notably absent from this list, however, are the wild, native fruits that are encountered archaeologically as funerary food offerings at Deir al-Ballas as well as at many other sites. Although the better preservation conditions in tombs have preferentially preserved these foods in the archaeological record, these foods were not restricted to funerary use and are now being recovered and identified from settlement areas at Deir al-Ballas (sycomore fig, dom, and dates, as well as the domesticated grape) (Shahat, 2020). Ethnographic evidence notes that wild fruits are still enjoyed by modern Egyptians. For instance, *nabq*, dom, and dates are collected and eaten by families in Upper Egypt; these fruits are also sold in local markets in Qurna and Luxor, Upper Egypt (Jensen and Shahat ethnographic observation, 2019), as well as in several parts of Cairo (S. Ikram, personal communication, 2020). Moreover, these species and many more (including domestic watermelon, grape, persea, desert date, and juniper) have been found are attested in the workmen's village at Amarna, in its refuse dump and animal pens, as well as the gardens of its private chapels (Stevens & Clapham, 2014: 158-161). In all, Stevens and Clapham present 60 plant species identified from Amarna, greatly enhancing our understanding of the breadth of plant use in popular contexts. Combined with the present study, we propose a more nuanced interpretation of the C_3 stable isotope signature by considering the archaeobotanical records that give a detailed view of what species contributed to the C_3 signature. The archaeobotanical analysis gives a contextualized understanding of the continuity and changes of the diet of C_3 taxa, which goes beyond domesticated wheat and barley to include a range of wild, local C_3 plants plus imported or newly introduced fruits such as pomegranate, watermelon, or juniper berries at this time in Egyptian history.

For the sources of C_4 isotopes in the Egyptian human evidence, a suggested origin is the consumption of animals fed with C_4 plants (Dupras *et al.*, 2001; Dupras & Tocheri, 2007; Touzeau *et al.*, 2013: 120) as the C_4 cereals (millet and sorghum) only were found regularly during the Roman period as a contributor to the human diet in Egypt (Touzeau *et al.*, 2013: 120). This archaeobotanical study from Deir al-Ballas invites us to think beyond domesticated cereals and the animals that consumed them to include wild plants and rhizomes that are regularly encountered in the archaeobotanical records when studied in detail. For example, a tiger nut rhizome from the Cyperaceae family was found in Deir al-Ballas in Tomb 244/255 with the watermelon. The tiger nut is a C_4 plant (Arnold *et al.*, 2016: 5) and bioarchaeologists should give it more consideration as a contributor to the ancient Egyptian diet. Its history of usage extends back to the Late Paleolithic in Egypt (Hillman *et al.*, 1989) and large quantities of tiger nuts have been found dating from the Predynastic (*e.g.*, Nag' al-Deir)[20] to the New Kingdom eras, found in the tombs of Kha and Merit[21] as well as the refuse pit beside the workmen's village at Amarna (Stevens & Clapham, 2014: 159). Thus, we respectfully would like to modify the statement made by Touzeau and colleagues, regarding sorghum and millet being the only C_4 plant sources encountered in Africa (2013: 115), to include wild species such as tubers and rhizomes such as tiger nut.

Conclusions

This study has confirmed that the carbon dates obtained for the botanical samples support the relative dating provided by the ceramics and other material culture artifacts such as scarabs found in the tombs at Deir el Ballas. Despite significant plundering of the cemetery over the years, the AMS carbon dates obtained from the plant remains align with the early-mid 18th Dynasty dates suggested by the material culture, thus authenticating that the archaeobotanical samples were part of the original grave goods. While the plant remains that were found within ceramic containers in the tombs were not likely to be intrusive, it was particularly important to conduct independent testing in the case of the watermelon seeds from Tomb 244 or 255 that were not associated with any particular vessel to demonstrate whether or not they dated to the same time period as the rest of the tomb assemblage. Additionally, the dates obtained for the dom sample from Tomb 105 informed us that this tomb dated to the late Eighteenth-early 19th Dynasty; the absence of

20 Baskets full of tiger nut rhizomes were excavated at Nag' al-Deir by Reisner and Lythgoe (Shahat, 2019).

21 Among funerary contexts, *Cyperus esculentus* (n=8) is attested in the tomb of Kha, where it was found in a box along with 5 almonds (Shahat, unpublished botanical analysis report, Museo Egizio, Turin, 2019).

material culture from this context had made dating the tomb impossible without this archaeobotanical data.

The juniper sample produced ^{13}C results that point to the plant having grown in water-stressed conditions, presumably outside of Egypt, and especially outside of the Nile river valley. However, understanding whether the pomegranate and watermelon grew locally in a process of indigenization or they were imported to the site (either from within or outside of Egypt) is the subject of further rigorous stable isotope analysis of oxygen and strontium for identifying their water source and the region in which they grew.

Describing the identified species of archaeobotanical remains found in archaeological contexts is important, but it does not substitute for the importance of conducting stable isotope analysis on both the archaeobotanical materials and on the human remains as well. Each analysis contributes to a different set of information that complements the others. When added to evidence from human remains, stable isotope analysis can provide a broader picture regarding the extent these foods contributed to the diets of different individuals in the cemetery and whether the individual buried was a local person or an immigrant. However, we did not have the opportunity to conduct stable isotope on the human bodies as we lack information on where Reisner stored the human remains that he excavated at Deir al-Ballas.

In summary, this variety of food offerings combining local wild and domesticated plants as well as imported fruits, most of which are isotopically C_3 plants, invite bioarchaeologists to be more broad-minded about their interpretations of carbon stable isotope values from human teeth and bones (*e.g.*, Touzeau *et al.*, 2013: 122). The archaeobotanical analysis is thus of great importance to better understand the extent of the repertoire of C_3 plants that contributed to the diet that resulted in these isotopic values.

This study of the food offerings from the non-elite Cemetery 1-200 at Deir al-Ballas provides evidence of the complexity of Egyptian plant use in the diet for the residents that we can conclude combined native wild plants, domesticates, as well as the rare imported species. While stable isotope analysis on human teeth, bones, and hair confirms the continuity of a generally C_3-based diet, archaeobotanical materials provide a more refined view of the "menu" of plant taxa that constituted the basis of these isotope results. This menu was not fixed but rather shows both continuity and changes over Egyptian history. Moreover, even though C_4 plants were a minor contributor to the diet, the identification of their archaeobotanical remains gives us an idea of what C_4 plants may have made to this contribution, such as the tiger nut rhizome. Further research on additional native and imported species to determine whether they were C_3 or C_4 plants may help to clarify the picture of the ancient Egyptian diet even further and contribute to our understanding of how Egyptian foodways were shaped and reshaped by Egypt's social history of cultural interactions.

Bibliography

Araus, J., J.P. Ferrio, J. Voltas, M. Aguilera & R. Buxo. 2014. Agronomic Conditions and Crop Evolution in Ancient Near East Agriculture. – Nature Communications 5: 3953, Online, DOI: 10.1038/ncomms4953, www.nature.com/naturecommunications.

Arnold, E.R., G. Hartman, H.J. Greenfield, I. Shai, L.E. Babcock & A.M. Maeir. 2016. Isotopic Evidence for Early Trade in Animals between Old Kingdom Egypt and Canaan. - PloS one, 11 (6), p.e0157650.

Aston, B. 1994. Ancient Egyptian Stone Vessels: Materials and Forms. Studien zur Archäologie und Geschichte Altägyptens. – Cairo/Heidelberg, Deutsches Archäologisches Institute Abteilung Kairo/Ägyptologisches Institut Universität Heidelberg.

Aston, D. 2012/2013. Radiocarbon, Wine Jars and New Kingdom Chronology. – Ägypten und Levante 22/23: 289-315.

Aston, D. 2003. New Kingdom Pottery Phases as Revealed Through Well-Dated Tomb Contexts. In: Bietak, M. Ed. The Synchronisation of Civilisations in the Eastern Mediterranean in the Second Millennium B.C. II: Proceedings of the SCIEM 2000 – EuroConference, Haindorf, 2nd of May – 7th of May 2001. – Vienna, Verlag der Österreichischen Akademie der Wissenschaften.

Asouti, E., C. Kabukcu, C.E. White, I. Kuijt, B. Finlayson & C. Makarewicz. 2015. Early Holocene Woodland Vegetation and Human Impacts in the Arid Zone of the Southern Levant. - The Holocene 25 (10): 1565-1580.

Asouti, E., M. Ntinou & C. Kabukcu. 2018. The Impact of Environmental Change on Palaeolithic and Mesolithic Plant Use and the Transition to Agriculture at Franchthi Cave, Greece. - PloS ONE 13 (11), p.e0207805. https://doi.org/10.1371/journal.pone.0207805.

Bourriau, J. 1982. 52 Drinking Cup. In: Anonymous. Ed. Egypt's Golden Age: The Art of Living in the New Kingdom, 1550-1085 B.C. – Boston, Museum of Fine Arts: 78-79.

Bourriau, J. 1990. The Pottery. In: Lacovara, P. Ed. Deir el-Ballas: Preliminary Report on the Deir el-Ballas Expedition, 1980-1986. – Winona Lake, Eisenbrauns (American Research Center in Egypt Reports 12): 15-22.

Bronk Ramsey, C., M.W. Dee, J.M. Rowland, T.F.G. Higham, S.A. Harris, F. Brock, A. Quiles, E.M. Wild, E S. Marcus

& A.J. Shortland. 2010. Radiocarbon-Based Chronology for Dynastic Egypt. – Science 328: 1554-1557.

Chomicki, G. & S.S. Renner. 2015. Watermelon Origin Solved with Molecular Phylogenetics including Linnaean Material: Another Example of Museomics. – New Phytologist 205: 526-532.

Daressy, G. 1894. Notes et Remarques. – Recueil de travaux relatifs à la philologie et à l'archéologie égyptiennes et assyriennes 16: 44.

Dorman, P. 1988. The Monuments of Senenmut: Problems in Historical Methodology. – London, Kegan Paul International (Studies in Egyptology).

Dupras, T.L. & M.W. Tocheri. 2007. Reconstructing Infant Weaning Histories at Roman Period Kellis, Egypt using Stable Isotope Analysis of Dentition. – American Journal of Physical Anthropology 134: 63-74.

Eichler, S.S. 2000. Die Verwaltung des "Hauses des Amun" in der 18. Dynastie. – Hamburg, Helmut Buske (Studien zur altägyptische Kultur, Beiheft 7).

El-Bana, M., K. Shaltout, A. Khalafallah & H. Mosallam. 2010. Ecological Status of the Mediterranean *Juniperus phoenicea* L. Relicts in the Desert Mountains of North Sinai, Egypt. - Flora 205(3): 171-178.

Escudero A., S. Mediavilla & H. Heilmeier. 2008. Leaf Longevity and Drought: Avoidance of the Costs and Risks of Early Leaf Abscission as Inferred from the Leaf Carbon Isotopic Composition. – Functional Plant Biology 35 (8): 705-713.

Faure, G. & T.M. Mensing. 2005. Isotopes: Principles and Applications. – Chichester, John Wiley and Sons (3rd Ed).

Frood, E. 2010. Social Structure and Daily Life: Pharaonic. In: Lloyd, A.B. Ed. A Companion to Ancient Egypt. – Oxford, Blackwell: 469-490.

Gardiner, A.H. 1937. Late-Egyptian Miscellanies. – Brussels, Fondation Égyptologique Reine Élisabeth.

Galán, J.M. 2014. The Inscribed Burial Chamber of Djehuty (TT11). In: Galán, J.M., B.M. Bryan & P.F. Dorman. Eds. Creativity and Innovation in the Reign of Hatshepsut. – Chicago, University of Chicago Press (Studies in Ancient Oriental Civilizations 69) : 247-272.

Hartman G. & A. Danin. 2010. Isotopic Values of Plants in Relation to Water Availability in the Eastern Mediterranean Region. – Oecologia 162(4): 837-852.

Hayes, W.C. 1960. A Selection of Tuthmoside Ostraca from Deir el-Bahri. – Journal of Egyptian Archaeology 46: 29-52.

Hillman, G., E. Madeyska & J. Hather. 1989. Wild Plant Foods and Diet at Late Paleolithic Wadi Kubbaniya: The Evidence from Charred Remains. In: Wendorf, F., R. Schild, & A. E. Close. Eds. The Prehistory of Wadi Kubbaniya 2: Stratigraphy, Paleoeconomy, and Environment. – Dallas, Southern Methodist University Press: 162-242.

Holthoer, R. 1977. New Kingdom Pharaonic Sites: The Pottery. – Uppsala, Scandinavian Joint Expedition to Sudanese Nubia (The Scandinavian Joint Expedition to Sudanese Nubia 5).

Hope, C. 1987. Innovation in the Decoration of Ceramics in the Mid-18th Dynasty. – Cahiers de la Ceramique Egyptienne 1: 97-122.

Iacumin, P., H. Bocherens, A. Mariotti & A. Longinelli. 1996. An Isotopic Palaeoenvironmental Study of Human Skeletal Remains from the Nile Valley. – Palaeogeography, Palaeoclimatolology and Palaeoecolology 126: 15-30.

Jensen, V. 2019. The Cemeteries of Deir el-Ballas: Non-elite Burials of the 17th-19th Dynasties and Their Relationship to the Royal Palace. – Unpublished PhD Thesis, University of California, Berkeley.

Lacovara, P. 1981. The Hearst Excavations at Deir el-Ballas: The 18th Dynasty Town. In: Simpson W.K. & W.M. Davis. Eds. Studies in Ancient Egypt, the Aegean, and the Sudan: Essays in Honor of Dows Dunham. – Boston, Museum of Fine Arts: 120-124.

Lacovara, P. 1990. Deir el-Ballas: Preliminary Report on the Deir el-Ballas Expedition, 1980-1986. – Winona Lake, Eisenbrauns (American Research Center in Egypt Reports 12).

Lacovara, P. 1997. The New Kingdom Royal City. – London, Kegan Paul.

Lambers, H., F.S. Chapin III. & T.L. Pons. 2019. Plant Physiological Ecology. – Springer Science and Business Media, E-Book.

Lilyquist, C & R. Brill. 1993. Studies in Early Egyptian Glass. – New York, Metropolitan Museum of Art.

Loret, V. 1898. Le tombeau d'Aménophis II et la cachette royale de Biban el-Molouk. – Bulletin de l'Institut Egyptien 9: 98-112.

Merrillees, R. 1968. The Cypriote Bronze Age Pottery Found in Egypt. – Lund, Studies in Mediterranean Archaeology (Studies in Mediterranean Archaeology 18).

Minor, E. forthcoming. Faience Tiles from Deir el-Ballas and Kerma: New Evidence of Egyptian-Nubian Relations at the Foundation of the New Kingdom. In: Kiser-Go, D. & C. Redmount. Eds. Weseretkau "Mighty of Kas": Papers Submitted in Memory of Cathleen A. Keller.

Murray, M.A. 2000. Fruits, Vegetables, Pulses and Condiments. In: Nicholson, P. & I. Shaw. Eds. Ancient Egyptian Materials and Technology. – Cambridge, Cambridge University Press: 609-655.

Netolitzky, F. 1943. Nachweise von Nahrungs- und Heilmitteln in den Trockenleichen von Naga-ed-der (Ägypten). – Mitteilungen des Deutschen Instituts für Ägyptische Altertumskunde in Kairo, Erstes Ergänzungsheft: 5-33.

Niespolog, E.M, W.D. Sharp, C.A. Tryon, J.T. Faith, J. Lewis, K. Ranhorn, S. Mambelli, M.J. Miller & T.E. Dawson. 2020. Carbon, Nitrogen and Oxygen Isotope of Ostrich Eggshells Provide Site-scale Pleistocene-Holocene Paleoenvironmental Records of Eastern African Archaeological Sites. – Quaternary Science Reviews 230: 106142.

Olsson, I. 1986. Radiometric Methods. In: Berglund, B. Ed. Handbook of Holocene Palaeoecology and Palaeohydrology. – Chichester, John Wiley and Sons: 273-312.

Pearsall, D.M. & Hastorf, C. 2011. Reconstructing Past Life-Ways with Plants II: Human-Environment and Human-Human Interactions. In: Anderson, E.N., D. Pearsall, E. Hunn & N. Turner. Eds. Ethnobiology. – Hoboken, Wiley-Blackwell: 173-187.

Petrie, W.F. & J.E. Quibell. 1896. Naqada and Ballas. – London, Bernard Quaritch.

Reimer, P.J., E. Bard, A. Bayliss, J.W. Beck, P.G. Blackwell, C.B. Ramsey, C.E. Buck, H. Cheng, R.L. Edwards, M. Friedrich & P.M. Grootes. 2013. IntCal13 and Marine13 Radiocarbon Age Calibration Curves 0-50,000 years cal BP. – Radiocarbon 55 (4): 1869-1887.

Reisner, G.A. 1900. Hearst Expedition Notebook D-3. Unpublished field journal. – Boston, Museum of Fine Arts.

Reisner, G.A. 1905. The Hearst Medical Papyrus: Hieratic Text in 17 Facsimile Plates in Collotype with Introduction and Vocabulary. – Leipzig, J.C. Hinrichs (University of California Egyptian Archaeology 1).

Reisner, G.A. 1908. Naga ed-Der. – Leipzig, J.C. Hinrichs (University of California Egyptian Archaeology 2).

Renner, S.S., G. Chomicki, & W. Greuter. 2014. Proposal to Conserve the name *Momordica lanata (Citrullus lanatus)* (watermelon, *Cucurbitaceae*), with a Conserved Type, against *Citrullus battich*. – Taxon 63(4): 941-942.

Savage, S.H. 1998. AMS Radiocarbon Dates from the Predynastic Egyptian Cemetery, N7000, at Naga-ed-Der. – Journal of Archaeological Science 25: 235-249.

Schweinfurth, G. 1884. Further Discoveries in the Flora of Ancient Egypt. – Nature 29: 312-315.

Shahat, A. & V. Jensen. In press. Social Archaeology of Food at Deir el Ballas: an Archaeobotanical Study of the Non-Elite Cemetery Food Offerings. In: El Dorry, M.-A. Ed. Feeding a Civilisation: Food and Drink in Egypt and Sudan. – Cairo, Institut français d'archéologie orientale (Bibliothèque d'Étude. Cairo).

Shahat, A. 2019. Internal lab report of carbon dates and stable isotope baselines for botanical samples from Egypt held at the Phoebe A. Hearst Museum of Anthropology. (manuscript in author's possession).

Shahat, A. 2020. Unpublished field report, Deir el-Ballas expedition 2019-2020 (manuscript in author's possession).

Sharp, Z. 2017. Principles of Stable Isotope Geochemistry (2nd Ed). Published online doi: https://doi.org/10.25844/h9q1-0p82.

Shirley, J.J. 2014. The Power of the Elite: the Officials of Hatshepsut's Regency and Coregency. In: Galán, J.M., B.M. Bryan & P.F. Dorman. Eds. Creativity and Innovation in the Reign of Hatshepsut. – Chicago, University of Chicago Press (Studies in Ancient Oriental Civilizations 69): 173-246.

Stevens, C. & A.J. Clapham. 2014. Botanical Insights into the Life of an Ancient Egyptian Village: Excavation Results from Amarna. In: Stevens, C.J., S. Nixon, M.A. Murray, & D.Q. Fuller. Eds. Archaeology of African Plant Use. – Walnut Creek, Left Coast Press (University College London Institute of Archaeology 61): 151-164.

Szpak, P. & K.L. Chiou. 2019. A Comparison of Nitrogen Isotope Compositions of Charred and Desiccated Botanical Remains from Northern Peru. – Vegetation History and Archaeobotany: 1-12.

Thanheiser, U. 2004. Die Pflanzenreste. In: Hein, I & P. Jánosi. Eds. Tell el-Dab'a XI: Areal A/V, Siedlungsrelikte der Späten 2. Zwischenzeit. – Vienna, Verlag der Österreicchischen Akademie der Wissenschaften: 377-383.

Thompson, A.H., L. Chaix & M.P. Richards. 2008. Stable Isotopes and Diet at Ancient Kerma, Upper Nubia (Sudan). – Journal of Archaeological Science 35(2): 376-387.

Touzeau, A., R. Amiot, J. Blichert-Toft, J.P. Flandrois, F. Fourel, V. Grossi, F. Martineau, P. Richardin & C. Lécuyer. 2014. Diet of Ancient Egyptians Inferred from Stable Isotope Systematics. – Journal of Archaeological Science 46: 114-124.

Wasylikowa, K. & M. van der Veen. 2004. An Archaeobotanical Contribution to the History of Watermelon, *Citrullus lanatus* (Thunb.) Matsum. & Nakai (syn. *C. vulgaris* Schrad.). – Vegetation History and Archaeobotany 13: 213-217.

Wetterstrom, W. 1990. The Botanical Remains. In: Lacovara, P. Ed. Deir el-Ballas: Preliminary Report on the Deir el-Ballas Expedition, 1980-1986. – Winona Lake, Eisenbrauns (American Research Center in Egypt Reports 12): 25.

Animal Butchering Technology in Old and Middle Kingdom Egypt: Textual and Iconographic Evidence for the Shift from Stone to Metal Tools

Eléuterio Sousa & Haskel J. Greenfield

Introduction

There are very few Egypt-based studies that consider the overall transition from stone to metal-based technologies, particularly those dealing with quotidian activities, such as animal butchery. While some research on this topic has been conducted on meat processing in ancient Egypt (*e.g.* Ikram, 1995), there is very little consideration of the types of technologies used in the butchering process and their chronology of adoption. This lack of consideration has relevance for the larger issue of the origins of metallurgy and its influence on mundane aspects of daily life (*i.e.* food processing).

In this study, we summarise the existing literature on the butchery technologies during the Old Kingdom in order to answer the following question: when does the shift from a stone- to metal-based (*i.e.* bronze) tool technology take place for such mundane matters as meat processing? Even though a wealth of data have been collected, it is still not clear when utilitarian metallurgy appears and is widely adopted in Egypt for quotidian activities. Even though there is evidence for the use of copper metal objects beginning in the pre-Dynastic most or all copper tools appear to be prestige-related since they were used for the display of status by Egyptian elites (Odler, 2016).

Some scholars consider the transition from a stone- to a metal-based technology to have occurred during the Middle Kingdom (ca. 1975-1640 BC) since bronze tools only begin to appear at this time (Graves-Brown, 2010: 188; Ikram, 1995: 66-75). However, the presence of copper metal tools from the Early Dynastic and Old Kingdom (ca. 2680-2180 BC) suggests that a metal-based technology for animal butchery may have appeared earlier than has been suggested, at least to a certain level or specific activity within the society.

In this survey, we review the evidence for animal processing with flint and metal implements (*e.g.* knives), and summarise current knowledge on the introduction of bronze utilitarian metallurgy. These will provide insights on how widespread metal use was in daily life and if it was available to individuals from all social strata of ancient Egyptian society. This study is important for increasing our understanding of the relationship between tool type, the nature of the transition from stone to metal-based technology, and the relationship between status and resources.

Evidence of Animal Butchery

The evidence that ancient Egyptians killed, processed, and ate meat is extensive from the earliest periods onwards. In the Neolithic and Chalcolithic periods (Badarian and Naqada

In: Ikram, S., J. Kaiser & S. Porcier. Eds. 2021. *The Ancient Egyptians and the Natural World. Flora, Fauna, and Science.* – Leiden, Sidestone Press: 253-264.

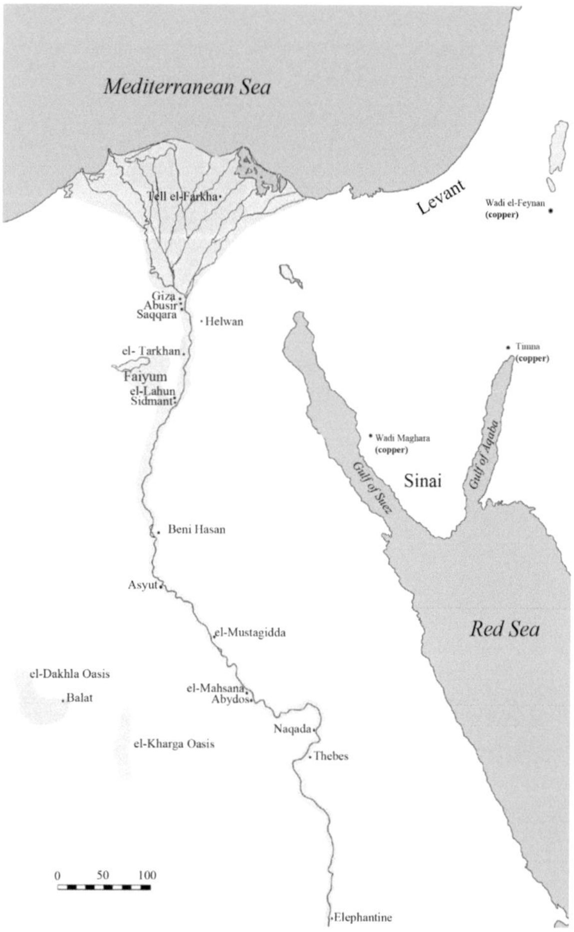

Figure 1. Map of Egypt with the names of locations mentioned in text.

I-II), the faunal remains recovered from a variety of sites demonstrate the importance of meat to ancient Egyptian culture (*e.g.* Köhler, 2010: 27; Wetterstrom, 1993). While there is limited iconography from the Early Dynastic period in general, it is only from the Old Kingdom onward that texts and iconography with evidence of butchery begin to appear. These are discussed next.

Two-Dimensional Butchery Scenes

Butchery scenes are frequently illustrated in tombs for the elite dating from the Old Kingdom onward. Although cattle are most commonly depicted in these scenes, a few cases of wild game, such as gazelle, have been recorded (Ikram, 1995; Strandberg, 2009: 103). An Old Kingdom limestone relief from the tomb of Niankhenesut, Saqqara (5th-6th Dynasty, ca. 2300 BC; AS5967, Tomb 67, Porter & Moss, 1981: 696), depicts men slaughtering cattle (or oxen) (Figure 2). In this scene, what appears to be a large knife is used to skin, chop, slice, and dismember the animal for food. In another Old Kingdom tomb, that of Iput (Porter & Moss, 1981: 396; Firth & Gunn 1926: 1-14) at Saqqara (6th Dynasty, ca. 2325-2150 BC), there is a depiction of butchers slaughtering what is possibly an ox – the sex is indeterminate because the sexual organs are not clearly depicted (Figure 3). Butchery scenes, such as those mentioned above, have successfully been used to understand the nature of animal processing in ancient Egypt and the different tool types used in butchering activities (Ikram, 1995).

Three-Dimensional Butchery Scenes

Three-dimensional models of animal butchering also exist from the late Old Kingdom through the early Middle Kingdom, and are useful in creating a picture of animal butchery in ancient Egypt. From the Old Kingdom, there is only a single example – a limestone painted figurine of a butcher slaughtering an ox (E10626, Oriental Institute Chicago) found in the tomb of Nykauinpu at Giza dating from the 5th Dynasty (ca. 2477-2466 BC; Figure 4; Porter & Moss, 1974: 300). The man holds a knife in preparation of slaughtering the animal. The figurine presents a rare example of three-dimensional depictions of butchery in the Old Kingdom since animal butchery scenes are generally represented only in two-dimensional reliefs in tombs during this period.

During the First Intermediate Period and Middle Kingdom, butchery scenes are more commonly represented by three-dimensional models of animal butchery. These are in addition to two-dimensional ones. These models include the well-known painted wooden models recovered from a variety of tombs. The wooden models became a common addition (or replacement) for the representation of two-dimensional tomb scenes from the Old Kingdom. The wooden models that portray animals being butchered with a large knife from the tomb of Meketre in Thebes (Tomb 280; Porter & Moss, 1960: 359-364) (ca. 11th-12th Dynasty) are some of the best-preserved (Figures 5-8). Similar models come from a variety of tombs in Middle Egypt (Beni Hassan and Asyut) and near the Faiyum (Sidmant), such as a model found in the tomb of Sebekhetepi at Beni Hasan (Tomb 723) represents a cow whose neck has been painted with a red line to denote its slaughter (BM EA41576 – Tooley, 1989: 46) (Figure 6). The knife is not preserved. Another example of a butchery model comes from the 12th Dynasty tomb of Khety (Tomb 366) at Asyut that shows butchers slaughtering cattle (Breasted, 1948: 37; Garstang, 1907: 105, 224; Strudwick & Dawson, 2016: 142, 145). A unique model of a butcher's workshop was also found in 11th Dynasty at Asyut (Tomb 7) (Breasted, 1948: pl. 33; Chassinat & Palanqe, 1911: 50, pl. XV). In other words, models of butchery scenes appear at the end of the Old Kingdom and continue into the Middle Kingdom.

There seems to be no correlation between the placement of daily life models in tombs and regional

Figure 2. Animal butchery scene. Painted limestone relief. Provenance: Tomb of Nianchenesut, Saqqara. Old Kingdom (5th-6th Dynasty). State Museum of Egyptian Art, Munich, Germany. ÄS 5970 © Osama Shukir Muhammed Amin FRCP(Glasg)/Ancient History Encyclopedia.

Figure 3. Animal butchery scene. Painted limestone relief. Provenance: Tomb of Idut, Saqqara. Old Kingdom (5th-6th Dynasty). © Werner Forman Art Resource.

Figure 5. Wooden model of a slaughterhouse. Painted wood. Provenance: Tomb of royal chief Meketre, Thebes. Reign of Amenemhat I, 12th Dynasty, Middle Kingdom. Metropolitan Museum of Art, New York. MET 20.3.10. © H. Greenfield.

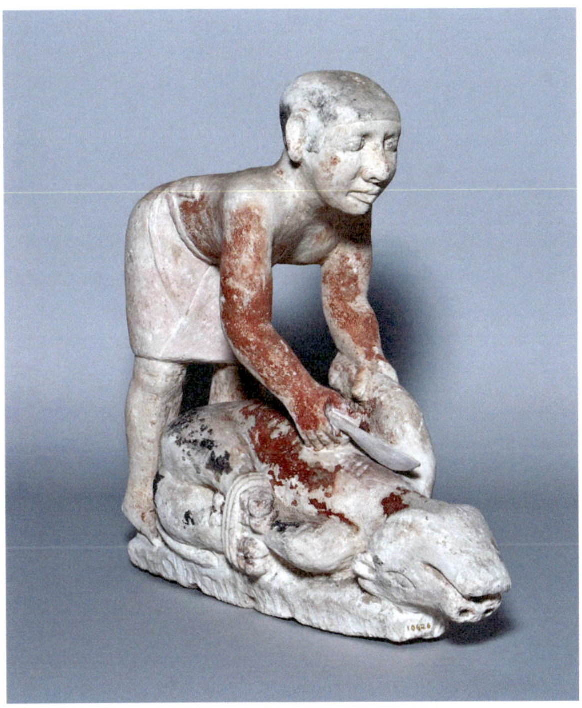

Figure 4 (left). Figurine of a man slaughtering a cow. Painted limestone. Provenance: Tomb of Nykauinpu, Giza. Old Kingdom (3rd Dynasty). Oriental Institute Museum, Chicago. E10626. © The Oriental Institute Museum Chicago.

Figure 6. Wooden model of a butchery activity. Painted wood. Provenance: Tomb of Sebekhetepa, Tomb 723, Beni Hasan. Middle Kingdom. British Museum, London. EA41576. © The British Museum.

religious practices. Models reflecting animal butchery scenes are found throughout Egypt. The prominence of animal butchery three-dimensional wooden models in funerary contexts reflects the ritual role that animal butchery played in ensuring the well-being of the deceased in the afterlife. The variation in artistic quality and style between models suggests that there was a range of artists working on these painted wooden models throughout Egypt and that production was not localised. Compare the aesthetic quality execution of the model in BM EA41576 (Figure 6) with that of MET 20.3.10 (Figure 5) – one is executed with greater aesthetic qualities than the other.

Written Evidence for Butchery

Written records can also be a line of evidence for understanding animal butchery and its place within early Egyptian society. There is clear evidence that butchery was a specialised profession already during the Old Kingdom. Literary evidence of the titles of individuals dating from the Old Kingdom suggests that local specialist butchers were hired for animal slaughter.

The early 3rd Dynasty tomb of Neferherenptah Fefi at Giza (G 8412; Porter & Moss, 1927: 253) contains two lintels that bear inscriptions with the title "Butcher of the slaughterhouse" that belong to his son, Itisen (Fisher, 1960: 171; Porter & Moss, 1927: 253). Perhaps the most exquisite examples of butchery titles from the Old Kingdom is found in the tomb of Irukaptah, also known as the "Tomb of the Butchers" (McFarlane 2000; Porter & Moss, 1927: 639) at Saqqara (5th-6th Dynasty) who held the title "Butcher of the Main Property of the Palace" *i.e.* "Head Butcher", "The Royal Overseer of the Butchers", "The Royal Acquaintance, the Head Butcher" (Rachewiltz, 1960). Itukaptah's tomb is beautifully carved with portraits of himself and his family and a well painted butchery scene illustrating men slaughtering an ox with black and white hide (Rachewiltz, 1960: 17-18).

A statue from Saqqara also dating to the 5th Dynasty from a member of the royal household has the title "Butcher of the slaughterhouse" and "Master Butcher" (Fisher, 1960: 171). The official Seshemnefer, who served under Pharaoh Unas (6th Dynasty) as the 'Overseer of the Tenants of the Great House', is also given the title 'Master Butcher' on his mastaba at Saqqara (Rice, 1999: 187; Saad, 1947: 56-57).

Khenemu (5th Dynasty) held the title "The Master Butcher of the Great House" as denoted by the inscription on the lintels of his tomb in the Western Cemetery at Giza (Tomb G2191; Morenz, 1998). The tomb of Neferherenptah (priest for the mortuary cult of kings Khafre and Menkaure, 5th-6th Dynasties) also contained statues of his family, one of which belonged to his son Tesen who is given the title "Butcher in the Palace Slaughterhouse" (Hassan, 1944: 279-87, figs. 143-50; Rice, 1999: 131). The titles given to these individuals demonstrate a link between the royal administration and animal butchery that places some butchers (or butchery administrators) within the upper strata of ancient Egyptian society.

Butchers also had titles that placed them within the lower strata of society (Fisher, 1960: 177; Ikram, 1995: 109). For instance, some individuals are simply given the title of butcher as is evident from tombs at the Old Kingdom cemetery of Qubbet al-Hawa (Vischak, 2015: 198). Nonetheless, the presence of 'butcher' as a title denotes the presence of a specialised 'craft' during the Old Kingdom.

The differences in titles clearly demonstrate a hierarchy in animal butchery activities. There is a clear social ranking among butchers as reflected in these titles, such as the "master butcher" who held a higher rank among butchers (Fisher, 1960; Ikram, 1995: 109-111). This is evident even in the wooden models. The wooden model slaughterhouse found in the tomb of Meketre (Figure 5) depicts two higher status individuals (holding sceptres) who oversee the slaughterhouse and the workers. One (perhaps the "Royal Overseer of the Butchers") claims his status by placing the sceptre onto the shoulder of a butcher at work. It is clear from the evidence that the administration of animal butchery in ancient Egyptian society was directly linked to social hierarchy, whereby higher ranked individuals supervise those from the lower social strata.

Deducing Technology from Iconography and Models

It is very difficult to reconstruct the nature of raw materials used in butchery from the iconographic and three-dimensional models. While many of the models have been well-preserved, which allowed for their original shape and colour to be maintained, the knife held by the butcher is missing in some of them (*e.g.* Figure 6). In these cases, the butcher's hand is empty, where once a model of the knife existed. This makes tool identification impossible. The raw material of the knife is not clear even when the knife is preserved in the model. Furthermore, it is not possible to do a metallurgical analysis since the model knives are made of painted wood (*e.g.* MMA 20.3.10).

In most butchery scenes, a large knife is used to slaughter and dismember the animal. It is difficult to use the shape and colour of the knife as an indication of its raw material (stone or metal) because both flint and metal tools are similar in shape. The knives are often painted a dark colour (brown or dark red) that cannot be used to help identify the raw material since it could be painted to reflect either the blood of the animal or the colour of the knife. Flint knives might have a light chocolate hue, copper would be light, and bronze knives would be even lighter in colour. However, such colouring is not seen in either the iconographic depictions or on the models. Hence, these painted models of animal butchery are not particularly useful in identifying if metal or stone were used to slaughter the animal. It is assumed that the Old Kingdom butchering tools were made of stone (Ikram, 1995) given that hard tin-bronze (10% tin) does not appear until the end of the 3rd millennium BC. Stone tools dominate the butchering assemblages during this period in neighboring regions (*e.g.* Greenfield, 2013). But the situation is quite different in the Middle Kingdom.

Flint Knives

The literature on the use of flint in animal butchery is exhaustive, particularly evidence on flint knapped knives (Figure 7) similar to those depicted in animal butchery scenes (*e.g.* Graves-Brown, 2010; Ikram, 1995; Kobusiewicz 2015; Skarzynski, 2017 for some recent sources). Flint knives are the most exquisite during the Neolithic when they were at the peak of their manufacture. Beautiful flint knives are found in a variety of Neolithic contexts ranging from settlements to private tombs. The shape of these knives varies and their typologies have been studied to better understand the evolution of tool types (Graves-Brown, 2010: 173-179; Ikram, 1995: 63-73; Kobusiewicz, 2015). In fact, the variety of sharp flint tools may imply that they were likely used for a variety of activities, including animal butchery.

During the Old Kingdom, the visual evidence on animal butchery suggests that curved knives were the preferred tool for animal butchery. These flint knives are 'flaked, with slightly curving backs, rounded or pointed tips and gently curving blades' (Ikram, 1995: 63). However, there is a variety of sharp flint tools from Archaic and Old Kingdom sites that are not depicted in these scenes that could have been used in butchering activities, such as slicing or filleting. These types are bifacial flint knives, fishtail knives, or bitruncated blades (Kobusiewicz, 2015: 13-15). Particularly, bitruncated blade tools or "razor blades" are extensively found in archaeological contexts and although their use is unclear, it is not difficult to imagine that these small, sharp tools were used in filleting and other butchery activities (Graves-Brown, 2010; Kobusiewicz, 2015; Skarzynski, 2017). Several examples of these have been recovered from tombs at Saqqara (Kobusiewicz, 2015: 78; Macramallah, 1940: 6-8, 10). A higher number of bifacial flint knives (22.26%) than bitruncated regular blades (6.44%) are found at Giza. In contrast, at other sites (such as Elephantine) the results are opposite – 26.96% of bitruncated regular blades and 7.83% of bifacial knives (Kobusiewicz, 2015: 48). The reason for this discrepancy may lie in that Giza was a higher status

Figure 7. Flint bifacial knife. Provenance: Tomb of King Khasekhemui. Abydos. Late 2nd Dynasty, Early Dynastic Period. MET 01.4.30. © H. Greenfield.

mortuary site where ceremonial knives might have been more frequent due to accessibility, affordability, and ritual (Schmidt, 1992: 87).

Even though the shape of the knife is ambiguous in allowing determination whether it is made of metal or stone, it is clear from Old Kingdom iconographic depictions of slaughtering and carcass processing scenes that they are flint or stone knives. In a few scenes, the knife is being sharpened and one can see the baton used for sharpening by hitting the tool in such a way as to cause the sharpening flakes to flow off of the flint knife (Graves-Brown, 2010: 176).

Symbolic Importance of Stone Knives

In almost all animal butchery scenes, the use of these larger knives gives the impression that they were the primary tool used in slaughtering and carcass processing activities. However, their large size would make it difficult to conduct some activities (skin and fillet), even on large animals (such as cattle), and more difficult to dismember, disarticulate, skin, and fillet the common smaller domesticates (*e.g.* sheep, goat, or pig). Perhaps the reason for the consistent depiction of large flint knives in the slaughtering or butchering of larger mammals (*e.g.* cattle and antelope) is that they hold a special place in the ideological relationship of ritualistic activities (such as butchering), and that they are a token of status and wealth (Graves-Brown, 2010: 173-175). Large flint knives reflect ritual purity and are symbolic in ancient Egyptian ideology. Their association with the slaughtering and processing of larger livestock reflects status, wealth, and religious significance (Barbash, 2013: 29-30; Maitland, 2018; Schmidt, 1992: 87).

Moreover, sharpening the flint knife, slitting the throat of the animal with the knife, and catching of the butchered animal's blood with basins as represented in animal butchery scenes collectively epitomize the important relationship between cattle slaughtering with flint knives in ancient Egyptian ideology (Graves-Brown 2010; Hayes 1953: 263; Kobusiewicz 2015: 59-60; Maitland 2018). Therefore, these large flint knives may not have been used in daily butchering activities, but as mediums of their symbolic importance for the elite in mortuary activities.

The relationship between prime-cuts of meat (cattle, oxen) and large flint knives is strengthened by examples where they were found together (Tillman, 1992: 197). Large flint knives were uncovered at 1st Dynasty Helwan among oxen bones (Graves-Brown, 2010: 174; Saad 1951: 10-11), t Early Dynastic Tell al-Farka where cattle ribs and a flint knife are found together (Cialowic 2009: 86-87; Graves-Brown 2010: 174) and at the 5th Dynasty Abusir 'Shrine of the Knife' of Raneferet where knife fragments and animal bone found together (Graves-Brown, 2010: 174; Svoboda, 1993; 2006; Vachala & Svoboda, 1989). Bifacial knives similar to those in animal butchery scenes have been found in domestic contexts, such as Block 3 at al-Mahâsna where a wealthy individual is found with a large amount of wild fauna and cattle remains that perhaps were slaughtered using the flint knife found beside them (Skarzynski, 2017: 71-73). Although there are a few domestic contexts where large curved bifacial flint knives are found (Graves-Brown, 2010: 174-175), their limited presence outside of ceremonial contexts supports the suggestion that they may have been limited to ceremonial use (Schmidt, 1992: 87).

Development of Metallurgy in Ancient Egypt

The earliest metal objects appear during the Neolithic Badarian period (ca. 4500-3500 BC) in Egypt (*i.e.* at al-Mustagidda in Grave 596 as copper beads, Muhly, 1999: 629; Scheel, 1989: 8). Copper objects (such as axes, adzes, chisels, razors, and the like) begin to appear in the Chalcolithic Naqada I-II cultures (ca. 4,000-3,200 BC) (Muhly, 1999: 629; Needler, 1984: 280-282, cat. 180-183) and continue into the Early Dynastic, Old Kingdom, and later periods (Kobusiewicz, 2015: 1-2; Odler, 2016). It has long been recognised that the transition from stone to metal occured during the 3rd and 4th millenniums in Egypt – *e.g.* Breasted (1912: 28) stated "*It is indeed the age of the slow transition from stone to copper.*"

Metal knives that resemble those used in animal butchery appear in the Early Dynastic period – *e.g.* 1st Dynasty necropolis of Tarkhan where both flint and copper knives were found in a number of tombs (Petrie, 1912; 1913; examples see UC17081; UC28612). Perhaps the most exquisite example is a copper knife with a curved blade found dating to the Early Dynastic Period 1st Dynasty at Abydos (Manchester Museum 6783) (Figure 8). These early examples are made of arsenical copper (Muhly, 1999: 630).

Excavations of the 6th Dynasty Tomb of Impy at Giza (G2391) recovered a large amount of copper tools, including copper knives (Maddin *et al.*, 1984) (for example, Museum of Fine Arts, Boston 13.3471). The metallurgical study of the copper implements (including knives) in the Tomb of Impy demonstrated that they were made of unalloyed copper (Muhly, 1999: 630). A recent study

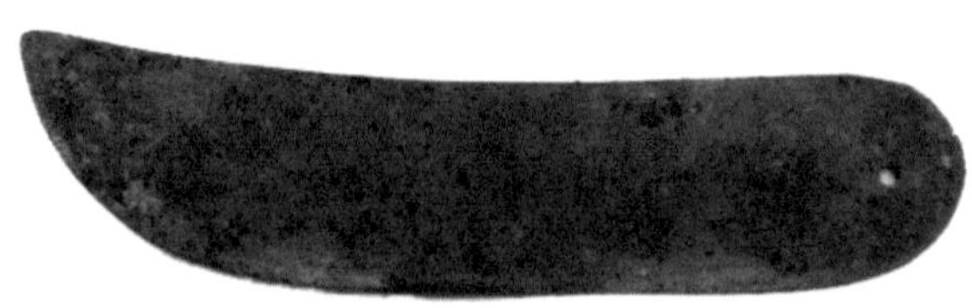

Figure 8. Copper knife. Provenance: Abydos. 1st Dynasty, Early Dynastic Period. Manchester Museum 6793 © Manchester Museum.

demonstrates that an astounding percentage (95%) of the sharp metal-based implements from the 3rd millennium BC that *potentially* may have been used in animal butchery (such as knives) come from elite, ritual, and ceremonial contexts, such as tombs, temples, and foundation deposits (Odler, 2016: 66-69). Very few specimens (5%) derive from settlement contexts where they may have been used in other, likely utilitarian, activities (Odler, 2016: 66-69). Whether these early copper knives are *actually* used in animal butchery for daily life or only for ceremonial purposes is unlikely given that copper is a relatively soft metal and would not hold an edge during the slaughtering or butchering processes.

Although the near absence of metal tools from settlement contexts may reflect their lack of use in daily life for activities, such as animal butchery, the phenomenon is complicated by the likely recycling of rare and expensive materials, such as copper or bronze, which can be melted down, repaired, and reused (Greenfield, 1999; 2013; Ikram, 1995: 67; Tillman, 1999: 314). An example of a copper production workshop area that reworked and recycled copper was found at the First Intermediate Period site of Balat in the Dakhla Oasis (Moeller, 2016: 242).

The presence of metal in wealthy burial contexts is indicative of a product that seems to have been reserved for the wealthier or higher classes of ancient Egyptian society. It cannot be determined as of yet whether the reuse of metal was a factor in hiding its distribution in antiquity. Metal may have been more widespread than previously believed, but this could only be investigated through proxy measures, such as microscopic butchering marks on bones (Greenfield, 2013).

Copper Production

Further evidence for metal as a highly desired product is reflected in its production. Evidence for metalworking in the Old Kingdom has been found at Buhen where a copper smelting production area containing three furnaces was excavated (Muhly, 1999: 630).

Metalworking in Old Kingdom Egypt was a specialised craft and the organisation of metal production overseen by the crown (Odler, 2016: 135-149; Roth, 1991). The organisation of metalworking is depicted in a number of scenes from the Old Kingdom. The best preserved examples are found in the tombs of Mereruka and Ti at Saqqara (Muhly, 1999: 630) (Figure 9). It is clear that metalworking is supervised by an official administrator who also may have overseen the main storage centre of raw materials (*e.g.* overseer of the treasury) (Odler, 2016: 30; Strudwick, 1985: 284; The Sakkarah Expedition, 1938: pl. 32-33).

Already in the 3rd Dynasty, "*the Egyptian court asserted direct control over the copper mines of south-western Sinai*" (Wengrow, 2006: 147). The royal monopoly on copper mining early in the Old Kingdom suggests an intent to produce metal in large quantities and control their distribution and use. The presence of mines east of the Nile Valley (Timna, Wadi al-Feynan, Wadi Maghara, etc.) (Bard, 1999: 1-2, 33; Feucht, 1999: 465; Shaw, 1998; Ward, 1999: 1031) and royal involvement in interregional trade demonstrates that metal was clearly being produced for the royal household and its retainers (Baud, 2010: 72; Mumford, 2010: 347; Tillman, 1999: 314). Royal involvement increased in the Middle Kingdom with the exploitation of mines by Amenemhat III (Leprohon, 1999: 54). In the New Kingdom, metal technology evolved to include the manufacture of pot bellows as a more efficient means of metallurgical production (Muhly, 1999: 631).

Both butchers and metal workers were integrated into the official administration since production was supervised by members of the royal household. Butchery and metal working were specialised tasks involving hard labours. For example, a Middle Kingdom text known as *Satire of the Trades* (*Instruction of Dua-Khety*) describes

Figure 9. Metal-workers weighing and smelting ore. Limestone relief. Provenance: Saqqara. 6th Dynasty, Old Kingdom. After The Sakkarah Expedition (1938: plate 30).

the hardships that metalworkers faced daily and their place within society. Khety states: *"I never saw a sculptor as envoy, nor is a goldsmith ever sent; but I have seen the smith at work; at the opening of his furnace; with fingers like claws of a crocodile. He stinks more than fish roe"* (Lichtheim, 2006: 186). This text provides a vivid image of the difficulties faced by metal workers. It also places metal workers into a lower class of society – individuals who would never be called to be present in the court.

Middle Kingdom and Later Stone and Metal Use

The development of metallurgy continued during the Middle Kingdom with the widespread of hard arsenic/tin-bronze tools alongside the continued use of flint in household activities (Graves-Brown, 2010; Ikram, 1995; Odler, 2016). Flint and copper tools are found together in a number of contexts, such as at Lahun (al-Lahun) in the Fayum (David, 1996: 168-170) (see objects 192, 205, 229 Manchester Museum). The site yielded both flint and (copper and bronze) knives that resemble in shape those used in animal butchery iconography and three-dimensional wooden models from the same time period. It is unclear if the Middle Kingdom bronze and copper knives were used for animal butchering for the population in general, restricted for use by higher status/wealthier individuals, or primarily for symbolic purposes. It is likely that given the abundance of stone tools during this period that they were the primary butchering technology for the lower strata of society.

By the time of the New Kingdom, bronze knives are found in abundance, and flint chipped stone flint tools begin to decline dramatically in frequency and quality (Graves-Brown 2010: 97). It is more than likely that by the New Kingdom, bronze had nearly replaced stone as the primary technology in animal carcass processing. Metalworking also continued to be a specialised craft as seen in the administrative document Papyrus P.BM 10068 which lists copper smiths as 5% of the total occupations that existed (Shaw, 2012: 135-138).

Discussion

Transition from Stone to Metal Knives

Although metal knives are present early in ancient Egypt and produced alongside flint knives from at least the Early Dynastic Period onward, the rate of change from stone to metal knives in mundane matters such as animal butchery is still unclear (Ikram, 1995: 65-69). Some authors attribute the transition from stone to metal for quotidian activities to the Middle Kingdom based on the decline in frequency of stone tools during this period (Graves-Brown, 2010: 97). The larger quantity of bronze objects recovered from New Kingdom sites suggests that the transition had already occurred. However, it is unclear how widespread the use of metal knives in the lower strata of society in each of these periods and the types of functions that they had. Whether metal knives were exclusively used in ceremonial contexts or in daily life is unknown as is its accessibility to all social classes of society.

A recent publication on the ideological significance of flint in Dynastic Egypt documents the declining frequency of flint during the Old and Middle Kingdoms by examining changes in quality, typology, technique of production, and most importantly (for our purposes) frequency over time (Graves-Brown, 2010: 97). The author documents a drastic decline in frequency of flint tools beginning in the Middle Kingdom with a near absence by the New Kingdom. This recent study corroborates earlier studies (Ikram, 1995: 65-69; Kobusiewicz, 2015: 13) and the decline is likely related to changes in metal production that would make it easier to mass produce at a resourceful rate and thus more affordable to a wider segment of society (Graves-Brown, 2010: 83-43; Rosen, 1996: 131, 153). Although metal knives existed in the Old Kingdom, there is no evidence on how widespread its use was during this time and if it was restricted to the wealthier members of society. It is clear from the decline of flint and increase of metal objects from the Old Kingdom into the Middle Kingdom that a socio-evolutionary change occurred that resulted in greater use of metal objects.

While the 3rd millennium is characterised by copper tools, bronze metal production appears to undergo a dramatic change in the Middle Kingdom based on the increased frequency of bronze metal objects. In combination with the decline in flint knives, it would suggest that the transition from stone to metal butchery technology occurs at this time. Nevertheless, the presence of copper knives from the third millennium begs the question as to whether they could have also been used in butchery activities.

Examining the Origins of Metallurgy through Animal Butchery

The replacement of stone, development of metallurgy and the impact on ancient Egyptian society cannot be completely understood simply by examining the material culture alone (specifically flint and metal tools). The metal (copper-based) record is biased through the limited depositional contexts and the likely reuse and melting down of metal objects. Hence, a new means needs to be developed to document and understand the shift from stone to metal technologies. Since ancient Egyptians domesticated, raised, and butchered animals for food, it is useful to examine the fauna that were slaughtered and butchered. The tools might have left marks on the bones that can be used to identify whether they were butchered with flint or metal tools.

Greenfield microscopically examined butchering marks on faunal remains from the southern Levant (Israel, Jordan, and Palestinian Territories). This method avoided the pitfalls of relying exclusively on the few metal objects that survive. Based on experiments conducted with modern tools and modern replicas of ancient tools (stone and metal), he demonstrated that there are microscopic diagnostic distinctions that allows grooves made by chipped stone to be differentiated from metal blades. The method uses both light optical and scanning electron microscopes (Greenfield, 1999; 2006). Application of his results to zooarchaeological materials from the southern Levant allowed for a quantitative assessment of the frequency of butchering marks made by flint and metal tools. These in turn provided an independent line evidence for the rate of adoption of metallurgy for animal butchering during the Bronze Age of the southern Levant. It is now clear that significant frequencies of butchering marks made by metal butchering implements did not appear until the Middle Bronze Age (Greenfield, 2013). This work has been largely validated as it was extended to other regions. Metal butchering implements only became the dominant technology in the Late Bronze Age across much of the Middle East and Europe (Greenfield, 2017; Marciniak & Greenfield, 2013). A recent study that examined animal bones from the workers settlement at Old Kingdom Giza suggests a similar metallurgical development for ancient Egypt (Sousa, 2020). However, in some regions, stone tools continue to be used through the Iron Age and even later periods (Greenfield & Marciniak, 2019). In consequence, one should conceptualise the Early Bronze Age (including 3rd millennium BC Egypt) as part of the long and continuous use of chipped stone tools that begin in the Palaeolithic. Given the diversity of rates and nature of adoption of metal from stone across the Old World, this analytical perspective can and should be applied to test theories about the development of metallurgy in ancient Egypt and to increase our understanding of the impact metal tools had in ancient Egyptian society.

Conclusion

Animal butchery in ancient Egypt was an important part of daily and ceremonial life for both the living and the dead. Previous research focused on animal butchering techniques, the types of animals butchered, and the types of knives used in the butchering process based on texts, iconographic reliefs, and models of animal butchery scenes (*e.g.* Ikram, 1995). However, the timing of the shift from stone to metal tools remains ambiguous and poorly understood. It is unclear how common metal knives were for butchery since the archaeological evidence is biased through recycling and toward funerary contexts which further distorts any understanding whether all levels of society had access to metal tools for more than elite funerary activities.

Given the published information summarised here, it may be suggested that the transition to metal knives for butchery was a complex process. It does not appear to be likely that animal butchery was conducted with metal knives with the first appearance of copper knives during the 3rd millennium BC. It is more likely that the transition took place a thousand years later during the Middle Kingdom when bronze knives first appear. Yet, we still do not know how frequent metal butchering implements were in the Middle Kingdom. It is likely that the transition away from chipped stone tools for butchering for all of Egyptian society occurred toward the end of the Bronze Age when a new and cheaper technology becomes widespread (*i.e.* iron). Flint knives continued to be used throughout the 3rd millennium BC in the southern Levant, but begin a long decline afterwards (Rosen, 1997). While the decline of flint in the Middle Kingdom may suggest a higher use of metal tools, the presence of copper metal knives in the Old Kingdom begs the question as to whether they were of a utilitarian nature for butchery activities. Future research will be conducted to answer these questions and fill in the gaps in the debate of the introduction and use of metal (copper) in mundane matters, such as animal butchering in ancient Egypt.

Acknowledgements

We would like to acknowledge the University of Manitoba, St. Paul's College (Near Eastern and Biblical Archaeology Laboratory), and the Social Sciences and Humanities Research Council for generously funding our research and participation in Bioarchaeology of Egypt 2019 Conference. Without their unstinting support, this research would not have been successfully carried out. Any errors are the responsibility of the authors.

Bibliography

Arnold, D. 1991. Amenemhat I and the Early Twelfth Dynasty at Thebes. – The Metropolitan Museum Journal 26: 5-48.

Barbash, Y. 2013. How Ancient Egyptians Viewed the Animal World. In: Bleiberg, E., Y. Barbash & L. Bruno. Eds. Soulful Creatures: Animal Mummies in Ancient Egypt. – Brooklyn, Brooklyn Museum in Association with D Giles Ltd.: 19-62.

Baud, M. 2010. The Old Kingdom. In: Alan, B. Ed. A Companion to Ancient Egypt. Vol. I. – Oxford, Wiley-Blackwell: 63-80.

Bard, K.A. 1999. Encyclopedia of the Archaeology of Ancient Egypt. – London, Routledge.

Binford, L. 1981. Bones Ancient Men and Modern Myths. – Toronto, Academic Press Inc.

Breasted, J.H. 1912. A History of Egypt From the Earliest Times to the Persian Conquest. – New York, Charles Scribner's Sons.
Breasted, J.H. 1948. Egyptian Servant Statues. – New York, Pantheon Books.
Chassinat, E.G. & C. Palanqe. 1911. Une Campagne de Fouilles dans la Nécropole d' Assiout. – Le Caire, l'Institut Français d'Archéologie Orientale (Mémoires pub. par les Membres de l'Institut Français d' Archéologie Orientale du Caire. Vol. 24).
Cialowicz, K.M. 2009. The Early Dynastic Administrative-Cultic Centre at Tell el- Farka. – British Museum Studies in Ancient Egypt and the Sudan 13: 83-123.
David, R. 1996. The Pyramid Builders of Ancient Egypt: A Modern Investigation of Pharaoh's Workforce. – London, Routledge.
Feucht, E. 1999. Jewelry. In: Bard, K. Ed. Encyclopedia of the Archaeology of Ancient Egypt. – London, Routledge: 462-465.
Fisher, H.G. 1960. The Butcher *Pḥ-r-ntr*. – Orientalia Nova Series 29, 2: 168-190.
Fisher, J.W. 1995. Bone Surface Modifications in Zooarchaeology. – Journal of Archaeological Method and Theory 2, 1: 7-68.
Frood, E. 2010. Social Structure and Daily Life: Pharaonic. In: Lloyd, A.B. Ed. A Companion to Ancient Egypt. Vol. I. – Oxford, Wiley-Blackwell: 469-490.
Garstang, J. 1907. The Burial Customs of Egypt. – London, Archibald Constable & Co.
Graves-Brown, C.A. 2010. The Ideological Significance of Flint in Dynastic Egypt. – Unpublished PhD Thesis, University College London Institute of Archaeology.
Golden, J. 2002. The Origins of the Metals Trade in the Eastern Mediterranean: Social Organisation of Production in the Early Copper Industries. In: Levy, E. & E.C.M. van den Brink. Eds. Egypt and the Levant. – London, Leicester University Press: 225-238.
Greenfield, H.J. 1999. The Origins of Metallurgy: Distinguishing Stone from Metal Cut Marks on Bones from Archaeological Sites. – Journal of Archaeological Science 26 (7): 797-808.
Greenfield, H.J. 2006. Slicing Cut Marks on Animal Bones: Diagnostics for Identifying Stone Tool Type and Raw Material. – Journal of Field Archaeology 31: 147-163.
Greenfield, H.J. 2013. The Fall of the House of Flint: A Zooarchaeological Perspective on the Decline of Chipped Stones for Butchering Animals in the Bronze and Iron Ages of the Southern Levant. – Lithic Technology 38 (3): 161-178.
Greenfield, H.J. 2017. The Spread of Productive and Technological Innovations in Europe and Near East: An Integrated Zooarchaeological Perspective on Secondary Animal Products and Bronze Utilitarian Metallurgy. In: P.W. Stockhammer & J. Moran. Eds. Appropriating Innovations. Proceedings of the Entangled Knowledge in Late Neolithic and Early Bronze Age Eurasia Conference, Heidelberg, 15th-17th January 2015. – Oxford, Oxbow: 50-68.
Greenfield, H.J. & A. Marciniak. 2019. Retention of Old Technologies Following the End of the Neolithic: Microscopic Analysis of the Butchering Marks on Animal Bones from Çatalhöyük East. – World Archaeology 51 (1): 76-103.
Hayes, W.C. 1953, The Scepter of Egypt: A Background for the Study of the Egyptian Antiquities in the Metropolitan Museum of Art. – New York, Metropolitan Museum of Art.
Ikram, S. 1995, Choice Cuts: Meat Production in Ancient Egypt. – Leuven, Peeters.
Janssen, J. 1975. Commodity Prices from the Ramesside Period: An Economic Study of the Village of Necropolis Workmen at Thebes. – Brill, Leiden.
Kobusiewicz, M. 2015. The Production, Use and Importance of Flint Tools in the Archaic Period and the Old Kingdom of Egypt. – Oxford, Archaeopress (Egyptology 12).
Köhler, C.E. 2010. Prehistory. In: Lloyd, A.B. Ed. A Companion to Ancient Egypt. Vol. I. – Oxford, Wiley-Blackwell: 25-47.
Leprohon, R.J. 1999. Middle Kingdom, Overview. In: Bard, K.A. Ed. Encyclopedia of the Archaeology of Ancient Egypt. – London, Routledge: 50-56.
Lichtheim, M. 2006. Ancient Egyptian Literature: The Old and Middle Kingdoms. Vol. 1. – Berkeley, University of California Press.
Macramallah, R. 1940. Un cimetière archaïque de la classe moyenne du people a Saqqarah. Fouilles a Saqqarah. – La Caire, Imprimerie Nationale, Boulaq (Service des Antiquites de l'Egypte).
Maddin, R., T. Stech, J.D. Muhly & E. Brovarski. 1984. Old Kingdom Models from the Tomb of Impy: Metallurgical Studies. – The Journal of Egyptian Archaeology 70: 33-41.
Maitland, M. 2018. Dirt, Purity, and Spatial Control: Anthropological Perspectives on Ancient Egyptian Society and Culture in the Middle Kingdom. – Journal of Ancient Egyptian Interconnections 17: 47-72.
Marciniak, A. & H.J. Greenfield. 2013. A Zooarchaeological Perspective on the Origins of Metallurgy in the North European Plain: Butchering Marks on Bones from Central Poland. In: Bergerbrant, S. & S. Sabatini. Eds. Counterpoint: Essays in Archaeology and Heritage Studies in Honour of Professor Kristian Kristiansen. – Oxford, Archaeopress (British Archaeological Reports, International Series 2508): 457-468
McFarlane, A. 2000. The Unis Cemetery at Saqqara I: The tomb of Irukaptah. – Warminster, Aris and Phillips.

Moeller, N. 2016. The Archaeology of Urbanism in Ancient Egypt from the Predynastic Period to the End of the Middle Kingdom. – Cambridge, Cambridge University Press.

Muhly, J. 1999. Metallurgy. In: Bard, K.A. Encyclopedia of the Archaeology of Ancient Egypt. – London, Routledge: 628-634.

Mumford, G.D. 2010. Settlements – Distribution, Structure, Architecture: Pharaonic. In: Lloyd, A.B. Ed. A Companion to Ancient Egypt. Vol. I. – Oxford, Wiley-Blackwell: 326-349.

Needler, W. 1984. Predynastic and Archaic Egypt in the Brooklyn Museum. – Brooklyn, Brooklyn Museum.

Odler, M. 2016. Old Kingdom Copper Tools and Model Tools. – Oxford, Archaeopress (Egyptology 14).

Olsen, S.. 1988. The Identification of Stone and Metal Tool Marks on Bone Artifacts. In: Olsen, S.L. Ed. Scanning Electron Microscopy in Archaeology. – Oxford, Archaeopress (BAR International Series 452): 337-360.

Petrie, F.W.M., G.A. Wainwright & A.H. Gardiner. 1912. Tarkhan I and Memphis V. – London, British School of Archaeology in Egypt.

Petrie, F.W.M. 1913. Tarkhan II. – London, British School of Archaeology in Egypt.

Petrie, F.W.M. 1916. Tools and Weapons. – London, British School of Archaeology in Egypt.

Pollio, C.J. 2009. Cut Mark Analysis of Protohistoric Bison Remains from EfPm-27 Utilizing the Scanning Electron Microscope. – Unpublished MA Thesis, University of Saskatchewan.

Porter, B. & R.L.B. Moss. 1927, 1960, 1974, 1981. Topographical Bibliography of Ancient Egyptian Hieroglyphic Texts, Reliefs, and Painting. Vol. I-III. – Oxford, Oxford University Press.

Potts, R. & P. Shipman. 1981. Cutmarks Made by Stone Tools on Bones from Olduvai Gorge, Tanzania. – Nature 291: 577-580.

Rachewiltz, de, B. 1960. The Rock Tomb of *Irw-K3-Pth*. – Leiden, Brill.

Rice, M. 1999. Who's Who in Ancient Egypt. – London, Routledge.

Rosen, S.A. 1996. The Decline and Fall of Flint. In: Odell, G.H. Ed. Stone Tools. Theoretical Insights into Human Prehistory. – New York, Plenum Publishing: 129-155.

Rosen, S.A. 1997 Lithics After the Stone Age: A Handbook of Stone Tools from the Levant. – Walnut Creek, CA, Altamira Press.

Roth, A.M. 1991. Egyptian Phyles In the Old Kingdom. – Chicago, Oriental Institute of The University of Chicago.

Saad, Z.Y. 1947. Royal Excavations at Saqqara and Helwan (1941-1945). – Le Caire, l'Institut Français d'Archéologie Orientale (Supplément aux Annales du Service des Antiquites de l'Egypte 3).

Saad, Z.Y. 1951. Royal Excavations at Helwan (1945-1947). – Le Caire, l'Institut Français d'Archéologie Orientale (Supplément aux Annales du Service des Antiquites de l'Egypte 14).

Scheel, B. 1989. Egyptian Metalworking and Tools. – Bloomsbury, Shire Publications.

Schmidt, K. 1992. Tell Ibrahim Awad: Preliminary Report on The Lithic Industries. In: Van den Brink, E.C.M. Ed. The Nile Delta in Transition 4th-3rd Millennium BC. Proceedings of the Seminar Held in Cairo, 21 – 24 October 1990, at the Netherlands Institute of Archaeology and Arabic Studies. – Tel Aviv, E.C.M. van den Brink: 79-96.

Shaw, I. 2012. Ancient Egyptian Technology and Innovation. – London, Bloomsbury.

Shaw, I. 1998. Exploiting the Desert Frontier: The Logistics and Politics of Ancient Egyptian Mining Expeditions. In: Herbert, E.W., A.B. Knapp & V.C. Pigott. Eds. Social Approaches to an Industrial Past The Archaeology and Anthropology of Mining. – London, Routledge: 242-258.

Sousa, E.L.A.D. 2020. On the Chopping Block: Animal Butchery Technologies in Old Kingdom Egypt. - Unpublished MA Thesis, University of Manitoba, Winnipeg.

Skarzynski, E.H.. 2017. Beyond Prestige: A Ritual Production Model for Stone Tool Specialization in Naqada Period Egypt. – PhD Thesis, University of Virginia.

Stager, L.E. 1992. The Periodization of Palestine from Neolithic through Early Bronze Times. In: Ehrich, R.W. Ed. Chronologies in Old World Archaeology. – Chicago, The University of Chicago Press (3rd Ed., Vol. I): 22-41.

Strandberg, A. 2009. The Gazelle in Ancient Egyptian Art Image and Meaning. – Sweden, Universitet Uppsala (Uppsala Studies in Egyptology 6).

Strudwick, N. 1985. The Administration of Egypt in the Old Kingdom. The Highest Titles and their Holders. – KPI, London.

Strudwick, H. & J. Dawson. 2016. Death on the Nile: Uncovering the Afterlife of Ancient Egypt. – London, D. Giles Limited.

Svoboda, J. 1993, Lithic Industries from Abusir, Lower Egypt. – Origini, 17: 167-218.

Svoboda, J. 2006. The King's Knives: Chipped Lithics from Raneferef's Mortuary Complex. In: Verner, M. Ed. Abusir IX. The Pyramid Complex of Raneferef. The Archaeology. – Prague, Academia, nakladatelstvi Academie ved ceske republiky: 502-512.

The Sakkarah Expedition. 1938. The Mastaba of Mereruka. – Chicago, The University of Chicago Press.

Tillmann, A. 1992, Die Steinartefakte des Dynastischen Ägypten, Dargestellt am Beispiel der Inventare aus Tell el-Dab'a und Qantir. – Unpublished PhD Thesis, University of Tübingen.

Tillman, A. 1999. Dynastic Stone Tools. In: Bard, K.A. Ed. Encyclopedia of the Archaeology of Ancient Egypt. – London, Routledge: 310-314.

Tooley, A.M.J. 1989. Middle Kingdom Burial Customs. A Study of Wooden Models and Related Material. Vol. I. – Liverpool, University of Liverpool.

Vachala, B. & J. Svoboda. 1989. Die Steinmesser aus Abusir. Ein Artefakt aus der Sicht von Zwei Historischen Disziplinen. – Zeitschrift für Ägyptische Sprache und Altertumskunde 116: 174-181.

Vischak, D. 2015. Community and Identity in Ancient Egypt. The Old Kingdom Cemetery at Qubbet el-Hawa. – Cambridge, Cambridge University Press.

Walker, P.L. & J.C Long. 1977. An Experimental Study of the Morphological Characteristics of Tool Marks. – American Antiquity 42 (4): 605-616.

Ward, W.A. 1999. Trade, Foreign. In: Bard, K.A. Ed. Encyclopedia of the Archaeology of Ancient Egypt. – London, Routledge: 1031-1034.

Wengrow, D. 2006. The Archaeology of Early Egypt: Social Transformation in North-East Africa, 10,000 to 2650 BC. – Cambridge, Cambridge University Press.

Wetterstrom, W. 1993. Foraging and Farming in Egypt: The Transition from Hunting and Gathering to Horticulture in the Nile Valley. In: Andah, B., A. Okpoko, T. Shaw & P. Sinclair Eds. The Archaeology of Africa: Food, Metals, and Towns. – London, Routledge: 165-226.

Anthropological Study of the Egyptian Mummy from the Pushkin State Museum of Fine Arts Using Computed Tomography

Sergey V. Vasilyev, Ekaterina B. Yatsishina, Ravil M. Galeev, Svatlana B. Borutskaya, Mikhail V. Kovalchuk, Olga A. Vasilieva, Olga P. Dyuzheva & Vadim L. Ushakov

Introduction

Paleoanthropological studies are traditionally based on skeletal remains. As a result of such complex investigations, anthropologists can determine the physical characteristics of individuals, calculate the proportions of the skeleton and the intravital body length, and determine the general robusticity of the bones. The research data makes it possible to understand the anthropological features of groups of people, the degree of homogeneity of populations, the genetic links with neighboring group, as well as to posit the origin of certain groups in different regions and to draw conclusions about migration processes in antiquity. The research study of human bone remains also allows to describe diet, illnesses, combat and domestic injuries that affected the skeleton. Biochemical study of bone tissue along with analysis of certain teeth features allow to assume the type of nutrition of people and the adequacy of food resources. Therefore, it is sometimes possible to indirectly learn about the living conditions of populations and the perspectives of anthropological groups in the demographic terms.

Such studies are traditional in paleoanthropology. However, in a number of cases research is complicated by mummification, natural or anthropogenic, which makes it impossible to directly access the bones of buried individuals non-destructively. In practice, there are large regions where paleoanthropological history is represented exclusively by mummified remains and mummies, most notably ancient Egypt. Often such anthropological materials obtained through excavation are stored or prepared for storage in museums and for exhibitions. The research study of bones in this case becomes impossible and paleoanthropological information about these individuals becomes lost to science.

In the last decades, the development of nuclear physics methods and technologies has led to the active use of radiations of different types in the study and conservation of cultural heritage. Modern methods of computed and magnetic resonance tomography are used for the non-invasive study of mummification techniques, the definition of procedures and ceremonies for mummification, the determination of biological sex and causes of death, the measurement of the anthropological characteristics, the forensic and medical examination of mummies, and the creation of three-dimensional stereolithographic

In: Ikram, S., J. Kaiser & S. Porcier. Eds. 2021. *The Ancient Egyptians and the Natural World. Flora, Fauna, and Science.* – Leiden, Sidestone Press: 265-274.

reconstruction (Aufderheide, 2003). Application of these methods provides possibilities to perform non-destructive research on mummified remains. Thus, such studies are conducted without the destruction of precious museum exhibits. Another important advantage of the computed tomography method is the ability to repeatedly examine mummies and to reconstruct the head/face no less effectively than on the bare skull. Development of 3D models allows us to visualize various pathologies and to conduct anthropometric studies of mummified remains (Partridge, 1996).

The first computed tomography of the mummified remains of a boy and a young woman were performed in 1977 by Derek Harwood-Nash from Toronto (Harwood-Nash, 1979). This study showed the promise of the computed tomography method. An example of a brilliant complex study of mummified remains by using the computed tomography method is the study of the so-called "Ice-man" or Ötzi. Multiple use of computed tomography allowed researchers to determine the probable cause of his death as an arrow tip was found in the soft tissues of the brachii (Nedden *et al.*, 1994)

Over time the number of mummies studied by computed tomography has increased significantly. Data on methods and techniques of embalming were reconstructed and supplemented by computed tomography. In some cases, it was possible to determine the causes of death of individuals (Cesarani *et al.*, 2003, Marx & D'Auria, 1988).

This paper presents the results of a paleoanthropological study of one of the nine Egyptian mummies stored in the Pushkin State Museum of Fine Arts (accession number I,1a №5301), using the data obtained on the computed tomograph of the National Research Center at the Kurchatov Institute. The stages and research methods are described in detail and characterized below.

In recent years, a Laboratory of Natural Science Methods in Humanitarian Sciences oriented to carry out complex analysis of museum pieces and archaeological artifacts was organized at the National Research Center Kurchatov Institute in Moscow. A complex study of the mummies from the Egyptian collection of the Pushkin State Museum of Fine Arts became a new stage in the development of this research.

Materials and Methods

The object of this study was mummy No. I,1a 5301 from the Department of the Ancient East of the Pushkin State Museum of Fine Arts (Figure 1). According to the inventory, this is the mummy of Khor-Kha, which dates to the 7th to 4th centuries BC. A grid of blue faience beads draped on top of the mummy of Khor-Kha is characteristic of Late Period Egyptian mummies (Ikram & Dodson, 1998).

The mummy in the sarcophagus is supine, with arms folded on the chest crosswise (right above the left). The

Figure 1. Mummy 5301. Courtesy of the Pushkin State Museum of Fine Arts.

right hand is extended, while the left hand is loosely curved as if it once held something. The legs are extended, knee joints are close together, and the bones of the feet are in contact, both feet slightly flexed.

X-ray CT of mummies was performed on a Biograph mCT40 (Siemens) positron emission tomography – computed tomograph (PET-CT). The spatial resolution of the cut thickness during scanning was 0.6 mm, the increment in reconstruction was 0.3 mm, and the spatial resolution in the scanning plane in reconstruction was 0.88 × 0.88 mm. Two modes of reconstructing images proceeding from 'raw' data were applied, which differ in the contrasting sensitivity to soft tissues and bone structures (Karlik *et al.*, 2007; Macková *et al.*, 2016; Panzer *et al.*, 2013; Posh, 2015).

In a brief description of the software for paleoanthropological research of mummies two principal points have to be mentioned. The first one includes the segmentation of bones, in other words, the separation of bone structures from the surrounding environment using preset filters, which 'bare' the skeleton without any difficulties. In the case of mummies this is a rather complicated operation, since bone density is often greatly altered by time, and the distal epiphyses of long bones especially are sometimes either fragmented or decayed. Additionally, mummies can contain artefacts that have densities that are similar to that of bone, and thus special software allowing for manual segmentation adjustment, and also 3D modeling software tools (Figure 2) are required. Secondly, the availability of measuring tools allowing for measurements in both linear and angular dimensions is necessary. In this case the main difficulty in performing measurements is the projection dimensions (the bone must be located in a strictly defined plane) and dimensions that do not have anatomical connection (for example, maximum and minimum diaphysis diameters).

In general, the functionality of the software used to visualize the raw medical images obtained from the computed tomography, Inobitech DICOM, allows for carrying out a full osteological study of the bones of mummies. Unfortunately, however, this functionality extends only to the postcranial skeleton. To our knowledge, none of the modern programs allow the full implementation of craniometric programs on 3D models derived from CT data. Therefore, craniological measurements were made on models printed on 3D printers with analog instruments.

Results

Sex and Biological Age

The sex was mainly determined by the morphology of the pelvic bones (Suchey, 1979) and the development of the relief on the cranium (Buikstra & Ubelaker, 1994). In particular, the triangular shape of the locus of the pelvic bones, the obtuse sub-angle and the obtuse angle of the large sciatic incision of the pelvic bones tell us that this particular individual is female. In addition, this is also indicated by the small mastoid of the cranium, a pointed upper edge of the orbit and a weak supraorbital relief (Buikstra & Ubelaker, 1994).

Age was determined by standard anthropological methods, namely, taking into account the degree of overgrowth (obliteration) of the sutures of the cranium (Acsadi & Nemeskeri, 1974), the signs on the pelvic bones (White & Folkens, 1991), the signs on other bones of the postcranial skeleton and the degree of abrasion of the teeth (Buikstra & Ubelaker, 1994). However, it should be noted here that anthropologists are limited to the assessment of biological age according to skeletal and odontological criteria. Biological age is a category that corresponds to the rates of growth and development, as well as physical indicators of aging. The relationship between indicators of biological age and chronological age is not always linear. Thus, the biological age of an individual, determined by a number of morphological (and also physiological) criteria may be younger or older than the actual chronological age, depending on the rate of growth or aging in a given organism.

In terms of the skull, the beginning of the overgrowth of the lower edge of the coronal suture is marked on the cranium of the individual, while the sagittal suture in the region of the obelion is not overgrown. Venous and sagittal sutures in the bregma area are open. The metopic suture is open. The wedge-occipital suture is overgrown. The general evaluation of biological age according to obliteration of the seams is 20-25 years (Buikstra & Ubelaker, 1994).

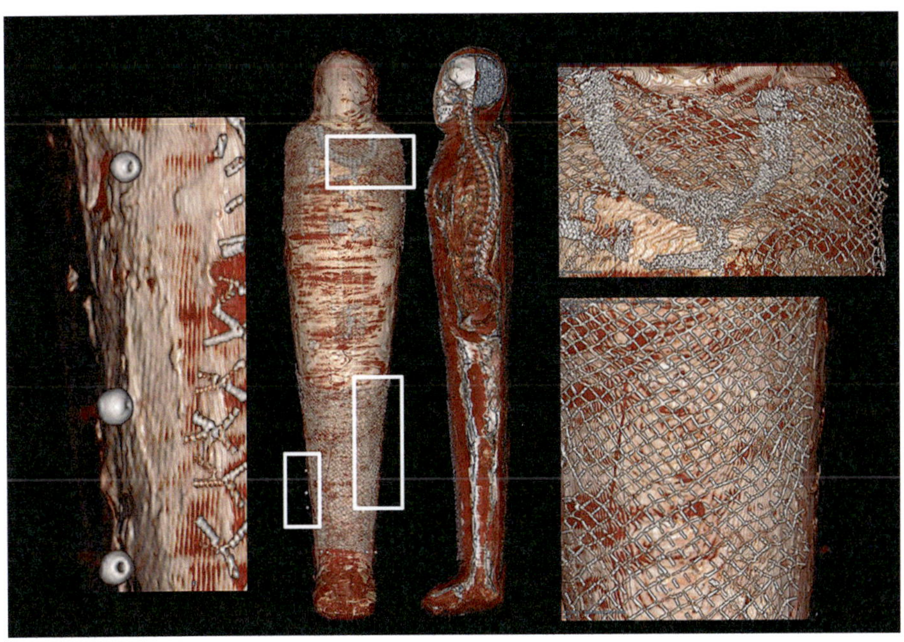

Figure 2. CT of Mummy 5301 and its details. By R.M. Galeev.

According to our observations, all the teeth of the individual had erupted, including the third molars. As far as we could see, there was no abrasion on the occlusal (chewing) surface of molars and premolars. The tops of the central and lateral incisors have been slightly erased. The general evaluation of biological age according to the dental abrasion is 20-25 years (Buikstra & Ubelaker, 1994).

Study of the pelvic bones shows that comparison of the patterns of the symphysial surface on both bones corresponds according to anthropological schemes to the age of 20-30 years (Brooks & Suchey, 1990). The auricular surfaces according to the degree of development correspond to the age of about 25 years (Lovejoy, 1985). Also iliac crests and sciatic mounds have already fused to the pelvic bones, which indicates the age of over 23-25 years (Ubelaker, 1989).

Thus, the features associated with the cranial sutures, teeth, and pelvic bones indicate the biological age of death of 20-25 years, most likely closer to the end of that range.

Craniological Study

The cranium of the mummy (Inv. No. I,1a 5301) is relatively well-preserved (Figure 3). The study was performed according to the classical craniological program (Alekseev & Debets, 1960; Martin & Saller, 1957).

The cranium of the mummy (Inv. No. I,1a 5301) has close to the average values dimensions of the horizontal circle through the ophryon and the transverse arch. The sagittal arch is large for a female, *i.e.*, the cranium is quite elongated.

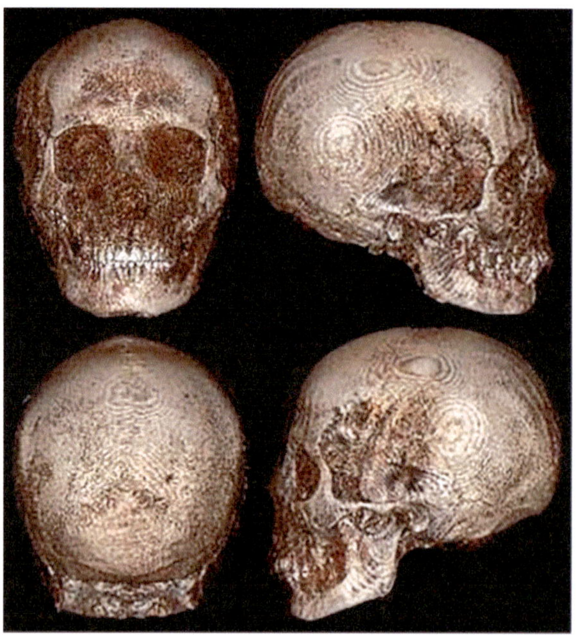

Figure 3. The skull of the mummy 5301. By R.M. Galeev.

The shape of the cranium when viewed from above is pentagonoid (pentagonal), the largest width of the cranium is shifted backward and falls to the rear third. Frontal and especially parietal tubers are quite strongly developed. It is their development that determines the pentagonal shape of the cranium. The cranium can be described as elongated and relatively wide – mesocranial. The altitude-longitudinal index has an average value and indicates the ortocrania. The altitude-longitudinal index indicates the metriocranial category of the cranium. Both indexes shows that the cranium is medium tall (Alekseev & Debets, 1960). The metope is straight and visually quite wide. Absolute sizes of the smallest and largest metope width are included in the category of large and very large widths. According to the metope-transversal index (average metope width) the cranium is mesogenic. The metope-zygomatic index is large. Due to sufficiently low index of the curvature of the frontal bone it can be concluded that its bend is quite strong. Development of the glabella area is estimated at two points on a six-point Broca scale (Alekseev & Debets, 1960). Superciliary arches (type II): there are barely noticeable elevations to the right and left from the glabella. The parietal tubers are located high. A relatively low index of curvature of the parietal bones indicates of a small radius of their curvature. Mammillary tubercles representing female characteristic have a length of less than 2 cm and evaluated with the index of 1. The nape is wide, it has a rounded shape. The upper nuchal lines (*linea nuchae superior*), which represent spindle-shaped massive structures, have a strong development, which is unusual in females.

The facial part of the cranium is narrow and relatively high, the upper facial index shows that it belongs to leptinous category (high lytic index). The angles of horizontal profiling belong to the category of very small angles, *i.e.* the face is sharply profiled even by the measures of the Europeid representatives. Prognathism of the face is weakly expressed, which is confirmed by the values of the face-pointing indicator (Flower index). The craniofacial vertical index has an average value for *Homo sapiens*, *i.e.* there is a tendency of a combination of a tall face and a sufficiently high cranium. The craniofacial transversal index is relatively small, which demonstrates a combination of a narrow face and a relatively wide cranium. The orbits are not high and relatively not wide (mesoconchic). The upper edge of the orbit is pointed. Supraorbital channels are not closed (in the cut form). The shape of the upper edge of the orbital socket is rectangular. In absolute terms the nose is high and relatively narrow (leptorinous), which is also confirmed by a nasal index. The nasal protrusion angle has an average value. Simotic and maxillo-frontal indexes are included in the average category, which indicates a not significant height of the noseband. Zygomaxillar area is narrow, gracile. The lower

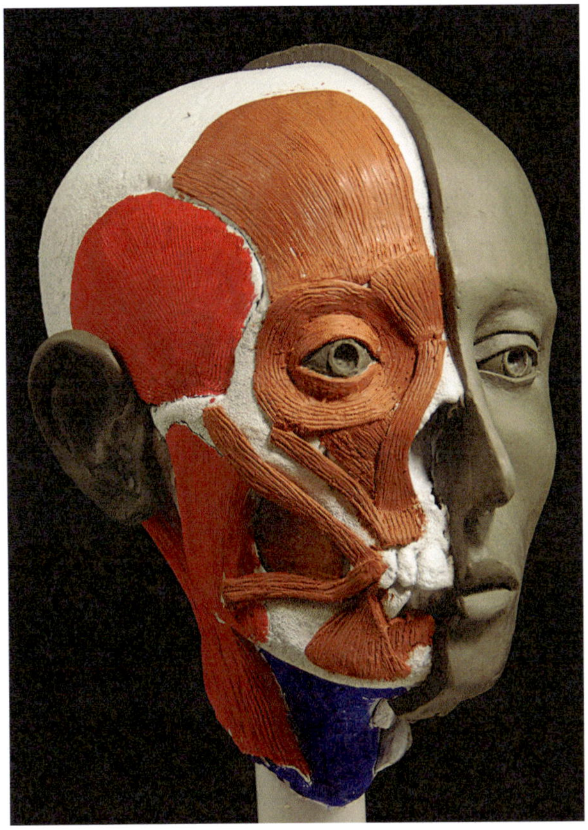

Figure 4. Stage of reconstruction of the face on the skull of the mummy 5301. By R.M. Galeev.

Figure 5. Reconstruction of the face on the skull of the mummy 5301, made by the method of Gerasimov. By R.M. Galeev.

edge of the piriform opening represents anthropina, that is, the lateral edges of the piriform opening directly pass into the lower edge, which has a sharp shape.

The facial reconstruction on the skull was carried out according to the method of M. M. Gerasimov (Gerasimov, 1955) (Figure 4, 5).

Osteological Study

The postcranial skeleton is well preserved and is almost complete (Figure 6). The bones of the postcranial skeleton were measured according to the standard osteometric program and the method of certain measurements of the bones of the skeleton was based on the rules described in the work of V.P. Alekseev "Osteometry" (Alekseev, 1966). The main osteometric form with the results of measurements of the postcranial skeleton is shown in Table 1.

Proportion of Limbs (Table 2)

The intermembral index corresponds to the shortened upper limbs in relation to the lower extremities. The brachial-femoral index is very high, which indicates a relative elongation, more precisely, a strong elongation of the brachii compared to the femora, or vice versa, a strong shortening of the femora in comparison with the brachii. The radial-shinbone index is small. Therefore, the relative shortness of the arms provides a significantly shortened antebrachium relative to the shin. The radial-brachial index shows the ratio of the proximal and medial segments of the arms. The value of the radial-brachial index is minimal for the individual studied within the range of the index variations of the modern human type. Therefore, one can speak of a strong relative shortening of the antebrachium in comparison with the brachii. The values of the absolute and relative lengths of the clavicles indicate a small size of the width of the brachii. The calculated width of the brachii is 31.4 cm, which corresponds to a very small value. The absolute width of the pelvis is not large. This observation points out some masculinity of this female. The pelvic index demonstrates that the pelvis of the individual is low enough, which, precisely, is typical for female.

Intravital length of the body determined using the formulas of Bunak (1961), Dupertuis and Hedden (1951) and Raxter (2008), which take into account the length of

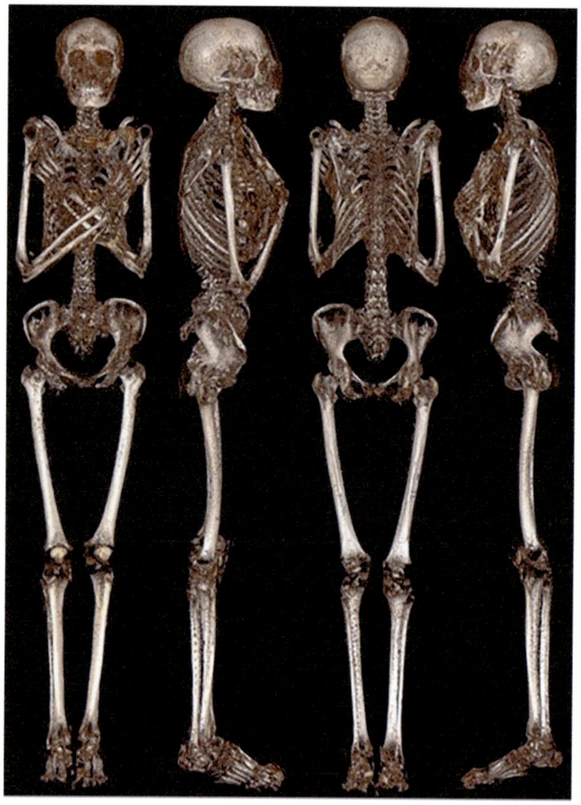

Figure 6. The skeleton of mummy 5301. By R.M. Galeev.

the femoral bone and shinbone, resulted in an estimated height of 158, 159 and 156 cm. (Bunak, 1961; Dupertuis & Hedden, 1951; Kaiser, 2018; Raxter, 2008; 2011; Zakrzewski, 2003).

Robusticity of the Skeleton (Table 3)

The clavicles and arm bones are gracile. The humeri have an average degree of flatness, while the deltoid tuberosity is well pronounced. The degree of robusticity for the femora was below average (world variation of the index according to Khrisanfova). The bones are well developed sagittally, and the rear pilaster is well expressed. In the upper layer, the femoral bones are expanded (or platimeric), the right bone is hyperplatimeric. The robusticity of the tibiae was below average world variation of the index (according to Khrisanfova) according to the index of the ratio of the circumference of the middle of the diaphysis to the length of the bone. The average degree of flattening of the bone in the middle part of the diaphysis could be also assumed.

Osteoscopy, a description of the degree of development of bone relief for attaching some of the most important muscles, was also carried out, using Fedosova's (1986) descriptions for the development of muscular relief. The phonetic description is based on Alekseev (1966). The muscular relief on the brachial bone has an average level

No. by Martin-Saller	Right in cm	Left in cm
Clavicle		
1. Length maximum.	129	130
6. Circumference middle diaphysis.	30	28
Humerus		
1. Length maximum.	301	299
2. Length.	298	297
3. Upper epiphyseal width	-	47
4. Lower epiphyseal width	53.8	54.5
5. Maximum width mid-diaphysis	21.2	20
6. Minimum width mid-diaphysis	14.8	14.2
7a. Circumference middle diaphysis	52.6	56
Radius		
1. Length maximum.	214.7	210.8
3. Minimum circumference diaphysis	36	33.8
Ulna		
1. Length maximum.	-	248
2. Physiological length.	-	225
3. Minimum circumference diaphysis.	29	30
Pelvis		
2. Maximum width pelvis	253	
1. Height pelvis	191.5	-
9. Height Ilium	125	-
15. Height sciatic bone	67.5	-
17. Length pubic bone	78	-
12. Width Ilium	144	-
Femur		
1. Length maximum.	416	420
2. Length in natural position.	413	417
6. Sagittal diameter mid-diaphysis	25.8	25.4
7. Transverse diameter mid-diaphysis	23.5	24.5
10. Upper sagittal diameter	35	33
9. Upper transverse diameter	26	25
8. Circumference middle diaphysis	79	76
Tibia		
1a. Length maximum.	356	354
1. Full length	349	348
8. Sagittal diameter mid-diaphysis	26.2	25.5
9. Transverse diameter mid-diaphysis	21.2	19.5
10. Circumference middle diaphysis	75	73

Table 1. Results of measurements of the postcranial skeleton (in mm).

Index	Right	Left
Intermembral	67.68	66.61
Brachial-femoral	72.88	71.65
Radio-brachial	71.33	70.55
Tibial-femoral	84.50	83.45
Radio-tibial	61.52	60.57
Clavicular-brachial	43.29	44.29
Shoulder width (cm)	31.4	
Shoulder-growth index	19.89	
Pelvis width (cm)	25.3	
Pelvic index	75.69	
Pelvic-growth index	16.02	
Pelvic-brachial	80.57	
Stature according to Bunak, Dupertuis and Hadden	157.9	

Table 2. Indices of proportions of limbs, pelvic and brachii measures, intravital length of the body (in cm).

Index	Right	Left
Robusticity clavicle (6/1)	23.26	21.37
Cross section diaphysis brachial index (6/5)	69.81	41.0
Robusticity radius (3/1)	16.77	16.03
Robusticity ulna (3/2)	14.04	13.78
Robusticity femur (8/2)	19.13	18.23
Pilastri femur index (6/7)	109.79	103.67
Strength femur index(6+7 / 2)	11.94	11.97
Platimery femur index (10/9)	74.29	75.76
Robusticity tibia (10/1)	21.49	20.98
Expansion mid-diaphysis tibia index (9/8)	80.92	76.47

Table 3. Indices of limb bones robusticity and strength (in cm).

of development. The supinator crest with the attachment of the brachioradial muscle is well expressed. This muscle flexes the antebrachium at the elbow joint and establishes the radial bone in the middle position between pronation and supination (position with freely lowered arms). On the radial and ulnar bones, the relief of the interosseous edge with the attachment of is the interosseous membrane, which giving rise to some muscles of the hand, is well developed. It should also be noted the presence of the radial roughness, *i.e.* the relief of the flexor of the arm in the elbow joint, that is the place of attachment of the biceps, which bends and pronates the arm.

Muscular relief on the femoral bone in general corresponds to that for females, especially taking into account the degree of development of greater trochanter. In the first place the presence of interstitial line, to which the iliofemoral ligament attaches, pulling the leg to the pelvis when walking, should be noted. However, the value of this relief, especially as a certain bone base (as well as in a number of other cases), strongly depends on the genetic component. In addition to the described structure, the gluteal tuberosity with the attachment of coxa extensor, the dorsal gluteal muscle and the femoral epicondyles with the starting position of third head of the triceps muscle of calf, which is of great importance for walking, running, jumping, as a significant flexor of the knee and foot, is weakly developed. The *linea aspera* and a pectineal line, where the muscles leading the femur to the median plane and the hamstrings biceps, involved in flexing the knee joint, are attached, have an average degree of development.

The degree of development of the relief on the tibiae corresponds to that on the femora. However, the level of the prominence of the shinbone tuberosity, as well as the level of the prominence of the greater trochanter on the femora, indicates the bones of a female. It is interesting to note that the soleus muscle line (third head of the triceps muscle of calf which is of great importance as a significant flexor of the knee and foot, as discussed above) on both tibiae is quite well developed.

In general, it can be concluded that the muscular relief of the upper limbs has an average level of development, as it should be for female engaged in some kind of physical labor associated with bending and pronating the arms in the elbow joint. Muscular relief of the bones of the legs is expressed quite well and indicates a significant load on the corresponding muscles.

Paleopathology

There are minor degenerative changes in the segment of cervical vertebrae C6-C7 in the form of unexpressed spondylosis (Figure 7). Otherwise, there are no signs of pathology on the skeleton.

Conclusion

The study has shown that the mummy belongs to a woman who died between the age of 20 and 25 years. Craniological studies suggest that she was of Mediterranean ancestry, characterized by dolichocrania and mesocrania narrow-facedness, a sharp horizontal profiling and a relatively high and narrow nose.

The observables of the postcranial, gracile, skeleton indicate a growth below the average for modern populations – 157.9 cm[1], a narrow pelvis, a small width of the brachii, an elongated brachii relative to the femor and a significantly shortened antebrachium.

1 Reported average stature for ancient Egyptian samples range from 150.8-155.9 cm for females, *cf.* Kaiser (2018: 225) and Raxter (2011: 124).

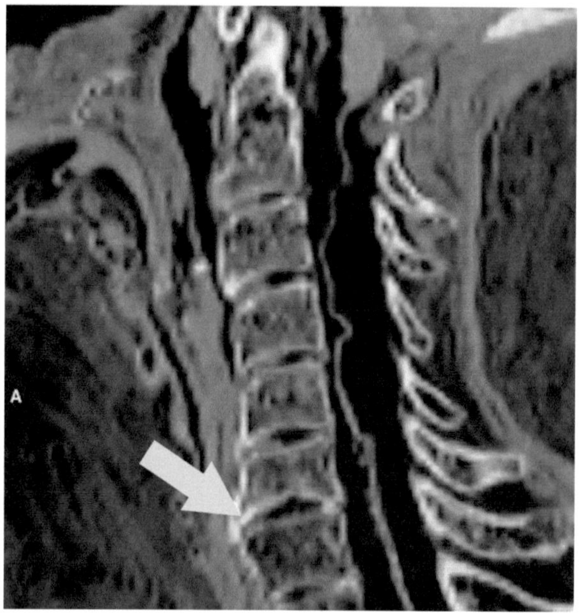

Figure 7. CT of the cervical spine of the mummy 5301. The unexpressed spondylosis in the form of anterior marginal osteophytes of C6-C7 vertebrae is determined (indicated by a white arrow). By R.M. Galeev.

Acknowledgement.

This study was supported by the Russian Foundation for Basic Research, project no. 17-29-04144 ofi_m.

Bibliography

Alekseev, V.P. & G.F. Debets. 1964. Craniometry. – Moscow, Nauka.

Alekseev, V.P. 1966. Osteometry. – Moscow, Nauka.

Acsadi, Gy. & J. Nemeskeri. 1974. History of Human Life Span and Mortality. – Current Anthropology 15 (4): 495-503.

Aufderheide, A.C. 2003. The Scientific Study of Mummies. – Cambridge, Cambridge University Press.

Brooks, S. & J.M. Suchey. 1990. Skeletal Age Determination Based on the Os Pubis: A Comparison of the Acsádi-Nemeskéri and Suchey-Brooks Methods. – Human Evolution 5: 227-238.

Buikstra, J.E. & D.H. Ubelaker. Eds. 1994. Standards for Data Collection from Human Skeletal Remains. – Fayetteville, Arkansas Archeological Survey Research Series (Vol. 44).

Bunak, V.V. 1961. Ratio of Segment Length and Total Body Length as Measured on Skeletons. – Questions of anthropology 7: 34-46.

Dupertuis, C. & J. Hadden. 1951. On the Reconstruction of Stature from Long Bones. – American Journal of Physical Anthropology 9: 15-53.

Cesarani, F., M.C. Martina, A. Farraris, R. Grilletto, R. Boano, E.F. Marochetti, A.M. Donadoni & G. Gandini. 2003. Whole-Body Three-Dimensional Multidetector CT of 13 Egyptian Human Mummies. – American Journal of Roentgenology 180: 597-606.

Fedosova, V.N. 1986. General Evaluation of the Development of the Mesomorphic Component According to Osteological Data (Osteological Method). – Questions of anthropology 76: 104-116.

Gerasimov, M.M. 1955. Facial Reconstruction from the Skull: Modern and Fossil Man. – Moscow, USSR Academy of Sciences Publishing House.

Harwood-Nash, D.F.C. 1979. Computed Tomography of Ancient Egyptian Mummies. – Journal of Computer Assisted Tomography 3: 768-773.

Ikram, S. & A. Dodson. 1998. The Mummy in Ancient Egypt: Equipping the Dead for Eternity. – London, Thames & Hudson.

Kaiser, J.E. 2018. Raising the Dead: The Bioarchaeology of the Saite and Roman Period Wall of the Crow Cemetery at Giza. – Unpublished PhD, Thesis Berkeley University.

Karlik, S.J., R. Bartha & K. Kennedy & R. Chhem. 2007. MRI and Multinuclear MR Spectroscopy of 3200-Year-Old Egyptian Mummy Brain. – American Journal of Roentgenology 189: 105-112.

Lovejoy, C.O., R.S. Meindl, T.R. Pryzbeck & R.P. Mensforth. 1985. Chronological Metamorphosis of the Auricular Surface of the Ilium: A New Method for the Determination of Adult Skeletal Age at Death. – American Journal of Physical Anthropology 68:15-28.

Marx, M. & S.D. D'Auria. 1988. Three-Dimensional CT Reconstruction of An Ancient Egyptian Mummy. – American Journal of Radiology 150: 147-149.

Macková, A., D. MacGregor, F. Azaiez , J. Nyberg & E. Piasetzky. 2016. Nuclear Physics for Cultural Heritage. – Nuclear Physics Division of the European Physical Society.

Martin, R. & K. Saller. 1957. Lehrbuch der Anthropologie 1. – Stuttgart, Gustav Fischer.

Meindl, R.S. & O. Lovejoy. 1985. Ectocranial Suture Closure: A Revised Method for the Determination of Skeletal Age at Death Based on the Lateral-Anterior Sutures. – American Journal of Physical Anthropology 68(1): 57-66. DOI: 10.1002/ajpa.1330680106

Nedden, D., R. Knapp, K. Wicke, W. Judmaier, W.A. Murphy, H. Seidler & W. Platzer. 1994. Cranium of a 5,300 Year Old Mummy: Reproduction and Investigation with CT Guided Stereolithography. – Radiology 193: 269-272.

Panzer, S., H. Gill-Frerking & W. Rosendahl. 2013. Multidetector CT Investigation of the Mummy of Rosalia Lombardo (1918-1920). – Annual Anatomy 195: 401-419.

Portridge, R.B. 1996. Faces of Pharaohs. Royal Mummies and Coffins from Ancient Thebes. – London, The Rubicon Press.

Posh, J.C. 2015. Technical Limitations on the Use of Traditional Magnetic Resonance Imaging in the Evaluation of Mummified Remains: A View from a Hands-On Radiologic Technologist's Perspective. – The Anatomical Record 298: 1116-1124.

Raxter, M.H., C.B. Ruff, A. Azab, M. Erfan, M. Soliman & A. El-Sawaf. 2008. Stature Estimation in Ancient Egyptians: A New Technique Based on Anatomical Reconstruction of Stature. – American Journal of Physical Anthropology 136(2): 147-155.

Raxter, M.H. 2011. Egyptian Body Size: A Regional and Worldwide Comparison. – Unpublished PhD Thesis, University of South Florida.

Suchey, J.M. 1979. Problems in the Aging of Females Using the Os Pubis. – American Journal of Physical Anthropology 51: 467-470.

Ubelaker, D.H. 1989. Human Skeletal Remains. Excavation, Analysis, Interpretation. – Washington, Taraxacum.

White, T.D. & P.A. Folkens. 1991. Human Osteology. – San Diego, Academic Press.

Zakrzewski, S.R. 2003. Variation in Ancient Egyptian Stature and Body Proportions. – American Journal of Physical Anthropology 121(3): 219-229.

Intentionally Burnt Human Remains from the Kom Ombo Temple Salvage Excavation

Afaf Wahba

Introduction

This paper presents preliminary anthropological research from the Ground Water Lowering Project (GWLP) in Kom Ombo (2018) carried out by CDM Smith and funded by USAID. The archaeological monitoring of the GWLP consisted of salvage excavation and recording of archaeological material encountered during the engineering project, and within tight time constraints. The site is known mostly for its Ptolemaic /Roman temple but includes settlement (and cemeteries) from other periods.

The archaeological monitoring revealed skeletal human remains of two different periods and in two different places within the site. The earliest group excavated, in area DTD16, dates to the Late Old Kingdom/First Intermediate Period. These six burials included both non-elite and more elaborate individual vaulted tombs and were part of a bigger cemetery that is, as yet largely unexcavated.

The focus of this paper is the second group of burials excavated at the site. These burials date to the Ptolemaic period and are represented by numerous inhumations of what appears to be intentionally burnt bones. At the time of publication, 48 individuals have been identified from an isolated structure, TP32 (Test Pit 32) to the east of the temple tell. Two chambers inside the structure were excavated, each containing the remains of several individuals whose bones showed signs of *in situ* burning.

Historical background

Kom Ombo, ancient *Nbt* or Ombos, was one of the most important cities of ancient Egypt. The site lies on the east bank of the Nile, approximately 48km north of Aswan, and is probably best known as the main cult center of the crocodile-god Sobek. The city became a Greek town in the Graeco-Roman period, controlling the trade routes from Nubia to the Nile Valley, and is notable for its temple, the current iteration of which was constructed during the Ptolemaic Period, mainly by by Ptolemy VI Philometor (180-47 BC) (David 1993: 99). The temple is unique in that it is built with dual axes and dedicated to two main gods and their divine triads, the eastern to Sobek, and the western to Horus the Elder (Morkot 2001). In addition to the temple, the site also includes settlement remains and cemeteries from other periods (Figure1).

Excavations at TP 32

The structure designated TP32 measured 5.83 m north-south, by 1.85 m east-west, by 0.96 m deep. It was composed of two chambers, one in the north and one in the south, delineated by five walls. It seems that the two chambers were originally for residential

In: Ikram, S., J. Kaiser & S. Porcier. Eds. 2021. *The Ancient Egyptians and the Natural World. Flora, Fauna, and Science.* – Leiden, Sidestone Press: 275-284.

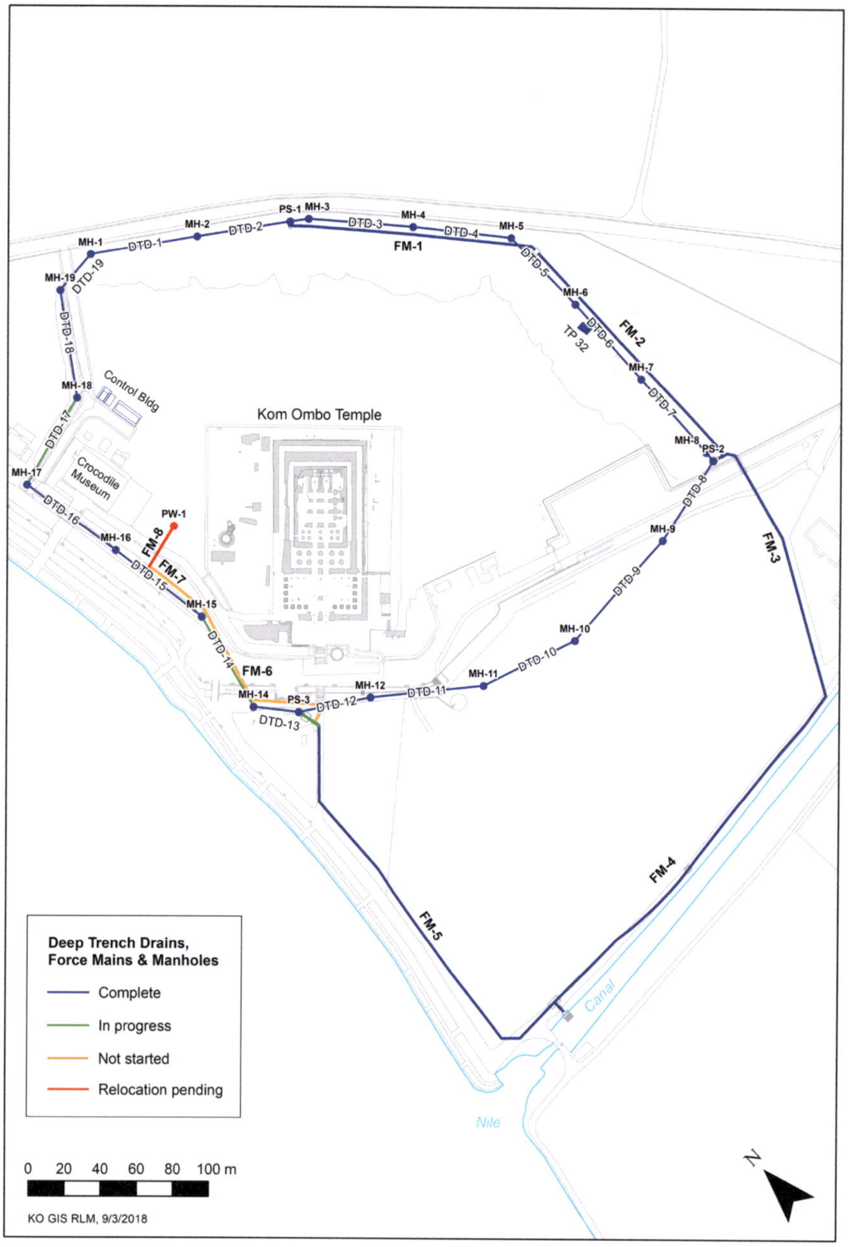

Figure 1. Map showing the two excavated areas (DTD16, TP32), in respect of the Kom Ombo temple and the Crocodile Museum. Map by M. Abd El Basset.

purposes and had not been designed specifically as a burial place. Both chambers had floors, layers of mud render, and the walls were covered with plaster and whitewash. The stratigraphic sequence from the excavation revealed six phases of domestic and funerary activity in this area (Figure 2).

During excavation of TP32, an assemblage of burnt human remains was recovered. The skeletal remains in the two chambers were either deposited in clearly defined pits, or rested directly on the floors of the two chambers (Figure 3). The relationship between material buried in the pits and the material on the floors within the assemblage was not always clear during excavation, though the ceramics recovered from the structure were exclusively Ptolemaic in date, suggesting that the human remains date to the same period.

The north chamber included a series of layers and features with commingled burnt bones, as well as one articulated inhumation in this chamber which pre-dated the burning event. One cut feature [1140] in the center of the chamber represented a phase of reuse comprised by a later deposit containing six almost complete adult skulls in addition to long bones and some vertebrae from individuals of varying ages. The long bones that appeared at the beginning of the excavation were laid around the edge of the feature, with the skulls placed in the center of

Figure 2. TP32 during excavation; north-facing.TP32 Constructions. Photograph by M. Badry.

Figure 3. North chamber. Feature [1140] during excavation; step 1 (top), step 2 (bottom), northwest-facing. Photographs by A. Wahba.

the pit on top of each other (Figure 3). A similar feature in the same chamber contained additional bones. However, no duplicate skeletal elements were recorded between the two deposits, suggesting that a) the skeletal materials from the two pits were drawn from the same group of individuals, and b) that the individuals were placed in these deposits post-skeletonization.

Evidence of articulated skeletal material and mummification in this chamber was only found in a cut [1140], where the bones showed signs of less burning with only traces of smoke and blackened or browned bones. The remains in the southern chamber were more extensively burnt, and completely fragmented and disarticulated.

The southern chamber was defined by three mudbrick walls. As in the north chamber, the walls of this space were covered with a layer of mud render which was covered with whitewashed plaster on the walls and floor. The degree of heat and level of cremation was much stronger in this chamber. It contained debris mixed loosely with sand and crushed limestone. All of these materials were mixed with human skeletal remains. The southern chamber displayed the same firing activity as the northern chamber but with harder, more calcified burnt sand and fewer human remains. The bones from this space were very strongly burnt and rendered fragmentary by the more intense fire that had caused the complete loss of organic materials and led to heavy discolouration of the bones (Figure 4).

Taphonomy of the TP32 Deposit

During excavation, the bones from TP32 were first examined to determine the state of preservation. The discolouration of the bones from the burning and the degree of fragmentation was noted. Information about the rate and shape of cracking resulting from exposure to fire was also observed in order to understand more about the burning process and to determine whether the bones were fresh or dry during the burning, whether burning took place *in situ*, and whether the burning signified a funerary ritual. Due to the commingling, every piece of bone was given a separate identifying number. All the observable traits were recorded separately for each anatomical

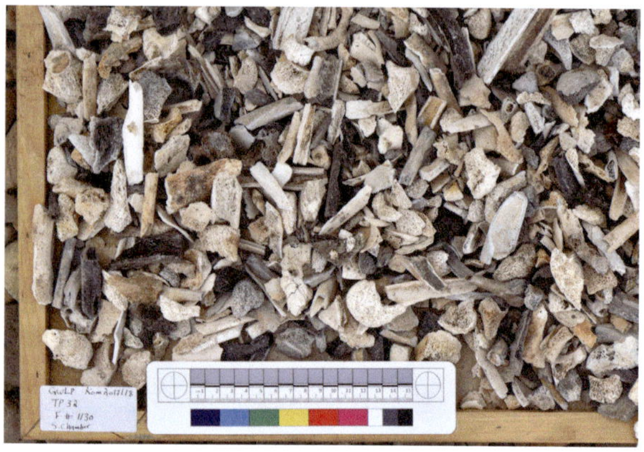

Figure 4. Fragmentation of the skeletal material from TP32, south chamber. Photograph by A. Gabr.

element. The human remains from the two chambers had been previously disturbed causing them to be fragmentary with the exception of one articulated burial in the north chamber, Burial 1, which predated the firing event.

A variety of peri-mortem events and post-mortem processes can be inferred through the study of bone colour, surface details, and shape and these may indicate the speed of burial or exposure to environmental factors such as humic acids, an oxygenated environment, burning or moisture. It can also be misleading since bones can change colour very quickly within and between sedimentary horizons. Colour also varies according to the type of bone and between young or old (Jalvo & Andrews, 2016).

Fresh, untreated bone has an ivory colour. A variety of agents, including grave inclusions, mortuary rituals, and dispositional environments may cause discolouration. Exposure to heat, whether accidental or as part of an interment procedure, causes systematic colour changes that provide information about the heat source and its intensity.

In the northern chamber, approximately 90% of the material had discoloured to brown or black from the effects of smoke, about 5% had discoloured to bluish gray and about 5% retained their natural colour. None of the bones had discoloured to white, which assumes a relatively low temperature of 200-300° C. In the south chamber about 75% of the material was discoloured to white, about 20% to blue-gray, and 5% to brown or black, which assumes as high a temperature as 800° C (Shipman *et al.*, 1984). The material recovered from the upper levels exhibits extensive discolouring to white, while the lower levels contained material with a high percentage of brown or black discolouration. Most of the taphonomical changes resulted from the burning process. The twisting, cracking and splitting that occurs under very high temperatures were common characteristics in the south chamber while no traces of these effects were seen in material from the north chamber.

The heat pattern in the two chambers, in addition to the colouring and discolouring of the skeletal materials, provides information about the condition of the bodies at the time of burning. This includes whether they were freshly buried or already dry and decayed. The bones from the northern chamber were not heavily burnt compared to those in the southern chamber. Therefore, the fire had not caused the complete loss of the organic portions of the bones. Sometimes some burnt organic soft tissue was found still attached to the bones. Bone discolouration from this chamber was smoked brown to black, indicating a low temperature of about 200° C (Shipman *et al.*, 1984), while the white, grey, bluish grey colours seen on the bones from the south chamber must have been caused by a much higher temperature of about 800° C (Buikstra & Ubelaker 1994: 95; Shipman *et al.*, 1984).

The heat pattern, together with the extent of the fire, can also be estimated from the condition of the floors and the wall plaster in the two chambers. The heavy burning process affected the floors and the plaster that covered the walls, but not to the same intensity in both chambers. Due to the different heat pattern and the sources of the fire and its location, the wall plaster in the two chambers had been found with differential patterning from the original white to black or red. The heat was more intense in the southern chamber where the white plaster had turned red. It had also partially fired the mudbricks from their unbaked state to a fired state. These findings complement the evidence from the skeletal material, where the bones from the south chamber were more extensively and heavily burnt than those in the north chamber.

A significant taphonomical feature were the cut marks found on three different pieces of bone. The first was a transverse cut mark on the posterio-lateral aspect of the distal third of a left femur. The second was on the shaft of an unsided femur. The third was four different cut marks on the right ramus of a mandible. A large cut mark was made on the posterior of the ramus in the masseteric tuberosity from the superior to the inferior direction. Two further deep cut marks were present on the anterior of the ramus and superior to the oblique line in the direction from anterior to posterior. The fourth cut mark was a shallow cut on the inferior of the condylar neck of the mandible in the direction from inferior to superior.

All of these cut marks were slicing cuts produced by objects harder than bone, and the agent can be either human action with a tool or weapon, movement of rock against bone during abrasion, or movement of the bone against a hard object (Jalvo & Andrews, 2016: 26). Furthermore, the marks had been inflicted when the bone was fresh. Twisting, cracking, cut marks, and articulated bones all combine to indicate that the material was fresh when it was disturbed and burnt.

Osteological Analysis

The skeletal remains were examined to estimate the age and sex of the individuals represented. Determination of sex was based on discriminatory characteristics of the skull and pelvis (Acsadi & Nemeskeri, 1970; Buikstra & Ubelaker, 1994). Estimation of age was based on several methods: degree of tooth wear (Brothwell, 1981) and ectocranial suture closure (Meindl & Lovejoy, 1985). Other methods used to determine age at death in this collection were age related changes at the pubic symphysis according to Todd (1920) and Suchey-Brooks (1990), and modal age-related changes to the auricular surface of the os coxae according to Lovejoy *et al.* (1985). Stages of bone development and bone fusion were used to assess the ages of juveniles/subadults (Schaefer *et al.*, 2009) according to long-bone length (Scheuer & Black, 2000), dental eruption and development (Ubelaker, 1989). Visible pathologies and taphonomic processes were recorded if present. All measurements were taken according to the guidelines in Buikstra and Ubelaker (1994) so that the data would be comparable to other similar samples in the future.

A total number of 48 individuals were excavated from the north and south chambers. The Minimum Number of Individuals (MNI) from the South Chamber was 12; seven adults and five juveniles based on the left ilium and ischium from different age categories. The North Chamber contained a minimum of 36 individuals, 18 adults and 18 juveniles. The MNI count of the North Chamber remains was based on the ilium bone from the left pelvis in addition to two distal epiphyses of the left femur for two juveniles, which added an additional two individuals to the ilium count. The MNI analysis can be summarized as follows (Table 1).

Even though the absolute number of individuals in the TP32 location was difficult to ascertain because of the high fragmentation of the material, almost half of the minimum number of individuals were non-adults (23/48, 48%). Among the adults, most individuals were above the age of 35, with young adults (20-25 years of age) representing the second largest age-group. The sample was fairly evenly distributed between the sexes; slightly more than half of the individuals for whom sex could be assessed with relative certainty were female. In general, then, the biological profile of the TP32 group is consistent with that of a preindustrial society with low life expectancy: mortality was high in early childhood, fell after approximately five years of age, and stayed relatively low through the teenage years. There was another peak in mortality among young to middle adults, but relatively few individuals reached old age (Scheidel, 2009).

Location	MNI #	Male	Male?	Female	Female?	Undetermined Sex	Juveniles
South Chamber	12	1		1	1	4	5
North Chamber	36	6	1	7	3	1	18
Total	48	7	1	8	4	5	23

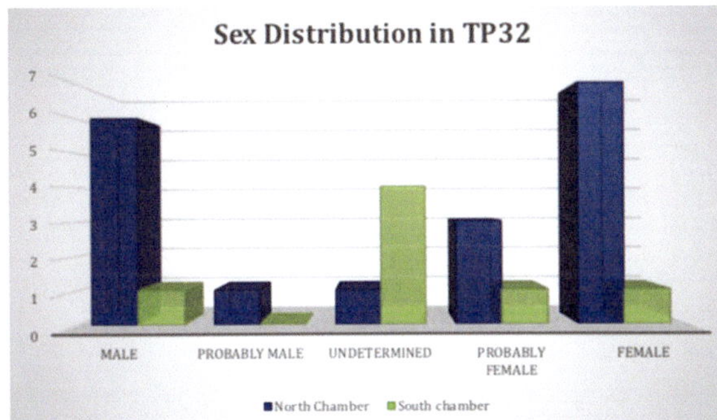

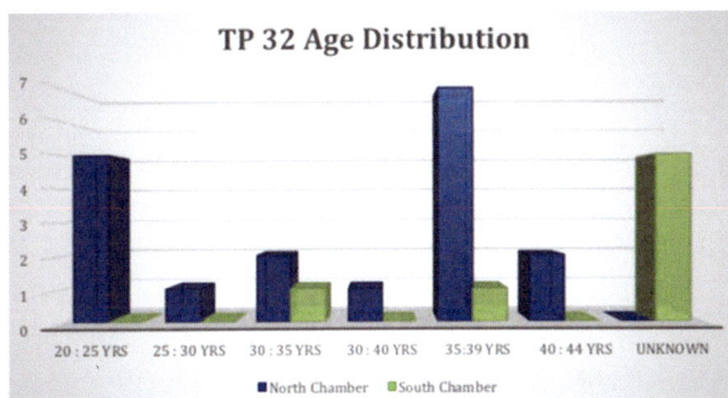

Table 1. Number of individuals, with estimated sex and age distribution.

While the sample size of the TP32 material was small, the pathological conditions observed on the bones can still give us some clues about the state of health among the population of Ptolemaic Kom Ombo. Lesions indicative of nutritional deficiencies, for example, were somewhat common in both the North and South Chambers, with both adults and juveniles exhibiting evidence of *cribra orbitalia* (slight, moderate, or severe), a skeletal stress marker exhibiting as pitting of the orbital roof that may be linked to nutritional deficiency (Walker *et al.*, 2009).

Several types of joint diseases were also noted amongst the skeletal remains discussed. All of the joint surfaces were examined looking for evidence of non-inflammatory conditions such as osteoarthritis and other degenerative joint diseases (DJD), as well as inflammatory conditions, rheumatoid arthritis and ankylosing spondylitis (Waldron, 2012; Ortner, 2003: 561). Several cases of degenerative joint disease were noted, such as osteoarthritis, osteophytic growth, new bone formation, and eburnation. Unsurprisingly, arthritic changes were more common in older adults than in younger individuals. Many studies have drawn inferences between osteoarthritis and functional stress in the lifestyles of past populations (Cohen & Crane-Kramer, 2007; Goodman & Martin, 2002; Goodman *et al.*, 1984; Larsen *et al.*, 2009), and some have gone so far as to suggest links between patterns observed in the distribution of osteoarthritis and specific activities (Hershkovitz *et al.*, 1996). However, a multitude of factors underlie the development of osteoarthritis. In addition to mechanical load, these include genetic predisposition, obesity, age, sex, and even diet and ethnic origin (Jurmain, 1999: 50-67; Waldron, 2012). Nevertheless, the fairly high levels of arthritic changes in the TP32 material suggest at the very least that this was a physically active population.

Discussion

The human skeletal remains which form the subject of this paper date to the Ptolemaic Period and were excavated from the Kom Ombo temple site. They consist of burnt skeletal material, much of which was composed of disarticulated remains. All but one were secondary burials showing signs of *in situ* burning represented by numerous inhumations of burnt bones in a domestic place. Mixed in with the skeletal remains were sherds of Ptolemaic era pottery, made according to Egyptian styles and techniques, Egyptian-style faience beads, and fragments of linen bandages. Taken together, this suggests that the individuals re-buried in the TP32 structure were originally buried according to native Egyptian funerary customs.

While cremation certainly existed as an option in Ptolemaic Egypt, the practice was generally limited to Alexandria in the north, a multicultural city founded on Hellenistic values which under Alexander's successors grew to the largest and most powerful metropolis in the known world, attracting scholars, scientists, philosophers, mathematicians, artists, and historians, and serving as the port to Europe. In the cemeteries of Alexandria, somewhere between 8-25% of the interments were cremations (Landvatter, 2013: 70). While the practice certainly went completely against the main tenet of native Egyptian funerary customs, that of keeping the body against the main tenet of Egyptian funerary practice, that of keeping the body intact for the afterlife, it was not limited only to the Greek population of Alexandria. Rather, it appears to have been used by many different immigrant groups to denote that they were *not* Egyptian, or possibly to signal allegiance to the new ruling elite. Dead treated this way were cremated on a funerary pyre, and subsequently placed in a distinct cinerary urn which was deposited either in singular, smaller tombs or larger communal hypogea (Landvetter, 2018).

An alternative hypothesis that may explain the burning of the remains in TP32 is that it is a reflection of the civil unrest and political strife that plagued Egypt from the late third century onward. In 205 BC, an independent state was proclaimed in the Thebaid, and the area was governed by two native rebellious kings for almost 20 years (Vandorpe, 2010). Texts from the period mention other signs of unrest: strikes, abandonment of and attacks on villages (Lloyd, 2000). The political turmoil also spelled trouble for the temples in the region: "ignorant rebels" disrupted the ongoing work on the temple of Edfu during the reign of Ptolemy IV (Pestman, 1995), and the temple at Medamud was destroyed (Vandorpe, 1995). It is certainly possible that similar attacks were made on the Kom Ombo temple, and that the bones in TP32 were burnt in the process.

Another possible explanation for the presence of burnt remains in TP32 is that the remains were disposed of in this manner to hide the evidence of tomb robberies in antiquity. Because of the significant amounts of wealth accompanying many Egyptian dead, tomb robberies were widespread in ancient Egypt, particularly during periods of civil unrest and difficulty. Several texts from earlier periods mention details of such robberies, including the Abbott Papyrus (BM 10221), Amherst Papyrus #7 (BM 10054) (Jankuhn, 1974: 1-4), and the Leopold II Papyrus in Brussels (Jankuhn, 1974: 69). These texts document the trials of robbers who had been caught in the act and put on trial for the pillaging of royal and elite burials in Thebes, and mention the burning of coffins, either inside the tomb or in other locations after the removal of the objects from the place of burial. However, the Theban robbers appear to have set fire not to the bodies, but to the coffins alone, presumably because it was easier to collect the gold used to decorate the coffins from the ashes of a fire than by chiseling it off the surface of the coffin itself (Strudwick, 2013). While there were no traces of coffins in the TP32 material, it is still possible that the bodies were collected from nearby cemeteries and subsequently set on fire to hide the evidence of pillaging activity. Such an interpretation has been put forth for other burnt deposits of burial items (which did not include bodies or fragmentary human remains) found in domestic contexts at Gurob (Gasparini, 2018: 301-305).

A final, and perhaps more plausible, explanation for the burning of the TP32 remains is that the building was used either as a communal burial place or as a storage location for bodies awaiting burial elsewhere, and that it accidentally caught on fire. The presence of an intact burial in the northeast corner of the building that predated the fire clearly shows that the area was already deemed appropriate for funerary use before the deposition of the additional skeletal material in the two chambers. Another possibility is that the TP32 structure functioned either as an embalming workshop or storage location for mummies by necropolis workers associated with the temple.

These workers are known from several preserved family archives of demotic texts. From the Ptolemaic period, the archives derive mainly from Memphis, Hawara, and Thebes (Cannata, 2009: 2), but there is no reason to believe that similar organizations did not exist elsewhere in Egypt. Though no embalming workshops have been definitively identified in the archaeological record, Aston (2003: 153-160) makes a convincing case for three possibilities, two in the Memphite necropolis, and one in Assasif, based on the high frequency of embalming caches found in the surrounding areas. All three examples were found in association with temples or mortuary monuments but had the appearance of domestic mud-brick buildings. In association with these buildings were also storage facilities used to store materials used for embalming, as well as bodies waiting to be embalmed and finished mummies ready for burial (Chapman, 2016: 136-140). Another group of necropolis workers that are known from texts to have stored bodies awaiting burial are the choachytes, whose roles were similar to that of an undertaker. The duties of the choachytes included both the provisioning of a burial place in the necropolis and subsequent care (*i.e.*, perpetuating the funerary cult of the deceased) as well as the storage and transport of mummies awaiting burial. The choachyte association in Thebes, for example, are known from texts to have owned a house for the express purpose of storing mummies waiting to be transported to the West Bank (Landvatter, 2013: 117).

Conclusion

This paper has aimed to explain the presence of burnt skeletal material in a structure excavated as part of the Kom Ombo ground water Lowering Project. Possible explanations considered were that the remains were cremated, that they were burnt in an attack on the temple due to civil unrest, that they were burnt intentionally in order to hide evidence of tomb robberies in antiquity, or that the TP32 structure was used either as a communal burial ground or a storehouse for embalmers or undertakers and caught fire accidentally. While the excavations at TP32 were limited in scope, the archaeological evidence does allow us to rule out some of these scenarios with relative certainty.

There were no traces of cinerary urns in the TP32 material. Rather, the material culture mixed with the skeletal material suggests that the individuals re-buried in TP32 were originally interred according to Egyptian funerary practice, and that their bodies were burnt *after* burial, not before. Thus, we can rule out intentional cremation as part of a funerary ritual as the reason for the burning of the remains.

Further, the burning in TP32 appears to have been an isolated event, as no other archaeological evidence of destruction were found during the recent excavations at Kom Ombo. If the fire was the result of a politically motivated attack on the temple, other indications of unrest at the site would probably be expected. Additionally, such a scenario still does not explain the presence of large quantities of burned remains in a place that had until then been used for residential and not burial purposes.

If the fire in TP32 was indeed set intentionally to hide the evidence of tomb robbing activities, the question remains as to why the job was not finished – while the skeletal material in the southern chamber was completely fragmented due to the fire, the bones in the northern chamber were still clearly recognizable as human, as the deposits included several complete human crania. There is also the question of from whence the bodies originally came, and why this specific area was chosen for the purpose of disposing of the remains. Nevertheless, it is certainly possible that a nearby and hitherto unknown Ptolemaic cemetery exists from which the erstwhile robbers collected their loot.

TP32 could also have functioned either as a communal burial place or as a storage location for bodies awaiting burial elsewhere, and accidentally caught on fire. That the area had been used as a burial space at one point in time is beyond doubt, since an intact burial that predated the fire was found in the northern chamber, and it is certainly possible that this practice continued. However, it is hard to explain the total incineration of the skeletal remains in the southern chamber of the building if the bodies were below ground at the time of fire.

A final explanation for the burning of the TP32 remains is that the bodies were above ground at the time of the fire because they were being stored in the structure awaiting final burial. The building could have been used as storage either for an embalming workshop, or for choachytes associated with the temple. Both types of storage facilities are known from contemporary texts, and are known to have existed in connection with temples in Thebes and Memphis. If, indeed, the TP32 structure was used to store numerous mummies above ground, the contents of the building would likely have been highly flammable, considering the many oils, unguents, and large quantities of linen used in the mummification process. Thus, an accidental fire appears to be the most likely explanation for the burning of the bones in TP32. However, complete excavation is needed to prove and support this idea, especially as we have evidence that the two chambers were extended further to the south.

Acknowledgements

For this work, I am grateful to Dr. Ayman Ashmaouy, Head of the Egyptian Antiquities Sector and Mr. Mohamed Abd

EL Badiaa, Head of the Upper Egypt Central Department. In Kom Ombo we are deeply indebted to Mr. Abd El Moniem Said, General Director of Aswan Antiquities. I am also indebted to my colleague Ahmed Gabr who helped me a lot, and all the other colleagues from Kom Ombo who were part of the team. The work would not have been possible without the great help and support from Ms. Freya Sadarangani and Ms. Ana Tavares, and the entire CDM team. I am deeply indebted to Dr. Jessica Kaiser for her advice and reading of this contribution.

Bibliography

Acsádi, G & J. Nemeskéry. 1970. History of Human Life Span and Mortality. – Budapest, Akadémiai Kiadó.

Aston, D.A. 2003. The Theban West Bank from the Twenty-fifth Dynasty to the Ptolemaic period. In: Strudwick, N. & J.H. Taylor. Eds. The Theban Necropolis: Past, Present and Future. – London, British Museum Press.

Brooks, S. & J.M. Suchey. 1990. Skeletal Age Determination Based on the Os Pubis: A Comparison of the Acsádi-Nemeskéri and Suchey-Brooks Methods. – Human Evolution 5: 227.

Brothwell, D.H. 1981. Digging up Bones. – Oxford, Oxford University Press.

Buikstra, J.E. & D.H. Ubelaker. Eds. 1994. Standards for Data Collection from Human Skeletal Remains. – Fayetteville, Arkansas Archaeological Survey (Research Series 44).

Cannata, M. 2009. The Realia of Egyptian Burial Practices in the Ptolemaic Period (332-30 BC). – Unpublished PhD Thesis, Oxford University.

Chapman, S.L. 2016. The Embalming Ritual of Late Period through Ptolemaic Egypt. – Unpublished PhD Thesis, University of Birmingham.

Cohen, M.N. & G.M.M. Crane-Kramer. Eds. 2007. Ancient Health: Skeletal Indicators of Agricultural and Economic Intensification. – Gainesville, University Press of Florida.

David, R. 1993. Discovering Ancient Egypt. – London, Michael O'mara Books.

Gasperini, V. 2018. Tomb Robberies at the End of the New Kingdom: The Gurob Burnt Groups Reinterpreted. – Oxford, Oxford University Press (Oxford Studies in Egyptian Archaeology).

Goodman, A.H. & D.L. Martin. 2002. Reconstructing Health Profiles from Skeletal Remains. In: Steckel, R.H. & J.C. Rose. Eds. The Backbone of History: Health and Nutrition in the Western Hemisphere. – New York, Cambridge University Press.

Goodman, A.H., D.L. Martin, G.J. Armelagos & G. Clarke. 1984. Indications of Stress from Bone and Teeth. In: Cohen, M.N. & G.J. Armelagos. Eds. Paleopathology and the Origins of Agriculture. – Orlando, Academic Press.

Hershkovitz, I., L. Bedford, L.M. Jellema & B. Latimer. 1996. Injuries to the Skeleton Due to Prolonged Activity in Hand-to-Hand Combat. – International Journal of Anthropology 6: 167-178.

Jalvo, Y.F & P. Andrews. 2016. Atlas of Taphonomic Identifications. – London, Springer.

JanKuhn, D . 1974. Bibliographie der Hieratischen und Hieroglyphischen Papyri. – Wiesbaden, Harrasowitz (Göttinger Orientforschungen, IV. Reihe Ägypten, Band 2).

Jurmain R. 1999. Stories from the Skeleton: Behavioral Reconstruction in Human Osteology. – Amsterdam, Gordon and Breach.

Landvatter, T.P. 2013. Identity, Burial Practice, and Social Change in Ptolemaic Egypt. – Unpublished PhD Thesis, University of Michigan.

Landvatter, T.P. 2018. Identity and Cross-cultural Interaction in Early Ptolemaic Alexandria: Cremation in Context. In: McKechnie, P. & J.A. Cromwell. Eds. Ptolemy I and the Transformation of Egypt, 404-282 BCE. – Leiden, Brill (Mnemosyne Supplements): 199-234.

Larsen, C.S, P.L. Walker, R. Steckel & O.W. Sciulli. 2009. History of Degenerative Joint Disease in Europe: Inferences about Lifestyle and Activity. Abstract: Poster presented at the Annual Meeting of the American Association of Physical Anthropologists, Chicago. American Journal of Physical Anthropology 138 (S48): 172.

Lloyd, A.B. 2000. The Ptolemaic Period (332-30 BC). In: Shaw, I. Ed. The Oxford History of Ancient Egypt. – Oxford, Oxford University Press: 388-413.

Lovejoy, C.O., R.S. Meindl, R.P. Mensforth & T.J. Barton. 1985. Multifactorial Determination of Skeletal Age at Death: A Method and Blind Tests of Its Accuracy. – American Journal of Physical Anthropology 68: 1.

Lovejoy, C. O., , R.S. Meindl, T.R. Pryzbeck & R.P. Mensforth. 1985. Chronological Metamorphosis of the Auricular Surface of the Ilium: A New Method for the Determination of Adult Skeletal Age at Death. – American Journal of Physical Anthropology 68: 15.

Meindl, R.S. & C.O. Lovejoy. 1985. Ectocranial Suture Closure: A Revised Method for the Determination of Skeletal Age at Death Based on the Lateral-Anterior Sutures. – American Journal of Physical Anthropology 68: 57.

Morkot, R. 2000. Kom Ombo. In: Redford, D.B. Ed. The Oxford Encyclopedia of Ancient Egypt. Vol. 3. – Oxford, Oxford University Press: 248-250.

Ortner, D.J. 2003. Identification of Pathological Conditions in Human Skeletal Remains. – San Diego. Academic Press.

Pestman, P.W. 1995. Haronnophris and Chaonnophris: Two Indigenous Pharaohs in Ptolemaic Egypt (205-186

B.C.). In: Vleeming, S.P. Ed. Hundred Gated Thebes. – Leiden, Brill: 101-137.

Schaefer, M., S. Black & L. Scheuer. 2009. Juvenile Osteology: A Laboratory and Field Manual. – Amsterdam, Elsevier Academic Press.

Scheidel, W. 2009. Population and Demography. In: Erskine, A. Ed. A Companion to Ancient History. – Malden, MA, Wiley-Blackwell: 134-145.

Scheuer, L. & S. Black. 2000. Development and Aging of the Juvenile Skeleton. In: Cox, M. & S. Mays. Eds. Human Osteology in Archaeology and Forensic Science. – London, Greenwich Medical Media: 9-22.

Shipman, P., G. Foster & M.J. Schoeninger. 1984. Burnt Bones and Teeth: An Experimental Study of Colour, Morphology, Crystal Structure and Shrinkage. – Journal of Archaeological Science II: 307-325.

Strudwick, V. 2013. Tomb Robberies at the End of the New Kingdom. – Oxford, Oxford University Press.

Todd, T.W. 1920. Age Changes in the Pubic Bone. I. The White Male Pubis. – American Journal of Physical Anthropology 3: 285-334.

Ubelaker, D.H. 1999. Human Skeleton Remains: Excavation, Analysis, Interpretation (3rd Edition) – Washington, DC, Taraxacum.

Ubelaker, D.H. 1987. Estimating Age at Death from Immature Human Skeletons: An Overview. – Journal of Forensic Sciences 32(5): 1254-1263.

Vandorpe, K. 1995. City of Many a Gate, Harbour for Many a Rebel: Historical and Topographical Outline of Greco-Roman Thebes. In: Vleeming, S.P. Ed. Hundred Gated Thebes. – Leiden, Brill: 203-239.

Vandorpe, K. 2010. The Ptolemaic Period. In: Lloyd, A.B. Ed. A Companion to Ancient Egypt. – Malden, MA: Wiley-Blackwell: 160-179.

Walker, P.L., R.R. Bathurst, R. Richman, T. Gjerdrum & V.A. Andrushko. 2009. The Causes of Porotic Hyperostosis and Cribra Orbitalia: A Reappraisal of the Iron-Deficiency-Anemia Hypothesis. – American Journal of Physical Anthropology 139(2): 109-125.

White, T.D., M.T. Black & P.A. Folkens. 2012. Human Osteology (3rd Edition). – Burlington, MA, Elsevier Academic Press.

Waldron, T. 2012. Joint Disease. In: Grauer, A.L. Ed. A Companion to Paleopathology. – Malden, MA, Wiley-Blackwell: 513-530.